Vibration Acoustics
Applied to
VVER-1200 Reactor Plant

Vibration Acoustics
Applied to
VVER-1200 Reactor Plant

G.V. Arkadov, Ph.D
Russian Society of System Engineering, Russia

V.I. Pavelko, Ph.D
Scientific Technical Center Diaprom, Russia

M.T. Slepov, Ph.D
Novovoronezh Nuclear Plant, Russia

Translated by: Gabor Por, Ph.D
Budapest University of Technology and Economics, Hungary

World Scientific

NEW JERSEY • LONDON • SINGAPORE • BEIJING • SHANGHAI • HONG KONG • TAIPEI • CHENNAI • TOKYO

Published by

World Scientific Publishing Co. Pte. Ltd.

5 Toh Tuck Link, Singapore 596224

USA office: 27 Warren Street, Suite 401-402, Hackensack, NJ 07601

UK office: 57 Shelton Street, Covent Garden, London WC2H 9HE

Library of Congress Cataloging-in-Publication Data

Names: Arkadov, G. V. (Gennadiĭ Viktorovich), author. | Pavelko, V. I. (Vladimir Il′ich), author. | Slepov, M. T. (Mikhail T.), author. | Por, Gabor, translator.

Title: Vibration acoustics applied to VVER-1200 reactor plant / G.V. Arkadov, Ph.D, Russian Society of System Engineering, Russia, V.I. Pavelko, Ph.D, Scientific Technical Center Diaprom, Russia, M.T. Slepov, Ph.D, Novovoronezh Nuclear Plant, Russia ; translated by: Gabor Por, Ph.D, Budapest University of Technology and Economics, Hungary.

Other titles: Vibroshumovai͡a diagnostika VVĖR. English

Description: Singapore ; Hackensack, NJ, USA : World Scientific, [2021] | Translation of: Vibroshumovai͡a diagnostika VVĖR. | Includes bibliographical references and index.

Identifiers: LCCN 2021005225 | ISBN 9789811234668 (hardcover) | ISBN 9789811234675 (ebook for institutions) | ISBN 9789811234682 (ebook for individuals)

Subjects: LCSH: Water cooled reactors--Testing. | Pressurized water

British Library Cataloguing-in-Publication Data

A catalogue record for this book is available from the British Library.

For any available supplementary material, please visit
https://www.worldscientific.com/worldscibooks/10.1142/12220#t=suppl

Desk Editors: Balasubramanian Shanmugam/Yu Shan Tay

Typeset by Stallion Press
Email: enquiries@stallionpress.com

Printed in Singapore

About the Authors

Gennady Arkadov has obtained PhD in Technical Sciences and Economics from the National Research Nuclear University (MEPhI) – Obninsk Institute for Nuclear Engineering (OINPE) (Obninsk, Kaluga region). Mr. Arkadov has 30 years of work experience in nuclear power engineering. He worked for JSC "All-Russian Scientific and Research Institute for Nuclear Power Plants Operation" (JSC "VNIIAES", State Atomic Energy Corporation "Rosatom") and rose through the ranks from an engineer to the General Director (2006–2014). In 2014, Mr. Arkadov was appointed as Vice-president of the Russian System Engineering Society. His research interests include technical diagnostics, serviceable life, life cycle management, and system engineering.

Vladimir Pavelko has obtained PhD in Technical Sciences from the National Research Center of the Russian Federation <<Kurchatov Institute>> (Moscow). Mr. Pavelko specializes in reactor plant diagnostics. In 1972, Mr. Pavelko graduated from Moscow Institute of Electronics and Mathematics, and in 1978 from the Faculty of Mathematics and Mechanics of Lomonosov State University. In 1972,

Mr. Pavelko was sent to work for the Kurchatov Nuclear Power Institute (currently known as NRC "Kurchatov Institute") where he was in charge of the laboratory of reactor plant diagnostic methods. Since 1996, he has been Deputy Director General for Research at JSC "STCD". Mr. Pavelko has more than 100 scientific publications. He is the co-author of a monograph on reactor plant diagnostics. He has 12 inventor certificates and 4 patents of invention.

Michael Slepov has obtained PhD in Technical Sciences from the State Scientific Centre of the Russian Federation — Leypunsky Institute for Physics and Power Engineering (Obninsk, Kaluga region). Mr. Slepov graduated from Obninsk Institute of Nuclear Power Engineering in 1992, and in 1998, he took a post-graduate course in RF State Scientific Center Leipunsky Institute of Physics and Power Engineering. Mr. Slepov has more than 25 years of work experience in nuclear power engineering. He is currently in charge of Novovoronezh NPP technical diagnostics department. Mr. Slepov participated in the virgin start of WWER-1200 unit. His research interests lie in reactor plants noise diagnostics.

Foreword by Scientific Translation Editor

I have been asked by the authors of this book to check the English translation. I am a member of that Hungarian school which has been mentioned by the Lieu as initiating the systematic noise diagnostics in nuclear reactors. Personally I graduated from Russia; therefore I had been chosen to lectorate the translated term text. My aim was to make the book easy to read for a specialist, who is used to certain term both in noise diagnostics and in the field of VVER-type reactors.

When we delivered a complex noise diagnostics system for Kalinin NPP in 1991, we had, on a contemporary technical level, a well-advanced, half-automated diagnostic system KAZMER and KARD. I believe we left a positive impression on the Russian specialists. However, I was astonished reading this book, how far they have advanced in this field in the past few decades. Noise and fluctuation diagnostic of NPPs has become from an experiment-based practice to real science, which is based on control theory and on long line theory modeling the dynamic effects. The diagnostics in their hand has become from a timid, experience-driven knowledge to a well-established model mathematically supported by differential equation which helps a lot in understanding the components of the measured spectral estimates. This is what you can find in the first half of the book, which could serve bravely as a textbook for universities.

Moreover, we can learn from the second half of the book that diagnostics of NPPs have reached a higher technical level today.

The application of knowledge gained from diagnostics experiments affects the planning of the main coolant loop in NPPs. I was surprised, that in order to assist the damping of the standing waves, significant changes have been introduced in the planning of a VVER-1200 NPP in comparison to a VVER-440 or a VVER-1000. I strongly suggest everyone who is interested in diagnostics of NPPs analyzing measurable signals in the primary loops to read this book. It is also an important book for constructors of primary loop components.

I beg for pardon from everyone who thinks we did not find the best technical words in the translation. PWR reactors in general and VVER particularly have a rather big bibliography today. That latter came from several different translations, therefore sometimes different terms have been used for the same object. We were trying to find those term which are the most accepted and make the text easily understandable for the reader, while preserving the way of thinking of the original Russian text as much as possible.

I am very satisfied with the results achieved by the authors of this book. I really hope that these developments do not hit a dead end at this level. I am looking forward to further success in this field.

Gabor POR, PhD
Professor Emeritus
Candidate of physical sciences

Preface

Technical diagnostics occupies a strong position in the global engineering community and it is included in the standards and recommendations for nuclear power plants in both design and operation. All foreign nuclear power plants in operation are to some extent equipped with tools of technical diagnostics of reactor plants either from the very beginning or in the course of modernization. Regardless of an I&C diagnostic architecture, whether it is the local systems of Framatome design, or the Westinghouse centralized systems, diagnostic algorithms are universal.

The Russian operating company JSC Concern Rosenergoatom pays a great deal of attention to development of technical diagnostics tools. Over the past 20 years, almost all Russian power units have been equipped with these. This has contributed to both improvement of operational safety and ensuring the reliability and life extension of nuclear power plants in operation.

This book is useful for further development of automated control systems for nuclear power plants of any type as well as for design of domestic I&C systems that meet requirements of modern international standards. The developed algorithms and approaches can be an integral part of the comprehensive mode diagnostics associated with new possibilities of centralized processing of digital information. I hope that the engineers and companies — JSC Atomenergoproekt, JSC OKB Gidropress, NRC Kurchatov Institute, JSC RASU — involved in design and engineering will appreciate this monograph.

For my part, I would particularly like to express my gratitude to the people who laid the groundwork for design and equipment of VVER nuclear power plants with domestic diagnostic systems and fostered a keen interest in diagnostic approaches, as one of the main areas of ensuring safe operation of complex process facilities. They include Abagyan Armen Artovazdovich and Kudryavtsev Boris Konstantinovich as well as all the engineers and scientists with whom I had a chance to work together in VNIIAES from 1989 to 2014. JSC VNIIAES has a special role in establishing the subject of this monograph in Russia.

Arkadov G.V.
President of Russian Institute of
System Engineering

Thoughts on Diagnostics Inspired by Reading the Monograph "Vibration Acoustics Applied to VVER-1200 Reactor Plant" by Arkadov, G.V., Pavelko, V.I. and Slepov, M.T.

In my youth, at the Kurchatov Institute in the early 70s of the last century, great efforts were put into pilot full-scale testing of reactor core. Critical assemblies of nuclear reactor core were tested. Neutron flux measurements were carried out in critical assemblies and the results were applied to estimate the main physical parameters of reactor cores: the reactivity coefficients, control rod performance, neutron cross-sections, etc. The right physicists carried out these kinds of activities according to established programs and procedures. However, there were also "wrong" physicists who measured not the neutron flux itself, but its fluctuations, that is, small deviations from the mean value, which in amplitude were less than 1% of the average or a so-called reactive noise. "Wrong" physicists were tolerated, but nobody took them seriously. The former were said to have measured while the others only happened to have calculated. They were not much of an intrusion, and the Kurchatov Institute could afford to support digging in the noise, too.

Now, 45 years later, in the references of a heavy tome with the title "Vibration Acoustics..." I find familiar names of "wrong" physicists. Their early publications are referred to as classic or fundamental papers on the diagnosis of reactors. Both vibrations and acoustics of circulation circuits are monitored via the reactor neutron field, and complex interpretation of neutron noise measurements allows extracting additional diagnostic information.

The monograph systematically presents the mathematical apparatus of reactor vibration acoustics, and I would like to emphasize the practical aspect of the case. This mathematical apparatus is used in reactor diagnostic systems that have become operational attributes. No new nuclear power plant unit can be commissioned and operated without them. These are strict international requirements for NPP safety. Russia could not have successfully entered the international nuclear power plant market without presenting a reactor diagnosis system as part of the NPP design. It is the newly built Unit 1 of Novovoronezh NPP-2 VVER-1200 that is equipped with advanced diagnostic systems of a new generation to a higher degree than any other VVER unit. Those systems comprised operating experience of diagnostic systems not only of Russian nuclear power plants, but also of foreign Russia-supplied nuclear power plants: Bushehr NPP (Iran), Tianwan NPP (China) and Kudankulam NPP (India). On the above-mentioned foreign nuclear power plants, Russian diagnostic systems managed to detect and localize at an early stage the so–called diagnostic events, namely, precursors of the development of the reactor malfunction. Russian systems proved to be not only competitive with the systems of well-known reactor companies – EDF, Framatome, Westinghouse – but also more capable in diagnosis of domestic VVER reactors. Our systems come with pre-generated start-up databases of diagnostic information. That is, they contain already known spectral portraits of the normal vibration state of the reactor plant.

Besides, it is quite important that the new diagnostic data were very quickly obtained on the new first power unit of the kind. This type of reactor is planned to be the reference for the future foreign nuclear power plants of Russian design. Moreover, this reactor will be equipped with a full range of diagnostic systems currently tested on Unit 1 of Novovoronezh NPP-2. As a Scientific Director of the AES-2006 design, I have firsthand knowledge of the problems of diagnostics systems implementation. It is with great difficulty that complex, high-end technology diagnostic systems become operational practice. The systems are slowly filled with diagnostic knowledge, but our foreign demanding customer will undoubtedly require it. Nevertheless, for the first time, the authors initiated unique measurements as early as at the stages of commissioning the first VVER-1200 unit, including the stage of low power testing and on to the stage of commercial operation.

I have known the authors for quite a long time and in this sense, I am a biased critic. I know that all three authors made themselves noted in promotion of Siemens diagnostic systems for VVER-440-type reactor plants. It happened in the mid-90s. After that, since 2005, all three began to develop domestic diagnostic systems for VVER-1000. Then there were years of numerous implementations of these systems on new and operating VVER-1000 units in Russia and abroad. The monograph summarizes the experience of all types of VVER applied to interpretation of diagnostic information obtained in a new VVER-1200 power unit.

Consistent reading of the 15 chapters of the monograph has led me to the following conclusions.

The fundamental approach according to which the vibration acoustics differential equations got lined up in an increasing difficulty order is obvious: the harmonic oscillator equation with generalizations in the form of motion and wave equations, with generalizations to all types of partial differential equations. Just sheer enumeration of the names of the equations with the names of great physicists and mathematicians such as Helmholtz, d'Alembert, Lagrange, Laplace, Navier, Kirchhoff, Stokes, Fresnel, Fourier, Euler indicates that the monograph is built on a serious mathematical apparatus. At the same time, the authors insist on the so-called "engineering narrative", illustrating the above equations with applied examples or simplifying the task by analogy methods.

The material presented in the monograph is undoubtedly relevant as at the time of writing this text the first VER-1200 unit has only been operating on the design parameters for a year. The monograph claims to become a desktop book for engineers and technical specialists one way or another involved in NPP operation and, in particular, with monitoring and diagnostics of the equipment of future VVER-1200 units. It systematically presents the mathematical apparatus of vibration acoustics, and it is a kind of reference guide to vibration acoustics methods for students, lecturers and researchers not necessarily involved in reactor engineering.

There is a clear focus of the authors on the practical result. The empirical regularities connecting vibration diagnostic parameters with the power unit operating mode, for example, the frequency of standing waves with the temperature of the coolant, are derived. On the other hand, the phase characteristics of the signals from pressure pulsation sensors help restore distribution patterns of amplitudes of the main standing waves types. In addition, the effect of synchronization of all MCPS oscillations at the frequencies of standing waves was unexpectedly discovered. The proposed new methods are based on actual multichannel measurements of different physical parameters, so that the feasibility of the methods is beyond any doubt. The monograph discusses measures to reduce the vibration loads on the primary circuit at the expense of structural changes that are obviously possible to implement in the future after bench testing the ideas themselves.

The authors once again demonstrate how simple electrical analogies, which are more than a century old, can lead to the correct answer. At the same time, they warn against the typical errors of such simplification. The analogies also apply a hierarchy according to which they start with circuits in lumped parameters and proceed into the theory of the electromagnetic field of ultrahigh frequencies.

The complexity of the mathematical description of elastic dissipative media is demonstrated in full. It is known that the methods of viscosity accounting by the Navier–Stokes equation have their limitations and are not always confirmed by experimental dependences. However, the authors managed to show that in this case simple diagnostic features may prove to work such as the linear phase in a subband of frequencies, carrying the measurement information about the viscosity of the coolant.

The authors highly appreciate the diagnostic potential of the commissioning measurement systems of the first VVER-1200 unit while admitting that there are still open issues of their integration with standard stationary diagnostic systems.

In general, the monograph is a valuable scientific and technical paper in all its aspects. I predict a high index of scientific citation to the monograph and a lively interest in all countries with nuclear power.

V.G. Asmolov
The Advisor to the General
Director of the State Corporation
"Rosatom",
Doctor of Technical Sciences
Professor

In Lieu of a Preface — Historical Journey and a Bibliographic Guide

It is not just a book in front of you, but a monograph. This means that besides new hypotheses, new methods and concepts, it provides extensive bibliographic material. Usually references are grouped either chronologically or alphabetically, and placed at the end of the book or after each chapter. We decided to use the bibliography as an introduction to reactor vibration acoustics and a brief historical journey. With this in mind, the reference list is grouped by topic, followed by comments on each group. Total number of references is 410 and it is huge, so trying to read them one by one is a most difficult and ineffective occupation. It is best to read this monograph first. This is one of its purposes. The second goal is to present systematically the entire broad range of reactor vibration acoustics methods focusing on typical errors. The third one, an obvious purpose, is to present in an accessible form the authors' new ideas with possible applications not only for reactor vibration acoustics.

Vibration acoustics is a synthetic science, and the form of some of its differential equations is similar to the corresponding equations of theoretical mechanics, electrical engineering, the theory of long lines, the theory of electromagnetic field and optics [1–14].

Vibration acoustics, as a science, has long been carrying an "applied" epithet [15–46]. However, we should not forget about its theoretical heritage and fundamental nature, set out in remarkable monographs and articles [6–10, 47–56], which a researcher of applied science should have “at hand”. Vibration acoustic effects observed in the main circulation circuit (MCC) are mostly described already in classical works on the acoustics of elastic media [15, 17, 21–27, 30–36, 38, 39, 45]. It only remains to take into account the features of the reactor plant (RP).

Vibration acoustics is only based on two classes of differential equations. The first class is ordinary differential equations of motion, the second class is wave partial differential equations [57–62]. The narrative in this monograph is structured in this order illustrating VVER-1200 vibration acoustics by specific examples. There are two classes of differential equations, and there are dozens of their types. In each case, the simplest equations of these two classes get generalized. Substantiating the analytical form of the equation, its order, determining the forcing effects, a very wide range of differential equations types can be obtained, and wave equations are generalized to all types of mathematical physics equations.

In addition to differential equations, the mathematical tools of vibration acoustics include the Theory of Functions of a Complex Variable and the Theory of Stochastic Processes [63–70]. Not all technical colleges offer such mathematical courses; it is the so-called university level of mathematical education. However, there are brilliant examples in development of theoretical electrical engineering by engineers with higher but non-university education. Theoretical electrical engineering [14] is based on the same mathematical tools as vibration acoustics, but there are many known examples of electrical engineering developed on the engineering level but not only at a high scientific one.

The next topic in order, but not in importance, a limitless and self-sufficient one is signal processing. If it is only studied theoretically, one can happen to fail to get to specific applications. We carefully selected such books [70–83], in which the mathematical tools are of the so-called engineering level and can be quickly mastered by graduates of technical colleges, despite the fact that these books set out the spectral-correlation mathematical tools of the Theory of Stochastic Processes, including multidimensional random ones.

Specific vibration acoustic applications can be found in radio electronics, construction, hydrology, oceanography, petrochemistry, sound recording, aircraft and automotive construction, etc., and in reactor engineering, obviously, too. Vibration acoustics is included into technical diagnostics as one of its applied areas, which should follow the general principles regardless of the object of diagnosis. Vibration acoustics, as the most important element of reactor diagnostics, is also an element of the general diagnostic theory already developed as an independent scientific area [84–107].

The development of domestic reactor diagnostics went its own special way largely determined by the research of the dynamics of reactor plants in normal operation and emergency modes. The specifics of the Russian school in reactor diagnostics and, in particular, vibration diagnostics of the reactor are collected in references [108–133]. It is in reactor engineering that neutron kinetics, thermal physics and hydrodynamics triumph; and vibration acoustics stands aside. Vibration, as one of the many factors of equipment wear, is not considered by any reactor plant designs (and rightly not considered!) the main factor of equipment wear-out. The acoustics of the circulation circuits are not taken into account when justifying the reactor plant safety in any domestic or foreign design. At the same time, there are known unsuccessful designs of reactor plants in which vibration acoustic effects led to equipment failures or to its premature wear-out [113, 115, 117, 119, 120, 122, 132, 133].

The use of vibration acoustics in reactor engineering grows extremely slowly, perhaps due to departmental barriers, the complexity of the mathematical tools, the science-intensive diagnostic systems, the lack of regulatory requirements governing the acoustic and vibration effects in the MCC and the necessity of their continuous monitoring. Reactor plant equipment has specific features that require an individual approach. Significant weight and dimensions of vibrating structural elements, the unique dynamics of fluid flow paths and huge values of the forces causing vibration make the vibration factor stand out as a special issue in analysis of reliability and service life of a reactor plant [134–150].

The huge efforts of the General Designer on experimental, computational and full-scale justification of VVER vibration resistance and strength are presented in the references [151–171]. Before start of these works, a significant foreign statistics of vibration incidents at various nuclear power plants related to abnormal vibrations of the internals and fuel assemblies have been collected. Vibration anomalies on the first VVER-440 power units were also registered. The General Designer during commissioning of the first VVER-1000 power unit (Unit 5 of Novovoronezh NPP), carried out unprecedented full-scale vibration measurements.

Comprehensive computational and experimental justifications of vibrational strength were conducted by the VVER RP General Designer together with the experts of the Mechanical Engineering Research Institute of the Russian Academy of Sciences named after A.A. Blagonravov [172–177]. Nonetheless, efforts to justify the vibration strength of the reactor plant only by calculations have failed so far [171]. Moreover, this has not been only a domestic problem.

The works performed [134–177] can be attributed to the first phase of development of vibration acoustics. Complex mathematical tools were not always justified by experimental studies. Theoretically, the tasks were set very far from applications, for example, the problems of turbulent flows around the shells, based on which it is extremely difficult to justify the full-scale measurements of the RP equipment vibration and pressure pulses of the coolant in the MCC. Such activities shall be carried out continuously, therefore, increasing the modern systems of reactor measurements (regular monitoring systems, commissioning measurement systems (CMS) and stationary diagnosis system); it is also necessary to continuously develop computational and analytical methods of vibration acoustics.

We could not help mentioning diagnostic activities on rotary equipment [178–187] as an allied area of technical diagnosis. Even more so that the main source of vibration acoustic energy is the main circulation pump set (MCPS). Diagnostics of rotary equipment favorably differs from other areas of technical diagnostics by the presence of *a priori* numerical values of diagnostic thresholds in absolute physical units.

Thus, two following areas of vibration acoustic applications to reactor engineering shall be indicated: design and operation. A special role is given to the very first operation stage of the reactor plant, the commissioning test stage. The flow of vibration acoustic diagnostic information received by special CMS can hardly be overestimated [188–200]. It is necessary to make the starting vibration acoustic image of the reactor plant and to "teach" the stationary diagnostic systems.

Vibration calculations are confirmed by a positive operating experience of NPP units. This experience has also allowed to establish criteria of extensive vibration measurements during commissioning on the new VVER units. The exhaustion of the vibration strength during the operation of the power unit shall be constantly monitored due to the uniqueness of the vibration state of each power unit and the individual history of its operational vibration loading. Vibration measurements should accompany the entire life of the reactor plant from the period of completion of its installation to its decommissioning. It is possible to obtain reliable empirical regularities of the RP equipment vibration by its measurements carried out for a long time and on a sufficiently large fleet of NPP units.

The current strength standards require the impossible, namely, to specify the amplitude and frequency of vibrations. In view of their *a priori* uncertainty, strictly speaking, the application of strength standards in this form could not serve as justification for the design vibration strength. Comprehensive operational vibration monitoring of the RP equipment is only possible with implementation of stationary systems of vibration noise diagnostics.

We have tried to go from the simplest differential equations of motion and vibration acoustics to the equations applicable in the interpretation of multichannel full-scale reactor measurements. Thus, such an important area of vibration acoustics as numerical simulation was shot out. In diagnostic applications, it is equally important to have a result, albeit in a simplified but an analytical form. Therefore, the monograph has addressed analytical methods. To make up for the above, the references [201–218] show the main achievements of numerical simulation of MCC vibration and acoustics.

At the same time, we sought the simplest mathematical formalism: In this monograph, to understand the process of deducing a formula, it is enough to cover the course of mathematical analysis of a technical college. We avoided vector-matrix equations (where possible) in principle in order to simplify the presentation. We sought to consider spatially one-dimensional cases (where possible) to present the physical essence of the phenomenon. If desired, the reader can engage in multi-dimensional generalizations. We also sought to give a mathematical conclusion not only to new ideas, but also to already well-known physical methods. In order to understand the limitations applicable to one or another method, it is enough to trace its mathematical conclusion. Unfortunately, applied vibration acoustics in its numerous similarities with mechanics and electrical engineering suffers from abundant errors due to the incorrect application of known formulas.

In presenting new ideas, we embraced a known one: "There is nothing more practical than a good theory." The authorship of this statement is attributed to dozens of scientists. We are pleased to know that of the mentioned scientists there are some brightest minds in vibration acoustics — G. Kirchhoff and N.N. Bogoliubov.

Problems of reactor vibration acoustics were often presented as ones of reactor noise [219–332]. Historically, regular reactor measurements of the first designs of both domestic and foreign reactor plants did not contain vibration acoustic measuring channels. Vibration acoustic information was mainly extracted from ex-core and in-core neutron measurement channels, and they were used for entirely different purposes. The noise (fluctuating) component of the signal was isolated from regular neutron measurements. This scientific area in reactor engineering was generalized by the term "reactor noise", and quite effective international scientific congresses were devoted to it (SMORN – "Specialist Meeting on Reactor Noise" [219–284]). This is how the Organizing Committee of the conferences positions itself [220]: SMORN is an international symposium devoted to the study and applications of reactor noise in order to monitor, diagnose and improve knowledge of physical processes in nuclear reactors. The SMORN Symposium aims to bring together both professionals and operators of monitoring systems currently installed in many nuclear reactors around the world. The first seven conferences were held: in 1974 in Casaccia (Italy), in 1977 in Gatlinburg (USA), in 1981 in Tokyo (Japan), in 1984 in Dijon (France), in 1987 in Munich (Germany), in 1991 in Gatlinburg (USA) and in 1995 in Avignon (France). If the first conferences were devoted to research and methods for determining the core physical parameters of research and power reactors using noise technology, the subject of the latest activities covered development and application of methods of supervision and diagnostics at commercial nuclear power plants.

Conferences 2–4 were particularly effective (the first conference was of an academic nature and was entirely devoted to sub-critical noise studies of non-energy multiplying media). At that time, noise studies were not yet commercialized, and scientists from different countries openly discussed development of new noise methods, bringing them to perfection as an analytical form of a physical model. The leaders can be enumerated by their names: Thie J.A., Wehling H.-J., Kosaly G., Wach D., Uhric R.E., and Saito K.

Then came the first expensive industrial systems of early noise diagnosis [236, 262, 264, 266, 276, 279, 284], implemented at various nuclear power plants with declared functionality and signal processing methods that were undisclosed for competitive reasons. Statistical information on use of diagnostic systems at nuclear power plants in different countries has now become practically useful. For example, information was provided for about as many as 54 power units of the French nuclear power plants [278], or the entire scope of the noise diagnostics data from Germany (when it still existed there), the USA, Belgium, Hungary, Finland was summed up [274, 275, 279–284]. Unfortunately, since 2003 (after SMORN-8) and at the time of publication of this book, none of the nuclear power states has taken on the difficulties of organizing and conducting the SMORN-9.

Reactor-noise studies of Hungarian authors evolved into an entire scientific school [207, 229, 241, 244, 252, 253, 255, 256, 261, 268, 270]. They flourished during the period of successive launches of four VVER-440 units of Paks NPP, starting from 1983. Further operation of these units led to development of the Hungarian noise diagnosis systems. The Hungarian school ceased to exist after 1991 and Hungarian researchers spread to various international scientific centers.

Despite the fact that the leadership in reactor noise research belongs to foreign researchers, we will specifically highlight the domestic pioneer works on reactor noise of the time [109, 121, 125, 145, 206, 212]. The acoustics of the VVER main circulation circuit as one of the main sources of vibration and, at the same time, a source of barometric neutron reactivity effects, is the subject of a number of research works of Russian authors [134, 135, 137, 145, 153]. For example, the pipeline vibrations were analyzed taking into account acoustic standing waves of the designed level [135, 137].

In countries with a developed nuclear power, usually a triad of scientific, design and operating companies is formed. In the Russian VVER nuclear power industry, it has the following form: the General Designer (JSC OKB "Gidropress"), the Scientific Supervisor (NRC Kurchatov Institute) and the operating company (JSC Concern Rosenergoatom that includes as branches all the Russian nuclear power plants in Russia). A specialist, who happened to work in one of these organizations, as fate would have it, should attach himself or herself to either a scientific or a design or an operating school, which does not always correspond to his or her natural talent. Three authors of this monograph, employed by three different companies of the industry, have complemented each other (as it turned out later) for a quarter of century, participating in numerous joint diagnostic projects. The interaction, however, has not always been cloudless and been accompanied by fundamental disputes that only improved the final product.

In addition to this triad, there is also a university science [333–353]. On the one hand, the higher education institutions constitute a kind of conveyor producing engineers and graduate students for companies of the above triad, who then become directors, chief engineers, scientific supervisors and scientists or engineers. On the other hand, university science is outside the Rosatom circuit, which creates departmental barriers to distribution of reactor measurement data. Honorable professors remain on their "conveyor", and not only give lectures, but also engage in scientific activities in the conditions of information hunger (the complete absence or extremely scarce flow of measured information from NPP units in operation.

Academic science is in a similar situation, with the only difference that its potential is sometimes addressed by the Chief Designer and Scientific Supervisor of VVER projects. In these cases, the most powerful scientific and technical breakthroughs are observed [172–177].

The insidiousness of vibration is that when its amplitude is small enough, its contribution to the exhaustion of the equipment life can be neglected, but vibrations can play a decisive role in resonant vibration phenomena. Resonant excitation of vibrations, of course, is considered unacceptable by any RP design. Acoustic shock effects are also not allowed in the coolant circulation circuits, such as hydraulic shock [170]. Yet, they do happen. The property of unpredictable growth of vibration amplitudes at resonance phenomena makes vibrations a permanent factor in our studies. Either the wear processes caused changes in the natural frequencies of equipment vibrations and they coincided with the frequency of the driving force, or there was external unpostulated driving force, or some structure was operating in the stationary operation mode with constant but high and uncontrolled vibration amplitudes, and the damage is fatigue-based.

This is a typical list of issues arising in the search for the causes of failures. There is a continuous flow of publications on vibration threats [120, 149, 311, 324, 343, 348–353].

In addition, many of the thousands of measurements that accompany operation of a nuclear reactor can be interpreted through the laws of vibration acoustics. From these measurements additional diagnostic information about vibration states of reactor plant assemblies and thermal state of its circuit fluids is obtained [228–231, 235, 238, 240–245, 250–259, 261, 263, 283, 289, 294, 313, 315, 318, 332]. Simultaneous measurements of ex-core and in-core neutron noise with signals of vibration sensors or pressure pulse sensors of pressure pulsations (DPD) promise new technological breakthroughs in reactor measurements. Moreover, it will be on-line noise measurements, indistinguishable in their functionality from traditional reactor deterministic measurements, for example, temperature or pressure measurements.

The modest contribution of the authors has also spinned off into a separate topic [354–385]. Working in different organizations, but closely communicating with each other, they covered by their scientific interests almost all the problems of reactor vibration acoustics. In our defense, we can only make it a point to note that the authors, working quite long in the area of applied reactor diagnostics, are "infected" with a very productive phenomenological approach. In accordance with it, at first, a sustainable experimental dependence is observed, and only then, a mathematical model is built, which does not necessarily go back to the fundamental physical laws. For example, as the basic premise a delay of one signal relative to another can be declared. Thus was born a well-known phenomenological theory of globality–locality of the neutron field of the boiling water reactor, in which the delay was declared as a consequence of the steam motion together with the flow of coolant [301]. Spectral features of the signals are first presented, then they demonstrate a stable physical phenomenon, for example, a linear dependence of the resonance frequency on the coolant temperature, and then, as a rule, mathematical conclusions follow. Although there are cases of the reciprocal sequence of the original application of the equations and, in particular, the Navier–Stokes equations, and the subsequent experimental confirmation.

The monograph most often demonstrates the spectral characteristics of the signals received at the new VVER-1200 power unit

(Unit 1 of Novovoronezh NPP-2) for two reasons. The first VVER-1200 power unit is equipped with the maximum number of standard stationary diagnostic systems, which allowed almost continuous flow of high quality information. The second VVER-1200 power unit is planned as a standard reactor designed to replace decommissioned nuclear power units in Russia and to supply abroad. Inevitably, there will be problems of personnel training, including diagnostics, and this monograph may be of significant help (in the humble hope of the authors). Of course, we used the experience gained at numerous reactor plants of other VVER reactor plant designs (VVER-1000, VVER-440). In addition, the software tools of long-term authorship are used, which remain "behind the scenes". However, the quality of spectral estimation shows the uniqueness of methods and algorithms of these software tools. They are integrated into a universal software package called "WING", which contains all the necessary range of methods for processing both arbitrary signals and reactor signals.

In conclusion, we have collected references to domestic and foreign regulatory documents on vibration, shock, reactor measurements and monitoring and diagnosis systems [386–410], thereby demonstrating the gap between theory and implemented practice.

One should not consider the bibliographic references presented by the authors of the monograph to be an exhaustive one though they are way or another related to vibration acoustics. Only significant works are included, according to the authors' subjective point of view, so we apologize in advance to the specialists whose works remained outside the scope of our attention.

We also hope that this book will be useful for college professors of not only reactor engineering colleges, but also for other technical universities in development of a course of lectures on vibration acoustics, and for students of these colleges for their exams.

Writing a monograph is a laborious and painstaking process, which would not have been possible without the work of our numerous predecessors, so we would like to mention the Russian experts who have made a practical contribution to development of vibration acoustics of reactor plants not only by the references to the publications. The authors also considered it important to mention their colleagues, highly qualified specialists, interactions with whom largely determined the authors' position on a particular application issue of control and diagnosis of reactor plants. This scientific and technical potential is distributed among the companies of the industry:

NRC "Kurchatov Institute": Kalinushkin A.E., Kaplar E.P., Korneev B.S., Semchenkov Y.M., Ustinov V.S., Shikalov V.D.

JSC OKB "GIDROPRESS": Berkovich V.Y., Bogachov A.V., Mineev S.A., Piminov V.A., Khairetdinov V.U.

Rosatom SC: Asmolov V.G., Lokshin A.M., Pavlov D.V.

JSC NIKIET: Strelkov B.P., Strelkov P.B.

JSC Rosenergoatom Concern: Berezanin A.A., Gutsev D.F., Katsman A.Í., Lovtchev V.N., Teterin Y.P.

Joint Stock Company "State Scientific Centre of the Russian Federation – Institute for Physics and Power Engineering: Dvornikov P.A., Kudryaev A.A., Kupriyanov V.M., Shvetsov D.M., Fedotovskikh V.S.

JSC SNIIP: Burtsev I.Y.

Mechanical Engineering Research Institute of the Russian Academy of Sciences: Kaplunov S.M., Makhutov N.A.

JSC VNIIAES: Verezemsky V.G., Davidenko N.N., Lupishko A.N., Potapov V.V., Taratunin V.V., Tereshonok V.A.

JSC RASU: Antipov S.I., Michailov M.N.

JSC "Atomtechenergo": Blokhin V.N.

JSC NITI named after Alexandrov: Klimov Y.T., Pankin A.M.

JSC STCD: Zhidkov S.V., Kalinin A.N., Kopiev S.Y., Kotelnikov V.I., Krylov K.A., Matveev V.P., Finkel B.M.

JSC KVANT PROGRAMM: Lazarev N.A.

JSC Atomenergoproekt: Yegorov S.V., Sarkis G.G., Topchiyan R.M.

Obninsk Institute for Nuclear Power Engineering of NRNU MEPhI: Leskin S.T., Skomorokhov A.O., Trofimov A.I.

Nizhny Novgorod State Technical University named after Alexeyev: Znyshev V.V., Ovchinnikov V.F., Smirnov L.V.

Moscow Power Engineering Institute: Gorbatykh V.P., Proskuryakov K.N.

Bauman Moscow State Technical University: Perevezentsev V.V., Solonin V.I.

Development of vibration and noise diagnostics required not only scientific research and development activities in various research institutes and design centers but also experiments carried out directly at the sites of NPPs with VVER reactor plants. The authors see it as a pleasant duty to thank the highly qualified specialists of Russian nuclear power plants who have made the greatest contribution to development of this very difficult area of reactor engineering:

Kozhin M.L., Titov S.I. — **Kola NPP;**
Muravsky V.B., Rumyantseva N.G. — **Kalinin NPP;**
Adamenkov A.K. — **Rostov NPP.**

The authors express their special and sincere gratitude to the specialists of employees of JSC STCD:

Svezhintseva P.Y. and Chernenko V.B. for assistance in processing of the experimental material;
Belov A.I. for critical comments contributing to the improvement of the book;
Kovalev E.V., Nikolaeva I.S., Tolokova P.A., Turaletova O.I. for their help in editing and organization of work for publication.

We are grateful for valuable feedback from Voronin L.M. We also thank the management and employees of JSC VNIIAES (General Director Tukhvetov F.T.) for their support.

Moreover, the monograph would not be complete without the practical part. The authors are grateful for the assistance in arrangement of activities on Unit 1 of Novovoronezh NPP-2, its director PhD Povarov V.P., Chief Engineer Fedorov A.I. and First Deputy Chief Engineer Vitkovsky S.L.

We sincerely admire the professionalism of the Novovoronezh NPP personnel who managed to plan and conduct a unique series of experiments for a limited time. These materials are the basis for the chapters devoted to vibration acoustics issues of the VVER-1200 reactor plant. It is necessary to mention personally:

Shevtsov S.I., the developer of the original measurement circuits, and the indispensable supervisor of the activities;

Petryanik V.L. and Tabunschik V.V., who performed the entire experimental part at a high professional level.

It would be wrong not to mention the staff of the journal *Izvestiya Vuzov* of the Nuclear Energy of Obninsk Institute for Nuclear Power Engineering of NRNU MEPhI represented by the Chief Editor Doctor of Science Kazansky Y.A. and Executive Secretary of the editorial board Mironovich V.L. It is on the pages of this journal and with their direct involvement that materials published that served as the basis for some chapters of the monograph were first published.

Abbreviations

ADS	absolute displacement sensor
AED	asynchronous electric drive
AFL	amplitude-frequency locus
ASPD	autospectral power density
ASW	acoustic standing wave
CCF	cross-correlation function
CM	commissioning measurements
CMS	Commissioning Measurement System
CSPD	cross-spectral power density
DCD	direct charge detector
DS	dynamic system
EPS	emergency protection system
FA	fuel assembly
FE	fuel element
FEM	finite element method
GOST	Russian national standard
HET	heat exchange tubes
HF	high frequency
HFF	low-frequency filter
IC	ionization chamber
ICND	in-core noise diagnostics
KNPP	Kalinin NPP
LD	logarithmic decrement
LDS	leak detection system
LF	low frequency
LFF	low-frequency filter

LL	long line
LPDS	Loose Part Detection System
LSM	least square method
MAR	multidimensional autoregressive
MAT	mean acoustic trajectory
MCC	main circulation circuit
MCP	main circulation pump
MCPL	minimum controllable power level
MCPS	main circulation pump set
ND	nominal diameter
NPP	Nuclear Power Plant
NVNPP	Novovoronezh NPP
PDMS	Pipeline Displacement Monitoring System
PFL	phase-frequency characteristic
PPS	pressure pulse sensor
PTU	protective tube unit
RC	reactor cavity
RDS	relative displacement sensor
RMS	root-mean-square value
RP	reactor plant
SG	steam generator
SHF	super high frequency
SLDS	Secondary Leak Detection System
SO	scheduled outage
TDS	Technical Diagnostic System
TEH	thermoelectric heater
TNPP	Tianwan NPP
VMS	Vibration Monitoring System
VVER	water-cooled power reactor
WCH	water chemistry

Contents

Chapter 1

Equations of Motion of Dynamic System

1.1 Single Degree of Freedom Dynamic Systems

The analytical description of mechanical vibrations is based on the equations of motion of dynamic systems [17, 21–29, 31–37, 40–45, 55, 97–100, 102, 103, 106]. Under the dynamic system, we understand an object that has the concepts of state and the law of its change defined. A basic dynamic system is described by an ordinary second-order linear differential equation:

$$M \cdot \frac{d^2x(t)}{dt^2} + C \cdot \frac{dx(t)}{dt} + K \cdot x(t) = f(t) \tag{1.1}$$

It describes the motion $\ll x(t) \gg$ of the mass $\ll M \gg$ connected to a fixed support by a spring with a stiffness $\ll K \gg$ due to a force $\ll f(t) \gg$. The dynamic system (DS) has the property of energy dissipation represented by a parameter called damping — "C".

Equation (1.1) is a balance of forces

$$f_{\mathrm{M}}(t)+f_{\mathrm{C}}(t)+f_{\mathrm{K}}(t) = f(t)$$

The external force $f(t)$ causes:

- inertial force f_{M}, giving the acceleration to the body with a mass M;
- damping force f_{C}, arising due to friction in motion proportional to the velocity of the body through the C coefficient;

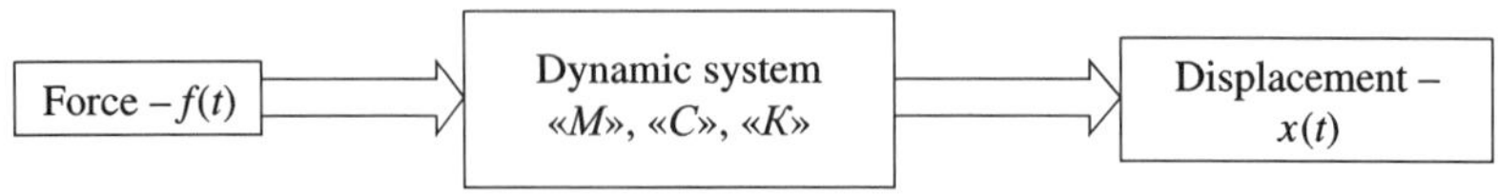

Figure 1.1: Presentation of a dynamic system using the input–output diagram.

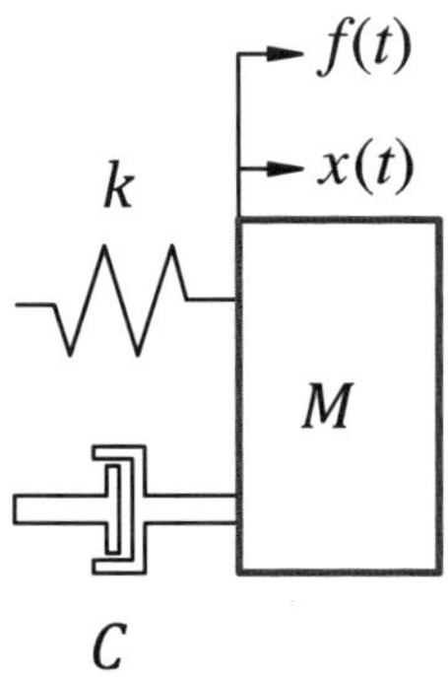

Figure 1.2: Schematization of a dynamic system with a single degree of freedom spring pendulum.

- elastic reaction force f_K, proportional to the actual displacement through the K coefficient (Fig. 1.1).

The three above forces $f_M(t), f_C(t), f_K(t)$ constitute reaction to, respectively, inertia, dissipation and stiffness of a dynamic system. In a static case when the displacement is constant in time x_0, Eq. (1.1) is reduced to the definition of the concept of stiffness $\ll K \gg$: $K \cdot x_0 = f_0$. The graphical equivalent of Eq. (1.1), consisting of a spring, damper and mass, is shown in Fig. 1.2.

1.2 Simplest Analytical Model of Oscillations of Main Equipment of VVER Heat-Exchanging Loop

This section discusses a two-degrees of freedom dynamic system or the so-called two-mass model. As a roughest approximation, one mass means the main circulation pump (MCP) set weight — M_1 and the other means the SG weight — M_2 (1.2). The balance of forces for each mass gives a system of two differential equations similar to Eq. (1.1):

$$\begin{cases} M_1 \cdot \dfrac{d^2x_1}{dt^2} + (C_1 + C_3)\dfrac{dx_1}{dt} - C_3 \cdot \dfrac{dx_2}{dt} + (K_1 + K_3)\,x_1 - K_3 \cdot x_2 = f_1 \\ M_2 \cdot \dfrac{d^2x_2}{dt^2} + (C_2 + C_3)\dfrac{dx_2}{dt} - C_3 \cdot \dfrac{dx_1}{dt} + (K_2 + K_3)\,x_1 - K_3 \cdot x_1 = f_2 \end{cases} \tag{1.2}$$

After the Laplace transformation,[1] we come to algebraic equations:

$$\begin{cases} x_1[M_1 \cdot p^2 + (C_1 + C_3)\,p + (K_1 + K_3)] - x_2[C_3p + K_3] = f_1 \\ x_2[M_2 \cdot p^2 + (C_2 + C_3)\,p + (K_2 + K_3)] - x_1[C_3p + K_3] = f_2 \end{cases}$$

or

$$\begin{cases} x_1[z_1 + C_3p + K_3] - x_2[C_3p + K_3] = f_1 \\ -x_1[C_3p + K_3] + x_2[z_2 + C_3p + K_3] = f_1 \end{cases}$$

where z_1^{-1} is a transfer function of Dynamic System DS_1, z_2^{-1} is a transfer function of Dynamic System DS_2.

The condition that the determinant $D = 0$ allows finding the resonant frequencies

$$(z_1 + I) \cdot (z_2 + I) - I^2 = 0, \quad \text{where } I = C_3p + K_3$$

or

$$\frac{z_1 \cdot z_2}{I^2} + \frac{1}{I}(z_1 + z_2) = 0$$

or

$$\frac{1}{z_1} + \frac{1}{z_2} = -\frac{1}{I}$$

or

$$H_1(p) + H_2(p) = -\frac{1}{I(p)} = -\frac{1}{C_3p + K_3} \tag{1.3}$$

[1] *Note*: In steady state case Fourier transformation can be used, which leads directly to Eq. (1.5).

At zero damping, for the relationship of DS1 and DS2 ($C_3 = 0$), the expression (1.3) stops being complex

$$\frac{\frac{1}{M_1}}{p^2 + \frac{R_1}{M_1}} + \frac{\frac{1}{M_2}}{p^2 + \frac{R_2}{M_2}} = -\frac{1}{K_3}$$

or after replacing the Laplace variable p with an imaginary frequency $p = j \cdot \omega$, we have:

$$\frac{M_2}{\omega^2 - \omega_{01}^2} + \frac{M_1}{\omega^2 - \omega_{02}^2} = \frac{M_1 \cdot M_2}{K_3} \tag{1.4}$$

An illustration of this relationship is given in Fig. 1.3.

For the purpose of practical interpretation, we should note that instead of two resonances — ω_{01}, ω_{02} formed separately of DS_1 and DS_2, a two-mass DS would manifest from four to one resonance depending on the stiffness of the connection of the two DS.

Figure 1.4 shows a "soft" connection of the two DS corresponding to a spring with low rigidity — K_3 and a high level of compliance — $\frac{1}{K_3}$. In this case, we have one pair of resonant frequencies near ω_{01} and the other one near ω_{02}. With a "stiff" connection, four resonances are observed at a considerable distance from each other. With a "very stiff" connection, a system of two masses can only have one high-frequency resonance.

A connecting element with a stiffness K_3 simulates a U-shaped pipeline. The above qualitative result indicates that by changing only

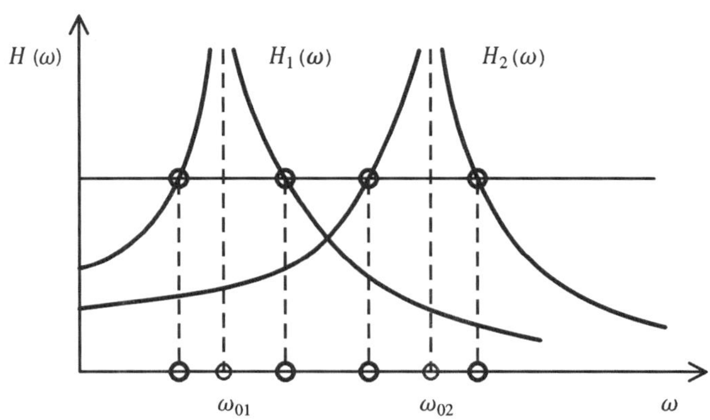

Figure 1.3: Natural frequency splitting effect.

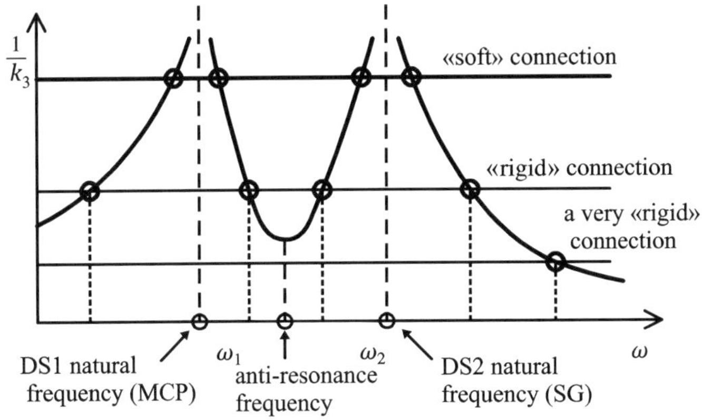

Figure 1.4: Three levels of stiffness of two DS connection.

its stiffness characteristics (for example, when designing a loop) and leaving the "cold" and "hot" legs unchanged, it is possible to obtain resonant vibration frequencies of the loop in a very wide frequency range. In particular, only by changing one U-shaped pipeline, it is possible to "dilute" the unwanted coincidence of the frequencies of the acoustic standing wave (ASW) and MCPS with the natural frequencies of the loop equipment. The local minimum in the vibration displacement spectrum in Fig. 1.4, called anti-resonance, can be selected so that its frequency coincides with the MCP set rotational frequency. The most natural way would be absorbing vibration forced by the MCP set. We emphasize that we are not talking about the MCP set natural oscillation but the MCP set rotational frequency $\mathrm{F_{MCPS}}$.

Figures 1.5 and 1.6 show equivalent schemes of the two-mass model and the balance of forces for each mass. It is obvious that this one-dimensional model (with a single space coordinate — "x") is very rough, suitable for qualitative interpretation of the vibration effects of the VVER loop. The real stiffness of the pipeline cannot be described by a single number — K but set by a matrix $[K]$, which determines the longitudinal, transverse and torsional vibrations of a basic dynamic system. When the set of basic DS is connected to a real structure, strictly speaking, it is impossible to interpret its fluctuations in terms of longitudinal, transverse or any other oscillation types. However, by isolating a single resonance from the set observed in the experiment, one can make assumptions in favor of

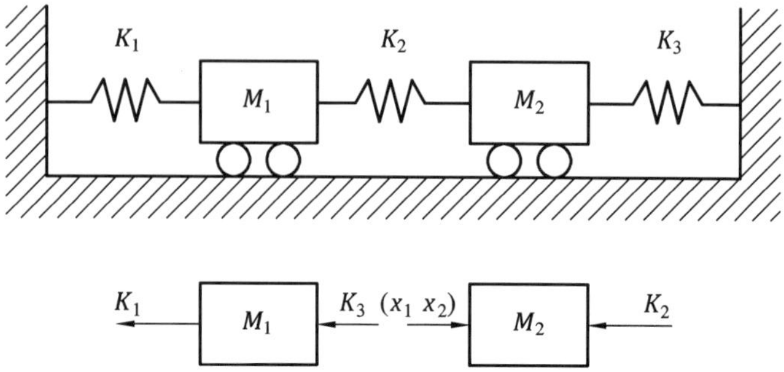

Figure 1.5: Schematization of free oscillations of a two-mass vibration model without damping.

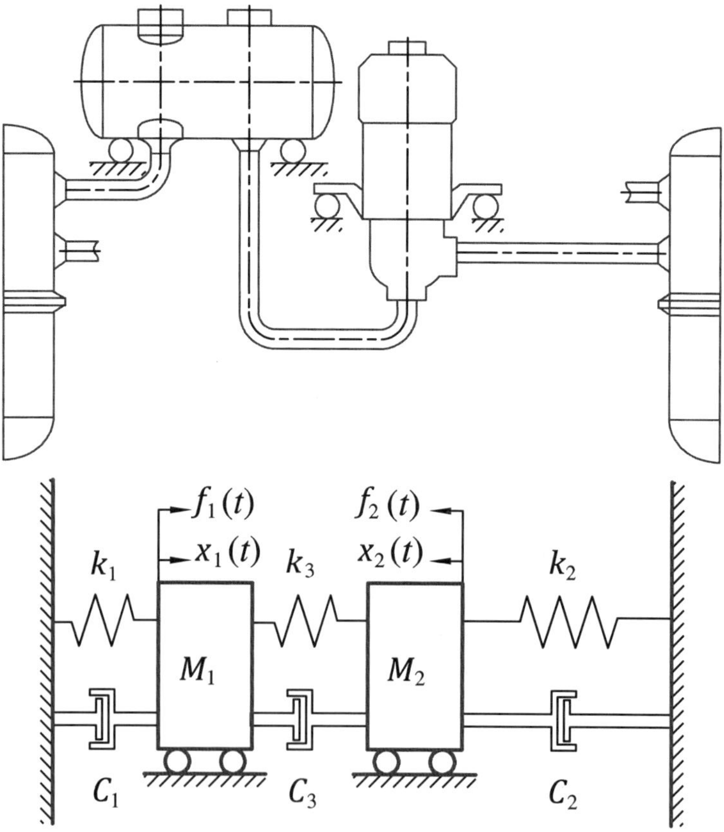

Figure 1.6: Schematization of the loop of a two-mass DS with three stiffnesses and three dampings.

the dominance of any one type of oscillations using the two-mass model discussed.

That means, this model is developed not to adequately describe all the observed resonances, but to find the causes of the resonances at a qualitative level.

In obtaining the expression (1.3) damping in DS1 and DS2 was neglected not quite correctly. It is also incorrect to carry out real levels $\frac{1}{K_3}$ in the complex plane. In the following, it is demonstrated in terms of complex transfer functions what exactly is the neglect. Differential equations of two-mass DS written in the time domain

$$\begin{cases} M_1 \cdot \ddot{x}_1 = -K_1 \cdot x_1 + K_3 \left(x_2 - x_1\right) \\ M_2 \cdot \ddot{x}_2 = -K_3 \left(x_2 - x_1\right) - K_2 \cdot x_2 \end{cases}$$

in the frequency domain in terms of two transfer functions are converted to the form:

$$H_1 \left(p\right) + H_2 \left(p\right) = -\frac{1}{K_3},$$

which, after multiplying by the complex-conjugate equation

$$H_1^* \left(p\right) + H_2^* \left(p\right) = -\frac{1}{K_3}$$

is written like this:

$$\left|H_1 \left(p\right)\right|^2 + \left|H_2 \left(p\right)\right|^2 + H_1 \left(p\right) \cdot H_2^* \left(p\right) + H_2 \left(p\right) \cdot H_1^* \left(p\right) = \frac{1}{K_3^2}$$

Squared absolute value of the transfer function $\left|H\left(p\right)\right|^2$ is usually larger than the cross-summand $\left|H_1\left(p\right) \cdot H_2^*\left(p\right)\right|$. The curve $\left|H\left(p\right)\right|^2$ has one hump and the curve $\left|H_1\left(p\right) \cdot H_2^*\left(p\right)\right|$ has two humps and when the resonance frequency ω_1 is separated from the second resonant frequency ω_2 by a significant "frequency" distance $\omega_1 \ll \omega_2$, the local maximum of one curve is multiplied by the small values of the "tail" of the other curve.

If $H_1\left(p\right) = H_2\left(p\right)$, then the related oscillations of two identical DS are not a resonance excitation, but an increase in the amplitude of the oscillations and an increase in the frequency distance between the two resonances (Fig. 1.7).

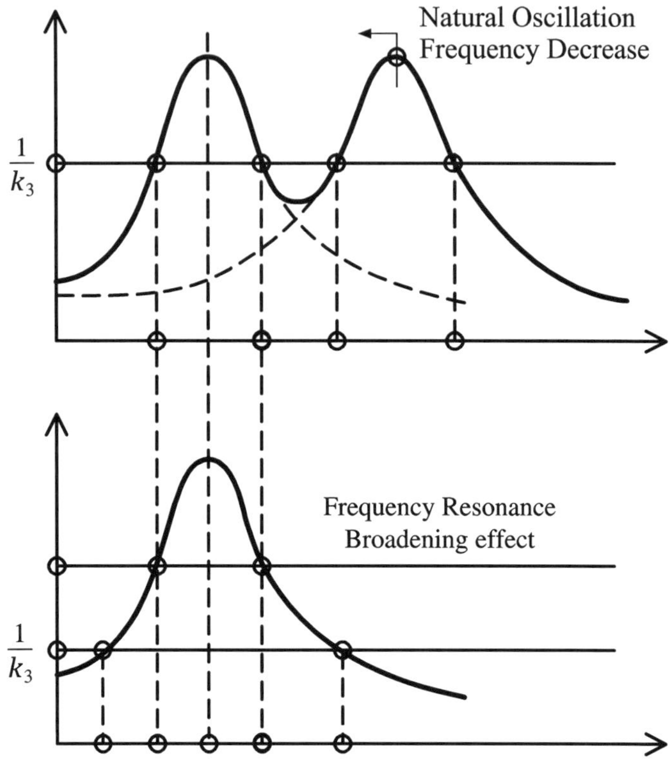

Figure 1.7: Coincidence of natural frequency of SG and MCP set when the "hot" leg has been prolonged.

The original system (1.2) in the frequency domain has the form:

$$\begin{bmatrix} x_1 \\ x_2 \end{bmatrix} \cdot \begin{bmatrix} -M_1\omega^2 + K_1 + K_3 & -K_3 \\ -K_3 & -M_2\omega^2 + K_2 + K_3 \end{bmatrix} = \begin{bmatrix} K_1 \cdot x_0 \\ K_2 \cdot x_0 \end{bmatrix} \tag{1.5}$$

or

$$\begin{bmatrix} x_1 \\ x_2 \end{bmatrix} \cdot \begin{bmatrix} z_1(\omega) + K_3 & -K_3 \\ -K_3 & z_2(\omega) + K_3 \end{bmatrix} = \begin{bmatrix} K_1 \cdot x_0 \\ K_2 \cdot x_0 \end{bmatrix}$$

$$z_1(\omega) = K_1 - M_1\omega^2$$

$$z_2(\omega) = K_2 - M_2\omega^2$$

Then we have determinants:

$$\Delta = (z_1 + K_3)(z_2 + K_3) - K_3^2$$

and

$$\Delta_{x1} = \begin{vmatrix} K_1 \cdot x_0 & -K_3 \\ K_2 \cdot x_0 & z_2 + K_3 \end{vmatrix} = K_1 \cdot x_0 \cdot z_2 + K_3 \cdot K_1 \cdot x_0 + K_3 \cdot K_2 \cdot x_0$$

$$= x_0[K_1 \cdot z_2 + K_3(K_1 + K_2)],$$

$$\Delta_{x2} = \begin{vmatrix} z_1 + K_3 & K_1 \cdot x_0 \\ -K_3 & K_2 \cdot x_0 \end{vmatrix} = K_2 \cdot x_0 \cdot z_1 + K_3 \cdot K_2 \cdot x_0 + K_3 \cdot K_1 \cdot x_0$$

$$= x_0[K_2 \cdot z_1 + K_3(K_1 + K_2)]$$

Accordingly, the SG and MCP set vibration displacements shall be written in the following form using Kramer's method as solutions of Eq. (1.4):

$$x_1 = \frac{\Delta_{x1}}{\Delta} = x_0 \frac{K_1 \cdot z_2 + K_3(K_1 + K_2)}{(z_1 + K_3)(z_2 + K_3) - K_3^2}$$

$$x_2 = \frac{\Delta_{x2}}{\Delta} = x_0 \frac{K_2 \cdot z_1 + K_3(K_1 + K_2)}{(z_1 + K_3)(z_2 + K_3) - K_3^2}$$

The vibration displacements of the first mass will be zero ($x_1 = 0$), when

$$z_2 = -\frac{K_3\,(K_1 + K_2)}{K_1} = -K_3 - K_3 \cdot \frac{K_2}{K_1}$$

$$K_2 - M_2\omega^2 = -K_3\left(1 + \frac{K_2}{K_1}\right)$$

Hence,

$$\omega_{1*}^2 = \omega_{20}^2 + \frac{K_3}{M_2}\left(1 + \frac{K_2}{K_1}\right) = \omega_{20}^2 + \frac{K_2 \cdot K_3}{M_2 \cdot K_2}\left(1 + \frac{K_2}{K_1}\right)$$

$$= \omega_{20}^2[1 + \iota_{2K} + \iota_{1K}]$$

At a frequency greater than the second natural frequency, the so-called anti-resonance (zero amplitude) of the first mass will be observed.

Similarly, $x_2 = 0$, when

$$z_2 = -\frac{K_3\left(K_1 + K_2\right)}{K_2} = -K_3 - K_3 \cdot \frac{K_1}{K_2}$$

$$K_1 - M_1\omega^2 = -K_3\left(1 + \frac{K_1}{K_2}\right)$$

$$\omega_{2*}^2 = \omega_{10}^2 + \frac{K_3}{M_1}\left(1 + \frac{K_1}{K_2}\right) = \omega_{10}^2 + \frac{K_1 \cdot K_3}{M_1 \cdot K_1}\left(1 + \frac{K_1}{K_2}\right)$$
$$= \omega_{10}^2[1 + \iota_{1K} + \iota_{2K}],$$

the anti-resonance is as many times higher than the first resonance.

The ratio of vibration displacements of the first and second mass depends on the ratio of the three DS stiffnesses and the two resonant frequencies:

$$\frac{x_1}{x_2} = \frac{K_1 \cdot z_2 + K_3\left(K_1 + K_2\right)}{K_2 \cdot z_1 + K_3\left(K_1 + K_2\right)} = \frac{K_1}{K_2} \cdot \frac{z_2 + K_3\left(1 + \frac{K_2}{K_1}\right)}{z_1 + K_3\left(1 + \frac{K_1}{K_2}\right)}$$

$$= \iota_K \cdot \frac{z_2 + K_3\left(1 + \frac{1}{\iota_R}\right)}{z_1 + K_3\left(1 + \iota_R\right)} = \iota_R \cdot \frac{K_2 - M_2\omega^2 + K_3\left(1 + \frac{1}{\iota_R}\right)}{K_1 - M_1\omega^2 + K_3\left(1 + \iota_R\right)}$$

$$= \iota_K \cdot \frac{K_2\left(1 - \left(\frac{\omega}{\omega_{02}}\right)^2\right) + K_3\left(1 + \frac{1}{\iota_K}\right)}{K_1\left(1 - \left(\frac{\omega}{\omega_{01}}\right)^2\right) + K_3\left(1 + \iota_K\right)}$$

$$= \frac{1 - F_{02}^2 + \iota_{2K}\left(1 + \frac{1}{\iota_K}\right)}{1 - F_{01}^2 + \iota_{1K}\left(1 + \iota_K\right)}$$

$$= \frac{1 - F_{02}^2 + \iota_{1K} + \iota_{2K}}{1 - F_{01}^2 + \iota_{1K} + \iota_{2K}},$$

where $\frac{K_1}{K_2} = \iota_K$ is the stiffness ratio; $F_{01} = \frac{\omega}{\omega_{01}}; F_{02} = \frac{\omega}{\omega_{02}}; \iota_{1K} = \frac{K_3}{K_1}; \iota_{2K} = \frac{K_3}{K_2}$.

Hence, the vibration displacements:

$$x_1 > x_2, \quad \text{if} \quad 1 - F_{02}^2 + \iota_{1K} + \iota_{2K} > 1 - F_{01}^2 + \iota_{1K} + \iota_{2K}$$
$$F_{01} > F_{02}$$

or

$$\frac{\omega}{\omega_{01}} > \frac{\omega}{\omega_{02}} \rightarrow \omega_{02} > \omega_{01}$$

If $\omega_{0\mathrm{SG}} > \omega_{0\mathrm{MCPS}}$, then for vibration displacements it is usually $X_{\mathrm{MCPS}} > X_{\mathrm{SG}}$ (observed in practice).

Let us imagine that the "hot-leg" pipeline is lengthened with the other loop parameters remaining constant (DS_1 and DS_2 parameters). Let the "hot-leg" pipe be elongated until the SG and MCP natural frequencies ω_{MCPS} and ω_{SG} are equal. What is the danger of such a coincidence? It is not as dangerous as the coincidence of the natural frequencies of the DS with the frequency of the external force, at which vibration amplitudes grow infinitely. Coincidence of ω_{MCPS} and ω_{SG} cannot be called a condition of resonant excitation of SG and MCP set oscillations. The coincidence of the transfer functions $H_{\mathrm{SG}}(\omega)$ and $H_{\mathrm{MCPS}}(\omega)$ leads to a doubling of the vibration amplitude (but no more than that, in contrast to the resonance excitation), the effect of broadening the resonance frequencies and reducing their number (Fig. 1.8).

Under certain conditions, the two-mass oscillatory system leads to a favorable vibration effect of "damping" vibrations (Fig. 1.4). In addition to the full-scale two-mass scheme described above, the anti-resonance frequency can be found by a simplified scheme.

In this simplified scheme, the frequency of anti-resonance is $\omega_0 = \sqrt{\frac{K_3}{M_2}}$, only determined by the DS_2 resonance properties (the second mass is the SG mass).

Knowledge of natural frequencies of single-mass DS and stiffness of the connection allows you to find the resonant oscillation

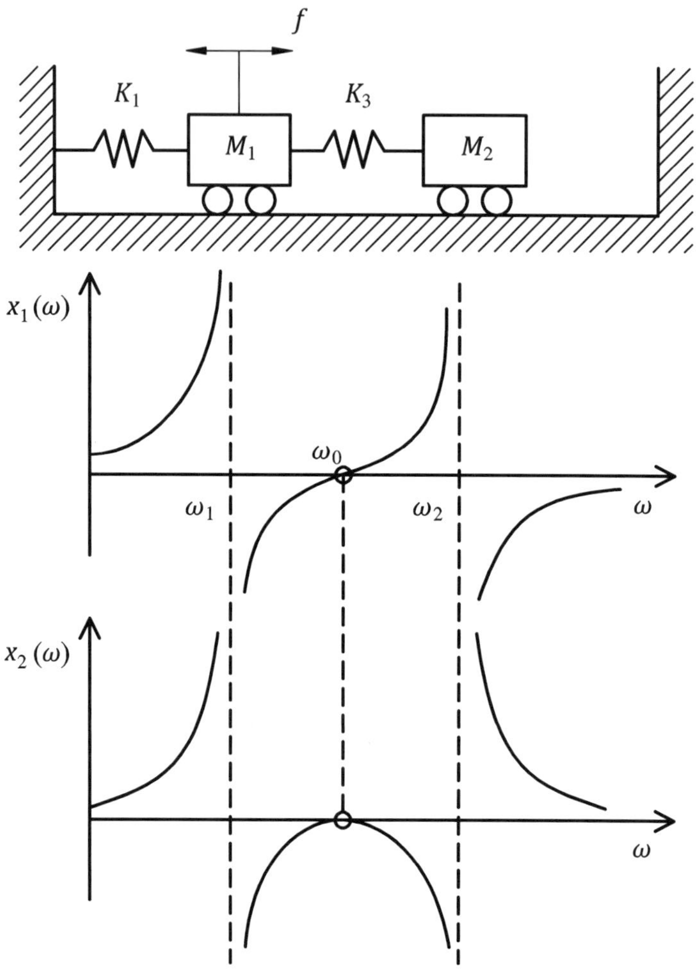

Figure 1.8: Idealized scheme of two-mass vibration dampener with two stiffnesses.

frequencies of two-mass DS. To do this, use the so-called Wien's diagram (Fig. 1.9), which has a complex analytical form and a simple graphic interpretation.

For practical cases, high-quality graphical representations shown in Fig. 1.10 are also useful.

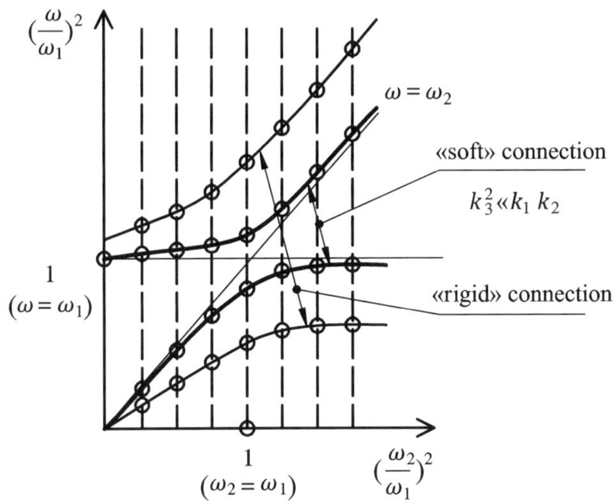

Figure 1.9: Natural frequencies of two-mass DS depending on the natural frequencies of two single-mass DS. Wien's diagram.

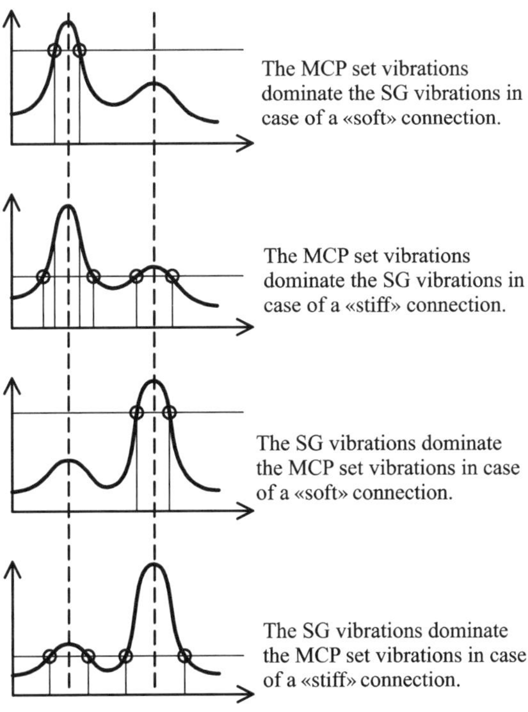

Figure 1.10: Cases of dominance of the vibrations of one DS over the other for different stiffness connections.

1.3 Points of Application of External Driving Vibration Forces

A system with two degrees of freedom is usually excited by applying force to one of the masses. In the case of VVER, this is unacceptable. One should never neglect at least two sources of excitation:

(1) MCP set is a centered (applied to the concentrated MCP set mass) polyharmonic force;
(2) ASW is a distributed force and each ASW type has its own spatial distribution.

Since the "cold-leg" and "hot-leg" pipe connections of the loop are located in close proximity to each other, the following assumptions can be taken as the boundary conditions: a two degrees of freedom suspension system makes the same oscillations — x_0 — on the left and right as shown in Fig. 1.11.

That is, excitation of oscillations occurs from the side of the casing, and not the vibrations of the casing itself but from the coolant pressure pulsations. Figure 1.12 shows points of application of driving forces.

In the model representation F_1 is replaced by an equivalent excitation by moving the point of "suspension" (x_0). Significantly, F_1 works the same on the SG side and on the MCP set side. The U-shape return does not have supports (either movable or stationary), so it is assumed that there is no damping. In the equivalent circuit in Fig. 1.13, damping is only introduced in $\mathrm{DS_{MCPS}}$ and, $\mathrm{DS_{SG}}$.

Due to the vibrations of the "suspension", both masses are synchronously excited, but by different forces. SG is excited by more forces than the MCP set since $k_1 > k_2$. Applying force to the mass

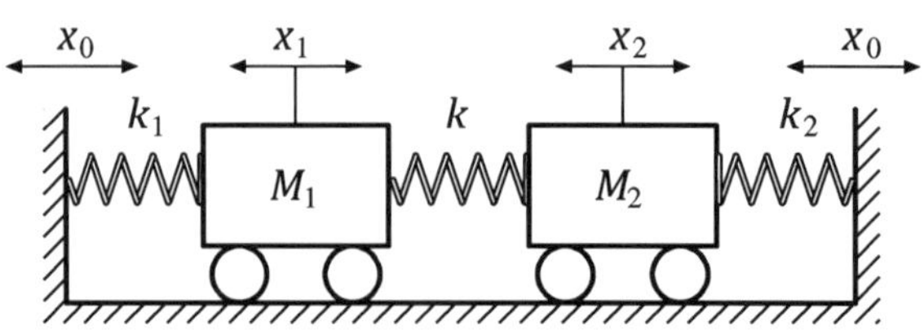

Figure 1.11: A two-mass model with the same excitation of the suspension of two spring pendulums.

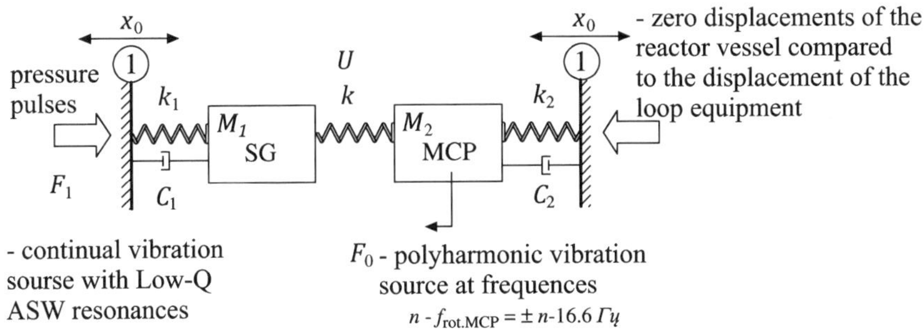

Figure 1.12: Excitation of two suspensions DS_1 and DS_2 and a single mass (MCP set).

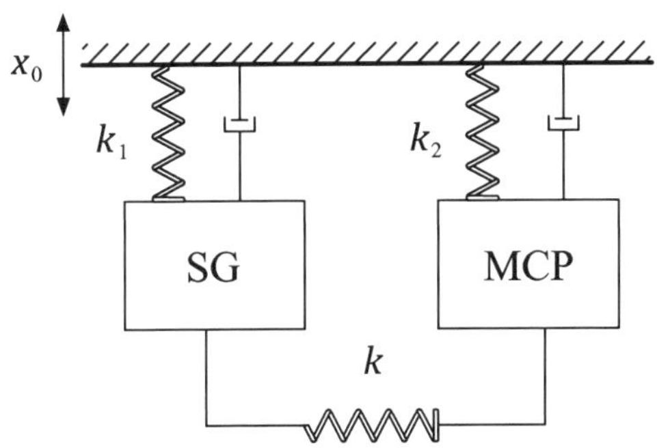

Figure 1.13: Equivalent circuit of a loop with two damping parameters.

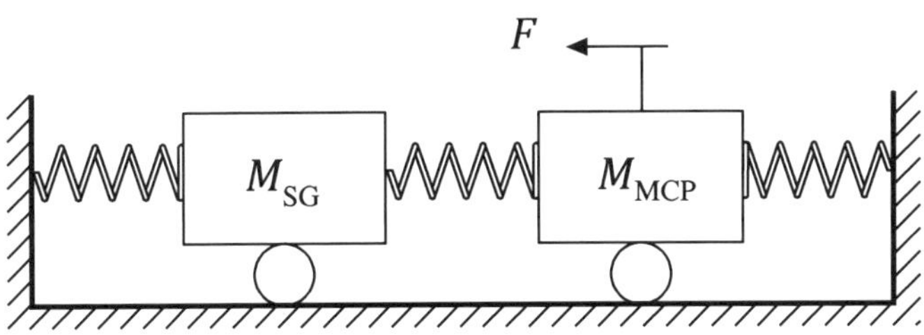

Figure 1.14: Application of external polyharmonic force to M_{MCPS}.

(Figs. 1.14 and 1.15) is equivalent to applying another force to the pendulum suspension (Figs. 1.14 and 1.15) $(k \cdot F)$.

Application of one and the same force to the suspensions of both masses (Figs. 1.15 and 1.16) is equivalent to application of two different forces to the two masses $k_1 \cdot F_0$ and $k_2 \cdot F$. The vibrations

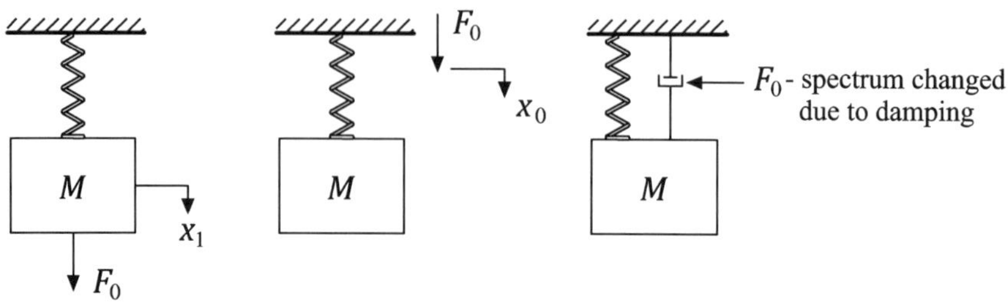

Figure 1.15: Application of external force.

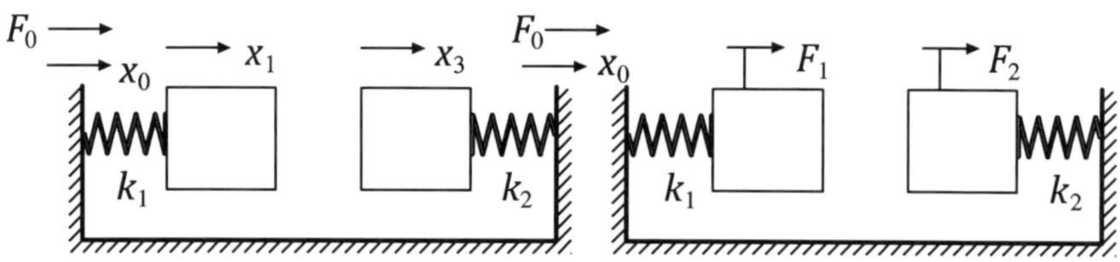

Figure 1.16: Application of one and the same force to the suspensions of both masses.

of the MCP set as an unbalanced mass at the rotational frequency can be described by a forced force acting through a damping mechanism (Fig. 1.15). This model leads to very complex nonlinear effects bringing about harmonics and sub-harmonics of the rotational frequency of the MCP set, as well as change in the quality factor of their resonances.

1.4 Numerical Estimates of the Ratio of MCP and SG Natural Oscillation Frequencies

Numerous analytical methods for calculating the absolute values of the DS natural frequencies are, on the one hand, the most difficult problems in terms of complexity of calculations. On the other hand, based on some schematization of the real DS design, the obtained frequencies are questioned, from the point of view of discrepancy of the model from the real DS.

Trust in the results of such calculations only occurs after verification procedures. Besides, if analytical methods allow a physical interpretation of the results received, since they explicitly give the dependence on the parameters of interest, the calculation methods, most often, do not allow to get to the cause of the change in the frequency of resonances. This so-called inverse problem is the main one in diagnostic applications.

The formulas for the natural frequencies of two-mass systems include the natural frequencies of the corresponding one-mass (with one degree of freedom) systems. As a first approximation, it makes sense to hypothetically break the links between one-mass systems and estimate their natural frequencies. Then they will undergo changes in the superposition of relationships, but will be those guidelines that will allow accepting the numerical values of frequencies obtained from the computational model with confidence.

For diagnostic interpretations, it is sometimes more profitable to find not the absolute values of natural frequencies, but their dependence (in particular, their ratios). Indeterminacy having constant character should be reduced, increasing this way the reliability of the ratio obtained. At the same time, verification of model representations on full-scale experimental data will not suffer from excessive regulation of the process of comparing one with another.

Without resorting to multiple finding of matrix determinants of the characteristic equation, the reduction of energy balances allows to find, at least, the first lowest natural frequency of the DS oscillation. Rayleigh's energy method is based on the law of energy conservation

$$E = E_K + E_P = \text{const}$$

at any time of the DS free oscillations where E_K is kinetic energy and E_P is potential energy. In particular, at a certain time $E = E_{K\max}$ and $E_P = 0$, and at another time $E = E_{P\max}$ and $E_K = 0$. Hence, from the law of energy conservation:

$$E_{K\max} = E_{P\max}$$

Transverse oscillations of a rod (a beam) at its natural frequency ω_0 can be presented as follows:

$$x\,(z,t) = x(z) \cdot \sin(\omega_0 t + \varphi_0)$$

where $x\,(z,t)$ is the displacement of the point of the rod in the coordinate "z" a the time "t"; $x\,(z)$ — the curved axis of the rod or the shape of the rod oscillations (Fig. 1.17).

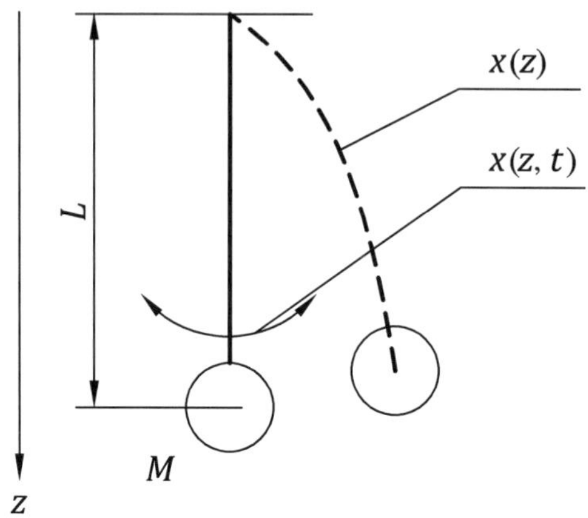

Figure 1.17: Rod oscillation model.

The kinetic energy of the oscillations in this case is written in the form

$$E_K = \frac{1}{2}\int_0^L m(z)\left[\dot{x}(z,t)\right]^2 dz = \frac{1}{2}\omega_0^2 \cos^2(\omega_0 t + \varphi_0)\int_0^L m(z)\, x^2(z)\, dz$$

where $m(z)$ is the rod mass distributed over a single spatial variable.

Hence,

$$E_{K\max} = \frac{1}{2}\omega_0^2 \int_0^L m(z)\, x^2(z)\, dz$$

Potential deformation energy of the rod bending

$$E_P = \frac{1}{2}\int_0^L k(z)\, M^2(z,t) dz = \frac{1}{2}\int_0^L \frac{M^2(z,t)}{E\cdot J(z)} dz = \frac{1}{2}\sin^2(\omega_0 t + \varphi_0)$$
$$\times \int_0^L E\cdot J(z)\left[\ddot{x}(z)\right]^2 dz$$

Hence,

$$E_{P\max} = \frac{1}{2}\int_0^L E\cdot J(z)\left[\ddot{x}(z)\right]^2 dz$$

where E is Young's modulus; $J(z)$ is the rod inertia moment.

After equating $E_{K\max} = E_{P\max}$, we have:

$$\omega_0^2 = \frac{\int_0^L E \cdot J(z)\left[\ddot{x}(z)\right]^2 dz}{\int_0^L m(z)\, x^2(z)\, dz} \tag{1.6}$$

When there is a concentrated mass at the end of the rod, "M", this formula becomes more complicated

$$\omega_0^2 = \frac{\int_0^L E \cdot J(z)\left[\ddot{x}(z)\right]^2 dz}{M \cdot x^2(L) + \int_0^h m(z)\, x^2(z)\, dz} \tag{1.7}$$

As a shape of the rod oscillations $x(z)$, a static deflection of the beam under impact of the force applied in a point with the coordinate "L" is taken.

$$x(z) = \frac{F}{6 \cdot E \cdot I}(3L \cdot z^2 - z^3)$$

After integration in the expression (1.6), we have

$$\omega_0^2 = \frac{140}{11} \cdot \frac{E \cdot I}{m_0 \cdot L^4} \tag{1.8}$$

where I, m_0 are constants not depending on the z and, respectively, the average moment of inertia and the average weight of the rod.

The dependence (1.8) $\omega_0 \sim \frac{1}{L^2}$ is valid for oscillations of a rod with a non-zero own mass distributed over the lowest first form of oscillations, and there is no point mass at the end of the rod.

As follows from (1.7), the formula for the oscillation frequency of a massless rod with a point mass "M" at its end will be

$$\omega_0 \sim \frac{\text{const}}{\sqrt{M \cdot x(L)}} \tag{1.9}$$

Or, after substitution in

$$x(z) = \text{const} \cdot z^2 (3L - z)$$

we have

$$x(L) = \text{const} \cdot L^3$$

Thus

$$\omega_0 = \frac{\text{const}}{\sqrt{M \cdot L^3}} \tag{1.10}$$

Here the development of Rayleigh energy dependences have been deliberately repeated so explicitly, with an accuracy to a constant, to produce the formulas that work out the ratio of frequencies. In practice, the following formulas are often not correctly applied referring to reference books:

$$\omega_0 \sim \frac{1}{\sqrt{L}}, \sim \frac{1}{L}, \sim \frac{1}{L^2}$$

Multi-mass vibration DS, united by complex rigid connections, are characterized by the number of resonances exceeding the number of masses. This effect of frequency multiplication leads to the fact that some frequencies will be in the "genetic" relationship. In particular, it can be harmonic, sub-harmonic and combinational relationships. If some physical phenomenon occurs, which shifts the frequency of any resonance, other related frequencies will move synchronously with it. This shift of a set of resonance frequencies is a diagnostic feature of this particular physical phenomenon.

A basic DS described above has a single scalar bending stiffness, determined by the length of the pipeline, L, causing the natural frequencies of such DS:

$$\omega_0 \sim \frac{1}{\sqrt{L^3}}$$

Other, non-bending types of vibrations, for example, longitudinal, have a different dependence on the length of the pipeline:

$$\omega_0 \sim \frac{1}{\sqrt{L}}$$

torsional types have another one, etc.

Already for a single-mass DS, different types of vibrations can be matched with the stiffness matrix, and hence, getting a series of natural frequencies. Therefore, the "manual" interpretation of the set of resonance frequencies even of a single spectral characteristic is extremely difficult. The appropriate software research tools should support the empirical methods of interpretation. With their help, from large archives of vibration measurements, diagnostic signs that are stable in time are obtained.

Let's neglect the connection of the MCP and SG masses in the form of a U-shaped pipeline. As a primitive DS, simulating the MCP set oscillations we assume cantilever with the MCP set point mass "M_{MCPS}". Considering the MCP set motor, for VVER-1000 it will be $\sim 140\,t$. The rod length matches the length of the "cold-leg" pipeline $L_{\text{MCPS}} = 16.58\,\text{m}$.

As a primitive DS, simulating the SG oscillations we assume a cantilever with the MCP set point mass "M_{SG}". Considering the mass of the water from the Main Circulation Circuit and the secondary circuit of the reactor plant for VVER-1000 $M_{\text{SG}} = 353\,t$, and the length of the "hot-leg" pipeline $L_{\text{SG}} = 8.16\,\text{m}$ (almost twice as short L_{MCPS}).

Here are deliberately given numerical values so that the reader, having, perhaps, other data or data of another main circulation circuit (MCC) design, is able to repeat such elementary calculations. It is obvious that neither full-scale analytical nor computational models provide such a possibility. Moreover, usually, the actual method is presented as a "black box", not subject to disclosure.

It is easy to take into account the mass of water in the "cold-leg" and "hot-leg" pipelines, but being in the numerator and denominator, these parameters do not significantly affect the frequency ratio. Thus,

$$\frac{\omega_{\text{SG}}}{\omega_{\text{MCPS}}} = \sqrt{\frac{M_{\text{MCPS}}}{M_{\text{SG}}}} \cdot \sqrt{\left(\frac{L_{\text{MCPS}}}{L_{\text{SG}}}\right)^3} = \sqrt{0.48} \cdot \sqrt{(2.03)^3}$$
$$= 0.69 * 2.89 = 1.99$$

The SG, as an oscillatory system, has twice the natural oscillation frequency than the natural oscillation frequency of the MCP set. There are prejudices that the steam generator, as an element significantly exceeding the weight of the MCP, has a lower natural frequency. In fact, the ratio of length of "cold-leg" and "hot-leg" pipelines is much more critical than the ratio of mass.

Realizing that a rough estimation method of the natural frequency ratio was used, we emphasize that ω_{SG} is practically the second harmonic ω_{MCPS}. This is not quite a favorable frequency ratio of the elements of the two-mass oscillatory system. The increased vibration amplitudes of one element more easily cause increased vibrations of another element when their natural frequencies are in a harmonic ratio.

1.5 On Modal Analysis in the Simplest Case of Oscillations of Connected Masses

In the previous section, in the example of a two-mass vibration model the different effects of resonant frequency multiplication were shown. Two single-mass DS with resonant frequencies ω_{01} and ω_{02}, connected by a single rigid element, can generate up to four resonant frequencies. In this case, frequencies ω_{01} and ω_{02} come into motion, and new resonances are formed, the location of which is not limited on the frequency axis.

It is obvious that for a multi-mass structure, such as the MCC, one can expect unpredictably many resonance frequencies observed at various points. Even *a priori* knowledge of the natural resonance frequencies of all the MCC items $\omega_{0n}(n = 1, 2, 3, \ldots, N)$ does not guarantee knowledge of the origin of all the observed resonances.

MCC decomposition on N basic dynamic systems (it is better to have DS with one degree of freedom) already contains some model assumptions. Overlapping the relationships between basic DS, that is, MCC modeling as a whole, is the next stage of model assumptions.

The solution of the characteristic multidimensional equation of free oscillations is most often associated with the so-called procedure of a diagonalized matrix of natural values. In terms of the dominant component analysis, this corresponds to the linear transformation of the observed space of vibration displacements into a space with a smaller dimension, in which incoherent oscillations of a small mass number occur. This is the main target purpose of modal analysis.

Figure 1.18 shows the spectrum of vibration displacements consisting of four resonances. Elementary actions on it are decomposition on four one-resonance spectra. Moreover, the location of the resonances remains unchanged. It is obvious that such decomposition, strictly speaking, is incorrect.

As an example, Fig. 1.19 shows some random DS consisting of eight DS with one degree of freedom. Each basic DS connects to other DS according to the actual design.

After the modal analysis, we will have an equivalent circuit consisting of eight (or fewer) DS with one degree of freedom unrelated to each other (Fig. 1.20). Three parameters of each DS, (M, K, C), will be, generally speaking, different from the initial DS.

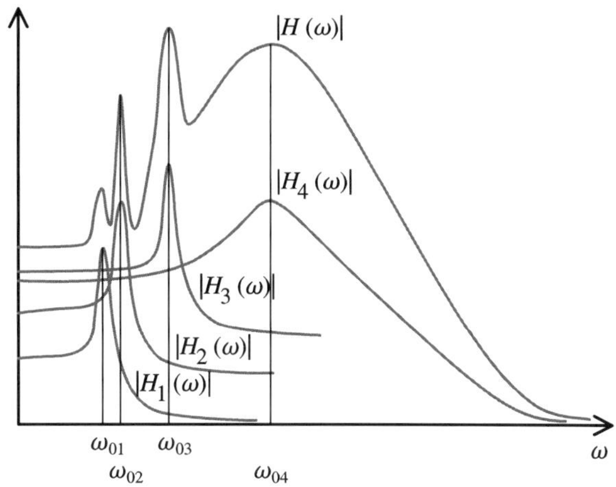

Figure 1.18: Experimentally determined transfer function modulus, $H(\omega)$ and four transfer function moduli of four DS with one degree of freedom.

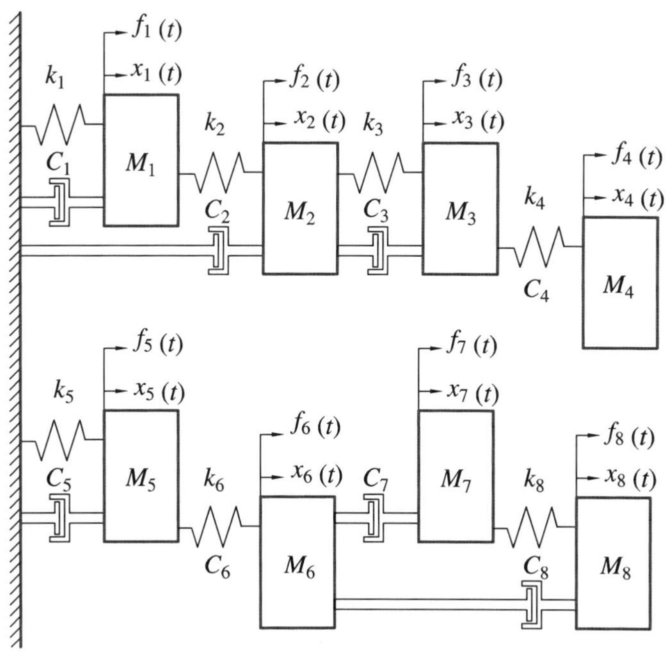

Figure 1.19: Schematization of the structure of eight point masses in the form of eight DS with one degree of freedom each. The relationships between the DS are superimposed in accordance with their interaction in the real structure.

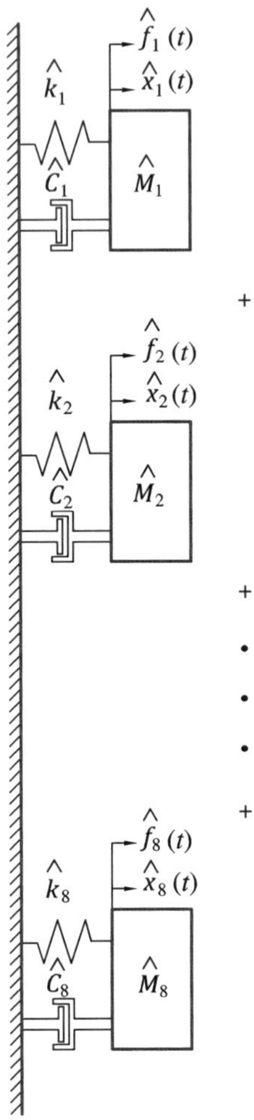

Figure 1.20: Equivalent representation of the DS shown in Fig. 1.19, in the form of disconnected DS with one degree of freedom.

Because of such solutions of the multidimensional problem for eigenvalues, one can get a series of natural frequencies (tone) of the structure that characterize the entire structure globally and generally speaking cannot be "tied" to any one of its elements. However, for the VVER-1000, the lowest type of MCP oscillations at a frequency of 2.6 Hz has the maximum vibration displacement at all four MCP. Therefore, it's called the natural frequency of the MCP oscillations.

For the same reason, the next resonance frequency of the MCC oscillations, 4.6 Hz is called the natural frequency of the SG oscillations, etc.

Thus, in practice, as the first iteration, an incorrect attempt is always made to represent the oscillations of a complex structure in the form of a "direct" sum of oscillations of disconnected masses.

Chapter 2

Parameters of Basic Dynamic System

2.1 DS Dynamic Stiffness and Transfer Function

Parameters of dynamic systems are the subject of an extensive fundamental research [21, 25–27, 29, 36, 39, 43, 45, 48, 64, 67, 95, 97–101, 106]. Let's consider the simplest case of DS with a single degree of freedom.

After the Laplace transformation, the differential equation (1.1) results in the following algebraic equation:

$$(M \cdot p^2 + C \cdot p + K) \cdot X(p) = F(p)$$

or

$$K(p) \cdot X(p) = F(p)$$

where p is complex variable and $K(p)$ is a polynominal:

$$K(p) = M \cdot p^2 + C \cdot p + K$$

The Laplace transformation in $p = j\omega$ goes into Fourier series transformation:

$$(-M \cdot \omega^2 + jC \cdot \omega + K) \cdot X(\omega) = F(\omega)$$

$K(p)$ or $K(j\omega)$ second degree polynominal is characterized by static stiffness of "K" and by coefficients for the second (acceleration) and the first (velocity) derivatives of displacement, "M" and "C", respectively. Therefore, $K(p)$ is called dynamic stiffness. If you consider force $f(t)$ as an external impact, and $x(t)$ — as a response, then the polynominal $K(p)$ with its three coefficients – "M", "C", "K" — provides a complete definition of the dynamic system properties. One of the typical tasks of identification of DS is using the input impact $f(t)$ and the output response $x(t)$ to estimate the numerical values of DS parameters "M", "C", "K".

In the Laplace representation, the polynomial $K(p)$ is usually the product form:

$$K(p) = M \cdot (p^2 + 2\sigma \cdot p + \omega_0^2) \tag{2.1}$$

where σ is a relative damping and ω_0 is a natural frequency.

The polynominal $K(j\omega)$ in the Fourier representation is often reassigned with the polynominal $K(\omega)$

$$K(\omega) = -M \cdot \omega^2 + jC \cdot \omega + K = K \cdot (1 - F^2 + 2\alpha j \cdot F) \equiv K \cdot Z(F)$$

where $\alpha = \frac{\sigma}{\omega_0}$ is a non-dimensional damping factor and $F = \frac{\omega}{\omega_0}$ is a non-dimensional frequency.

Every time it is necessary to stipulate that the polynominals $K(p)$ and $K(\omega)$ are not the same. The confusion is caused by identification with frequency — ω of the Laplace variable $p = j\omega$, which is a purely imaginary value. In both representations, the stiffness dimension (N/m = kg/s^2) is the dimension of polynominals $K(p)$ and $K(\omega)$.

The static stiffness "K" in the Fourier representation $K(\omega) = K \cdot Z(F)$ is allocated explicitly; in the Laplace representation $K(p) = M \cdot (p^2 + 2\sigma \cdot p + \omega_0^2)$ the explicit factors are the mass and the polynominal with dimensions (Hz^2).

Non-dimensional frequencies $F < 1$ will be called before-resonance frequency, and $F > 1$ — after-resonance frequency. Non-dimensional composite polynominal

$$Z(F) = 1 - F^2 + 2\alpha j \cdot F$$

has a modulus

$$|Z(F)| = \sqrt{(1 - F^2)^2 + 4\alpha^2 F^2}$$

and a phase

$$tg\psi(F) = \frac{2\alpha F}{1 - F^2}$$

Table 2.1: Polynominal representations with Laplace and Fourier variables and normalized frequency.

$K(p)$	$M \cdot (p^2 + 2\sigma \cdot p + \omega_0^2)$
$K(\omega)$	$-M(\omega^2 - j2\sigma \cdot \omega - \omega_0^2)$
$Z(F)$	$(1 - F^2 + 2\alpha j \cdot F)$

Three variables p, ω, F interpreted as the corresponding frequencies define their own polynominal type $K(\ldots)$ (Table 2.1).

Inverse function $H(p) = K^{-1}(p)$ of dynamic stiffness means a transfer function between force and displacement:

$$X(p) = H(p)F(p)$$

$$H(p) = \frac{1}{K(p)} = \frac{1/M}{p^2 + \frac{C}{M} \cdot p + \frac{K}{M}}$$

A transfer function $H(p)$ becomes canonical after transition to dynamic factors $\sigma = C/2M$, $\omega_0\sqrt{K/M}$ or, respectively, to a relative damping — σ and a natural frequency — ω_0:

$$H(p) = \frac{1/M}{p^2 + 2\sigma \cdot p + \omega_0^2} \tag{2.2}$$

Parameters σ and ω_0 have frequency dimension (Hz). If the value ω_0 is interpreted as a natural frequency of the dynamic system oscillations, the dimensions of the relative damping factor "σ" are better to interpret as the inverse seconds (Hz = 1/s). The damping factor "σ" defines damping of the dynamic system oscillations after it was affected by the impulse (δ) force. When only frequency factors are introduced, the dynamic stiffness (denominator of the transfer function) will be

$$K(p) = M \cdot (p^2 + \Omega_1 \cdot p + \omega_0^2) \tag{2.3}$$

where $\Omega_1 = 2\sigma = C/M$.

The polynominal $p^2 + \Omega_1 \cdot p + \omega_0^2$ defines a two-dimensional space for its parameters (σ, ω_0) or (Ω_1, ω_0), while the dynamic stiffness $K(p)$ and the transfer function $H(p)$ define three-dimensional space for the parameters (M, C, K). Both coordinates (Ω_1, ω_0) have the same dimension — (Hz). All three physical values have their own dimensions in the "mass-damping-stiffness" space (M, C, K). Zeros of the denominator or poles of the transfer function do not uniquely determine the transfer function, although they depend on three DS parameters, setting an ambiguous correspondence $(\sigma, \omega_0) \Longleftrightarrow (M, C, K)$.

When the dynamic system is excited by a white noise, the force spectrum is a constant in the frequency form: $F(p) = \text{const}$ and delta function $\delta(t)$ in the time domain (ideal shock). After multiplying the equation

$$K(p) \cdot X(p) = F(p) = \text{const} \tag{2.4}$$

by the complex-conjugate equation

$$K(p)^* \cdot X(p)^* = F(p)^* = \text{const} \tag{2.5}$$

with subsequent averaging, we will obtain an expression for the spectral power density of displacements — $S_x(p)$

$$|K(p)|^2 S_x(p) = \text{const}_1 \tag{2.6}$$

With an accuracy to a constant we have a coincidence of the auto power spectral density (APSD) of displacement $S_x(\omega)$ with a square of the transfer function: $S_x(\omega) \equiv \frac{\text{const}_1}{|K(p)|^2} = \text{const}_1 |H(\omega)|^2$.

Thus, while measuring displacements which typically refer to a random process and moving to the APSD of displacement, we obtain information on the object of study as a dynamic system. At the same time, all the information on DS is hidden in three parameters "M", "C", "K" of one complex frequency function $H(\omega)$. This is one of the ways to solve the problem of DS identification, which obviously refers to stochastic (random) methods.

2.2 Global Properties of the Dynamic System

2.2.1 *Modulus and phase of the transfer function*

The transfer function is a real function for two special cases $\alpha = 0$ and $\alpha = 1$. At $\alpha = 0$ (Fig. 2.1),

$$H(\omega) = -\frac{1/M}{\omega^2 - \omega_0^2}$$

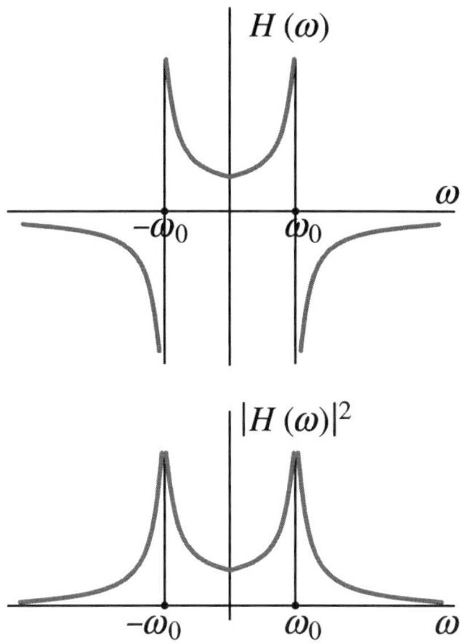

Figure 2.1: Transfer function and square of its module at $\alpha = 0$.

Here, $H(\omega)$ for positive frequencies consists of two sections of monotonic growth, and the function $|H(\omega)|^2$ consists of section of monotonic growth (pre-resonant) and monotonic decrease (post-resonant).

At $\alpha = 1$ (Fig. 2.2)

$$H(\omega) = -\frac{1/M}{\omega^2 + \omega_0^2}$$

there is a monotonic function of the frequency in each semi-axis.

Here, the greater the attenuation DS introduces, the higher is the frequency of impact. The minus before the function $H(\omega)$ refers to the phase relations. In the first case ($\alpha = 0$) — we have a common phase to the natural frequency and phase opposition after the natural frequency. In the second case ($\alpha = 1$) — there is always phase inversion: in any frequency, the displacement will be half a period behind the perturbation.

In both cases, the transfer functions are even functions, which were repeatedly used to justify omission of the negative frequencies. The transfer function is often doubled as $2H(\omega)$ in the positive semi-axis to preserve the integral properties, and this results in different conclusions in different literature sources.

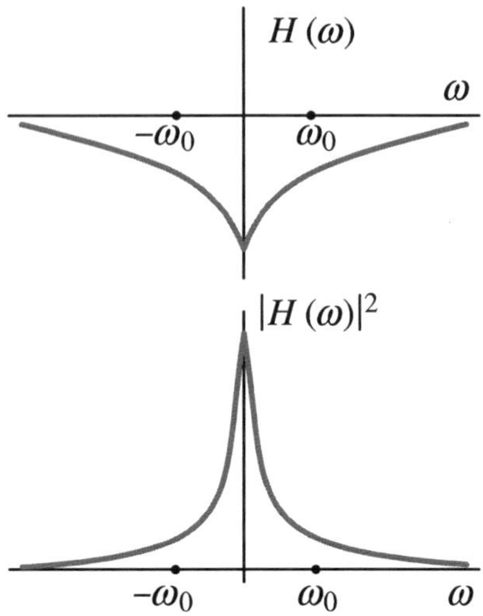

Figure 2.2: Transfer function and square of its module at $\alpha = 1$.

If we have an idealized case of non-dissipative system of free oscillations at $\alpha = 0$ with a specific natural frequency, then at $\alpha = 1$ in the natural frequency there is no maximum of the transfer function, but at the same time the natural frequency does not deteriorate at zero frequency. The frequency of the maximum value of the transfer function becomes a zero frequency. The case of $\alpha > 1$ or the case of the so-called strongly damped DS differ from each other only in the damping constant of the transient process when there is no oscillation in it.

Three parameters — M, K, C — defining the dynamic object itself are conservative parameters. In most diagnostic applications people seek to find their numerical values as the diagnostic features. They are often assumed to be stationary in time. In this case, the object under research is called a stationary object. All characteristics concluded solely from these parameters are called global parameters (including natural frequencies and Q-factors). Characteristics depending on initial state of the dynamic systems are the local characteristics.

Being a complex function, the transfer function can be represented in exponential form using its absolute value and phase.

$$H(\omega) = |H(\omega)|e^{-j\psi(\omega)}$$

The modulus and the phase are obtained from the imaginary and real parts of expression (2.2)

$$|H(\omega)| = \frac{1/M}{\sqrt{(\omega^2 - \omega_0^2)^2 + 4\omega^2\sigma^2}}$$

$$\psi(\omega) = -arctg\frac{2\omega\sigma}{\omega^2 - \omega_0^2} = \begin{cases} 0, & \omega = 0 \\ -\frac{\pi}{2}, & \omega = \omega_0 \\ & -\pi, \omega \Longrightarrow \infty \end{cases}$$

The transfer function modulus refers to an even function, but it is not symmetric with respect to the vertical line passing through the natural frequency: the left and right parts of the absolute value of the transfer function decrease monotonly, but according to their own laws (Fig. 2.3).

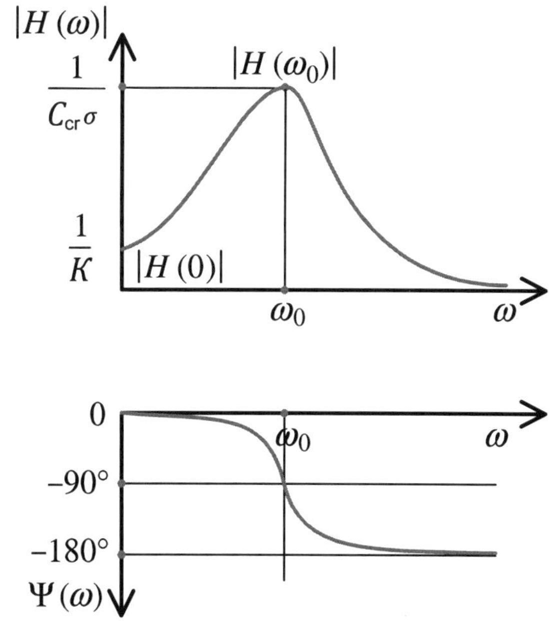

Figure 2.3: Absolute value and phase of the transfer function with its special points.

The transfer function has characteristic values in specific points:

$$H(\omega) = \begin{cases} \dfrac{1}{K}, & \omega = 0 \\ -\dfrac{j}{\sigma C_{\text{cr}}} = -\dfrac{j}{C\omega_0} = -\dfrac{j\sqrt{M}}{C\sqrt{K}}, & \omega = \omega_0 \\ -\dfrac{1}{M\omega^2}, & \omega \Longrightarrow \infty \end{cases}$$

The "$-j$" factor in this expression is equivalent to the phase shift to 90° at $= \omega_0$, and the "-1" multiplier (negative sign) is equivalent to the opposite phase taken at infinitely high frequencies.

If a certain mass "M" is a component of DS, then it can be estimated at a frequency that is much higher than the resonant one $\tilde{f} \gg F$, in accordance with the value of the transfer function at this frequency.

$$M \approx \frac{1}{H(\tilde{f}) \cdot \tilde{f}^2}$$

If a certain spring with "K" stiffness is a DS component, then its value can be also measured, but at a zero frequency (the so-called static experiment)[1]

$$K = \frac{1}{H(0)}$$

In these two experiments, the transfer function shall also be available in absolute physical units, i.e. (m/N) to obtain mass and stiffness in absolute physical values. The absolute value and phase of the transfer function for the normalized frequency take the following form:

$$|H(F)| = \frac{1/K}{\sqrt{(1 - F^2)^2 + 4\alpha^2 F}}$$

$$tg[\psi(F)] = \frac{2\alpha F}{1 - F^2}$$

The phase includes information on delay of response (displacement) from impact (force). Close to zero Hertz, when the impacts are

[1]*Note*: It is only an extrapolated value of the transfer function (remark of the editor).

slow, DS "manages" to track them almost without a delay and the phase is close to zero. DS response range remains almost unchanged in frequency. DS response is close to the opposite phase in case of extremely high frequencies. The phase shift is minus 90^0 at a natural frequency. This property of the phase response along with the maximum property of the transfer function modulus is often used to find the numerical value of DS natural frequency.

DS instantly (without delay) increases (reduces) impact of the external force up to $\ll 1/K \gg$ times close to zero Hz. So, all DS properties are defined by its stiffness only: neither mass nor damping matter at a zero frequency. A static deflection of the beam under impact of the force is the simplest illustration of this property. As the expression $\omega_0 = \sqrt{K/M}$ shows, a natural DS frequency referring to DS dynamic parameter at a constant mass can be defined by such a simple static (without excitation of oscillations) experiment. The transfer function area close to the zero Hz is called a stiffness line (Fig. 2.4). When DS stiffness changes, this area of the logarithm of the transfer function almost additively shifts along the ordinate axis while the curve shape remains unchanged.

The "$1/C$" damping has the greatest impact on the maximum of the transfer function amplitude. When DS mass changes, this high frequency domain of the logarithm of the transfer function almost additively shifts along the ordinate axis while the curve shape remains unchanged (Fig. 2.4).

DS is some kind of low-frequency filter. Near the zero Hz there is a transparency band with slightly monotonically increasing function and with minimal phase distortion, and at high frequencies there is the stop band in a form of monotonically decreasing function, finally the maximum amplitude and the biggest change in the characteristic appear close to the natural frequency. In this regard, while performing vibration diagnostics it is important to know from which side – to the left or to the right from the DS natural frequency – is the analyzed range of frequencies.

To identify changes in the three DS parameters — ("M", "C", "K") it is not always effective to consider only the neighborhood of the natural frequency. All three DS parameters contribute to this neighborhood, even though the stiffness determines the low-frequency range, while the DS mass dominates in the high-frequency range and the change of damping will most of all affect the amplitude

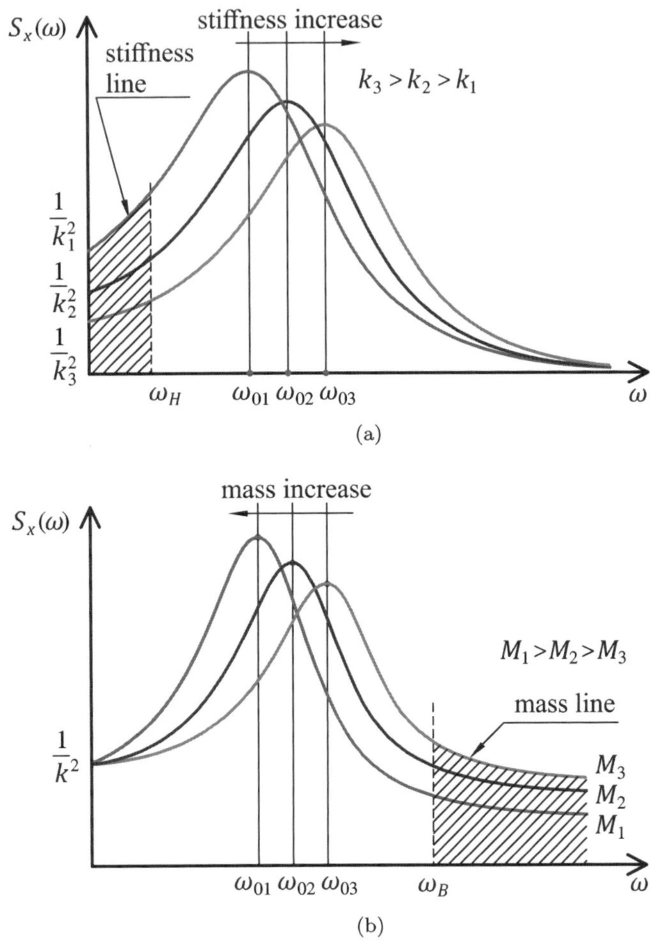

Figure 2.4: ASPSD of displacements (or a square of the transfer function modulus when exposed to white noise) when only DS stiffness (a) or mass (b) are changed.

of the transfer function in the neighborhood of the natural frequency (Fig. 2.5). In this case, the natural frequency itself will slightly vary from damping.

We will select a total square in the denominator to calculate the maximum of function $|H(\omega)|$ or function $|H(\omega)|^2$

$$|H(\omega)|^2 = \frac{1}{K^2\left\{\left[\left(\frac{\omega}{\omega_0}\right)^2 - (1 - 2\alpha^2)\right]^2 + 4\alpha^2(1 - \alpha^2)\right\}} \tag{2.7}$$

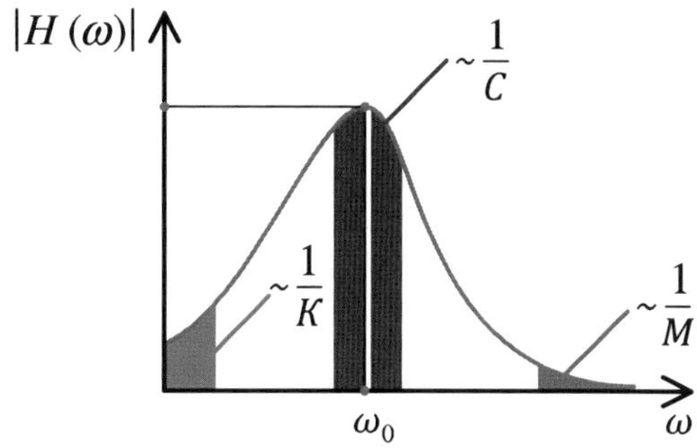

Figure 2.5: Dominant impact on the transfer function of stiffness (K) near the zero Hz, damping (C) about the natural frequency and mass (M) in the high-frequency range.

The denominator is represented as a sum of two non-negative summands, and only the first of them is frequency-dependent and becomes zero at a frequency

$$\omega_{\max} = \omega_0 \sqrt{1 - 2\alpha^2}$$

or

$$F_{\max} = \sqrt{1 - 2\alpha^2} \tag{2.8}$$

i.e. at a frequency $\omega_{\max}$ the maximum of the real function $|H(\omega)|^2$ is achieved (Fig. 2.6).

For $0 < \alpha < \frac{\sqrt{2}}{2}$, the formula for the maximum values takes the following form:

$$|H(\omega_{\max})|^2 = \frac{1}{4k^2\alpha^2(1-\alpha^2)}$$

$$|H(\omega_{\max})| = \frac{1}{2k\alpha\sqrt{1-\alpha^2}}$$

$$G(F_{\max}) = \frac{1}{2\alpha\sqrt{1-\alpha^2}}$$

$$|A(F_{\max})| = \frac{f_0}{2k\alpha\sqrt{1-\alpha^2}} \tag{2.9}$$

DS with a small damping ratio $\alpha = 0.1$ at the frequency of its maximum has amplification factor of more than five, increasing

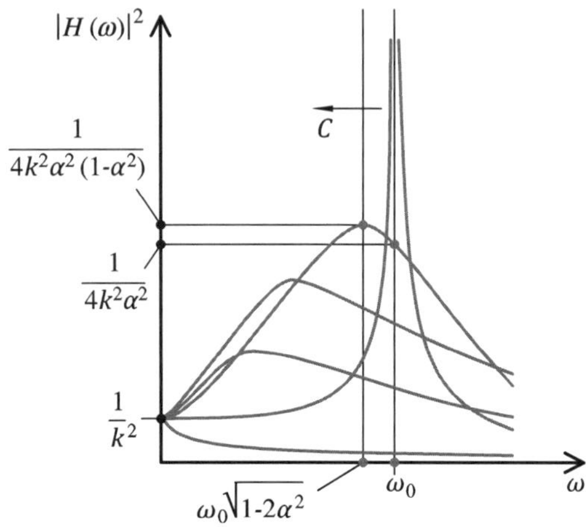

Figure 2.6: Square of the transfer function modulus in case of DS damping change.

α to $0,25$ it is more than two, at $\alpha = 0.5$ — around 1.4, and at $\alpha = \frac{\sqrt{2}}{2} \approx 0.707$ there is no amplification as the maximum is achieved at a zero frequency.

The value $\alpha = \frac{\sqrt{2}}{2}$ is a threshold value above which all transfer functions have a global maximum at a zero frequency. This explains the allocation to a separate class of the so-called aperiodic processes with $\frac{\sqrt{2}}{2} < \alpha < 1$. With such a high damping ratio, the DS solution still contains a periodic cofactor $\cos(\omega_0 t)$, and the monotone transfer function no longer has the so-called resonance at the natural frequency. This example illustrates a common misconception: if there is a sinusoidal factor with some frequency in the process recording, then the resonance at this frequency shall be observed in its spectrum.

The square of the absolute value of the dimensionless transfer function is a quadratic hyperbole depending on the damping ratio

$$|G(F,\alpha)|^2 = \frac{1}{(1-F^2)^2 + 4\alpha^2 F^2} = \frac{1}{C_1(F) + C_2(F)\alpha^2} \tag{2.10}$$

Function $|G(F,\alpha)|^2$ for the post-resonant frequencies $F > 1$ monotonically decreases in the parameter "α", i.e. the higher the damping, the greater the attenuation introduced. Such monotony is not

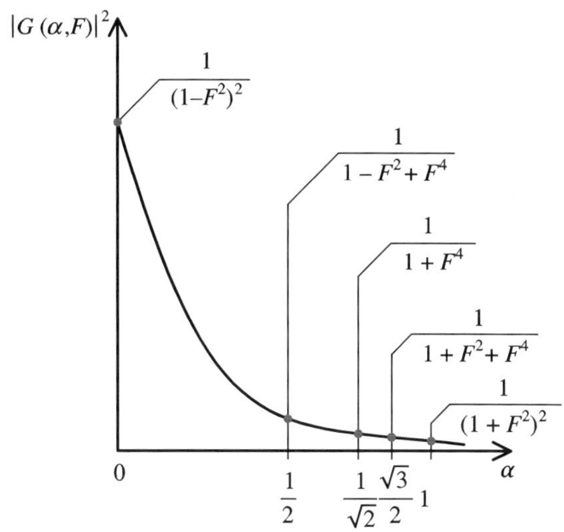

Figure 2.7: Square of the absolute value of the transfer function versus change of DS.

observed for the pre-resonant frequencies. Figure 2.7 shows the analytical form of the transfer function with typical values of the damping ratio.

In addition to the transfer function modulus, other functions which differ from each other by coefficient only are also used as an amplitude-frequency function and this causes some confusion. These functions are summarized in Table 2.2.

To obtain an overall displacement from the overall force it shall be multiplied by the frequency-dependent complex function:

$$x(t) = \frac{f(t)}{K} \cdot \frac{1}{-F^2 + 2j\alpha F + 1}$$

A dimensionless function of the dimensionless frequency

$$Z(F) \equiv (1 - F^2) + 2j\alpha F = U(F) + jV(F) \tag{2.11}$$

is also a complex function

$$U(F) + jV(F) = x_0 \cdot \frac{1}{A(F)} \cdot e^{-j\psi(F)} \tag{2.12}$$

depending on frequency functions $A(F)$, $\psi(F)$, where $x_0 = \frac{f_0}{K}$ — is the static displacement due to the static force f_0. Hence, an unknown

Table 2.2: DS amplitude-frequency functions applied in practice.

Name (dimensions)	Amplitude formula
DS growth ratio (dimensionless)	$\lvert G(F)\rvert = \frac{1}{\sqrt{(1-F^2)^2+4\alpha^2F^2}}$
DS frequency response (m)	$\lvert A(F)\rvert = x_0\lvert G(F)\rvert = \frac{f_0}{K}\lvert G(F)\rvert = f_0\lvert H(F)\rvert$
Modulus of DS transfer function (from force to displacement) (m/N)	$\lvert H(F)\rvert = \frac{\lvert G(F)\rvert}{K} = \frac{\lvert A(F)\rvert}{f_0}$

complex amplitude has a form depending on the normalized frequency which coincides with the one of DS transfer function multiplied by a constant.

$$|A(F)| = \frac{\frac{f_0}{k}}{\sqrt{(1-F^2)^2+4\alpha^2F^2}}$$

$$tg[\psi(F)] = -\frac{2\alpha F}{1-F^2}$$

Thus, the force – $|f|$ at a normalized frequency "F" causes DS displacement – $|x|$ through the coefficient $\frac{1}{k\sqrt{(1-F^2)^2+4\alpha^2F}}$. This frequency-dependent coefficient cannot be interpreted as a amplification, as it links two physical values of different dimensions (force and displacement). It shall be interpreted as a distributed calibration factor or transformation ratio with inverse dimension of stiffness — [m/N].

The DS amplification factor can be reasonably introduced to compare displacements at different frequencies. In this case, it is normalized to one at a zero frequency. Then we obtain that the amplification factor as a dimensionless frequency function (or amplitude factor distributed in frequency) also coincides with the form of the transfer function modulus with the accuracy to a constant:

$$|G(F)| = \frac{1}{\sqrt{(1-F^2)^2+4\alpha^2F^2}}$$
$$= \frac{K}{f_0}|A(F)| = K\cdot|H(F)|$$

The difference of functions $|G(F)|$, $|A(F)|$, $|H(F)|$ is intentionally emphasized, although all three functions have similar frequency dependence which causes a terminological confusion.

The modulus of the frequency amplification factor of the dynamic system $G(F)$ is transformed by means of identity substitutions as follows:

$$|G(F)| = \left|\frac{1}{\sqrt{(1-F^2)^2+4\alpha^2F^2}}\right| = \left|\frac{1}{1-F^2}\right| \cdot \left|\frac{1}{\sqrt{1+\left(\frac{2\alpha F}{1-F^2}\right)^2}}\right|$$

$$= |G_0(F)| \cdot \left|\frac{1}{\sqrt{1+[tg\psi(F)]^2}}\right| = |G_0(F)| \cdot |\cos\psi(F)|$$

where $G_0(F) = \frac{1}{1-F^2}$ is the frequency-dependent amplification factor of the dynamic system without damping.

In extreme cases with $F = 0$, $F \to \infty$, $|\cos\psi(F)| = 1$ and the damped DS does not differ from the undamped one (but only in these cases). In asymptotic cases, the amplification factor takes the following form (Fig. 2.8):

$$|G(F)| \approx \begin{cases} 1+F^2, & F \sim 0 \\ \dfrac{1}{2\,\alpha}, & F = 1 \\ \dfrac{1}{F^2}, & F \to \infty \end{cases}$$

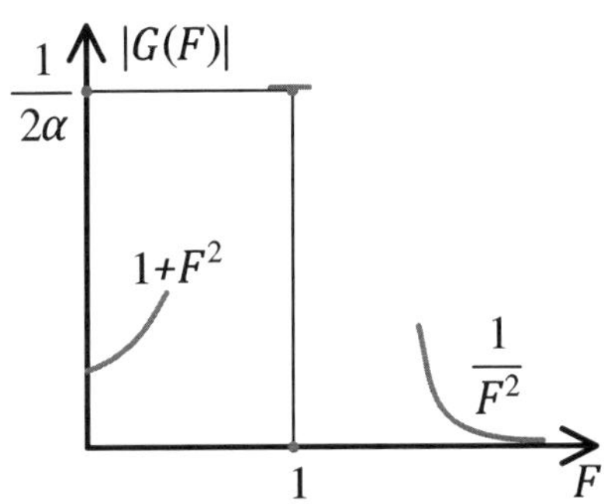

Figure 2.8: DS amplification factor in asymptotic cases.

2.2.2 *Deterministic and random applied force*

While interpreting the DS amplitude & frequency properties either a modulus — $|H(\omega)|$ or a square of the modulus — $|H(\omega)|^2$ of the transfer function are used due to the two possible fundamental properties of the applied force. When the external force is deterministic, the transfer function is used in cases when the DS perturbance is produced by a random force — a square of the modulus. In case of the random impact, the displacement obtained as a response is a random function (random process) being a result of transformation of the random impact into a random output process by the dynamic system with deterministic parameters ("M", "C", "K"). After the Laplace (or Fourier) transformation of the random process, $x(t)$, we will also have a random process in the frequency domain (spectrum), $X(p)$ although DS itself is a deterministic system — $H(p)$, which transforms the random function $F(p)$:

$$X(p) = H(p) \cdot F(p) \tag{2.13}$$

A random spectrum, $X(p)$, cannot be a characteristic, determined by DS. Averaging operations are required to achieve deterministic characteristics of the output signal. To do this, let's multiply Eq. (2.13) by the complex-conjugate equation and apply expectation functional (averaging operation) to both parts – $M[\ldots]$:

$$X^*(p) = H^*(p) \cdot F^*(p) \tag{2.14}$$

$$M[X(p) \cdot X^*(p)] = M[H(p) \cdot H^*(p) \cdot F^*(p)] \tag{2.15}$$

In the left-hand part of Eq. (2.15) determination of autospectral power density $S_x(\omega)$ of the random process $x(t)$ is given. In the right-hand part of Eq. (2.15), after multiplying the deterministic functions out of expectation function, we have:

$$S_x(\omega) = |H(\omega)|^2 \cdot S_F(\omega) \tag{2.16}$$

This expression consists only of deterministic functions, however, as well as Eq. (2.13) it describes DS. When DS is exposed to the white noise, there is a coincidence of the transfer function modulus square with APSD of displacements with an accuracy to a constant.

$$S_x(\omega) = |H(\omega)|^2 \cdot \text{const} \tag{2.17}$$

In the deterministic case, we have a linear relationship without averaging or a coincidence of the transfer function with a displacement spectrum with an accuracy to a constant

$$X(\omega) = H(\omega) \cdot F(\omega) \tag{2.18}$$

$$X(\omega) = \text{const} \cdot H(\omega) \tag{2.19}$$

Knowing the spectra $-X(\omega)$ and $F(\omega)$ in a deterministic case (2.19) or a spectral density $-S_x(\omega)$ in a stochastic case (2.17), the transfer function modulus and then its parameters "M", "C", "K" are found. An error often occurs when an equation of the quadratic spectral characteristics (2.17) is replaced by an equation of linear spectral characteristics (2.18) or vice versa.[2]

An imaginary component of the transfer function and/or a phase of the transfer function can be found by the known frequency functions of impact and response $F(\omega)$, $X(\omega)$. When the DS is exposed to a random force, the expression (2.6) does not contain information on delay, as there is no imaginary components or phase relations, which disappear after averaging to obtain APSD. To obtain phase relations, it is necessary to assess not only auto- but also cross-spectral characteristics of the processes $x(t)$ and $f(t)$. To do this, we will multiply both parts of Eq. (2.18) by the complex-conjugate function $X^*(\omega)$ and again apply the mathematical expectation functional:

$$M[X(\omega)X^*(\omega)] = M[H(\omega)X^*(\omega)F^*(\omega)] \tag{2.20}$$

or

$$S_x(\omega) = H(\omega)S_{xF}(\omega) \tag{2.21}$$

The left-hand part of this equation is a real function; the right-hand part consists of two complex functions, namely the transfer

[2]To avoid this today the APSD for deterministic case is also defined as the multiplication of the Laplace (or Fourier transfer to its complex conjugate (without averaging) (note of the editor).

function and cross-spectral density $S_{xF}(\omega)$, which means that their phase relations differ only by a symbol and the phases can be extracted both from the function $H(\omega)$ and from the function $S_{xF}(\omega)$: $\varphi_H(\omega) = -\varphi_{xF}(\omega)$. They are often mistakenly identified; the phase is described as a "cross-spectral density phase" or a "transfer function phase".

Thus, all information on delay of the random process $x(t)$ with regard to the random process $f(t)$ is contained in the cross-spectral density function. Note that the connection between APSD and CPSD (2.21) is linear but not quadratic as in case (2.17) between two CPSD.

2.2.3 *DS secular equation*

At a zero force (which is a specific case of white noise with zero amplitude) we have the equation of free oscillations or a secular equation

$$p^2 + 2\sigma \cdot p + \omega_0^2 = 0 \tag{2.22}$$

with roots being complex-conjugate numbers

$$p_1 = -\sigma + \sqrt{\sigma^2 - \omega_0^2} = \omega_0(-\alpha + j\sqrt{1 - \alpha^2}) \tag{2.23}$$

$$p_2 = p_1^* = -\sigma - \sqrt{\sigma^2 - \omega_0^2} = \omega_0(-\alpha - j\sqrt{1 - \alpha^2}) \tag{2.24}$$

where $\alpha = \frac{\sigma}{\omega_0}$ is a damping ratio, being a dimensionless parameter, which relates the oscillation damping parameter with dynamic system oscillation parameter. When they are equal, $\sigma = \omega_0$, two roots (p_1, p_2) also coincide with each other and become a totally real number $p_1 = p_2 = -\sigma$. The absence of the imaginary component in the roots of the dynamic stiffness corresponds to the absence of the dynamic system oscillations.

The concept of critical damping of the dynamic system — C_{cr} is introduced based on the condition of equality $\sigma = \omega_0$:

$$\alpha = \frac{\sigma}{\omega_0} = 1 = \frac{C}{C_{\text{cr}}} = \frac{C}{2\sqrt{\text{MK}}} \tag{2.25}$$

Hence, $C_{\text{cr}} = 2\sqrt{\text{MK}}$. The critical DS damping is such a threshold DS parameter crossing which results in qualitative change of the process properties $x(t)$. Response of DS with critical damping to δ perturbance will be a monotonically decreasing exponential process with a time constant equal to the inverse value of the damping ratio — $1/\sigma$. Q-factor of DS, and obviously the defining and damping facto, along with the DS oscillation properties are uniquely connected with the damping factor:

$$Q = \frac{C_{\mathrm{cr}}}{2C} = \frac{1}{2\alpha} \tag{2.26}$$

As another property of damping (in addition to the already introduced parameters C_{cr}, Q, α) a logarithmic decay ratio (LDR) is widely used in applications:

$$\delta = 2\pi\alpha \tag{2.27}$$

All the above dimensionless parameters one way or another define the scattering of the oscillatory energy by the dynamic system for a single oscillation period. It would seem that the logarithmic decay ratio is an excessive parameter since it coincides with the damping ratio with an accuracy to a constant. However, the logarithmic decay ratio has a simple visual physical interpretation and can be a basis for the direct estimation of DS damping properties. LDR is equal to the hyperbolic logarithm of the ratio of two neighboring maxima of the damped oscillatory displacement process after the impulse force (shock) impacted the DS: $\delta = \ln \frac{x_{\max(i)}}{x_{\max(i+1)}} = \frac{1}{n}, \ln \frac{A_{i+n}}{A_i}$ (Fig. 2.9).

Based on the fact that the expression for the critical damping — $C_{\mathrm{cr}} = 2\sqrt{\mathrm{MK}}$ includes a mass and stiffness, but does not include the dynamic system damping — "C", the DS can loose its oscillation character while increasing the mass and maintaining other fixed parameters or while increasing the stiffness and maintaining other fixed parameters.

Real structures tend to have much lower damping than the critical one ($C \ll C_{\mathrm{cr}}$), and so $\alpha \ll 1$, $Q \gg 1$. If special damping devices have not been introduced to the DS structure, then damping usually does not exceed 10% of the "C_{cr}".

So, $\sqrt{\mathrm{MK}}$ has the same dimension as the damping — "C" (kg/s); the ratio of C/M and K/C have frequency dimension (Hz), and their

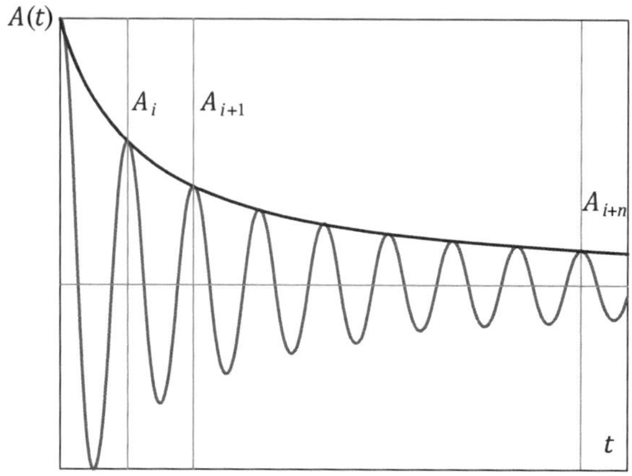

Figure 2.9: Free oscillations of a dissipative system with a single degree of freedom. A_i, A_{i+1} — envelope amplitudes coinciding with $x_{\max}(i)$ and $x_{\max(i+1)}$, respectively. A_{i+n} — envelope amplitude through "n" periods.

product is a square of the natural frequency:

$$\frac{C}{M} = \Omega_1 = \frac{\omega_0}{Q} \tag{2.28}$$

$$\frac{K}{C} = \Omega_2 = \omega_0 Q \tag{2.29}$$

$$\Omega_1 \Omega_2 = {\omega_0}^2 \tag{2.30}$$

According to (2.28)–(2.30), three frequencies $(\Omega_1, \Omega_2, \omega_0)$ are interdependent: being aware of two allows to define the third frequency. Three frequencies $(\Omega_1, \Omega_2, \omega_0)$ depending on three DS parameters (M, C, K) do not define them uniquely since these nonlinear equations are dependent on each other.

$$\Omega_1 = F_1(M, C, \_)$$

$$\Omega_2 = F_2(\_, C, K)$$

$$\omega_0 = F_3(M, \_, K)$$

Based on the triplet of numbers (M, C, K) it is possible to uniquely define the triplet of numbers $(\Omega_1, \Omega_2, \omega_0)$. However, in reverse there is no certainty.

With a low (less than critical) damping, the parameters Ω_1 and Ω_2 on the frequency axis are to the left and to the right from the natural

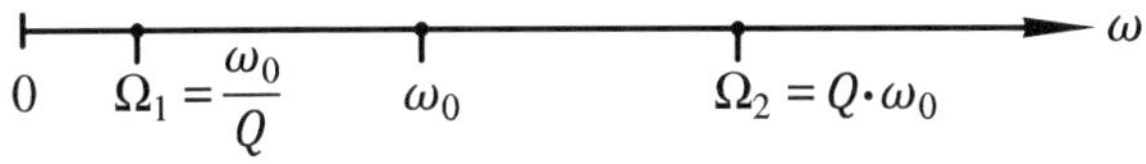

Figure 2.10: DS characteristic frequencies.

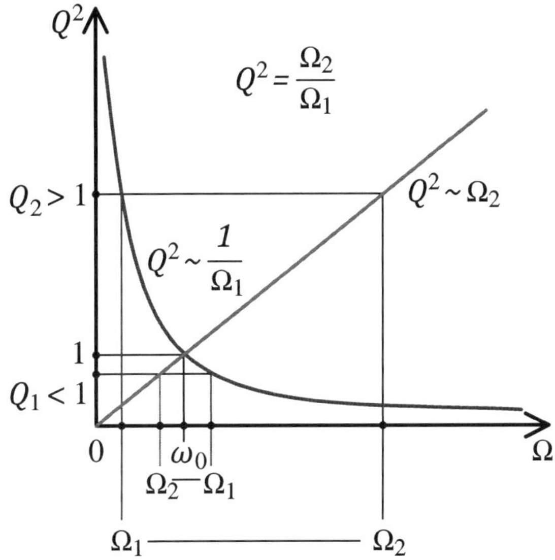

Figure 2.11: Dependence of Q-factor from characteristic frequencies.

frequency (Fig. 2.10), and only when $Q = 1$ all three frequencies coincide with each other, and the segment $[\Omega_1, \Omega_2]$ degenerates into a point.

Generally, the natural frequency ω_0, which is a geometric mean of the two characteristic frequencies, does not divide the segment $[\Omega_1, \Omega_2]$ into two halves. The Q-square is linearly dependent on the second characteristic frequency and hyperbolically dependent on the first one: $Q^2 = \frac{\Omega_2}{\Omega_1}$.

According to Fig. 2.11, the relative position of characteristic frequencies with respect to the natural frequency ω_0 is determined for a specific dynamic system with a known Q-factor.

The damping ratio as well as the Q-factor are determined by the ratios of the characteristic frequencies,

$$Q^2 = \frac{\Omega_2}{\Omega_1} = \frac{1}{4\alpha^2}$$

i.e. the characteristic frequencies contain information on both oscillatory DS properties and DS ability to introduce attenuation into the free oscillation process.

The two-dimensional space of the transfer function poles $H(p)$ – (σ, ω_0), represented by parameters of DS oscillation and attenuation parameters, and by the given equation $K(p) = 0$:

$$p^2 + 2\sigma \cdot p + \omega_0^2 = 0 \tag{2.31}$$

is equivalent in terms of amount of information on DS properties to the two-dimensional space represented by two frequencies (Ω_1, Ω_2) and given by the equation

$$p^2 + \Omega_1 \cdot p + \Omega_1\Omega_2 = 0 \tag{2.32}$$

or represented by two frequencies (Ω_1, ω_0) and given by the equation

$$p^2 + \Omega_1 \cdot p + \omega_0^2 = 0 \tag{2.33}$$

or represented by the damping ratio and natural frequency (α, ω_0) and given by the equation

$$p^2 + 2\alpha\omega_0 \cdot p + \omega_0^2 = 0 \tag{2.34}$$

The given equations of free oscillations (2.31)–(2.34) are identical to each other and determine the interdependence of three DS parameters "M", "C", "K" through its roots $p_{1,2}$ and contain information on the natural, the so-called global DS parameters, which are independent from the disturbing force parameters and initial DS oscillation conditions (local parameters).

Three DS parameters — "M", "C", "K" — are not only real numbers with a very specific physical meaning, but also always non-negative aka positive numbers. At the same time, the equation of free oscillations $p^2 + \frac{C}{M} \cdot p + \frac{K}{M} = 0$ has complex roots which in the frequency representation of this very equation in a form of

$$\omega^2 - \frac{C}{M} \cdot j\omega - \frac{K}{M} = 0 \tag{2.35}$$

shall be interpreted in the frequency language and using terms for DS oscillation and attenuation. It is usually said that neither negative nor imaginary frequencies are observed in nature, although we are talking about mathematical formalism. The secular equation makes it possible to represent a transfer function in the form of pole expansion (to the roots of the denominator) in the form of a sum

$$H(p) = \frac{A}{p - p_1} + \frac{A^*}{p - p_1^*}$$

The complex-conjugate parameters "A" and "A^*", or the so-called deductions define the DS amplitude properties:

$$A = \frac{1}{j2M\sqrt{\omega_0^2 - \sigma^2}} = \frac{1}{jC_{\text{cr}}\sqrt{1 - \alpha^2}},$$

and connect to the natural types of DS oscillations.

Taking into account the Laplace variable is a purely imaginary value (imaginary frequency) $p = j\omega$, we will obtain an expression for the transfer function as a complex-valued function

$$\begin{aligned} H(\omega) &= -\frac{1/M}{\omega^2 - \omega_0^2 - 2j\omega\sigma} \\ &= -\frac{1/M}{[(\omega^2 - \omega_0^2)^2 + 4\omega^2\sigma^2]}[(\omega^2 - \omega_0^2) + 2j\omega\sigma] \\ &= \text{Re}_H(\omega) + jIm_H(\omega) \end{aligned} \tag{2.36}$$

The denominator of the transfer function $H(\omega)$

$$\omega^2 - 2j\omega\sigma - \omega_0^2$$

has two complex-conjugate roots (two poles of the function $H(\omega)$)

$$\omega_{1,2} = j\omega_0\left(\alpha \mp j\sqrt{1 - \alpha^2}\right) \tag{2.37}$$

which become purely real $\omega_{1,2} = \mp\omega_0$ at $\alpha = 0$ (no damping) and purely imaginary $\omega_{1,2} = \mp j\omega_0$ at $\alpha = 1$ (critical damping). In intermediate cases — $0 < \alpha < 1$ — the solution to Eq. (2.35) is a damped oscillation process. Asymptotic expressions for the damping roots at characteristic $\omega_{1,2}$ have a form of

$$\omega_{1,2} = \left\{\begin{array}{ll} \omega_0, & \text{at } \alpha = 0 \\ j\omega_0, & \text{at } \alpha = 1, \\ \approx j\Omega_1 & \text{and } j\Omega_2, \text{ at } \alpha \gg 1 \end{array}\right\}$$

The roots of the secular equation at very high and very low damping are close to either purely imaginary or purely real numbers. It shall be recalled that the case of $\alpha \ll 1$ is the most common one in practice. Therefore, it is possible that roots of the secular equation are often considered falsely as equal and always real: $\omega_{1,2} = \omega_0 = \sqrt{K/M}$.

The real type roots of the secular equation in DS representation after the Laplace transformation is identical to the imaginary numbers of these roots in DS representation after the Fourier transformation due to the relationship of their variables $p = j\omega$. In some literature sources, the imaginary roots of the secular equation are interpreted as an evidence of the solution oscillation, in other sources — as an evidence of exponential decay of its amplitude. Transition from the complex variable "p" in Laplace transformation to the purely imaginary variable $p = j\omega$ in Fourier transformation causes discrepancies: the frequencies (2.37) shall be multiplied by the complex unit "j" to obtain the frequencies (2.23).

2.3 Solutions of DS Homogeneous Equation

2.3.1 *Free oscillations*

DS homogeneous equation

$$M \cdot \frac{d^2x(t)}{dt^2} + C \cdot \frac{dx(t)}{dt} + K \cdot x(t) = 0$$

with a zero damping

$$M \cdot \frac{d^2x(t)}{dt^2} + K \cdot x(t) = 0$$

written in the equivalent form

$$\ddot{x} + \omega_0^2 \cdot x = 0 \tag{2.38}$$

in addition to the standard zero solution $x(t) = 0$ has a solution in a form of the complex exponent:

$$x(t) = A \cdot e^{-j(\omega_0 t + \varphi)} \tag{2.39}$$

The correctness of the solution shall be checked by direct substitution of (2.39) to (2.38). It is often called a sinusoidal solution (which is not quite correct) since both imaginary and real components of the complex exponent, respectively,

$$\mathrm{Re}_x x(t) = \cos(\omega_0 t) \quad \text{and} \quad \mathrm{Im}_x x(t) = \sin(\omega_0 t)$$

are also solutions of Eq. (2.38).

At zero damping the remaining two DS parameters ("M" and "K") are compressed into a single constant $\omega_0 = \sqrt{K/M}$, which is the multiplicator of the displacement in the initial DS Eq. (2.38). Therefore, the square of the natural frequency ω_0^2 can be called a normalized stiffness. Normalizing by the mass, a transmission is made from the stiffness (kg/s^2) to the frequency square dimension (Hz^2).

The regularities of the simple solution for the non-dissipative DS (harmonic oscillator) are transferred to the case of DS with dissipation, if in the equation:

$$\ddot{x} + 2\sigma\dot{x} + \omega_0^2 \cdot x = 0 \tag{2.40}$$

we substitute

$$x(t) = y(t) \cdot e^{-\sigma t} \tag{2.41}$$

Then Eq. (2.40) is changed as

$$\ddot{y} + y \cdot (\omega_0^2 - \sigma^2) = 0 \tag{2.42}$$

or to a form without the summand with the first derivative similar to the equation of non-dissipative DS (2.38):

$$\ddot{y} + y \cdot \omega_C^2 = 0$$

where $\omega_C = \sqrt{\omega_0^2 - \sigma^2}$.

That is, the general solution of the DS equations without dissipation, where the substitution $\omega_0^2 \leftrightarrow \omega_C^2$ was made shall be multiplied by decreasing exponent $e^{-\sigma t}$.

Now instead of two particular solutions of non-dissipative DS $x_1(t) = \sin\omega_0 t$ and $x_2(t) = \cos\omega_0 t$ we have a fundamental system of solution for dissipative DS $x_1(t) = e^{-\sigma t} \cdot \sin\omega_c t$ and $x_2(t) = e^{-\sigma t} \cdot \cos\omega_c t$. Hence, a general solution in a complex representation is written as follows:

$$x(t) = C_1 \cdot e^{-\sigma t + j\omega_c t} + C_2 \cdot e^{-\sigma t - j\omega_c t} \tag{2.43}$$

The values of these two exponents were found to be complex-conjugate roots of the initial secular equation

$$p^2 + 2\sigma p + {\omega_0}^2 = 0$$

$$p_{1,2} = -\sigma \pm j\omega_0\sqrt{1-\alpha^2} = \omega_0\left(-\alpha \pm \sqrt{\alpha^2 - 1}\right) \tag{2.44}$$

with a frequency dimension ($p = j\omega$). For the dimensionless frequency $\left(F = \frac{\omega}{\omega_0}\right)$, the equation and roots take the form,

$$-F^2 + j2\alpha F + 1 = 0 \tag{2.45}$$

$$F_{1,2} = j\alpha \pm \sqrt{-\alpha^2 + 1}$$

Sign of discriminant of Eq. (2.44):

$$\Delta = C^2 - 4\text{MK} = C_{kp}^2(\alpha^2 - 1) \lesseqgtr 0, \quad C_{\text{cr}} = 2\sqrt{\text{MK}},$$

or a value of the dimensionless parameter α compared to the unit $0 \le \alpha \lesseqgtr 1$ define type of solution of the homogeneous Eq. (2.40). Unlike non-dissipative DS, the formula (2.43) contains three classes of functions. They can be monotonically decreasing, oscillating or having a single extreme point. The nature of the function cannot be changed in the given solution class ($\alpha < 1, \alpha = 1, \alpha > 0$) just by varying the constants C_1, C_2 (only by initial conditions):

$$x(t) = C_1 \cdot e^{p_1 t} + C_2 \cdot e^{p_2 t}, (p_1 \neq p_2) \tag{2.46}$$

or

$$x(t) = \begin{cases} e^{-\alpha\omega_0 t} \cdot \left[C_1 \cdot e^{\omega_0 t\sqrt{\alpha^2-1}} + C_2 \cdot e^{-\omega_0 t\sqrt{\alpha^2-1}}\right], \Delta > 0, \alpha > 1, \\ p_{1,2} = \omega_0\left(-\alpha \pm \sqrt{\alpha^2 - 1}\right) \\ e^{-\alpha\omega_0 t} \cdot (C_1 + C_2 \cdot t), \quad \Delta = 0, \quad \alpha = 1, \quad p_{1,2} = -\omega_0 \\ e^{-\alpha\omega_0 t} \cdot \left[C_1 \cdot \cos\left(\omega_0 t\sqrt{1-\alpha^2}\right)\right. \\ \quad \left. + C_2 \cdot \sin\left(\omega_0 t\sqrt{1-\alpha^2}\right)\right], \Delta > 0, \alpha > 1 \\ p_{1,2} = \omega_0\left(-\alpha \pm j\sqrt{1-\alpha^2}\right) \end{cases} \tag{2.47}$$

For example, if DS is highly damped ($\alpha > 1$), then no matter to what extent the mass deviates from equilibrium (x_0) and what the initial velocity (v_0) was, the displacement will monotonically tend to zero.

If DS is critically damped ($\alpha = 1, \Delta = 0$), the displacement will monotonically tend to zero after the global extreme value has been reached. In this case, the initial conditions affect the coordinates of extreme value (amplitude and time of the extreme value of displacement).

The decreasing oscillatory nature of the displacement cannot be changed at $\alpha < 0, \Delta < 0$ by changing the initial conditions.

The frequency ω_c at $\alpha > 1$ becomes a complex number

$$\omega_c = j\omega_0\sqrt{\alpha^2 - 1},$$

and the decay parameter of the exponent $\sigma = \alpha \cdot \omega_0$ is always a non-negative real number.

A general solution of the DS dissipative equation

$$\begin{aligned} x(t) &= C_1 \cdot e^{p_1 t} + C_2 \cdot e^{p_2 t} \\ &= e^{-\alpha\omega_0 t} \cdot \left[C_1 \cdot e^{j\omega_0 t\sqrt{1-\alpha^2}} + C_2 \cdot e^{-j\omega_0 t\sqrt{1-\alpha^2}}\right] \end{aligned} \tag{2.48}$$

is represented by two factors and only one of them contains information on initial conditions (C_1, C_2). It is obvious that the constants C_1 and C_2 determine the initial displacement amplitude $x(t)$ and their

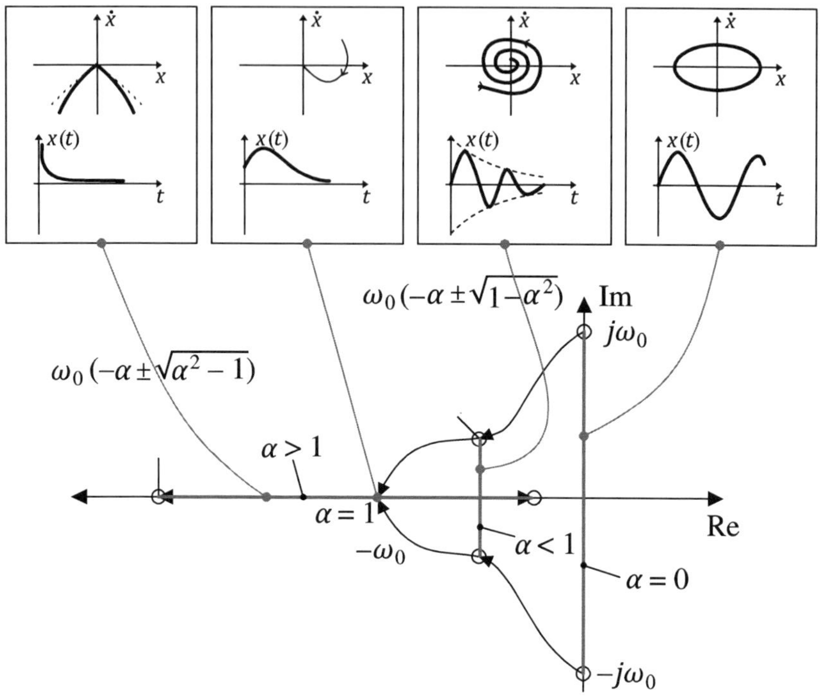

Figure 2.12: Correspondence of the secular equation roots and displacement as a function of time and trajectory on the phase plane.

relations in a form

$$\frac{C_1}{\sqrt{C_1^2 + C_2^2}}, \qquad \frac{C_2}{\sqrt{C_1^2 + C_2^2}}$$

determine the contribution of each of the two exponentials (2.48).

For a class of damped oscillating displacement (when $\alpha < 1$), the ratio of C_1 and C_2 determines the phase of oscillations. Figure 2.12 shows the correspondence of the secular equation roots, displacement and trajectory on the phase plane $(x, \dot{x})$.

The solution (2.48) has a simple asymptotic behavior in the form of complex and real exponentials:

$$x(t) \approx \text{const} \begin{cases} e^{-j\omega_0 t}, & \Big| \alpha \ll 1 \\ e^{-\omega_0 t}, & \Big| \alpha = 1 \\ e^{-\Omega_1 t}, \quad \text{and} \quad e^{-\Omega_2 t} & \Big| \alpha \gg 1 \end{cases} \tag{2.49}$$

The real solution will be as follows:

$$x(t) = \begin{cases} \text{const} \cdot e^{-\sigma \cdot t} \cdot \cos[\omega_0 \cdot t \cdot \sqrt{(1-\alpha^2)} + \varphi], & \alpha < 1 \\ \text{const} \cdot e^{-\sigma t}, & \alpha = 1 \\ \text{const} \cdot e^{-\sigma \cdot t} \cdot ch[\omega_0 \cdot t \cdot \sqrt{(1-\alpha^2)} + \varphi] & \\ \quad = \text{const} \cdot e^{\omega_0 \cdot t \left[\alpha - \sqrt{(\alpha^2-1)}\right]} + \text{const} \cdot e^{\omega_0 \cdot t \left[\alpha + \sqrt{(\alpha^2-1)}\right]}, & \alpha > 1 \end{cases}$$

In general, three DS parameters ("M", "C", "K") are compressed into two parameters $\sigma = C/2M$, $\omega_0 = \sqrt{K/M}$, which are again compressed into one factor before moving in DS equation – $\Omega = \omega_0\sqrt{1-\alpha^2}$. It can also be considered as a generalized normalized stiffness, but with damping included.

The non-zero damping not only affects the displacement amplitude, but also slightly shifts the natural frequency ω_0. To avoid a terminological confusion, the natural frequency hereinafter will be referred to as a frequency ω_0, but cases when the natural frequency refers to the offset value $\Omega = \omega_0\sqrt{1-\alpha^2}$ will be specified additionally.

All three DS parameters ("M", "C", "K") are included in the solution (in the displacement) as arguments of the complex exponent. When the complex exponent argument is a purely real number, we observe oscillation in the solution, and when it is a purely imaginary value, we have a monotonically damped function. In general, a complex argument specifies both.

Equations of DS in frequency representations:

$$\omega^2 - 2j\omega\sigma - \omega_0^2 = 0 \tag{2.50}$$

$$Z(F) = F^2 - 2j\alpha F - 1 = 0 \tag{2.51}$$

where $F = \frac{\omega}{\omega_0}$ is a normalized dimensionless frequency, and they have the following solutions:

$$\omega_{1,2} = j\omega_0 \left(\alpha \mp j\sqrt{1-\alpha^2}\right)$$

$$F_{1,2} = j\alpha \mp \sqrt{1-\alpha^2} \tag{2.52}$$

The normalized complex frequency of the later formula depends only on a single parameter, namely on damping ratio (Fig. 2.13). The complex frequency (2.52) in exponential representation

$$F \equiv |F|e^{-j\theta}$$

always has a unit modulus $|F| \equiv \sqrt{\text{Re}F^2 + \text{Im}F^2} = \sqrt{\alpha^2 + 1 - \alpha^2} = 1$ and a phase angle θ defined by the expression

$$tg\theta \equiv \frac{\text{Im}F}{\text{Re}F} = \frac{\alpha}{\sqrt{1-\alpha^2}}$$

By introducing two real variables $t = \alpha$, $s = \sqrt{1-\alpha^2}$ instead of a single complex number $F = j\alpha + \sqrt{1-\alpha^2}$, we obtain their relation in the form of a circle of a single radius $s^2 + t^2 = 1$ with a phase ratio $tg\theta = \frac{t}{s}$. The roots $F_{1,2} = j\alpha \mp \sqrt{1-\alpha^2}$ are uniquely related to the damping ratio — α, but due to $|F_{1,2}| = 1$ all the information on the parameter α is included in the complex frequency phase: $tg^2\theta = \frac{\alpha^2}{1-\alpha^2}$ or $\alpha = \sin\theta$, $0 \le \theta \le \frac{\pi}{2}$ (Fig. 2.13).

At $\alpha = 0$, the displacements $x(t)$ are a sine wave (Table 2.3); the dimensionless roots of the complex frequency equation (2.52) are the real $F_{1,2} = \pm 1$, i.e. they have a zero phase or antiphase, and the characteristic frequencies Ω_1 and Ω_2 have the extreme values of zero and infinity, respectively. In this case, their geometric mean which is a natural frequency $\omega_0 = \sqrt{\Omega_1 \Omega_2}$ is an uncertainty $(0 \cdot \infty)$ which

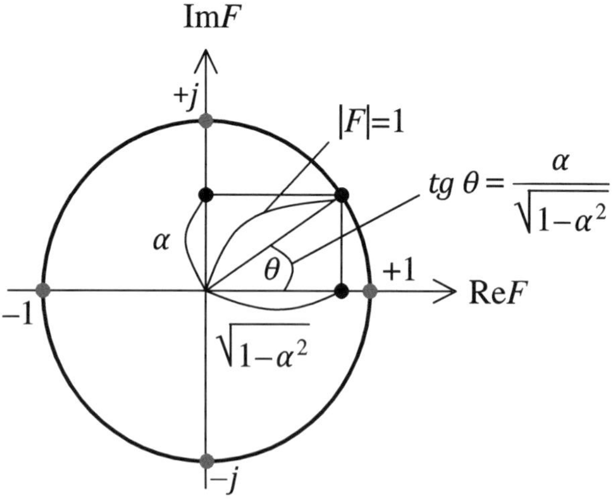

Figure 2.13: The roots of the secular equation on the complex plane for $\alpha \ll 1$.

can have any value on the positive half-axis of frequencies. At the same time, the Q-factor $Q = \sqrt{\frac{\Omega_2}{\Omega_1}}$ is an infinitely large value. This means that the oscillation amplitude on the resonant frequency is also infinitely high.

The properties of the DS solution at various damping are listed in Table 2.3.

Based on the considered hypothetical case ($\alpha = 0$), it can be concluded that in practice a normalized frequency at a zero damping is a complex number and the characteristic frequencies Ω_1 and Ω_2 are within a limited sub-band of frequencies, and Q-factor is a finite number.

In practice, the most commonly used half-open interval of the damping ratio change $\alpha \in (0, 1]$ is divided into three intervals according to the type of DS solution. A damped oscillatory type of the solution (row 2 of Table 2.3) will be observed for all small values α within the sub-band $\alpha \in \left(0, \frac{1}{2}\right]$. In this case, the complex frequency phase will be within a range $(0^0$–$30^0]$, and the Q-factor will always be more than one. It is the boundary value of $Q = 1$ which resulted in the allocation of this sub-band of the damping change into a separate sub-band. For all $Q > 1$, the characteristic frequencies Ω_1 and Ω_2 occupy the entire half-axis of the positive frequencies, $\Omega_1 \in (0, \omega_0)$ is before its natural frequency, and Ω_2 is after: $\Omega_2 \in (\omega_0, \infty)$.

The frequencies ω_0 and Ω_1 can be observed directly according to $x(t)$ at free DS oscillations. A single shock disturbance of DS results in free oscillations with an amplitude decreasing according to $e^{-\sigma t} = e^{\Omega_1 t/2}$, and the frequency of decreasing oscillations is approximately equal to ω_0. This is a basis for measuring the natural frequency and DS oscillation decrement in a single experiment.

It would seem that the frequency Ω_2 has no effect on the solution of the DS equation $x(t) = [e^{-\Omega_1 t/2}] \cdot e^{-j\omega_0 t\sqrt{1-\alpha^2}}$, but this solution can be represented in terms of Ω_2 and ω_0. A damping ratio α in the formula can be represented as terms of the characteristic frequency Ω_2 for an exact natural frequency $\Omega = \omega_0\sqrt{1-\alpha^2}$:

$$\Omega = \omega_0\sqrt{1 - \left(\frac{\omega_0}{2\Omega_2}\right)^2}$$

If the first characteristic frequency Ω_1 determines the attenuation of free oscillations, the second characteristic frequency Ω_2 determines

Table 2.3: Roots of the secular equation, characteristic frequencies and DS solutions at various damping ratios.

Relative damping ratio α	Q-factor $Q=\frac{1}{2\alpha}$	$\Omega_1=\frac{\omega_0}{Q}$, $\Omega_2=\omega_0 Q$	$F_{1,2}=j\alpha\mp\sqrt{1-\alpha^2}$ on the complex plane	Graphic representation of DS equation solution
$\alpha=0$	∞	$\Omega_1=0,\ \Omega_2=\infty$ 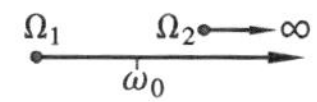	$F_{1,2}=\mp 1$ 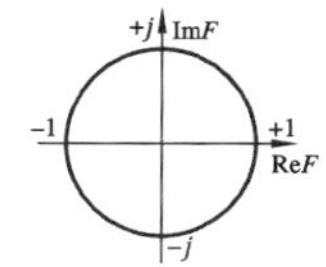	undamped 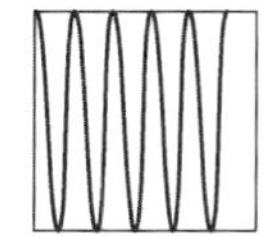
$0<\alpha<1/2$	$Q>1$	$\Omega_1<\omega_0,\ \Omega_2>\omega_0$ 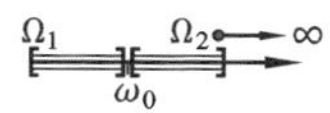	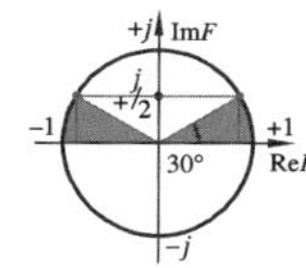	damped oscillatory 
$1/2<\alpha<\sqrt{2}/2$	$\sqrt{2}/2<Q<1$	$\omega_0<\Omega_1<\sqrt{2}\omega_0$ $\omega_0/\sqrt{2}<\Omega_2<\omega_0$ 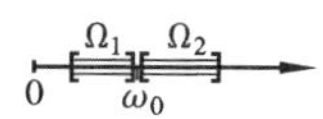	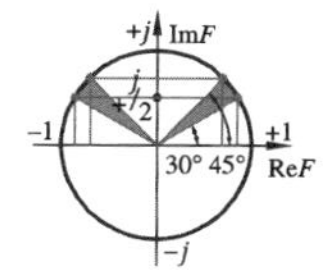	damped aperiodic
$\sqrt{2}/2<\alpha<1$ $\sqrt{2}/2<\alpha<1$	$1/2<Q<\sqrt{2}/2$	$\sqrt{2}\omega_0<\Omega_1<2\omega_0$ $\omega_0/2<\Omega_2<\omega_0/\sqrt{2}$ 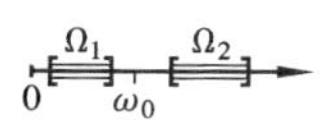	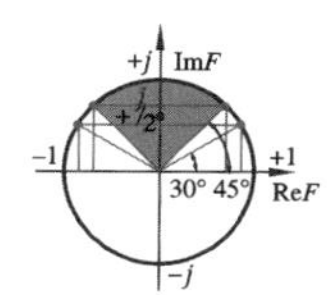	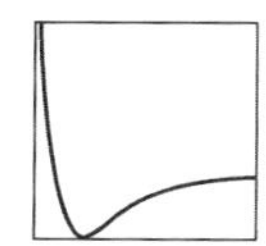

$\alpha = 1$	$Q = 1/2$	$\Omega_1 = 2\omega_0,\ \Omega_2 = \omega_0/2$ 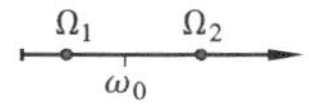	$F_{1,2} = j$ 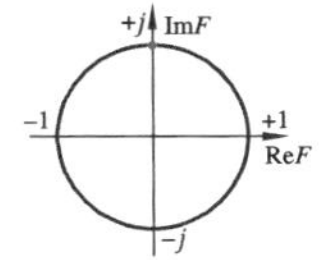	Damped monotonous $\alpha = 1$ 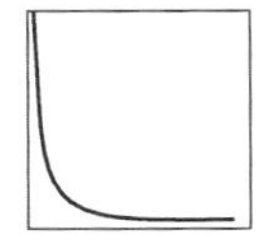
$\alpha > 1$	$Q < 1/2$	$\Omega_1 > 2\omega_0,$ $\Omega_2 < \omega_0/2$ $F_{1,2} = j(\alpha \mp \sqrt{\alpha^2 - 1})$ 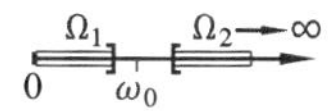	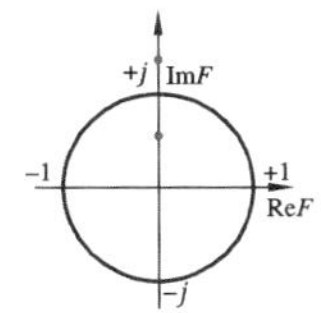	

the frequency shift of natural oscillation from the value ω_0. For $\alpha \ll 1$, we have an approximate expression $\Omega \approx \omega_0\left(1 - \frac{\alpha^2}{2}\right) = \omega_0 - \frac{\omega_0{}^3}{2\Omega_2^2}$ according to which the second characteristic frequency is not actually manifested in oscillations of weakly damped DS oscillations.

As soon as the characteristic frequencies Ω_1 and Ω_2 change their places in relation to natural frequency, i.e. $\Omega_1 > \omega_0$, and $\Omega_2 < \omega_0$, the DS solution shifts to a decaying aperiodic form (rows 3, 4 of Table 2.3). This boundary of the section corresponds to the Q-factor, which is equal to one: $Q = 1(\alpha = 1/2)$.

At $\alpha \in \left(\frac{1}{2}, \frac{\sqrt{2}}{2}\right]$, the characteristic frequencies are in the vicinity of the natural frequency, i.e. in adjacent sub-bands

$$\Omega_2 \in \left(\frac{\omega_0}{\sqrt{2}}, \omega_0\right) \quad \text{and} \quad \Omega_1 \in (\omega_0, \omega_0\sqrt{2}),$$

and a phase of the complex frequency – in sub-band (30°–45°). In this case, a small number of oscillation periods are visualized. In the solution of DS equation

$$x(t) = e^{-\sigma t} e^{-j\omega_0 t\sqrt{1-\alpha^2}} \tag{2.53}$$

the first factor — $e^{-\sigma t}$ – determines the damping of the amplitude, and the second — $e^{-j\omega_0 t\sqrt{1-\alpha^2}}$ – determines the oscillation properties of displacements, but only at $< \alpha < 1$. In other words, while the first factor is a real number, the second factor is a complex one, as the above separation of DS properties by factors is made. At values α close to one, but still less than one, the oscillation of displacement is very weak.

In practice, there is the case of DS oscillation when displacement after the first intersection of the equilibrium point returns to the equilibrium with almost zero amplitude. This is a case with a large but below critical damping at $\sqrt{2}/2 < \alpha < 1$ allocated into a separate row in Table 2.3. Visually, such an oscillation is perceived as if its decay occurred for one period with a single intersection of the equilibrium position. When the Q-factor crosses the boundary $Q = \sqrt{2}/2$, the frequency sub-bands with the natural frequencies cease to touch each other:

$$\Omega_2 \in \left(\frac{\omega_0}{2}, \frac{\omega_0}{\sqrt{2}}\right) \quad \text{and} \quad \Omega_1 \in \left(\sqrt{2}\omega_0, 2\omega_0\right),$$

At final value $Q = \sqrt{2}/2$, the damping ratio has the same value $\alpha = \sqrt{2}/2$, and the phase of the complex frequency is 45°.

When each of the two sub-bands of the characteristic frequency degenerates into a point

$$\Omega_2 = \frac{\omega_0}{2}, \quad \Omega_1 = 2\omega_0,$$

a critical damping occurs: $\alpha = 1$ $\left(\text{or } Q = \frac{1}{2}\right)$ — see the fifth row in Table 2.3. In this case, the complex frequency takes a single value, which is equal to the imaginary unit: $F = j$. The solution $x(t)$ eliminates the oscillating cofactor and there remains the only exponent defining the attenuation $x(t) = \mathrm{e}^{-\sigma t}$. It was present in the solution even before with the constant damping parameter $\sigma = \frac{-\Omega_1}{2}$, but only as the damping envelope of the displacement extrema (Fig. 2.12).

At $\alpha > 1$ — the characteristic frequencies again fall into the non-contiguous sub-bands: one sub-band is below the half natural frequency — $\Omega_2 < \frac{\omega_0}{2}$, the other is above the double natural frequency $\Omega_1 > 2\omega_0$. The roots of the complex frequency $F_{1,2} = j\alpha \mp \sqrt{1-\alpha^2}$ become purely imaginary. At $\alpha > 1$, the transfer DS function monotonically decreases with a frequency and has a global maximum at zero Hz. This does not mean that the natural frequency becomes zero, $\omega_0 = \sqrt{K/M}$ is a real number, and in practice it is always non-zero. But now there is neither global, nor local maximum of the transfer function at the natural frequency ω_0.

Now both factors of the solution (2.48) are the real numbers, i.e. both affect the displacement amplitude only, as there is no oscillation at all (the sixth row of Table 2.3). In this regard, it is not quite correct to define DS in terms of frequencies Ω_1, Ω_2, but it is more convenient to use their inverse values — $\tau_1 = \frac{1}{\Omega_1}$, $\tau_2 = \frac{1}{\Omega_2}$ as characteristic times of the amplitude decay. In case of heavily damped DS, the displacement shall be mainly determined only by the second characteristic frequency Ω_2. Indeed, the small value Ω_2 determines a large time constant of decay of the displacement process by its inverse value $\tau_2 = \frac{1}{\Omega_2}$, and on the contrary, the first characteristic frequency defines the rapidly attenuating process (Fig. 2.14).

At $\leq \alpha \leq 1$, the complex frequencies $F_{1,2}$, as the roots of the secular equation, are on the semi-circle of the unit radius in the complex plane (Fig. 2.15). As soon as solutions become real at $\alpha > 1$, they are located on the positive half-axis of the complex plane, and

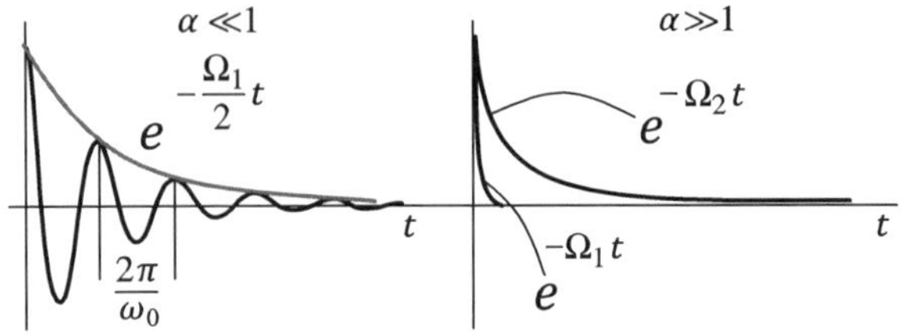

Figure 2.14: Free oscillations of the dynamic system with very low (the left graph) and very high damping after the shock impact.

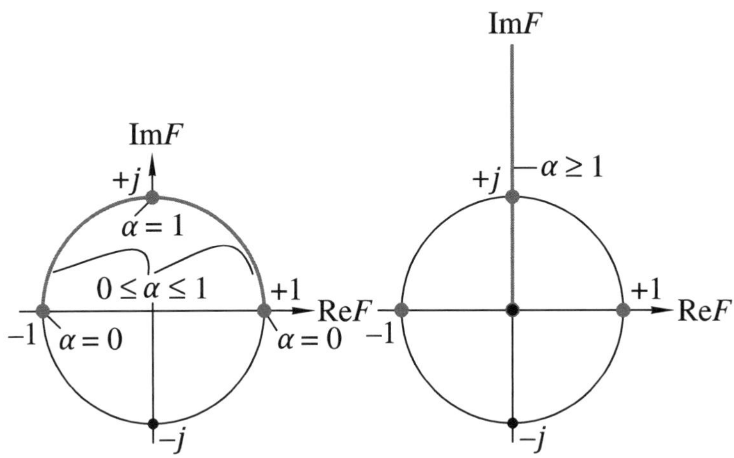

Figure 2.15: Location of the roots of DS secular equation at the complex plane with various relative damping (α).

their modules can take any values, not only number one, except for zero.

The specific global properties of DS are determined by the complex root of the secular equation, which cannot be the random values at the complex plane, but the values from the single circle and imaginary axis only, i.e. the imaginary and real parts (t, s) of the roots of the secular equation depend on each other.

DS displacements for very small and very large dampings can be written analytically as (Fig. 2.16)

$$x(t) \approx \text{const} \begin{cases} e^{-\frac{\Omega_1}{2}t} \cdot e^{-j\omega_0 t}, & \alpha \ll 1 \\ e^{-\Omega_2 t}, e^{-\Omega_1 t} & \alpha \gg 1 \end{cases} \tag{2.54}$$

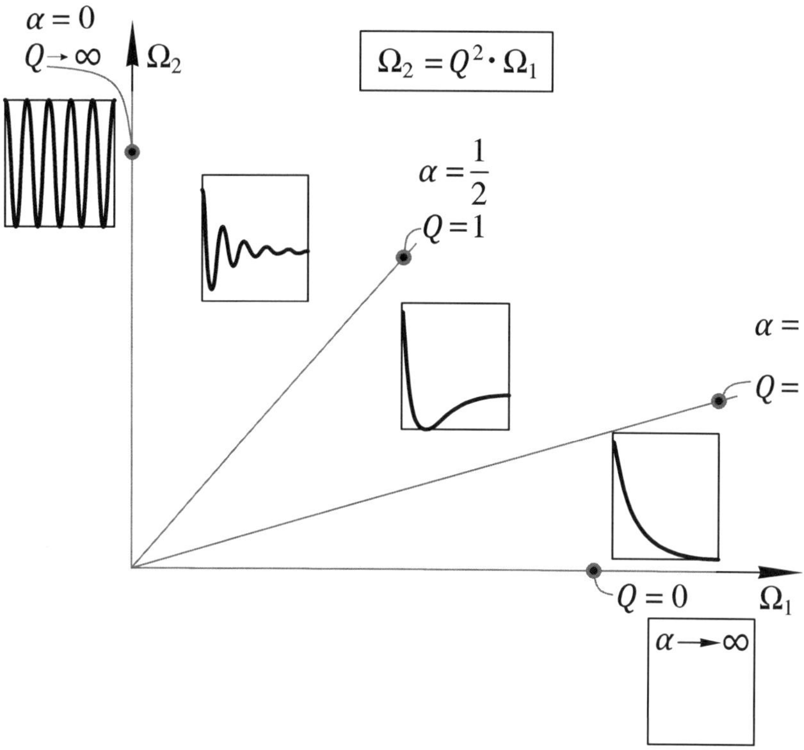

Figure 2.16: Dependency of DS natural frequencies through the Q-factor $\Omega_2 = Q^2\Omega_1$ and partitioning of the feature space into three sectors.

For strongly damped systems ($\alpha > 1$), there are two solutions of the DS equation at the same time. In practice, the rapid exponent $e^{-\Omega_1 t}$ is usually neglected, although at any time starting from $t = +0$ in the signal $x(t)$ there is information both on ω_0 and on σ. Any strongly damped DS has a non-zero natural frequency ω_0, and since $\omega_0 \ll \sigma$, we do not have to wait till the end of the transient exponentially $e^{-\Omega_1 t}$ to evaluate both parameters. If there is a need to estimate the DS parameters as soon as possible, then the parameter of the "fast" exponent near the zero time mark shall be estimated. In all other cases, it is easier to estimate the decay parameter of the "slow" exponent.

Secular DS equation expressed via the frequency factors Ω_1, Ω_2

$$p^2 + \Omega_1 \cdot p + \Omega_1\Omega_2 = 0,$$

corresponds to a point in two-dimensional parameter space (Ω_1, Ω_2), more precisely in the first quadrant of the coordinate system (Ω_1, Ω_2). Given that these two frequencies are connected by Q-factor via expression $\Omega_2 = Q^2\Omega_1$, we obtain partitioning of the first quadrant into three characteristic sectors: the oscillatory, aperiodic and damped solutions of the secular equation (Fig. 2.16).

At critical damping ($\alpha = 1$), there is no oscillation of the solution while the natural frequency remains unchanged: $\omega_0 = \sqrt{K/M} =$ const. In this case, the rate of DS tending to the equilibrium is maximal and is determined by the solution $e^{-\Omega_1 t/2}$ or the decay constant as a half of the first characteristic frequency $\Omega_{1/2} = \omega_0$. The oscillatory factor $\cos(\omega_0 t)$ is not degenerated into an identical unit, and there is no oscillation of the solution. The coincidence of the argument ($\omega_0 t$) of the function $\cos(\omega_0 t)$ with exponent value of the envelope $A(t) = e^{-\omega_0 t}$ is another sign of achieving critical damping ($\alpha = 1$).

2.4 Evaluating Parameters of Motion Equation Based on Spectral Measurements

To calculate the transfer function with dimension (m/N), two physical values — force and displacement — are to be measured. Both values can be measured with accelerometers. In order to obtain the displacement signal $x(t)$ from the acceleration signal $a(t)$, it is necessary to perform the double integration: $a(t) = \int v(t)dt$, $v(t) = \int x(t)dt$, where $v(t)$ is the velocity.

The accelerometer signal is proportional to the force in accordance with Newton's second law ($f(t) = a(t) \cdot M$), through the proportionality factor equal to the mass — M. The same accelerometer cannot perform both measurement functions at the same time. The accelerometer measures acceleration of the mass it is installed on, and another accelerometer installed on different mass measures the force, obtained when exposing the DS to that different mass. The easiest way to demonstrate this is by a well-known pulse excitation of DS oscillation by mean of a test hammer. To measure displacements of the studied structure, one can install accelerometers on it; the test hammer has a built-in accelerometer to measure the impulsing force. The signal of this built-in accelerometer is recalculated into the shock force. Hence, each accelerometer is installed on its own mass, while one mass is a DS component, and the other is a component of device producing the force.

In many cases, primary transducers (sensors) are based on a DS. Their operating frequency band can be located both before the natural DS frequency ($F = 1$) and after it. It is required that the transfer function in the operating frequency band must be uniform. The required uniformity of the frequency response is achieved by varying the damping properties of the sensor. Therefore, the upper frequency of the operating range for the sensor with extremely low damping shall be much lesser than for the natural DS frequency. The uniformity of the frequency response for DS with a higher damping ratio from the sub-band $\frac{1}{2} < \alpha < \frac{\sqrt{2}}{2}$, when the the resonance at the natural frequency is very low-attenuated ("blurred" in frequency), will improve and it will be possible to obtain quite wide operating frequency bands. As for extremely high damping $\frac{\sqrt{2}}{2} < \alpha < 1$, with non-resonant transfer function, the monotonic drop of the response frequency as the frequency increases will have the same hyperbolic character with a permanent decay constant for any sub-band of the operating frequencies.

The experimental studies of DS vibrations are based on the vibration displacements. They are obtained either by directly measuring with the primary displacement transducers, or by measuring the vibration velocity $v(t) = \frac{dx}{dt}$, or by measuring the vibration accelerations $a(t) = \frac{d^2x}{dt^2}$. In the latter two cases, the integration shall be implemented on the electronic level directly during measurements or on the mathematical level during processing of the measurements. Due to the high performance and relatively low cost, accelerometers are the most common vibration sensors. The acceleration obtained by means of double integration is converted into vibration displacements.

There is a misconception about the ability to measure both force and displacement by one accelerometer at the same time and thus find the DS transfer function $H(\omega) = \frac{F(\omega)}{X(\omega)}$, where there is a Fourier transformation of the force and displacement in the numerator and denominator, respectively. The measured acceleration $a(t)$ at a constant mass — "M" is proportional to the force $f(t) = M \cdot a(t)$

or in the frequency sector

$$F(\omega) = M \cdot A(\omega) = M \cdot \omega^2 \cdot X(\omega)$$

and displacement is calculated according to acceleration

$$X(\omega) = \frac{A(\omega)}{\omega^2}$$

Hence, the transfer function

$$H(\omega) = \frac{F(\omega)}{X(\omega)} = \frac{1}{M} \cdot \frac{1}{\omega^2}$$

is a quadratic hyperbole independent from any measurements of acceleration. This paradox is possible if we consider the force, proportional to acceleration, to be only the inertial force of the DS, and not the force applied externally to DS. Both displacement and force shall be measured via the independent channels within the experiment aimed at obtaining the DS transfer function. For example, when exposing the DS to excitation with the shock hammer, the impact force is measured by accelerometer built into the shock mechanism of the hammer. The other accelerometer installed on the DS structure produces acceleration signal as a reaction to the shock impact, which is then converted into displacement. Then the transfer function, which depends on two functions, will depend on two independent measurements.

In any experiment, the frequencies are obtained in absolute units — Hz. To obtain $H(\omega_0)$ in absolute units $[\frac{M}{N}]$, the measurements of force and displacement shall also be calibrated in absolute units.

When both force and displacement are stationary random processes, the transfer function can be obtained from their spectral densities — $S_x(\omega)$, $S_f(\omega)$:[3]

$$|H(\omega)| = \sqrt{\frac{S_x(\omega)}{S_f(\omega)}}$$

Thus, the problem of measuring parameters (M, C, K) in absolute physical units requires metrologically secured measurements of the minor fluctuations of displacement and force. At the same time, the measurements of time (frequency) are considered to be *a priori*

[3] At least its absolute value (remark of editor).

metrologically secured by the control (diagnostic) system itself. Metrological requirements for measurements of force and vibration displacements are unnecessary for measurements of the normalized parameters, ω_0 in the units of time (frequency) or dimensionless α, Q.

The equation of motion converted into a normalized form contains only two parameters: natural frequency ω_0 and damping ratio σ, or dimensionless normalized damping factor α and Q-factor $Q = \frac{1}{2\alpha}$. In the assumption of smallness of the damping parameter $\alpha \ll 1$, the natural frequency $\omega_0 \approx \sqrt{\frac{K}{M}}$ is the abscissa of the only maximum in APSD of the displacement sensor signal. The related damping parameters (σ, α or Q) are found from its width $\Delta\omega$ and/or phase of the transfer function "force–displacement" — $\varphi(\omega)$. We will introduce the following time constants as inverse of the characteristic frequencies Ω_1, Ω_2:

$$\tau_K = \frac{1}{\Omega_2} = \frac{C}{K} = \frac{\Delta\omega}{\omega_0^2} = \frac{1}{\omega_0 \cdot Q} = \frac{2\alpha}{\omega_0}$$

$$\tau_M = \frac{1}{\Omega_1} = \frac{M}{C} = \frac{1}{2\sigma} = \frac{1}{\Delta\omega} = \frac{Q}{\omega_0} = \frac{1}{2\alpha\omega_0}$$

$$\tau_0 = \sqrt{\tau_K \cdot \tau_M} = \frac{1}{\omega_0} = \sqrt{\frac{M}{K}}$$

where τ_M is a time constant of damping by mass, τ_K is a time constant of damping by stiffness,

Note that the product of $\tau_K \cdot \tau_M$ does not depend on damping. Time τ_0 is their geometric mean and it is connected to the natural frequency which in the first approximation does not depend on damping (the same as at $\ll 1$, $\omega_C \approx \omega_0$). From this point on, we will need approximations of low frequency $\omega \sim 0$, resonant frequency $\omega = \omega_0$ and very high frequency $\omega \gg \omega_0$, as well as approximation of the small damping ($\alpha \ll 1$). As shown above, the transfer function modulus in these three approximations will be the following:

$$|H(\omega)| = \begin{cases} \dfrac{1}{K}, & \omega \sim 0 \\ \dfrac{\sqrt{M}}{C\sqrt{K}}, & \omega = \omega_0 \\ \dfrac{1}{M\omega^2}, & \omega \gg \omega_0 \end{cases}$$

The normalized (with Hz-dimension) or dimensionless parameters of the oscillatory system are evaluated without calibration and verification of measurements by auto-characteristics of vibration displacement signals as follows: the natural frequency ω_0 and Q-factor (or damping ratio) σ, α.

Moreover, it is enough to have *a priori* knowledge of one system parameter in absolute units to restore two remaining parameters in absolute units as well. Most frequently, the mass is that parameter. It can be known from the design parameters of the system (linear or volumetric dimensions) and the reference density of materials. In addition, by performing the static experiments stiffness can be obtained (e.g. as in the classical experiment using the deflection of the beam under its own weight, its natural frequency can be determined).

The phase of the transfer function in these three frequency ranges has the following approximations:

$$tg\varphi(F) = \frac{2\alpha F}{1-F^2} \approx \begin{cases} 2\alpha F = 2\pi f \tau_{\text{delay}}, & F, \omega \sim 0, \text{ where } \tau_{\text{delay}} = \frac{C}{K} \\ \mp\frac{\pi}{2}, & F = 1 \text{ or } \omega = \omega_0 \\ \frac{C}{M\omega} = \frac{1}{2\pi f \tau_M}, & \alpha \ll 1, \quad \omega \gg \omega_0 \end{cases}$$

It is still assumed that $\omega_C \approx \omega_0$ and $\alpha \ll 1$. The phase near zero Hertz is a linear function of frequency, the slope of which depends on the stiffness time constant $\tau_K = \frac{C}{K}$. At low (pre-resonant) frequencies, the "C" damping is assessed by the phase characteristic in addition to assessment of stiffness of the transfer function modulus. At high frequencies, the phase is determined via the mass time constant, which also extends capabilities of assessing the system parameters by the transfer function modulus. At the resonant frequency, the phase is a constant $\mp\frac{\pi}{2}$. In comparison with the transfer function modulus, the situation here is reversed: at the resonance frequency the phase characteristic is a constant which does not contain any information on the system parameters. All three DS parameters are involved in the transfer function amplitude. You must set some frequency detuning in order to extract system parameters from the phase in vicinity of the resonant frequency.

2.5 Practical Aspects of Measuring Parameters of DS Transfer Function

2.5.1 *Assessment of small damping factor*

In practice, in weakly damped DS ($\alpha \ll 1$), the approximation of the transfer function modulus as a function being symmetric with respect to the vertical axis $F = 1$ is true. In this case, the numeric value of the damping ratio $\alpha \ll 1$ is extracted from the normalized transfer function, and there are specific points of the transfer function that contain either an abscissa or an ordinate, depending on the parameter $\alpha \ll 1$ (Fig. 2.17).

The most common method for assessing the damping ratio is based on the transfer function modulus. It is normalized by the value of the global maximum $\frac{G(F)}{G(1)}$ and the abscesses of this function's intersections with the level $\frac{1}{\sqrt{2}}$ are found. It is easy to show that with the accuracy to the second-order variables of smallness, one abscissa will be to the left of the natural frequency ($F = 1$) and equal to $1 - \alpha$, and the second abscissa is to the right of the natural frequency and equal to $1 + \alpha$ (Fig. 2.17):

$$\frac{G(1-\alpha)}{G(1)} = \frac{2\alpha}{\sqrt{1-(1-\alpha)^2 + 4\alpha^2(1-\alpha)^2}} \approx \frac{2\alpha}{\sqrt{4\alpha^2 + 4\alpha^2}} = \frac{1}{\sqrt{2}}$$

The same result will be for $F = 1+\alpha$ as well. This is a consequence of the approximate symmetry of the curve $G(F)$ in the neighborhood of the point $F = 1$. For sufficiently small damping, level $\frac{1}{\sqrt{5}}$ also has a similar property. This level crosses the curve $\frac{G(F)}{G(1)}$ in points $1 - 2\alpha$ and $1 + 2\alpha$. It is possible to further reduce this level, reaching frequencies $F = (1 \mp n\alpha), n = 1, 2, 3, \ldots$, while the parameter α is sufficiently small.

This level decrease has a limit, as $G(0) = 2\alpha$ and $G(2F) \approx \frac{2\alpha}{3}$, which breaches symmetry of the function $G(F)$.

Real structures have several vibrating masses which result in several resonances in the transfer function. Different resonances of one transfer function affect the others: the resonances widen and their central frequencies shift from the true value. The above method of assessing the parameter α will have the error not only due to the assumption of enough smallness of the parameter α, but also due to

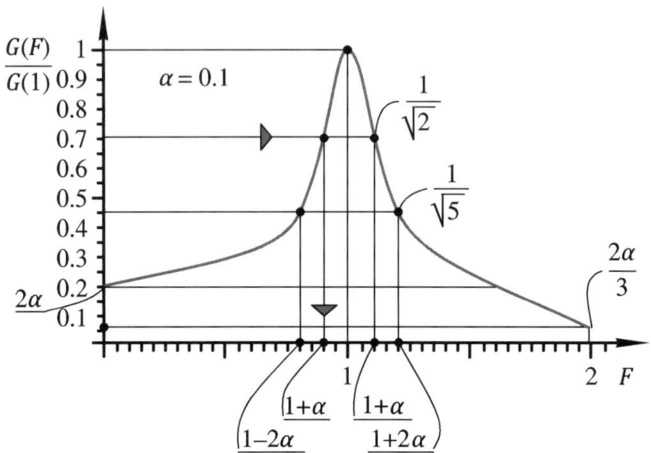

Figure 2.17: Close view of the transfer function modulus at small damping ratio.

the uncontrolled influence of the resonances on each other. As a necessary condition of the applicability of the presented method if the following equations are true,

$$\frac{G(1-\alpha)}{G(1)} \approx \frac{1}{\sqrt{2}}$$

$$\frac{G(1-2\alpha)}{G(1)} \approx \frac{1}{\sqrt{5}}$$

Note that there is no desired parameter α in the right parts of these equations. More rigid conditions of asymmetry show the insignificance of the impact of other resonances on the assessed one.

$$\frac{G(0)}{G(1)} = 2\alpha, \quad \frac{G(2)}{G(1)} \approx \frac{2\alpha}{3}$$

These equations contain a parameter α in the right parts, so they shall be used for testing, or for testing the validity of the method assumptions after assessing the parameter α.

According to $\alpha = \frac{\sigma}{\omega_0}$, the higher the resonance frequency, the more symmetrical it is (other factors being the same). The curve section to the left from the maximum is usually a gentle slope, to the right — steep slope. Approximation of the natural frequency to zero and the increase of damping makes the left and the right sections of the resonance curve increasingly distinct from each other.

Thus, by characteristic points of the transfer function modulus $|H(0)|$, $|H(\omega_0)|$ and ω_0, three DS parameters can be assessed in the first approximation: (M), (K), (C). If you know $|H(0)|$, ω_0, $|H(\omega_0)|$, then first the stiffness

$$K = \frac{1}{H(0)}$$

then mass

$$M = \frac{K}{\omega_0^2}$$

and then damping are found

$$C = \frac{1}{\omega_0 H(\omega_0)}$$

The curve $|H(\omega)|^2$ also has its own characteristic properties (Fig. 2.18). First, this is due to the damping factor σ (or "C" damping, and hence with logarithmic oscillation decrement, Q-factor and relative damping α). The half level of the function maximum $|H(\omega)|^2$

$$\frac{1}{2}|H(\omega_0)|^2 = \frac{1}{8 \propto^2 K^2}$$

is achieved (in the first approximation) on two frequencies: $\omega_0 \mp \sigma$. This implies the method of assessing the damping ratio $= \frac{C}{2M}$.

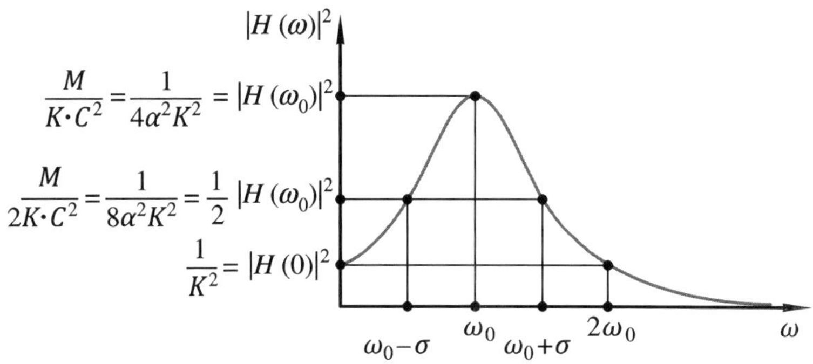

Figure 2.18: Characteristic points of the square of the DS transfer function modulus with one degree of freedom.

The given method is called the method of assessing according to 0.707 level (more precisely, according to level $1/\sqrt{2}$). At this level, not at the half one, the abscissa of function $|H(\omega)|$ coincides with the value $(\omega_0-\sigma)$. The function $|H(\omega)|^2$ is operated by the same simple method called the half-height method. When assessing the parameter α, sometimes confusion occurs with functions $|H(\omega)|$, $|H(\omega)|^2$ or with levels $1/\sqrt{2}$ and $1/2$.

Thus, the point of spectral space $\left[(\omega_0-\sigma), \frac{1}{2}|H(\omega_0)|\right]$ on the curve $|H(\omega)|$ also refers to a specific one. Based on this point and knowing two resonance parameters ω_0 and $|H(\omega_0)|$, the third resonance parameter — Q-factor $Q=\frac{\omega_0}{2\sigma}$ (or damping ratio $\alpha=\frac{\sigma}{\omega_0}$) is found.

It is obvious that to assess the relative damping "σ" according to the resonance form of the function $|H(\omega)|$ or $|H(\omega)|^2$, these function must be strictly monotonic in the frequency range $(\omega_0-\sigma,\omega_0)$ and have no other vibration effects. This requirement is not always met; there are often two different vibration effects with close natural frequencies and both transfer functions affect each other.

While narrowing the range $(\omega_0-\sigma,\omega_0)$ to the range (ω_1,ω_0), where $\omega_0>\omega_1>(\omega_0-\sigma)$, it is possible to achieve the function monotony $|H(\omega)|^2$ in this range and find the value proportional to the damping ratio by assessing the value ω_1 and the ratio $\frac{|H(\omega_0)|^2}{|H(\omega_1)|^2}$. In this case, the found abscissa ω_1 will depend on all DS parameters $([M],[K],[C])$, and not only on σ.

The curves $G(F)$ with the resonance amplitude reduced to identity are shown in Fig. 2.19. Also, level $1/\sqrt{2}$ and intersection points are indicated.

The difficulty of extracting small values of the parameter α, since the resonance itself has a very high Q-level (of small width at level $1/\sqrt{2}$), point $F=1$ and $F=(1-\alpha)$ have very close frequencies. On the one hand, the method error is reduced, while α is also decreased. On the other hand, it can be increased due to the small number of discrete spectral lines in the segment $(1-\alpha,1)$. The same curves are shown with normalization $\frac{G(F)}{G(0)}$ in Fig. 2.19. Their amplitude at the resonance frequency dramatically increases when α decreases according to the law $G(1)=\frac{1}{2\alpha}$. The level $1/\sqrt{2}$ is growing accordingly. At $\alpha=0.05$, $G(1)$ is 10 times more than $G(0)$, at $\alpha=0.01$–50 times, and the abscissa $1-\alpha$ and 1 can be indistinguishable. In practice,

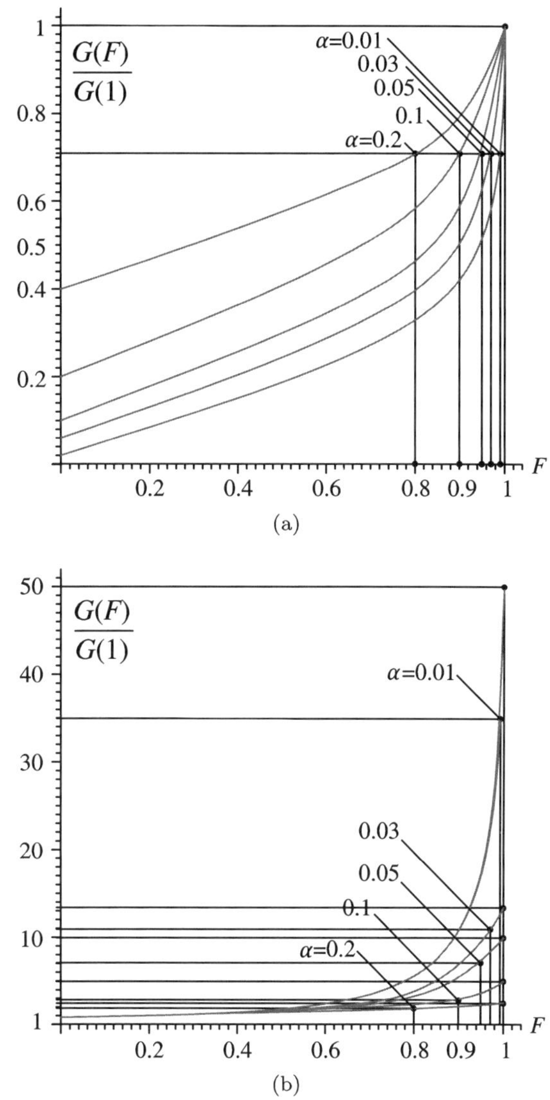

Figure 2.19: DS transfer functions, normalized by their maximum value at low damping (a) and the same transfer functions, normalized by the values in zero Hertz (growth ratio, b). The graphs show level 0.707 based on which the damping ratio is calculated.

when assessing the damping ratio, it is useful to present the transfer function in both normalizations.

2.5.2 *Assessment of arbitrary damping ratios*

For sufficiently high damping, the assumption on the approximate equality of the natural frequency and the maximum transfer function frequency is unfair: $\omega_0 \neq \omega_{\max}$. Figure 2.20 shows comparison of pre-resonance DS frequency area with different damping.

Table 2.4 shows a sequence of actions on assessment of ω_0, $\omega_{\max}$, α regardless of the value α and then all three DS parameters (M, C, K) according to the phase and modulus of the transfer function using a proprietary technology.

At the first step, the frequency ω_0 can be found where the phase takes the 90^0 value in the phase function of the transfer function (phase delay of displacement from the force measured in degrees). The frequency is determined in absolute units (Hz). The maximum global frequency $\omega_{\max}$, is also determined in absolute units (Hz) from

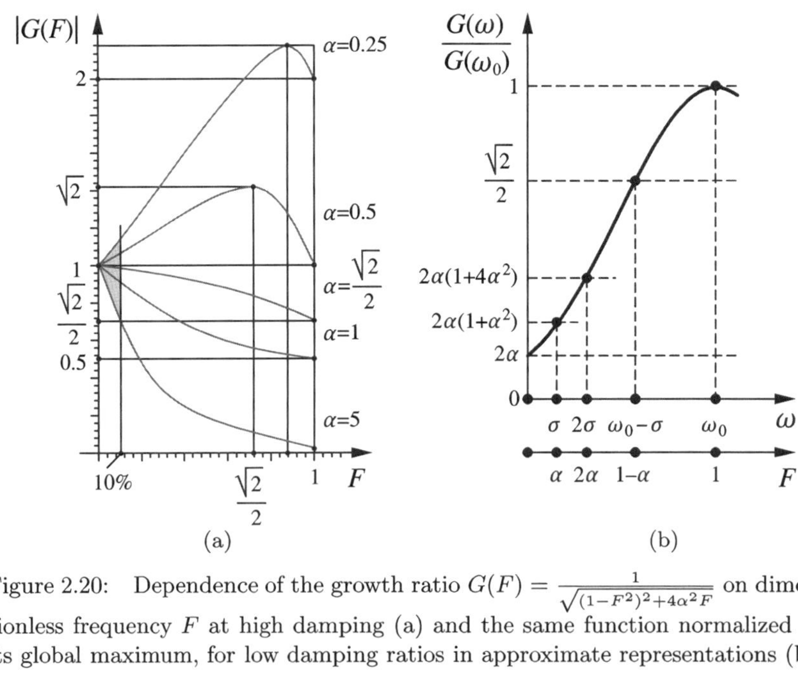

Figure 2.20: Dependence of the growth ratio $G(F) = \frac{1}{\sqrt{(1-F^2)^2+4\alpha^2 F}}$ on dimensionless frequency F at high damping (a) and the same function normalized by its global maximum, for low damping ratios in approximate representations (b).

Table 2.4: DS parameters assessment.

No. of step	Assessed parameter	Formula (assessment condition)
1	Natural frequency, ω_0	$\varphi(\omega_0) = \pi/2$
2	Maximum frequency, $\omega_{\max}$	$\lvert G(\omega)\rvert \leftarrow \max$
3	Dimensionless maximum frequency, $F_{\max}$	$F_{\max} = \omega_{\max}/\omega_0$
4	Damping factor, α	$\alpha = \sqrt{\dfrac{1 - F_{\max}^2}{2}}$
5	Stiffness, K or mass, M	$K = \dfrac{1}{H(0)}$ or $M \approx \dfrac{1}{H(\widetilde{f}) \cdot \widetilde{f}^2}$
6	Stiffness, K or mass, M	$\omega_0 = \sqrt{K/M}$
7	Value of the transfer function at natural frequency	$H(\omega_0) = \dfrac{1}{2K\alpha}$
8	Damping, C	$C = \omega_0 H(\omega_0) = \omega_{\max} H(\omega_{\max})$
9	Critical damping, C_{cr}	$C_{\text{cr}} = \dfrac{C}{\alpha}$
10	Stiffness, K	$K = \dfrac{C_{\text{cr}}}{2}\omega_0$
11	Mass, M	$M = \dfrac{C_{\text{cr}}}{2}\omega_0$

the displacement amplitude–frequency function (amplitude dependence of displacements on the frequency); then one has to check condition if $\omega_0 > \omega_{\max}$. The resulting accuracy of assessing the DS parameters will depend on the number of discrete frequency values in the interval $[\omega_0, \omega]$. It is obvious that the lesser the assessed damping, the more the frequency resolution that is required.

The next two rows of Table 2.4 (Nos. 3 and 4) transition to dimensionless parameters which are insufficient to assess the DS parameters (M, C, K) in absolute physical dimensions. The ultimate objective of the experimental assessments can be the damping ratio, i.e. α and the natural frequency, ω_0. Based on these assessments the relative damping $\sigma = \alpha\omega_0$ can be obtained and then based on definition $\sigma = C/2M$ — the absolute damping — "C" is obtained if only *a priori* the "M" mass is known. In practice, the DS mass is usually known based on independent weighing or elementary analysis of DS structural materials. Then it is easy to calculate stiffness: $K = \omega_0^2 M$. This provides calculation of all three DS parameters (M, C, K).

If DS mass is *a priori* unknown, it can be measured within DS by the formula $M \approx \frac{1}{H(\tilde{f})\cdot\tilde{f}^2}$ by exciting DS at a very high frequency — $\tilde{f}$. On the one hand, the higher the excitation frequency compared to the natural frequency, the more accurate the presented approximation. On the other hand, while the frequency increases, the value of the measured transfer function decreases and this may worsen the accuracy of assessing the mass. A static experiment on determination of stiffness $K = \frac{1}{H(0)}$ is preferable to vibration measurement of the mass, if possible. Therefore, the fifth row of Table 2.4 is given as an alternative: either mass, or stiffness. One of these parameters is enough to finally calculate all three DS parameters. The sixth row of Table 2.4 shows the natural frequency used to obtain either DS mass or DS stiffness.

Until now, the extreme properties of the transfer function were used: its values in 0 Hz and at a very high frequency. The transfer function has other specific points. In particular, these are frequencies ω_0 and $\omega_{\max}$ where the conditions specified in row eight of Table 2.4 are met and after which the damping (C) can be assessed.

$$C = \omega_0 H(\omega_0) = \omega_{\max} H(\omega_{\max})$$

Knowledge of the transfer function $H(\omega)$ in only one point ω_0 allows to calculate the DS damping, "C". Formulas for obtaining *mass* (M) *or stiffness* (K) from damping are given in the last two rows of Table 2.4.

For the four characteristic values of the damping ratio $\left(\alpha = 0, \frac{1}{2}, \frac{\sqrt{2}}{2}, 1\right)$, the biquadrate polynomial $[(1 - F^2)^2 + 4\alpha^2 F^2]$, defining the poles of the transfer function, has the frequency parameters as listed in Table 2.5.

According to Table 2.5, with an increase of the damping factor the transfer function maximum frequency decreases and reaches zero already at $\alpha = \frac{\sqrt{2}}{2}$, and not at $\alpha = 1$, when the oscillating factor in solution of the DS equation is extinct. At the maximum frequency $F_{\max} = \sqrt{1 - 2\alpha^2}$, the phase progression $\psi(F_{\max})$ is

Table 2.5: The polynomial roots $[(1-F^2)^2+4\alpha^2F^2]$ or frequency of the transfer function modulus maximum at characteristic damping.

α	0	$\frac{1}{2}$	$\frac{\sqrt{2}}{2}$	1
$(1-F^2)^2+4\alpha^2F^2$	$(1-F^2)^2$	$1-F^2+F^4$	$1+F^4$	$(1+F^2)^2$
$F_{1,2}$	∓ 1	$\frac{1}{2}\mp j\frac{\sqrt{3}}{2}$	$\mp j$	$\mp j$
$F_{\max}=\sqrt{1-2\alpha^2}$	1	$\frac{\sqrt{2}}{2}$	0	$\mp j$

calculated according to the definition $tg\psi = \frac{2\alpha F}{(1-F^2)}$:

$$tg[\psi(F_{\max})] = \frac{\sqrt{1-2\alpha^2}}{\alpha} \tag{2.55}$$

so, we have dependence of the transcendent phase from the maximum frequency (Fig. 2.20).

$$tg[\psi(F_{\max})] = \frac{F_{\max}}{\alpha} \tag{2.56}$$

It turns out that at the maximum frequency of the transfer function modulus we can obtain the damping ratio value according to the phase-frequency response:

$$\alpha = \frac{1}{\sqrt{\{tg[\psi(F_{\max})]\}^2+2}} \tag{2.57}$$

After excluding damping from expressions (2.55) and (2.56), we will obtain

$$tg^2[\psi(F_{\max})] = \frac{2F_{\max}^2}{1-F_{\max}^2}$$

Thus, the knowledge of the dimensionless frequency $F_{\max}$ allows to calculate the damping factor and phase at this frequency and, on the contrary, calculate the damping (Fig. 2.21) from the phase at the maximum frequency:

$$F_{\max} \leftrightarrow \alpha \leftrightarrow \psi(F_{\max})$$

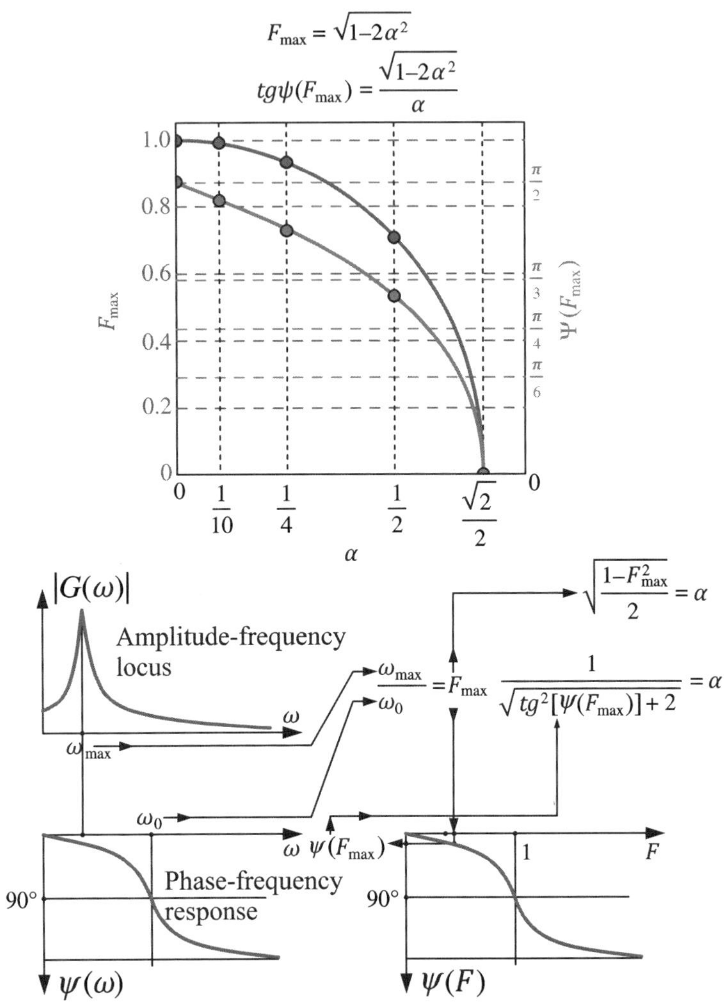

Figure 2.21: Amplitude-frequency function and phase-frequency response at the natural frequency and at the maximum frequency, formula for calculating the damping ratio.

Having obtained the value ω_0, from the phase-frequency response and value ω_{max} from the amplitude-frequency function, and by making transition to dimensionless frequency F_{max}, we can calculate the damping ratio α. We can return to phase-frequency response from the value F_{max}, and obtain the damping ratio α from the value $\psi(F_{max})$. In both cases, the phase-frequency response, amplitude-frequency function and $F_{\max}$ are used, but calculation results α will obviously differ. Assessment errors α will be different as well. All depends on the specific type of phase-frequency response, amplitude-frequency function and absolute value of parameter α.

Dimensionless parameters $F_{\max}$ and $\psi(F_{\max})$ are interrelated with all known dimensionless parameters of the damping factor (Q-factor, logarithmic oscillation decrement, damping factor) and can be equal to them, characterizing the properties of DS oscillation and damping.

In practice, a method of assessing parameters α only by amplitude-frequency function, when it is assumed that $F_{\max} \approx 1$, is common; the parameter α is assessed by the resonance width. This method is internally inconsistent: The assumption $F_{\max} \approx 1$ is valid for very small values α, but the resonance is too small for them. The "wide" resonances are common for high damping, but assumption $F_{\max} \approx 1$ is invalid for them and this can also result in significant error.

The value $\alpha = \frac{\sqrt{2}}{2}$ is a threshold for methods based on the resonance parameters. At $\alpha > \frac{\sqrt{2}}{2}$, all frequency functions have a global maximum at a zero frequency. This explains the allocation to a separate class of the so-called aperiodic processes with $\frac{\sqrt{2}}{2} < \alpha < 1$.

In practice, the weakly damped DS ($\alpha \ll 1$) the approximate representation of the transfer function modulus as a function being symmetric with respect to the vertical axis $F = 1$ is true. In this case, the numeric value of the damping ratio $\alpha \ll 1$ is extracted from the relative transfer function, and there are specific points of the transfer function that contain either an abscissa or an ordinate, depending on the parameter $\alpha \ll 1$.

In case of erroneous identification of $F_{\max}$ with the natural frequency $F = 1$, a reverse conclusion on absence of information on DS damping at frequency $F_{\max}$ is made. The formula (2.57) is particularly effective for highly damped DS with $\alpha \in \left[\frac{1}{2}, 1\right]$, when $F_{\max} \ll 1$. In other cases, a high-resolution spectral analysis shall be performed to distinguish values $F_{\max}$ and $F = 1$.

If DS is a filter with a transparency band $[0, F_{\max}]$, which is not an optimal strategy for building it, and requirements for a minimum phase incursion in this frequency sub-band are imposed, then the closer α is to $\frac{\sqrt{2}}{2}$, the smaller will be the phase incursion. At $\alpha = \frac{\sqrt{2}}{2}$, the sub-band $(0, F_{\max})$ is reduced to single point.

When external disturbing force has an infinite spectrum of frequencies as in the case of a step-wise force, the DS response in a form of displacement will also have the same infinite frequency spectrum. DS refers to a linear system which does not produce new frequency components but only changes the amplitude and phase of perturbation applied to the input. In the frequency language: DS is a linear passive filter which imposes its weight on each spectral component of the disturbing force.

2.6 Coolant Pressure Fluctuation as an External Vibration Force Perturbation

Direct measurement of the force causing vibrations is an extremely rare situation in practice. The distributed source of the force in MCC are pulsations of coolant pressure. They are measured by local point sensors for pressure pulsations. Thus, the pair "force–displacement" is replaced here by the measuring pair "PPS–accelerometer".

Equipment vibrations of MCC are being imaged in the form of multi-resonance spectral characteristics of vibration sensor signals (e.g. accelerometers). Let us assume that some harmonic vibration source has a frequency of forced oscillations ω_1. Now we will replace the multi-resonance spectrum with a single-resonance spectrum, which corresponds to a single equivalent spring pendulum with the natural frequency ω_0. The frequency ω_1 not necessarily coincides with the natural frequency ω_0. The coincidence of these frequencies, i.e. a resonance excitation, will be obtained automatically as a result of the effect of the given method.

Let us also suppose that all other resonances do not have significant effect on the frequency sub-band $(\omega_0 - \Delta\omega, \omega_0 + \Delta\omega)$, i.e. it is assumed that the single vibration effect dominates in this sub-band (Fig. 2.22).

Three parameters of the pendulum are interconnected by the known equation of the forced oscillations:

$$M \cdot \ddot{x}(t) + C \cdot \dot{x}(t) + K \cdot x(t) = f_0 \cdot e^{j\omega_1 t} \tag{2.58}$$

Pressure pulsations of coolant at a single frequency ω_1 can serve as a harmonic force perturbation $f(t) = f_0 \cdot e^{j\omega_1 t}$.

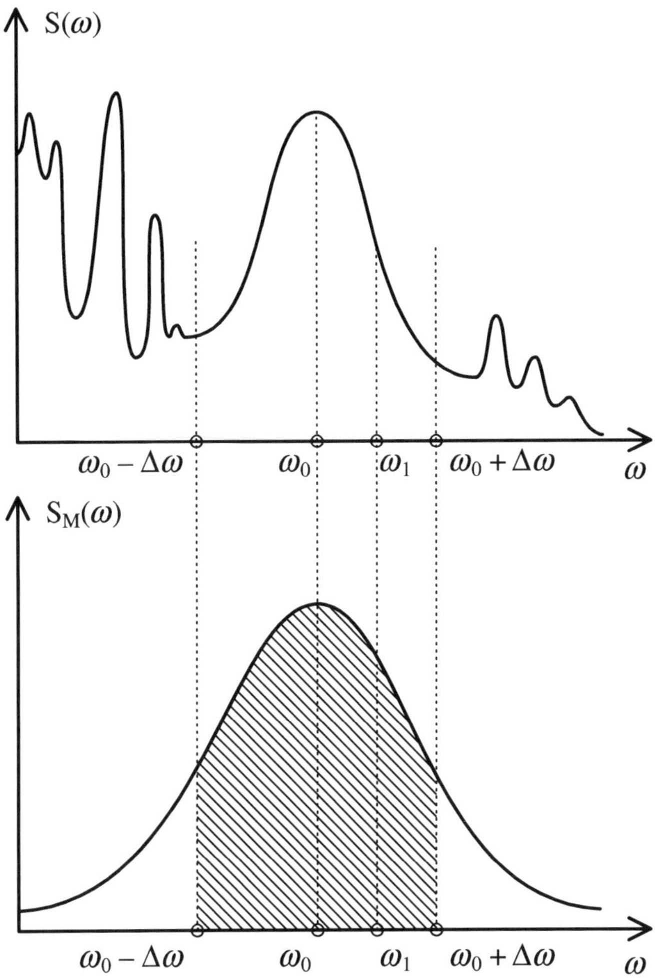

Figure 2.22: Multi-resonance spectral density of signal power of the vibration displacement sensor and single-resonance spectral density of equivalent spring pendulum power in frequency sub-band $[\omega_0 - \Delta\omega, \omega_0 + \Delta\omega]$.

Such frequency can be assigned both to the rotational frequency of MCP — ω_{MCP} — and to the central frequency of any ASW — ω_{ASW}. The response displacement will also be at a frequency of the disturbing force ω_1 (but not at the natural frequency ω_0), so the solution of the equation will be as follows:

$$x(f) = x_0 \cdot e^{j\omega_1 t}$$

After its substitution in Eq. (2.58), we will have the following:

$$(-\omega_1^2 \cdot M + j\omega_1 C + K) \cdot x_0 = f_0$$

with the known transfer function

$$H(\omega_1) = \frac{1/M}{-\omega_1^2 + j2\sigma \cdot \omega_1 + \omega_1^2} \quad (2.59)$$

The previous section shows that the natural frequency ω_0 and damping factor α (or σ) can be obtained from the autospectral power density of displacement sensor signal $S_x(\omega)$ at the same frequency under the assumption that the disturbing force is a white noise in the vicinity of the natural frequency: $[\omega_0 - \Delta\omega, \omega_0 + \Delta\omega]$. Then the square of the transfer function modulus will be proportional to APSD — $S_x(\omega)$:

$$|H(\omega)|^2 = \frac{S_x(\omega)}{S_p(\omega)} = \text{const} \cdot S_x(\omega)$$

The transfer characteristics of the pressure pulsation signal (PPS) and accelerometer signal will provide additional information on damping, but already at perturbation force frequencies and not only at natural vibration displacement frequencies.

The presence of the linear phase dependence on frequency with respect to the perturbance force implies a "pure" delay effect of vibration displacements with the delay time τ_{delay}, proportional to the slope of the linear phase dependency of the "PPS–accelerometer" pair of signals. Let us show this in an analytical way. The phase at the forced oscillation frequency ω_1

$$\text{arc}tg[\varphi(\omega_1)] = \frac{2\alpha\frac{\omega_1}{\omega_0}}{1 - \left(\frac{\omega_1}{\omega_0}\right)^2}$$

can be presented by the Taylor series

$$\text{arc}tg[x] = x - \frac{x^3}{3} + \frac{x^5}{5} - \cdots \quad (2.60)$$

where $x = \frac{2\alpha\frac{\omega_1}{\omega_0}}{1-\left(\frac{\omega_1}{\omega_0}\right)^2}$, and for $|x| > 1$, it is a rapidly convergent series with inverse power functions:

$$\text{arc}tg[x] = \frac{\pi}{2} - \frac{1}{x} + \frac{1}{3x^3} - \frac{1}{5x^5} \cdots \quad (2.61)$$

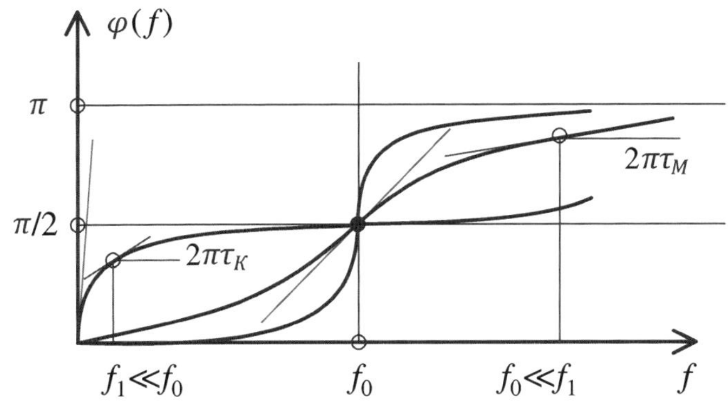

Figure 2.23: Phase-frequency response "force–displacement" with frequencies of perturbation forces — f_1 and natural oscillation — f_0.

If frequency ω_1 is near zero Hertz and $\omega_1 \ll \omega_0$, σ, then limited to the first term of decomposition (2.60), we have the following:

$$\varphi(\omega_1) \approx 2\alpha\frac{\omega_1}{\omega_0} = \tau_K\omega_1 = 2\pi\tau_3 f_1 \quad \text{where } \omega_1 = 2\pi f_1$$

This linear dependence of the phase means that there is a delay of value $\tau_3 = \frac{2\alpha}{\omega_0} = \tau_K$ of the vibration displacement from the force perturbation (Fig. 2.23).

That is, the time constant of the oscillatory system by stiffness τ_K has a physical meaning of the time delay of displacements from the effect of force at low frequencies.

If $\frac{\omega_1}{\omega_0} \gg 1$, then limited to two terms of decomposition (2.61), we have the following:

$$\varphi(\omega_1) \approx \frac{\pi}{2} + \frac{1}{2\alpha} \cdot \frac{\omega_1}{\omega_0}$$

or

$$\varphi(\omega_1) - \frac{\pi}{2} \approx \omega_1 \cdot \tau_M = 2\pi\tau_3 f_1$$

This linear dependence of the phase to the right from the value $\frac{\pi}{2}$ means that there is a delay of $\tau_3 = 2\alpha\omega_0 = \tau_M$ of the vibration displacements from the disturbing force perturbation value, i.e. the time constant of the oscillatory system by mass τ_M has a physical meaning of the time delay of displacements at high frequencies (Fig. 2.23).

Evaluation of the phase characteristic slope $\varphi(f_1) = \text{const} \cdot f_1 + \cdots$

$$\text{const} \cdot f_1 \approx \begin{cases} 2\pi\tau_K, & f_1 \ll f_0, \quad f_1 \approx 0 \\ 2\pi\tau_M, & \qquad\qquad f_1 \gg f_0 \end{cases}$$

or delay time

$$\tau_3 \approx \begin{cases} \tau_K, & f_1 \ll f_0, \quad f_1 \approx 0 \\ \tau_M, & \qquad\qquad f_1 \gg f_0 \end{cases}$$

by the phase characteristic allows to evaluate either one time constant – τ_K, or another – τ_M – depending on the ratio of the forced oscillation frequency ω_1 and the natural frequency of the system ω_0.

Both parameters (delay time and slope of the phase characteristic) are functions of two system parameters (M, S or K, C), i.e. they are diagnostic features defining in particular the power dissipated by the system. The frequencies of the standing waves ω_{ASW_i} and harmonic rotational frequencies of RCP $n \cdot \omega_{\text{MCP}}$ are the natural discrete frames of reference, setting the discrete forces with regard to frequency, which are always present in the circulation loop.

Thus, at the natural frequency of structure oscillation, the natural frequency and relative damping are evaluated from APSD of a single vibration displacement sensor. And from the mutual characteristics of the "PPS–accelerometer" signal pair, one can get diagnostic information on system parameters including relative damping at frequencies of perturbation forces. Thus, the valuable diagnostic vibration information for the entire frequency range is obtained in passive mode (without perturbance of natural noise fields of the reactor) from each resonance feature of auto- and cross-spectral characteristics. Interpretation of the phase function on frequency between accelerometer signals and PPS signals as a delay is a proprietary methodology.

Example. During the heating-up of VVER-1200 unit in multichannel mode, the signals of PPS and accelerometers were synchronously recorded. Three pressure pulsation sensors PPS were installed in each loop of VVER-1200 reactor plant:

- on "hot" pipeline — PPSh;
- on U-pipeline — PPSU;
- on "cold" pipeline — PPSc;

and two accelerometers:

- on U-pipeline — AccelU;
- on "cold" pipeline — Accelc.

Any change of coolant temperature in one or another direction results in shift of ASW resonances along the frequency axis, while its increase causes a shift of the resonances towards the beginning of the axis (zero Hertz) and accordingly the decrease of the coolant temperature results in movement in the opposite direction. Such resonances in Fig. 2.24 are marked by inclined straight lines (top-bottom temperature increase process). Harmonics of the rotational MCP frequency are marked by vertical lines. They are obviously permanent when coolant temperature changes and this is a sign of "mechanical" nature of these maxima. The monotonous decrease of ASW resonance frequency with increasing coolant temperature is a test for their identification. From the experimental data (Fig. 2.24) based on the group of coherence function, the following ASW are obtained for this measurement range of coolant temperature (nature and mechanism of generation of specific ASW data are described in the last chapters of the book):

- first loop — ASW1loop — (7.5–8.4) Hz;
- first vessel — ASW1v — (10.9–12.1) Hz;
- second loop — ASW2loop — (14.4–15.6) Hz;
- fourth loop — ASW4loop — (27.6–33.2) Hz.

To clarify this, we will consider only the first resonance, generated by ASW1loop. The right column of the graphs in Fig. 2.24 consists of fragments of the relevant coherence and phase functions in the narrow sub-band of ASW1loop frequencies. In the top line of the graphs, coherence between PPSh1 and PPSc1 at ASW1loop frequencies are close to one for the signals of pressure pulsation sensors at significant distance from each other. The phases are the zero constants in this frequency band for all coolant temperature values. This means that ASW1loop exhibits a powerful global effect which affects the entire loop. The half-wave length of this ASW almost coincides with the length of the loop. In a sense, its effect can be considered as a point effect: forces generated at ASW1loop frequency unidirectionally affect all points of the loop.

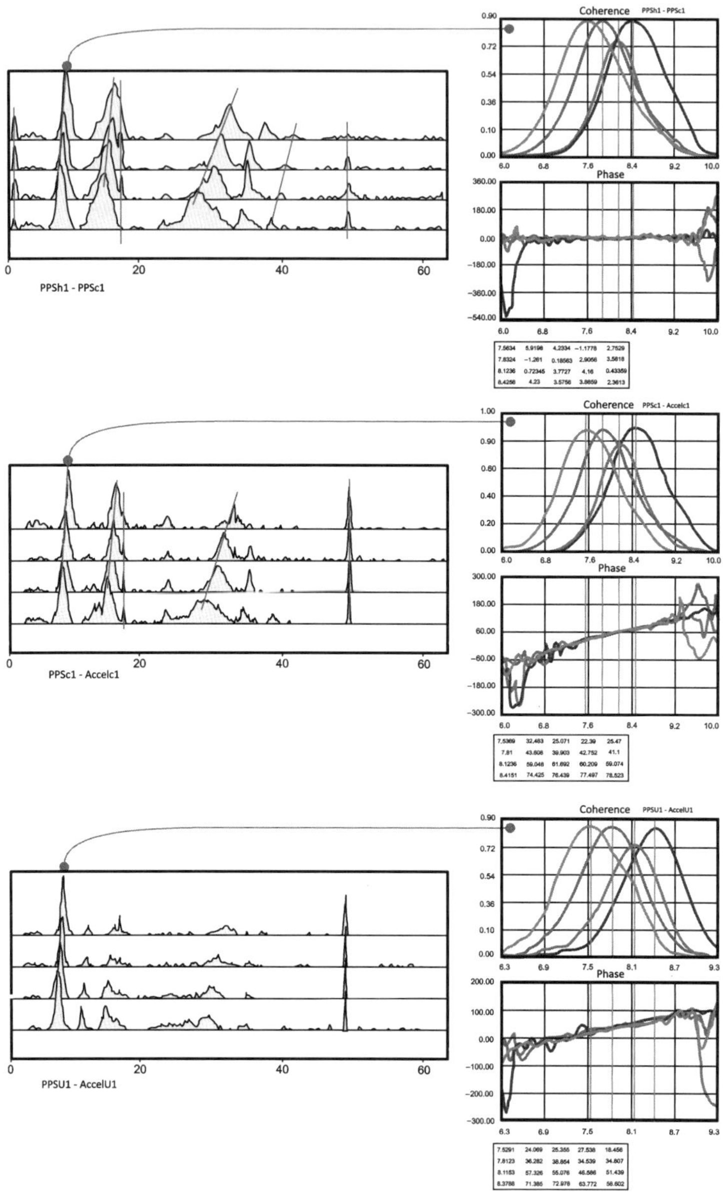

Figure 2.24: Groups of coherence and phase functions during unit heating-up: PPSh1 — PPSc1 — upper third, PPSh1 — Accelc1 in the middle, PPSU–AccelU — lower third.

In the next line of the graphs in Fig. 2.24, the "PPSc" signal is replaced by the "Accelc" signal in "PPSh–PPSc" signal pair. The last two sensors are near each other, but the sensitive component of PPS is in contact with the coolant, and the sensitive component of accelerometer is in contact with the outer side of the pipeline wall. However, the coherence function "PPSh–Accelc" does not actually differ from the previous coherence function "PPSh–PPSc". That is, the impact force ASW1loop, which is proportional to pressure pulsations in PPSh installation point, is also proportional and in-phase to the other force, generated due to impact of the same ASW1loop in PPSh (or Accelc) installation point. This is the point impact of ASW1loop on the pipeline walls.

A significant difference from the previous case is the linear slope in the phase function. Moreover, this slope does not change in conditions of changing coolant temperature and it is approximately 55°/Hz, which corresponds to a delay of 0.15 s (55°/Hz: 360°). If the entire loop as an oscillatory system is substituted as a rough approximation by a single spring pendulum with parameters (M,S,K), then the obtained delay time is an evaluation of the time constant $\tau_K = \frac{C}{K}$, but not the time constant $\tau_M = \frac{M}{C}$. Indeed, with the increase of coolant temperature the coolant mass within the loop decreases; therefore, the parameter τ_M must change, and hence, the phase slope should change as well. The temperature dependences of stiffness and damping are much weaker than the temperature dependence of mass.

The signal PPSh in "PPSh–Accelc" signal pair can be replaced by the PPSc signal; in this case, the sensors are near each other and the location of estimated functions is in the lower third of the graphs in Fig. 2.24. Coherences and phases with the same slope of about 55°/Hz remained almost unchanged. Replacement of the remote PPSh by the nearby PPSc has changed nothing. Generally, the delay time of first signal relative to another signal is typically associated with transfer of perturbance from one spatial point where the first sensor is installed to the other spatial point where the other sensor is installed. However, there is no transfer here. The obtained linear phase dependence on frequency in "PPS–Accel" signal pair is the phase of the "force–displacement" transfer function at ASW1loop frequency as external dependence of perturbance force on frequency. This phase defines delay of vibration displacements of the pipeline wall in relation to the global coolant pressure pulsations in this pipeline.

Linear phase dependencies are observed in "PPS–Accel" signal pairs and at all other ASW frequencies. For example, the ASWvessel changes the sign of its amplitude in the loop. As a result, an antiphase is observed (Fig. 2.25) and the phase characteristic of "PPSc–Accelc" has been significantly changed. However, the error of delay evaluation increases due to decrease of coherence. This demonstrates the fact that other physical uncorrelated vibration sources not associated with ASW emerge. The restrictive requirements of the developed method are not fulfilled. With an increasing ASW frequency, not only does coherence (Fig. 2.25) decrease, but also the linear phase becomes dependent on the coolant temperature.

Now we can conclude that the natural frequency of oscillations at high frequencies becomes lower than the perturbation force frequency and the evaluated delay time is $\tau_M = \frac{M}{C}$.

It is necessary to warn about possible errors of evaluation for τ_K, τ_M. The method itself is based on several assumptions, including the main one – replacement of multi-mass oscillatory system with a single spring pendulum, which is already quite a rough approximation. It must correspond to a stand-alone resonance within a range at natural oscillation frequency, so that the next adjacent resonance is the forced oscillation resonance, and all other resonances do not actually affect it. Otherwise, a phase offset error occurs. Therefore, the developed method is better to use for diagnostic purposes by assessing the phase slope as a diagnostic feature, which reveals the oscillatory system parameters.

When two signals of different physical natures (PPS and accelerometer) are used to assess the mutual characteristics, it is also possible to make a systematic hardware error in the phase evaluation. The error occurs because the method of obtaining a useful signal in each instrument channel distorts the phase in its own way. At the same time, the requirements for uniformity of the amplitude in the operating frequency band within each instrument channel are strictly fulfilled. For example, the inductive type PPS receive a useful

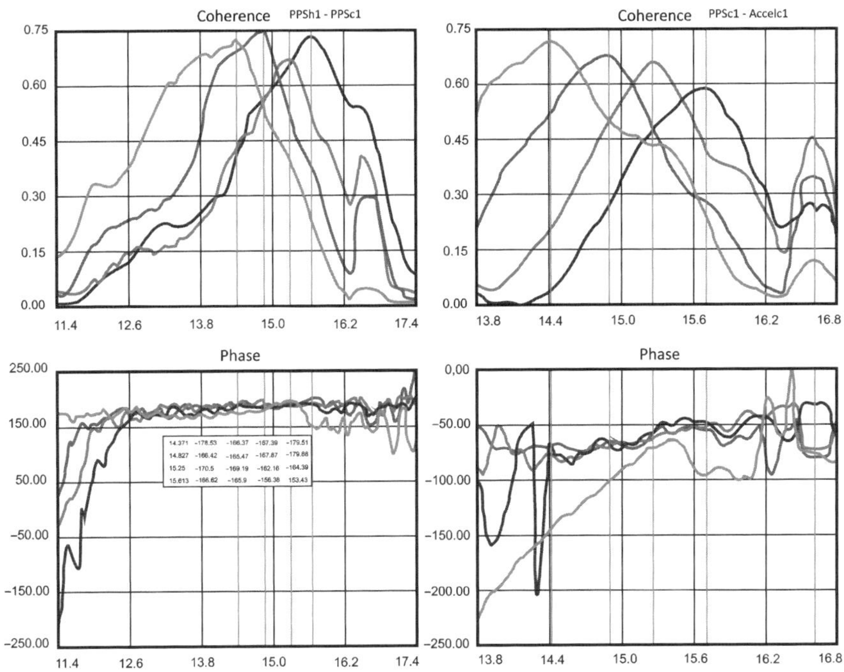

(a) (b)

Figure 2.25: Groups of coherence functions (upper line of the graphs) and relevant phases (lower line) for "PPSh–PPSc" (a) and "PPSc–Accelc" (b) signals pair.

signal of pressure pulsations in the frequency band of (0–200) Hz after demodulation of the carrying 5 kHz frequency signal. The electronic PPS route contains many active and non-linear components, which introduce an abnormal delay at each frequency of the operating range.

When the phase is evaluated by two signals of the same type (e.g. "PPS–PPS") with identical measuring equipment, then the individual "wrap" of the phase in each channel coincides with and is subtracted from others when calculating the phase.

Vibration accelerations in passive accelerometer are "immediately" converted into a charge on the sensitive component without any delay. Then it is subject to amplification, conversion (e.g. double integration to obtain displacement from acceleration), normalization, transmission through a long line, etc. As a result, phase distortion also occurs. This does not mean that in practice it is impossible to use the phase evaluation in "PPS–Accel" signals pair. To achieve this, it is necessary to obtain a hardware phase-frequency characteristic for this mutual analysis and then make the appropriate correction at each frequency.

Chapter 3

Vibration Diagnostic Features

3.1 Dynamic System with One Degree of Freedom as a Diagnostic Model

Conversions of vibration measurements into monitoring diagnostic features are covered in several references [25, 29, 37, 46, 55, 59, 97, 99, 100, 101, 103, 106] and regulatory documents [388, 392, 394, 396, 402–406]. In this chapter, we consider the simplest DS model again.

Each of the three resonance parameters (ω_0, $H(\omega_0)$, Q) may be seen as a diagnostic feature (diagnostic parameter), for which diagnostic thresholds are defined. When a threshold is exceeded, this is a diagnostic event indicating a malfunction with a certain degree of reliability. Diagnostic thresholds for each of the three parameters may be defined according to the empirical rule: natural fluctuations of the diagnostic feature in the "normal" state shall stay within the diagnostic threshold. If it is known *a priori* that the object is in the operable condition, selection of the diagnostic threshold becomes a purely strategic task when there is no physical model of the object and, moreover, there is no model demonstrating how the malfunction appears and develops.

Thus, one of the four main strategic tasks of diagnostics — malfunction detection — may be solved formally. Obviously, this kind of approach is not optimal. The second diagnostic task — malfunction localization — may also be solved with a formal approach, or the so-called localization to the nearest sensor location. Such an approach also has its drawbacks and is sometimes unacceptable even for rough estimations. Anyway, if we want to solve all four diagnostic tasks at once, which may be conventionally identified by the key words "detection", "localization", "cause" and "forecast", we will need a non-formal physical diagnostic model.

The analytical conversion of the three experimentally defined parameters of one spectral characteristic resonance, (ω_0, $H(\omega_0)$, Q), into physically interpreted diagnostic features is the way to build a physical diagnostic model. An example of such conversion is the transformation of a measurement space into a diagnostic feature space, with three dynamic system parameters ("M", "C", "K") (Fig. 3.1) used as diagnostic features. It does not mean that a measurement cannot be assigned a diagnostic feature, e.g. the measurement of vibration displacement as the function of time, without any processing, may have the upper and the lower diagnostic thresholds, and the maximum admissible vibration displacement may be justified by the design-basis structural constraints.

In this case, the conversion of a measurement space into a diagnostic feature space is the identical transformation of the space "in itself".

Figure 3.1 shows conventionally a three-dimensional space with axes (ω_0, $H(\omega_0)$, Q) as a measurement space, and a three-dimensional space with axes ("M", "C", "K") as a diagnostic space.

Spectral approach is the number one in vibration diagnostics, with spectral functions (most commonly, spectra) measured directly by spectrum analyzers, avoiding the stages of measurement and time series interpretation. That is why frequency functions are related to the measurement space, which is actually incorrect. First, vibration signals are measured (vibration displacement, velocity and acceleration) as the function of time, and then the simplest vibration diagnostic tasks are solved in terms of time, by estimating the root-mean-square (RMS) value and comparing it with the threshold value. Such an approach is widely used for vibration diagnostics of rotary mechanisms. Moreover, for rotary mechanisms, GOST defines the diagnostic thresholds for RMS values of vibration velocity in absolute units, with no statistical estimation for a specific diagnostics object. These thresholds also have a statistical origin, but they were obtained not as mean values in time, but as mean values for identical rotary equipment. It has been proven in experiments that different rotary equipment grouped according to their electric motor power have approximately the same RMS values of vibration velocity in

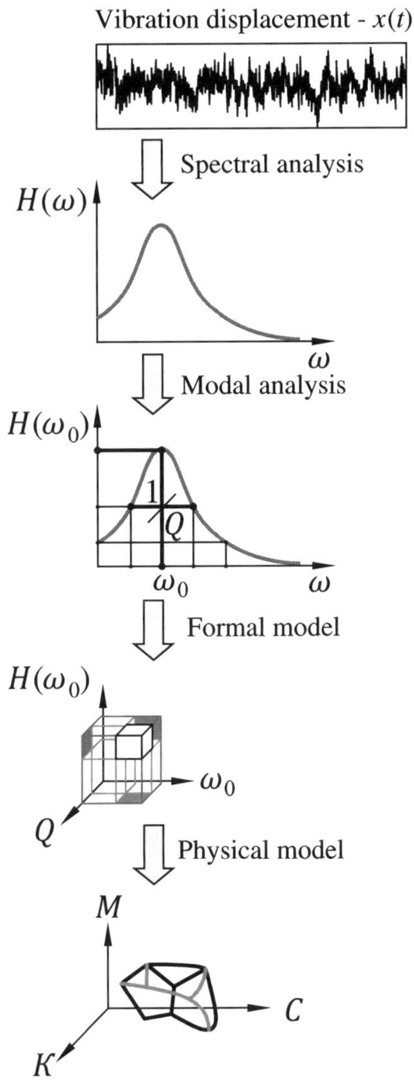

Figure 3.1: Serial conversion of vibration measurements into a physical diagnostic model based on dynamic system modeling.

the operable regime, i.e. vibration velocity thresholds in absolute physical values, nevertheless, have more formal character not based on a physical model.

A unique thing about rotary equipment diagnostics based on the absolute thresholds of vibration velocity RMS values is that the diagnostic information is focused on the rotational frequency and its harmonics, and its change is most often related to the only physical reason: imbalance of the rotating mass. Vibration diagnostics of reactor equipment, including MCP vibration diagnostics, is much more complicated, as it is based on the vibration dynamics of inherent characteristics of various RP components (SG, MCP, pipelines, reactor vessel, RI, FA) under harmonic forces — of mechanical origin from MCP, of acoustic origin from ASW, and of hydrodynamic origin from the coolant flow.

RMS value can also be expressed in terms of frequency: RMS value $= \sqrt{\int_0^\infty d\omega\, |H(\omega)|^2}$, which enables us to consider the spectral space as a measurement space right away. Obviously, this space is obtained as a result of the spectral analysis of multi-channel vibration signals, which are actually multidimensional.

Another space is obtained following the so-called modal analysis. This is a representation of an object's dynamics in terms of its natural frequencies and natural modes of oscillation. One of the features of a natural mode of oscillation is its amplitude at a natural frequency $H(\omega_0)$. The oscillation mode of a certain type of oscillation is characterized by a space coordinate, which means that its reconstruction is related to the issue of multi-channel measurements. Here, when we consider the single channel of vibration measurement, $H(\omega_0)$ is understood as the amplitude of a spectral function in the coordinate of the sensor location and, particularly, the amplitude of a transfer function at a natural frequency.

Apart from the widely known modeling issues, such as discretization of an object (splitting it into constituent parts), or the validity of assumptions on linearity or stationarity, etc., modal analysis involves some issues which may be difficult to solve. Actually, all types of natural oscillations contribute to each and every resonance feature of any spectral characteristic of displacement. On the one hand, for a complete modal analysis, an object has to be excited with all types of oscillations. On the other hand, we will need a sufficient number of sensors in order to detect the contribution of each type of oscillation and reconstruct the motion of any point of a dynamic object. In particular, a sensor located in the nodal point of a certain type of oscillation will fail to detect any displacement caused by this type of oscillation. This is why modal analysis cannot be complete *a priori*; its quality largely depends on the qualification of people and is achieved in iterations, through trial and error. So, it may be needless to require a complete modal analysis.

3.2 Vibration Diagnostic Spaces

The upper and the lower thresholds for each parameter (ω_0, $H(\omega_0)$, Q) define a regular rectangular prism in this three-dimensional space which represents a normal state of the diagnostics object, or the so-called "norm" (Fig. 3.2). Any other point in space outside this prism, i.e. a point in a complimentary space, is "abnormal" with regard to that prism. Such space, in its turn, may include subspaces, each defining a certain malfunction.

As seen in Chapter 2, for the transfer function modulus, an approximate relation between DS parameters and resonance parameters is expressed in the following way:

$$\begin{cases} \omega_0 = \sqrt{\dfrac{K}{M}} \\ H(\omega_0) = \dfrac{\sqrt{M}}{C\sqrt{K}} \\ Q = \dfrac{\sqrt{KM}}{C} \end{cases} \tag{3.1}$$

Considering that the resonance parameters (ω_0, $H(\omega_0)$, Q) are known, the three DS parameters ("M", "C", "K") can be easily derived from this system:

$$\begin{cases} M \approx \dfrac{Q}{H(\omega_0)\omega_0^2} \\ C \approx \dfrac{1}{H(\omega_0)\omega_0} \\ K \approx \dfrac{Q}{H(\omega_0)} \end{cases} \tag{3.2}$$

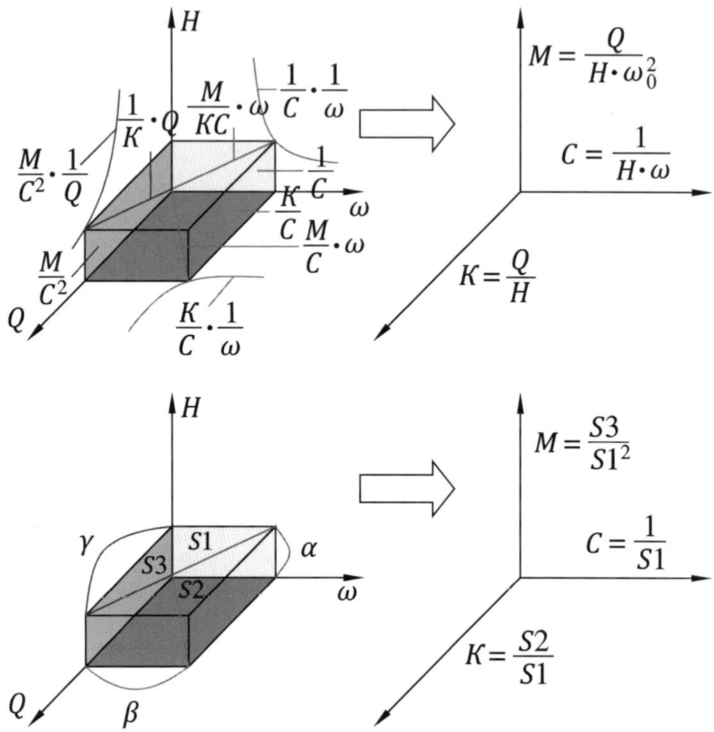

Figure 3.2: Conversion of a three-dimensional formal diagnostic space with measurements (ω_0, $H(\omega_0)$, Q), obtained from the resonance parameters, into a physical diagnostic space of dynamic system parameters ("M", "C", "K"). Point-to-point transformations (the upper lines of the graphs). Transformations of areas (the lower lines of the graph).

Thus, the expressions (3.1) and (3.2) define the nonlinear conversions of a three-dimensional measurement space into a three-dimensional space of diagnostic features and back:

$$(\omega_0, H(\omega_0), Q) \leftrightarrow (\text{“}M\text{”}, \text{“}C\text{”}, \text{“}K\text{”}).$$

The upper and the lower diagnostic thresholds for each of the three diagnostic parameters ("M", "C", "K") will also define a rectangular prism, but in their own space. If a "normal" set is defined by a rectangular prism in the measurement space (ω_0, $H(\omega_0)$, Q) after it is converted into the space of diagnostic features ("M", "C", "K"), it will not become part of any other rectangular prism. The same is true for inverse conversion.

That is, among the points of the measurement space which characterized the diagnostics object as an operable one, there will now be points corresponding to an inoperable object. This demonstrates that the notion of a "norm" is conditional and non-absolute — it depends on what is considered a "diagnostic threshold".

Diagnostic thresholds for each coordinate of the space ("M", "C", "K") may be justified physically, e.g. the mass threshold ("M") may be related to the loss of coolant or coolant leakage from the MCC. As the coolant is heated up and its density decreases, the mass contained in the constant volume also decreases. Correspondingly, the natural frequency of equipment oscillation and ASW frequency will be the diagnostic features related to the change of mass.

The damping threshold may be related to rolling friction of SG and MCP movable bearings, which means, to the state of their surfaces. The stiffness threshold may be related to MCC pipelines losing their elastic properties as a result of metal aging under radiation, thermal and mechanical impact, i.e. with the lifetime changes.

Thus, each of the three DS parameters is given a physically interpretable meaning of a diagnostic feature. The study of its trends in time with further interpolation to the intersection with the diagnostic threshold solves the issue of prediction.

Two other diagnostic tasks named "cause" and "localization" may be solved after we find out which of the three diagnostic features ("M", "C", "K") has exceeded its threshold, e.g. natural frequency may change due to two physical reasons: due to the change of mass and/or stiffness (in the first approximation, as natural frequency actually depends on damping as well). If the natural frequency of oscillations has changed — $\omega_0 = \sqrt{K/M}$ — with stiffness K remaining the same, possibly it may be caused by coolant leakage from MCC. If the natural frequency remains the same and the oscillation amplitude — $H(\omega_0) \approx \frac{\sqrt{M}}{C\sqrt{K}}$ — is decreases, it may be due to the wear of equipment movable bearings, represented by the parameter "C", etc.

Diagnostic thresholds in space (ω_0, $H(\omega_0)$, Q) may also have certain physical meaning, e.g. the diagnostic threshold for vibration displacement change may be defined by DS structural features as a threshold, meaning that a certain pre-defined gap between the parts shall not be exceeded.

The amplitude of oscillations may change due to the following three physical reasons: due to the change of mass and/or stiffness and/or damping — $H(\omega_0) \approx \frac{\sqrt{M}}{C\sqrt{K}}$.

When a diagnostic feature exceeds the amplitude threshold (e.g. when the vibration velocity RMS value threshold defined in GOST is exceeded), it gives no answers to the following question: which diagnostic parameter ("M", "C", "K") caused the diagnostic event?

The point with the coordinates (ω, H, Q) (hereinafter, for brevity, instead of "ω_0" we use symbol "ω", instead of $H(\omega_0)$ symbol "H") by its projections defines the characteristic interrelations between the DS parameters. From the expressions (3.1, 3.2) it is easy to show that in each coordinate plane (H, ω), (Q, ω), (H, Q) there are the following linear and quadratic dependencies with the DS parameters as coefficients (Fig. 3.3 and Table 3.1).

It turns out that the measurements of $I = (H, \omega, Q)$ as the coordinate of $\ll I \gg$ space are dependent on each other, i.e. the $\ll I \gg$ measurement space itself cannot be considered a principal component space, the main property of which is the independence of coordinates from each other. The same is true for the diagnostic features space $P = (M, C, K)$. Now, the measurements (H, ω, Q) of I space play the role of the coefficients of P space.

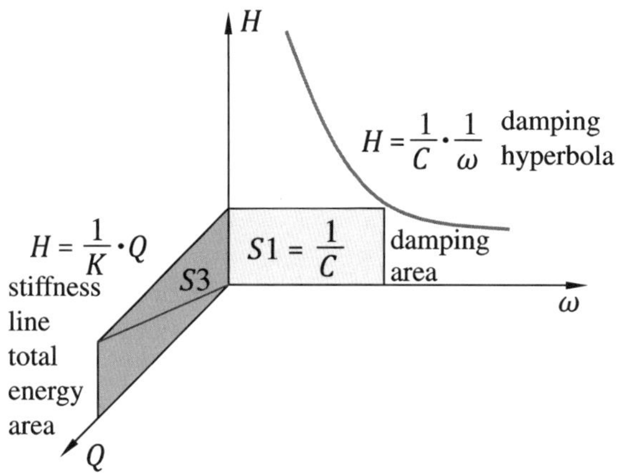

Figure 3.3: Areas space "S": *with* $S1$ — damping coordinate axis, [$S1$, $S2$] — stiffness plane with the stiffness line, [$S1$, $S3$] — mass plane with the mass parabola.

Table 3.1: Properties of coordinate planes (H,ω), (Q,ω), (H, Q) in space (ω, H, Q).

Coordinate plane	Linear dependency	Hyperbolic dependency	Area of a rectangle in the prism base	Direction tangent
(H,ω)	$H=\frac{M}{KC}\cdot\omega$	$H=\frac{1}{C}\cdot\frac{1}{\omega}$	$S1=\frac{1}{C}$	$tg\alpha=\frac{H}{\omega}=\frac{M}{KC}$
(ω,Q)	$\omega=\frac{C}{M}\cdot Q$	$\omega=\frac{K}{C}\cdot\frac{1}{Q}$	$S2=\frac{K}{C}=K\cdot S1$	$tg\beta=\frac{\omega}{Q}=\frac{C}{M}$
(Q, H)	$Q=H\cdot K$	$Q=\frac{M}{C^2}\cdot\frac{1}{H}$	$S3=\frac{M}{C^2}=M\cdot S1^2$	$tg\gamma=\frac{Q}{H}=K$

Independence of coordinates is important for the interpretation of modal analysis results. If the spectral characteristic under consideration has N (more than one) resonances, it is expanded into a direct sum of N single-resonance spectral characteristics, each modeled by its own DS (its own spring pendulum), only if the so-called generalized coordinates are independent. In particular, for such expansion it is necessary and sufficient to represent damping as a linear combination of mass and stiffness [99]:

$$C=\alpha\cdot M+\beta\cdot K \tag{3.3}$$

This ratio reduces the dimensionality of diagnostic features space from three to two dimensions.

As seen from Table 3.1, the easiest way to get the information on damping "C", is to take it from the coordinate plane (H,ω) due to a simple hyperbolic dependency, $H=\frac{1}{C}\cdot\frac{1}{\omega}$, and a simple expression for the prism base area, $S1=\frac{1}{C}$. If we build a diagnostic feature as a product of the two measured values $H(\omega_0)\cdot\omega_0$ and see how it changes in time, we will find that the only cause of such change is the change in damping, "C".

The easiest way to get the information on stiffness is to take it from the coordinate plane (H, Q) due to a simple linear dependency $K\approx\frac{Q}{H}$. The formulas for mass, "M", do not have such simple representation in the coordinate planes. The mass information is not concentrated in any single coordinate plane; it is distributed across the whole volume of the prism in quite a complicated way. For instance, mass "M" is included in the expressions for all three coordinate axes:

$$\begin{cases} \omega_0 = \sqrt{\dfrac{K}{M}} \\ H(\omega_0) = \dfrac{\sqrt{M}}{C\sqrt{K}} \\ Q = \dfrac{\sqrt{KM}}{C} \end{cases}$$

while in the diagnostic features space, all three coordinate axes are dependent on the resonance amplitude, "H":

$$\begin{cases} C \approx \dfrac{1}{H\omega} \\ M \approx \dfrac{Q}{H\omega^2} \\ K \approx \dfrac{Q}{H} \end{cases}$$

The given formulas for interrelation of the two noted spaces define the so-called point-to-point transformations (the point in one space is converted into a point in another space, although this is not a one-to-one transformation). Let us define one more diagnostic features space related both to the initial measurement space (ω_0, $H(\omega_0)$, Q) and the previously introduced diagnostic features space ("M", "C", "K"). Quadratic transformations from the coordinate planes (H, ω), (Q, ω), (H, Q) form the areas space — "S" $= (S1, S2, S3) = (H \cdot \omega$, $Q \cdot \omega$, $H \cdot Q)$, which is also a diagnostic features space, just like the DS parameters space, due to the following interrelations (Fig. 3.4):

$$\begin{cases} S1 = H \cdot \omega = \dfrac{1}{C} \\ S2 = \omega \cdot Q = \dfrac{K}{C} = K \cdot S1 \\ S3 = Q \cdot H = \dfrac{M}{C^2} = M \cdot S1^2 \end{cases}, \quad \begin{cases} M = \dfrac{S3}{S1^2} \\ K = \dfrac{S2}{S1} \\ C = \dfrac{1}{S1} \end{cases}$$

Now, all three coordinates ("M", "C", "K") are dependent on area $S1$. Coordinate "C" depends only on coordinate $S1$, coordinate

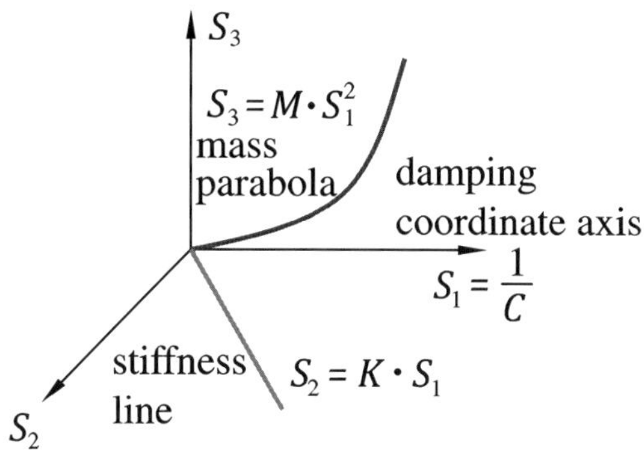

Figure 3.4: DS parameter planes: $[\omega_0,\ H(\omega_0)]$ — damping plane with the damping hyperbola, $[Q,\ H(\omega_0)]$ — stiffness plane with the stiffness line, $[S1,\ S2]$ — stiffness plane with the stiffness line, $[S1,\ S3]$ — mass plane with the mass parabola.

"K" depends on two coordinates, "$S1$", "$S2$", coordinate "M" depends on three coordinates, "$S1$", "$S2$", "$S3$". The information on

- damping, "C", is taken from axis "$S1$" and/or from plane (H, ω),
- stiffness, "K", is taken from plane $(S1, S2)$ and/or from plane (H, Q),
- mass, "M", is taken from plane $(S1, S3)$ and/or from all three axes (H, ω, Q).

The three area parameters $(S1, S2, S3)$ may also be considered as coordinate axes (Fig. 3.4) physically interpreted just the same way as the three DS parameters ("M", "C", "K"). Damping area $S1 = \frac{1}{C}$ is transformed into a damping axis "$S1$". In the coordinate plane $(S1, S2)$, there is the stiffness line $S2 = K \cdot S1$. In the coordinate plane $(S1, S3)$, there is the mass parabola $S3 = M \cdot S1^2$.

Apart from that, the coordinate plane (H, Q), or the coordinate axis $S3$, has a meaning in terms of energy. Indeed, the total energy of oscillations of a dynamic system comprise the potential energy accumulated in the damped spring and the kinetic energy of mass motion (Fig. 3.3):

$$\begin{aligned} E &= \frac{M}{2}\left(\frac{dx}{dt}\right)^2 + \frac{K}{2}x^2 = \frac{1}{2}M\left[\left(\frac{dx}{dt}\right)^2 + \omega_0^2 x^2\right] \\ &= \frac{M}{2}\omega_0^2|H(\omega_0)|^2 = \frac{M}{2C^2} = \frac{1}{2}HQ \end{aligned}$$

Total energy can be a diagnostic feature with the quadratic metric. The decrease of the DS total energy in time is the process of energy imbalance, meaning that there is a new consumer of oscillation energy, e.g. the jamming of equipment movable support. It can be interpreted as a "disengagement" of resonance: its amplitude decreases while the resonance broadens at the same time.

3.3 Integral Parameters of Resonance in a DS with One Degree of Freedom

The following parameters may serve as spectral vibration characteristics:

- spectrum modulus $|X(\omega)|$ for deterministic vibration displacements;
- APSD — $S_x(\omega) = M[|X(\omega)|^2]$ for random effects on the DS and random vibration displacements;
- ratio of APSD to CPSD modulus $= |H(\omega)|$, as an estimator of the transfer function modulus;
- square of transfer function modulus for random vibration displacements $|H(\omega)|^2$.

Actually, each point of function $|H(\omega)|$ or function $|H(\omega)|^2$ depends on only three DS parameters: (M, C, K). Certain transformations of a transfer function lead to simple dependencies applicable in vibration diagnostics. In particular, the area of a rectangle with the sides $|H(\omega_{\max})|$ and $\omega_{\max}$ is equal to the inverse damping (Table 3.1):

$$|H(\omega_{\max})|\omega_{\max} = 1/C$$

i.e. is the invariant of DS mass and stiffness. The same is true for a rectangle with the sides $|H(\omega_0)|$ and ω_0:

$$|H(\omega_0)|\omega_0 = 1/C$$

while the natural frequency of a non-dissipative system $\omega_0 = \sqrt{K/M}$, on the contrary, is the invariant of damping.

In a point of "frequency–amplitude" spectral space – $(\omega_0, |H(\omega_0)|)$ – the dependency of coordinates is such that their product depends on only one DS parameter. $|H(\omega_{\max})|$ and $|H(\omega_0)|$ values, just like $\omega_{\max}$ and ω_0 values, differ little from each other at low damping (when α values are close to zero). However, without simplifications of function $|H(\omega)|$, the following equalities are satisfied at any damping:

$$|H(\omega_{\max})|\omega_{\max} = |H(\omega_0)|\omega_0 = 1/C$$

Table 3.2 includes the integral transformations of functions $|H(\omega)|$ and $|H(\omega)|^2$, which are equal in their meaning but play different roles in the assessment of DS parameters. "Minus" sign in the table is for DS parameters which are invariant, i.e. which have no impact on this integral transformation.

Invariants have a selection capability, e.g. if some diagnostic feature is invariant to both mass and stiffness, but is changing in time, the root cause for this is the DS damping changing in time.

The integral characteristics of APSD resonances as certain areas under the resonances may highlight specific time features of a DS.

3.4 Relative Vibration Diagnostic Thresholds

A directly measured parameter or a function performed over a measurement may be put in a diagnostic range as an allowance. This means that the lower and the upper diagnostic thresholds are defined by taking a certain step up and down from an average value of the parameter or the function of the operable object.

A diagnostic threshold with a random physical dimension $\triangle P$, set by the diagnostic feature, P_0, may be compared with another diagnostic threshold for another diagnostic feature if they are normalized and made dimensionless in the following simple way:

$$P_0 \mp \Delta P = P_0\left(1 \mp \frac{\triangle P}{P_0}\right) = P_0(1 \mp \triangle_P)$$

where $\triangle_P = \frac{\triangle P}{P_0}$ is the normalized relative dimensionless diagnostic threshold, as a fraction of the stationary normal value, $P_0 \cdot \triangle_P$ is usually set in per cent: $\triangle_P \ll 100\%$. Under *a priori* uncertainty, relative diagnostic thresholds for different diagnostic features are usually set equal to each other. The $\triangle_P$ value is so small that any nonlinear

Table 3.2: Invariants of modulus and squared modulus of transfer function.

Transfer function modulus $\|H(\omega)\|$	**Transfer function squared modulus** $\|H(\omega)\|^2$
Damping rectangle. Invariant of mass and stiffness.	Not invariant to any DS parameter.
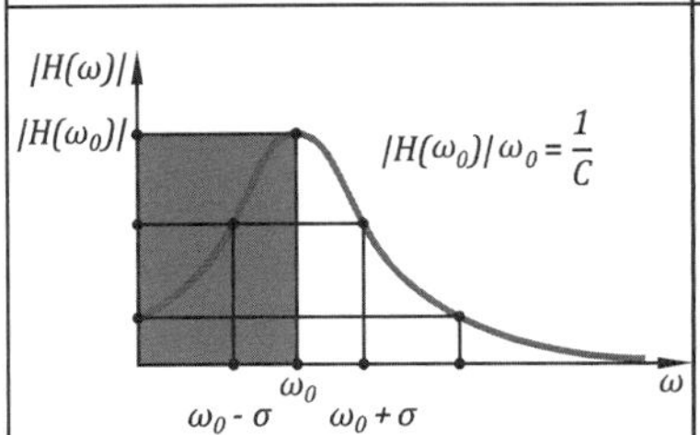	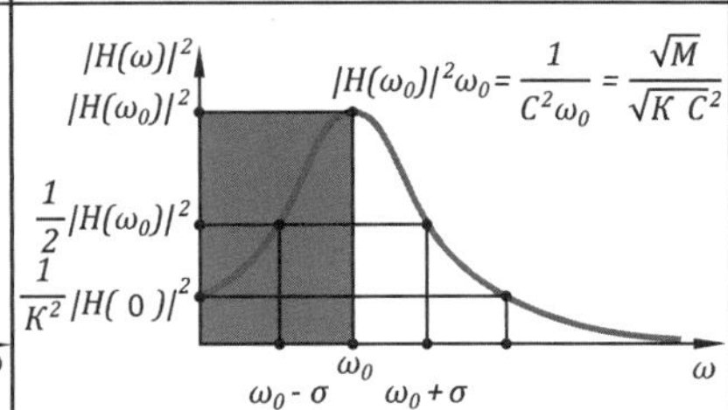
M: -, C: +, K: -	M: +, C: +, K: +
Vertical rectangle of critical damping. Damping invariant.	Mass invariant.
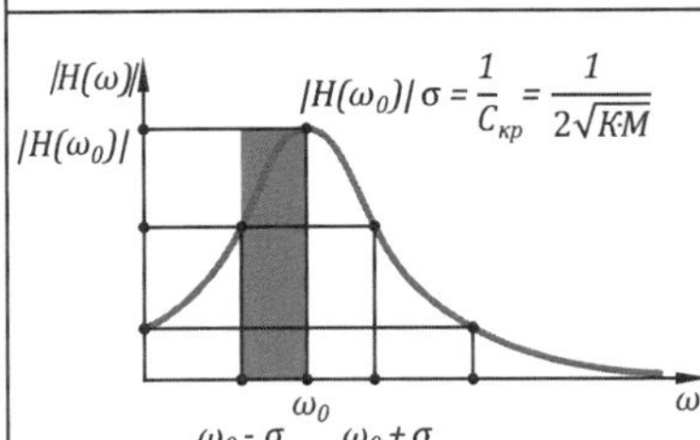	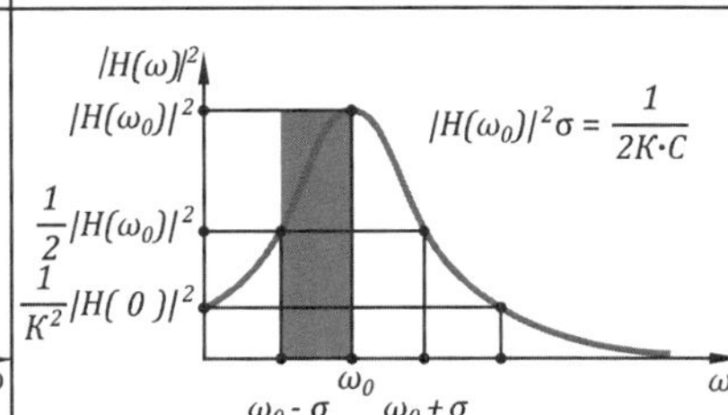
M: +, C: -, K: +	M: -, C: +, K: +
Horizontal rectangle of critical damping. Damping invariant.	Damping invariant.
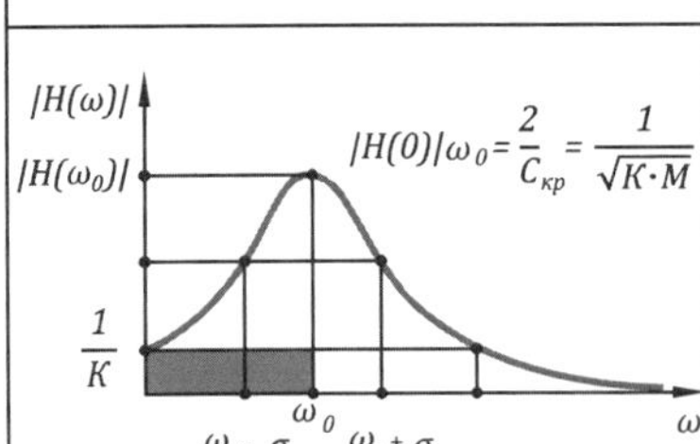	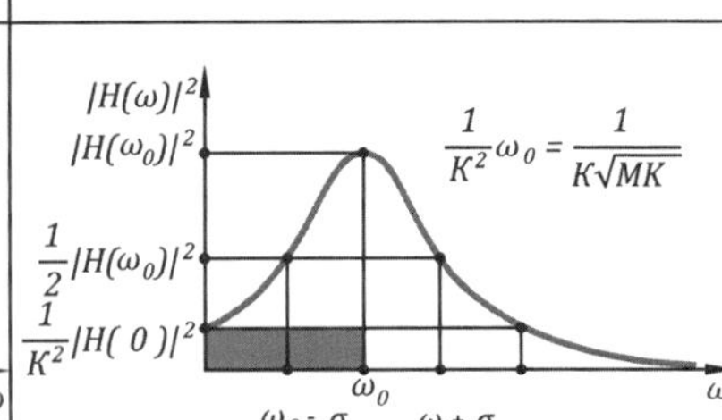
M: +, C: -, K: +	M: +, C: -, K: +

(*Continued*)

Table 3.2: (*Continued*)

<table>
<tr><td>Not invariant to any DS parameter·</td><td>Mass invariant·</td></tr>
<tr><td>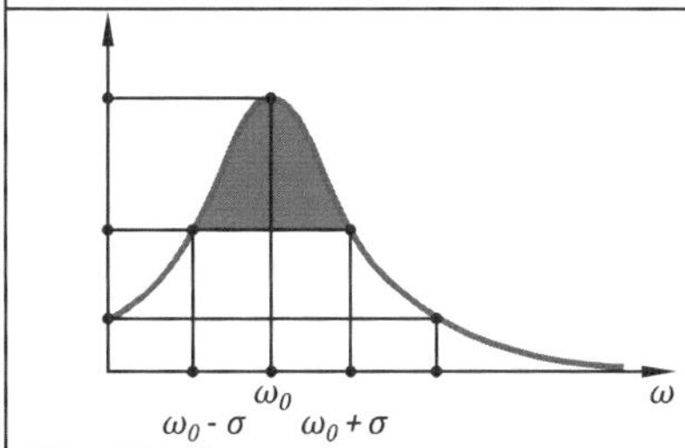

<table><tr><td>M</td><td>C</td><td>K</td></tr><tr><td>+</td><td>+</td><td>+</td></tr></table></td>
<td>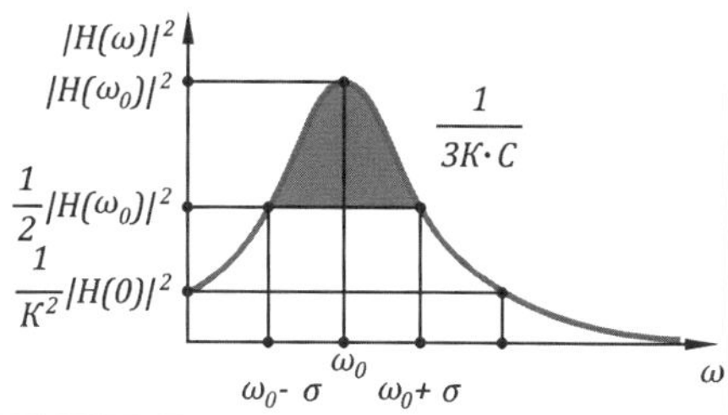

<table><tr><td>M</td><td>C</td><td>K</td></tr><tr><td>-</td><td>+</td><td>+</td></tr></table></td></tr>
</table>

transformations of diagnostic features may be linearized, e.g. in case of a small change of resonance frequency,

$$(\omega + \triangle\omega)^2 = \frac{K + \triangle K}{M + \triangle M}$$

the following approximation is true:

$$\omega^2(1 + \triangle_\omega)^2 \approx \omega^2(1 + 2\triangle_\omega) \approx \frac{K}{M}(1 \ + \triangle_K)(1 - \triangle_M)$$
$$\approx \frac{K}{M}(1 + \triangle_K - \triangle_M)$$

Thus, for relative thresholds we have

$$\triangle_\omega \frac{1}{2}(\triangle_K - \triangle_M) \tag{3.4}$$

Let us consider the case when the value of a random relative threshold equals to some constant, (say, one $\triangle_P = 1$, or any other, it does not matter in this case), meaning that the diagnostic object has gone beyond the normal limits. Then, a malfunction (or abnormality) identified by the value of diagnostic feature — $\triangle_K = 1$, with the same, unchanged value of feature $\triangle_M = 0$, will give half value of the relative threshold $\triangle_\omega = 1/2$, i.e. in one case the object will be considered inoperable, while in another case — with the only diagnostic feature, by natural frequency — will be considered operable. Two relative thresholds $\triangle_K$ and $\triangle_M$, with different signs in expression (3.4), may compensate each other, leaving the relative threshold $\triangle_\omega$ equal to zero. In this case, when the stiffness and mass of the object are monitored separately, the monitoring is more sensitive than the one based on natural frequency only as a single diagnostic feature.

Let us match three features in the measurement space (ω_0, $H(\omega_0)$, Q) with the three features in the diagnostic space ("M", "C", "K"). For this, we need to linearize the interrelations:

$$\begin{cases} \omega_0 \approx \sqrt{\dfrac{K}{M}} \\ H(\omega_0) \approx \dfrac{\sqrt{M}}{C\sqrt{K}} \\ Q \approx \dfrac{\sqrt{KM}}{C} \end{cases} \Longrightarrow \quad \begin{aligned} \triangle_\omega &= +\frac{1}{2}\triangle_M + \frac{1}{2}\triangle_K \\ \triangle_H &= +\frac{1}{2}\triangle_M - \triangle_C - \frac{1}{2}\triangle_K \\ \triangle_Q &= +\frac{1}{2}\triangle_M - \triangle_C + \frac{1}{2}\triangle_K \end{aligned}$$

and the interrelations:

$$\begin{cases} M = \dfrac{Q}{H\omega^2} \\ C = \dfrac{1}{H\omega} \\ K = \dfrac{Q}{H} \end{cases} \Longrightarrow \quad \begin{aligned} \triangle_M &= -2\triangle_\omega - \triangle_H + \triangle_Q \\ \triangle_C &= -\triangle_\omega - \triangle_H \\ \triangle_K &= -\triangle_H + \triangle_Q \end{aligned}$$

The matrix of the first system

$$A = \begin{matrix} -1/2 & 0 & 1/2 \\ 1/2 & -1 & -1/2 \\ 1/2 & -1 & 1/2 \end{matrix}$$

is related to the matrix of the second system, as direct and inverse matrices $AA^{-1} = E$

$$\mathrm{A}^{-1} = \begin{matrix} -2 & -1 & 1 \\ -1 & -1 & 0 \\ 0 & -1 & 1 \end{matrix}$$

which enables us to put down the ratio between the measurement space and the diagnostic features space in the vector-and-matrix form:

$$[I] = \begin{bmatrix} \triangle_\omega \\ \triangle_H \\ \triangle_Q \end{bmatrix} = A \begin{bmatrix} \triangle_M \\ \triangle_C \\ \triangle_K \end{bmatrix} = A[P] \tag{3.5}$$

$$[P] = \begin{bmatrix} \triangle_M \\ \triangle_C \\ \triangle_K \end{bmatrix} = A^{-1} \begin{bmatrix} \triangle_\omega \\ \triangle_H \\ \triangle_Q \end{bmatrix} = A^{-1}[I] \tag{3.6}$$

or

$$[I] = A[P] \quad \text{and} \quad [P] = A^{-1}[I]$$

When vector $[I] = \begin{bmatrix} \triangle_\omega \\ \triangle_H \\ \triangle_Q \end{bmatrix}$ takes on the value $\begin{bmatrix} 1 \\ 0 \\ 0 \end{bmatrix}$, it means that a malfunction (abnormality) was identified by the first diagnostic feature (natural frequency), when $[I] = \begin{bmatrix} 0 \\ 1 \\ 0 \end{bmatrix}$, a malfunction was identified by the transfer function amplitude, when $[I] = \begin{bmatrix} 0 \\ 0 \\ 1 \end{bmatrix}$, the malfunction was identified by Q-factor. Following the matrix actions according to the formula (3.6), we will see that in the diagnostic features space

vector $[I] = \begin{bmatrix} 1 \\ 0 \\ 0 \end{bmatrix}$ corresponds to vector $[P] = \begin{bmatrix} -2 \\ -1 \\ 0 \end{bmatrix}$

vector $[I] = \begin{bmatrix} 0 \\ 1 \\ 0 \end{bmatrix}$ corresponds to vector $[P] = \begin{bmatrix} -1 \\ -1 \\ -1 \end{bmatrix}$

vector $[I] = \begin{bmatrix} 0 \\ 0 \\ 1 \end{bmatrix}$ corresponds to vector $[P] = \begin{bmatrix} 1 \\ 0 \\ 1 \end{bmatrix}$

This means that abnormal frequency in the measurement space leads to two abnormal diagnostic features — abnormal mass and abnormal damping; abnormal spectral amplitude leads to the abnormality of all three diagnostic features; abnormal Q-factor leads to two abnormal diagnostic features — abnormal mass and abnormal stiffness.

In the inverse direction, in case of abnormalities in the diagnostic features space:

$$\text{vector } [P] = \begin{bmatrix} 1 \\ 0 \\ 0 \end{bmatrix} \text{ corresponds to vector } [I] = \begin{bmatrix} -\frac{1}{2} \\ \frac{1}{2} \\ \frac{1}{2} \end{bmatrix}$$

$$\text{vector } [P] = \begin{bmatrix} 0 \\ 1 \\ 0 \end{bmatrix} \text{ corresponds to vector } [I] = \begin{bmatrix} -0 \\ -1 \\ -1 \end{bmatrix}$$

$$\text{vector } [P] = \begin{bmatrix} 0 \\ 0 \\ 1 \end{bmatrix} \text{ corresponds to vector } [I] = \begin{bmatrix} \frac{1}{2} \\ -\frac{1}{2} \\ \frac{1}{2} \end{bmatrix}$$

When mass crosses its diagnostic threshold, all three measurements (ω_0, $H(\omega_0)$, Q) will deviate from their average values, but none of them will cross their thresholds. The same is true for stiffness. Only in case of abnormal damping, spectral amplitude and Q-factor will exceed their thresholds. Thus, under the *a priori* uncertainty, switching from the measurement space to the diagnostic features space increases the sensitivity to any malfunction.

The volume of a rectangular prism, formed by the point (ω, H, Q) in the measurement space,

$$V_I = \omega \cdot H \cdot Q = \frac{\sqrt{M}\sqrt{K}}{C^2} = \frac{Q}{C} \quad \left[\frac{\text{m}}{\text{N} \cdot \text{s}}\right] \tag{3.7}$$

or the same volume obtained from the areas of the prism base

$$V_I = S1 \cdot Q = S2 \cdot H = S3 \cdot \omega$$

is the so-called potency of a set included in the prism. The volume of another rectangular prism, formed by the point (M, K, C) in the diagnostic features space, obviously, is not equal to the volume V_I:

$$V_P = M \cdot K \cdot C = \frac{Q^2}{H^3 \omega^3} = C^3 Q^2 \quad \left[\frac{(\mathrm{N} \cdot \mathrm{s})^3}{\mathrm{m}} \right] \tag{3.8}$$

Two prisms are not transformed one into another by linear transformations and are physically represented by different dimensions. In formal language, this means that the efficiency of a set in the volume of one prism does not match the efficiency of a set in another prism. The prism volume as a diagnostic feature is applicable for identification of malfunctions according to the "OR" scheme, as when any diagnostic feature exceeds its relative diagnostic threshold, the prism volume will also be abnormal in any space:

$$\triangle_{V_I} = \triangle_{\omega} + \triangle_H + \triangle_Q$$
$$\triangle_{V_P} = \triangle_M + \triangle_K + \triangle_C$$

The two volumes have different dependency on Q-factor and damping, with the volume of the prism in the diagnostic features space, as seen from expressions (3.7) and (3.8), being much more sensitive to the change in damping and/or Q-factor than the volume in the measurement space:

$$\triangle_{V_I} = \triangle_Q - \triangle_{\mathrm{C}}$$
$$\triangle_{V_P} = 2\triangle_Q + 3\triangle_{\mathrm{C}}$$

By transforming the volumes, we can come to the dependencies on only one of the parameters:

$$V_I^3 \cdot V_P = Q^5$$
$$\frac{V_I^2}{V_P} = \frac{1}{C^5}$$

which are especially sensitive to the change in Q-factor or damping.

For any rectangular prism, the following interrelations of area and volume are true

$$S1 \cdot S2 \cdot S3 = V_I^2$$

as well as the interrelations for tangent slopes:

$$\mathrm{tg}\alpha \cdot \mathrm{tg}\beta \cdot \mathrm{tg}\gamma = 1$$

These dependencies make the space of areas $(S1, S2, S3) = (H\omega, \omega Q, QH)$ degenerate, in a sense that coordinate axes $S1$, $S2$, $S3$ are interdependent by definiteness, that is why of the three numbers ($S1$, $S2$, $S3$) one cannot unambiguously correspond to the three (H, ω, Q). The volume of a rectangular prism V_S in the areas space ($S1$, $S2$, $S3$), which is quadratically interrelated with the prism volume in the measurement space $V_S = V_I^2$, does not give any additional diagnostic information compared to the measurement space.

The space of relations $Z = (Z1, Z2, Z3)$ is also degenerated by definition:

$$\left(Z1 = \frac{S_1}{S_2} = \frac{H}{Q} = tg\gamma,\ Z2 = \frac{S_2}{S_3} = \frac{\omega}{H} = tg\alpha,\ Z3 = \frac{S_3}{S_1} = \frac{Q}{\omega} = tg\beta\right).$$

Obtained from the measurement space (H, ω, Q) and/or from the areas space $(S1, S2, S3)$ it always has a unit volume of the prism $V_z = 1$, i.e. "Z" space — the area ratio space, or the tangent slope space — has the potency of a set invariant to the measured values. Such invariants also have certain diagnostic value. Their time invariance proves that the adopted diagnostic model is true. When a malfunction appears, which makes this model false, the invariance no longer exists.

The normalized thresholds in "Z" space are interrelated with the normalized thresholds in the measurement space — ≪I≫ via a degenerate matrix:

$$\left\{\begin{matrix} 1 & 0 & -1 \\ -1 & 1 & 0 \\ 0 & -1 & 1 \end{matrix}\right\}$$

or

$$\triangle_{Z_1} = \triangle_H + 0 - \triangle_Q$$

$$\triangle_{Z_2} = -\triangle_H + \triangle_\omega + 0$$

$$\triangle_{Z_3} = 0 - \triangle_\omega + \triangle_Q$$

As follows from the previous system, in case of any abnormality in the measurement space by one coordinate only, two diagnostic features become abnormal at once in the $(Z1, Z2, Z3)$ space. With abnormal spectral amplitude "H", two coordinates will be abnormal — $Z1$, $Z2$; with abnormal frequency "ω" — $Z2$, $Z3$ will be abnormal; with abnormal Q-factor — $Z1$, $Z3$ will be abnormal. The same conclusion based on the degeneracy can be arrived at for areas space "S".

The ways to build vibration diagnostic spaces and the related vibration diagnostic features and thresholds presented above are the authors' proprietary techniques.

3.5 Considerations on Diagnostic Feature Engineering in Local Diagnostics Systems

The four diagnostic tasks designated with the keywords "detection", "localization", "cause" and "forecast" may be solved using different diagnostic parameters (features).

The detection task is solved by building such a measurement functional at such a level (diagnostic threshold) which minimizes two detection errors: "probability of false alarm" and "probability of missed alarm". The time needed for detection is also important. The detection functional is most often based on the criteria of maximizing the "signal/noise" ratio, which is most often a likelihood ratio. The numeric values of the denominator (noise) are defined experimentally, by statistical inference, when the object is functioning at its operable state (background measurements). The signal parameters are set to *a priori*, based on the minimal detectable level above the noise level.

The parameters of a dynamic system considered above may be declared as "signals", and their noise characteristics may be defined *in situ* by long-term background measurements. The noise level of a spectral feature (amplitude, frequency, resonance Q-factor) depends on a number of factors:

- the physical mechanism of formation of spectral resonance (acoustic standing waves, vibration of equipment components at natural frequencies, forced vibrations at MCP harmonics);

- other physical phenomena within the operating range of frequencies (thermal hydraulic source of vibrations from the coolant side);
- hardware noise level, etc.

and cannot be postulated *a priori*.

Any transformation of a single measured parameter, like multiplying by a constant, exponentiation, taking logarithms, etc., adds no information on the diagnostic parameter, but can change the signal-to-noise ratio depending on the noise mechanism during the measurement. The increase of an exponential transformation of measurements when obtaining a diagnostic feature does not guarantee that the "signal/noise" ratio will be higher.

After a malfunction is detected as a result of comparing the diagnostic feature with the diagnostic threshold in order to solve one of the upcoming diagnostic tasks ("cause", "localization" and "forecast"), it is also necessary to calculate certain measurement functionals. They do not necessarily coincide with the diagnostic feature that had been used for the detection. While the first one may have a formal character and ensure early detection (high sensitivity to malfunctions, minimal detection time, high quality of detection), the other diagnostic functionals must be physically interpretable. At this stage, the diagnostic system plays the role of an evaluator (calculator) of such physical parameters as coolant flow rate in coolant leak detection systems, mass of a loose object (loose part detection system (LPDS)), coordinates (LPDS, LDS, VNDS), etc. The time factor is less important in this case: unlike the functional used for detection, evaluation accuracy is more critical here.

While LPDS and leak detection systems (LDSs) are mostly detection-oriented and they solve a binary task (coolant leak/no leak, loose object detected/not detected), VNDS always evaluates vibration parameters. There is always some vibration of components in any diagnostics object. Vibration is not an abnormality, and abnormal vibration (the one exceeding the preset threshold) is not a malfunction, it is just a predecessor of a malfunction. VNDS also becomes a detection system if we consider that abnormal vibration is another new phenomena, just like leakage for LDS, or a loose object for LPDS.

When building a diagnostic feature, it shall be taken into account that the increase in its dimension leads to higher variance. Under dimensions of a diagnostic feature, we understand here the number of independent measurements used for its calculation. An example of a one-dimensional diagnostic feature is the calculation of stiffness "K" by the measured value of transfer function amplitude at zero frequency "$H(0)$" — $K = \frac{1}{H(0)}$. For absolute deviations, we have:

$$(K + \triangle K) = \frac{1}{H + \triangle H} \approx \frac{1}{H}(1 - \triangle_H)$$

Thus, for relative thresholds the following is true:

$$\triangle_K \approx -\triangle_H$$

As a result, normalized variances of the diagnostic feature and that of the measurement will coincide in the first approximation: $\sigma^2_{\triangle_K} \approx \sigma^2_{\triangle_H}$.

An example of a two-dimensional diagnostic feature is the calculation of damping by the measured natural frequency and the value of transfer function at natural frequency: $C = \frac{1}{H \cdot \omega}$. The square normalized variance of the diagnostic feature in the first approximation will be equal to the sum of the square normalized variances of measurements: $\sigma^2_{\triangle_C} \approx \sigma^2_{\triangle_H} + \sigma^2_{\triangle_\omega}$. An example of a three-dimensional diagnostic feature — $M = \frac{Q}{H\omega^2}$, thus $\sigma^2_{\triangle_M} \approx \sigma^2_{\triangle_Q} + \sigma^2_{\triangle_H} + 4\sigma^2_{\triangle_\omega}$. It is easy to make a conclusion that the measurement variances are added directly if the formula for the diagnostic feature includes the measurements in the form of a quotient or a product raised to the first power, no matter where — in the numerator or in the denominator.

Let us consider the general case. Suppose the diagnostic feature "P" depends on "n" measurements $\ll I_1, \ldots, I_n \gg$:

$$P = F(I_1, \ldots, I_n) \tag{3.9}$$

Its minor change — $\triangle P$ is caused by a minor change of any number of measurements — $\triangle I_i$ (from one to "n"). Expanding the expression for the increment of the diagnostic feature into a Taylor series, we have:

$$P + \triangle P = F(I_1, +\triangle I_1 \ldots, I_n + \triangle I_n) \approx F(I_1, \ldots, I_n) + \sum_i \frac{dF}{dI_i} \triangle I_i \tag{3.10}$$

By extracting Eq. 3.9 from Eq. 3.10, we get the following expression for the absolute value of the diagnostic feature increment:

$$\triangle P \approx \sum_i \frac{dF}{dI_i} \triangle I_i = \sum_i A_i \triangle I_i \tag{3.11}$$

With independent measurements, we see that the variance of a diagnostic feature is the linear combination of measurement variances

$$\sigma_P^2 \approx \sum_i A_i^2 \sigma_i^2, \quad \text{where} \quad A_i = \frac{dF}{dI_i} \tag{3.12}$$

The error of each measurement is increased (decreased) by A_i coefficient before it becomes part of the diagnostic feature error. Thus, the more the measurements involved in the diagnostic feature engineering, the higher the absolute spread around its own average value. In order to get the normalized deviation, let us divide Eq. 3.11 by 3.10 and switch to variances again

$$\sigma_{\triangle P}^2 = \sum_i B_i^2 \sigma_i^2, \quad \text{where } B_i = \frac{dF}{dI_i} \Big/ F = \frac{d(lnF)}{dI_i} \tag{3.13}$$

The obtained variance is now dimensionless, but it is also a sum of variances of measurements with the weighting coefficients B_i^2; B_i, A_i coefficients in the expressions differ from each other just like a derivative differs from a logarithmic derivative. The contribution of each i measurement in the spread of the diagnostic feature is defined by the value

$$\vartheta_i = \frac{B_i}{\sqrt{\sum_j B_j^2}}$$

which represents the sensitivity of the diagnostic feature to this measurement (to the abnormality detected by this measurement).

Before, the diagnostic feature aimed at detection of a malfunction was built as a product of measurements or other diagnostic features. At the same time, its normalized deviation was represented in the first approximation as the sum of normalized deviations of each measurement. The same result will be obtained with a diagnostic feature built on the basis of a direct sum of measurements.

3.6 DS Resonance Parameters as a Function of Time

Following the long-term observation of vibrations of RP equipment with a large number of measurements, it is possible to trace how different diagnostic features are changing in time, including the resonance parameters — $\omega(t)$, $H(t)$, $Q(t)$ — which are, correspondingly, resonance frequency, amplitude at the resonance frequency, and Q-factor. Certain properties of those functions of time, both auto- and cross-properties (e.g. the properties of a cross-correlation function, CCF) may indicate the reasons why a resonance appears and how it changes in time. If a resonance, indeed, is of a vibrational origin, and its parameters may be satisfactorily approximated with the DS 3.1, then, by cross-correlation of the time functions of $\omega(t), H(t), Q(t)$, we obtain additional information on the changes.

From the three measured resonance parameters $(\omega,\ H,\ Q)$ via the expressions

$$\omega = \sqrt{\frac{K}{M}}, \quad Q = \frac{\sqrt{K \cdot M}}{C}, \quad H = \frac{\sqrt{M}}{C\sqrt{K}} \tag{3.14}$$

we get an estimation of the three DS parameters *(M, C, K)*:

$$K = \frac{1}{H(0)}, \quad M = \frac{K}{\omega_0^2}, \quad C = \frac{1}{\omega_0 H(\omega_0)}$$

This is the so-called absolute estimation of physical values, which requires measuring channels with proper metrological support. However, even for relative and dimensionless characteristics, without any special requirements to measuring channels, it is possible to find out the change of which DS parameter caused the shift in resonance parameters.

For example, the change of mass $M(t)$ in any direction will cause a change the time functions in opposite phase to each other in the following pairs: "frequency — $\omega(t)$ and Q-factor $Q(t)$", "frequency — $\omega(t)$ and amplitude $H(t)$", while a co-phased change will be seen in the time function pair "amplitude $H(t)$ and Q-factor $/Q(t)$".

It may be just a minor change of mass resulting from a standard process operation. For SG, an example of such operation is the feedwater and all kinds of blowdowns from the secondary circuit. The steam generation process, random in its nature, also causes fluctuations of feedwater mass, which leads to the change of SG mass.

The time function $s\omega(t)$ — frequency and $H(t)$ — resonance amplitude will be opposite in phase if the resonance changes due to stiffness — K change in time, as

$$\omega = \sqrt{\frac{K}{M}}, \quad H = \frac{\sqrt{M}}{C\sqrt{K}}$$

or mass M, etc. Table 3.3 shows for each pair of functions of time $\omega(t)$, $H(t)$, $Q(t)$ if they are co-phased or opposite in phase when one of the DS parameters (M, C, K) changes in any direction (M, C, K).

Figure 3.5 shows the curves obtained in one fuel cycle of VVER-1000 reactor plant in 230 measurements. In these 230 points, the functions of time $\omega(t)$ and $Q(t)$ demonstrate counter-phase behavior. As a result, their cross-correlation function demonstrates major anti-correlation down to the point "−0.93". As it has been shown above, this is typical for a resonance at the natural frequency of a mechanical DS.

In ASW resonances, resonance frequency and amplitude are usually co-phased (Fig. 3.6).

Correlation may be applied not only to the parameters of the same resonance, but also to any pair of parameters of resonances at different frequencies. For example, amplitudes of two resonances (Fig. 3.7). Significantly large positive correlation or co-phase means that these two resonances have one and the same source or a genetic relationship.

The time correlation of DS estimations — (ω, H, Q) — in order to confirm the adequacy of the vibration model with the parameters ("M", "C", "K"), and to identify the cause of the observed resonance, are the authors' proprietary technique.

Table 3.3: Correlation of pairs of functions of time under the change of DS parameters.

Changed DS parameter Pair of functions of time	M (mass)	C (damping)	K (stiffness)
$\omega(t), Q(t)$	Opposite in phase	—	Co-phased
$\omega(t), H(t)$	Opposite in phase	—	Opposite in phase
$H(t), Q(t)$	Co-phased	Co-phased	Opposite in phase

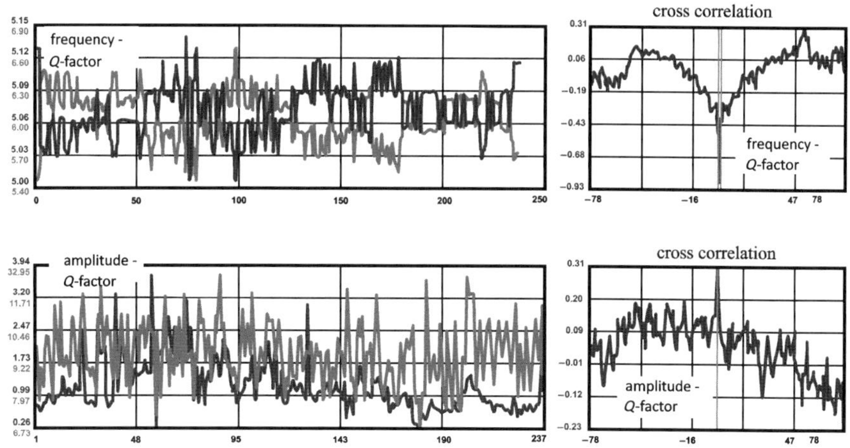

Figure 3.5: Estimated changes of resonance frequency and Q-factor of SG natural oscillations as functions of time, and their cross-correlation function – the top line of the graph. A pair of realizations $\omega(t)$, $Q(t)$ during the fuel campaign (on the left), and their counter-phase cross-correlation function (on the right). Estimated changes of amplitude and Q-factor $H(t)$, $Q(t)$ — lower line of the graph — with the co-phased cross-correlation function.

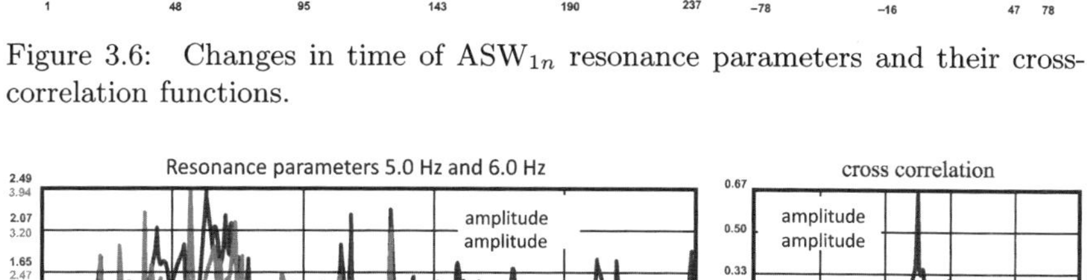

Figure 3.6: Changes in time of ASW_{1n} resonance parameters and their cross-correlation functions.

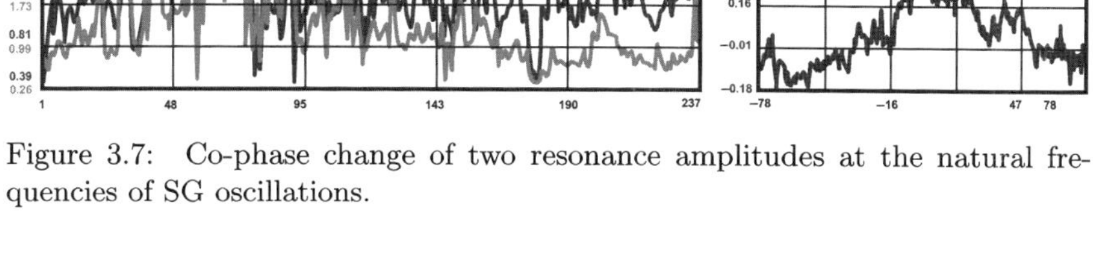

Figure 3.7: Co-phase change of two resonance amplitudes at the natural frequencies of SG oscillations.

Chapter 4

MCC Vibrations Modeling

4.1 Mathematical Model of Interaction Between Mechanical Dynamic System and MCC Pressure Pulsations Field

The modern software packages for numerical modeling of vibrations, together with the increasing computer performance, resulted in numerous vibration models not only for individual RP components, but also for the whole MCC [134, 135, 137, 139, 143, 148, 151, 152, 168, 201–215]. This important area of vibration study is very briefly covered by this chapter, as the authors are quite particular about it.

The applicability of a pre-designed computational unit of a numerical model to solve a particular vibration-related task is not always justified. Moreover, even the developer of a global numerical model sometimes does not understand such algorithms. Descriptions of software packages for numerical modeling or a modeling environment with extensive libraries of computational modules usually do not include algorithms in the analytical form. There are so many settings in the global model that it is impossible to shape a setting strategy which could lead to an unambiguous solution in such a multidimensional space of parameters. There have been some curious cases when experimental results were verified successfully based only on a variation of parameters of some complex model from another (non-vibrational) area. It does not mean that numerical modeling has no prospects at all. In our opinion, it needs more proof, including analytical methods.

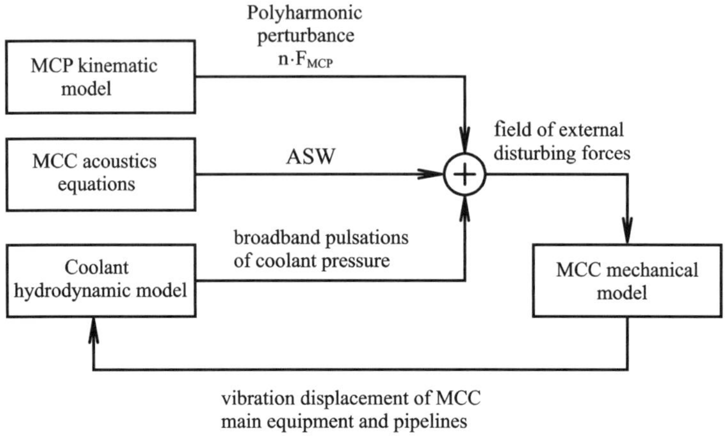

Figure 4.1: Structural diagram of MCC vibration acoustic model.

A vibration model should not be a "black box"; the multi-step process of getting the final numerical result shall be supported by intermediate, though sometimes simplified, but still analytical calculations. In the future, they will be really useful for diagnostic applications identifying the cause, localizing and forecasting the development of abnormal vibration.

MCC vibration acoustic properties are defined by the MCC mechanical structure and hydrodynamics of the MCC coolant. MCC acoustic properties, or, more generally, MCC hydrodynamic properties and the properties of MCC as a mechanical oscillating system, in the first approximation may be considered separately (Fig. 4.1).

MCC hydrodynamics can be traced back to two factors. First, to a number of spatially distributed but static forces which induce MCC vibrations due to acoustic standing waves (ASWs). This factor is described by acoustic equations.

Second, to a hydrodynamic broadband force with a continuous spectrum of pressure pulsations, which is described by an equation of fluid state, and, particularly, by Navier–Stokes equation. In the simplest case, this force is replaced by the so-called white-noise perturbance with the spatially-dependent amplitude MCC. These two forces are supplemented in each point of the MCC by polyharmonic forces appearing due to non-ideal MCP balancing.

Table 4.1: Characteristics of vibration acoustic sources in a standard VVER-1000 plant.

Sources	Frequencies	Resonance characteristics
ASW	$\begin{Bmatrix} \sim 6.60\,\text{Hz–ASW }\Pi_1 \\ \sim 9.00\,\text{Hz–ASW K}_1 \\ \sim 13.00\,\text{Hz–ASW K}_2 \end{Bmatrix}$	Low Q-factor, high-damping resonances
MCP vibrations	$\begin{Bmatrix} n \cdot 16.6\,\text{Hz — harmonics of MCP set} \\ p/q \cdot 16.6\,\text{Hz — sub-harmonics} \end{Bmatrix}$	High Q-factor, nearly damping-free polyharmonic spectra
MCC equipment vibrations at natural frequencies	{individually by loops}	Resonances with damping in a wide range of values

As a numerical example, the characteristics of vibration acoustic sources in a standard VVER-1000 plant are listed in Table 4.1.

Mechanical and hydrodynamic models in the elementary matrix form, regardless of damping and external forces, are represented by the following set of equations:

$$\begin{cases} ([K] - [M]\omega_n^2)\{X\} = 0 \\ \begin{Bmatrix} [M][0] \\ [M^{f_s}][M^P] \end{Bmatrix} \begin{Bmatrix} \partial^2 X \\ \partial^2 P \end{Bmatrix} + \begin{Bmatrix} [K][K^{f_s}] \\ [0][K^p] \end{Bmatrix} \begin{Bmatrix} X \\ P \end{Bmatrix} = 0 \end{cases} \tag{4.1}$$

where $[K]$ in the stiffness matrix of a mechanical DS, $[M]$ in the mass matrix of a mechanical DS, $\{X\}$ the displacement vector, ω_n in the DS natural frequencies, $[M^p]$ in the mass matrix of a hydraulic DS, $[K^p]$ in the stiffness matrix of a hydraulic DS, $[M^{f_s}]$ in the mass matrix of acoustic interaction, $[R^{f_s}]$ in the stiffness matrix of acoustic interaction, $\{P\}$ in the pressure vector of a hydraulic DS.

The set of equations (4.1) may be considered the first step of iteration where parameters of MCC vibration acoustic resonance are calculated (eigenvalues and eigenvectors). Such an approach does not consider the dissipation of vibration energy and the perturbances brought by an external field of forces. It is well known that minor damping ratios lead to a minor shift in natural frequencies of oscillations. The vector of external disturbing forces is taken into account in the second step of iteration, where the obtained MCC natural frequencies remain unchanged.

Modeling is often limited to solving homogeneous equations from the set (4.1) regardless of external forces. This is a weak point of numerical modeling. It is highly possible that a structure, being highly damped, exhibits almost zero vibrations at natural frequencies as one of the dominating vibration effects. They may have negligible amplitudes compared to the resonance amplitudes of external disturbing forces.

At the same time, acoustic standing waves, as external forces which induce vibrations, are natural oscillations themselves of the hydrodynamic fluid In this sense, calculation of ASW frequencies, as the solution of eigenvalue problem by means of homogeneous acoustic equations, is just a preparation for solving the task of considering external vibration forces. ASW, at least their lowest types, are always dominating in all kinds of vibration characteristics for both main equipment of the loops and MCC pipelines. The same is true for the harmonics of MCP rotational frequency. It is nearly impossible to get the ratio of amplitudes of resonances of spectral vibration characteristics *a priori*, or only from model calculations. In any case, numerical modeling of MCC vibrations must be supported by actual vibration measurements.

4.2 Finite Element Modeling

While VVER design was elaborated, it has always been supported by computational justifications and, particularly, by computational models of oscillations of MCC main equipment with the finite element methods (FEMs). FEM is a rapidly developing area of computational modeling of actual structures and, particularly, modeling of vibrations of VVER main circulation circuit. The models have a large number of elements (up to thousands) and, correspondingly, a large number of settings. As a result of a long-term verification of such models with actual measurements made at VVER RP and further reduction of dimensionality, the models become adequate for all measurement points. So far, vibration noise measurement data has been collected gradually, an iteration process has begun: "experiment → model verification → experiment → ...", and, in the result, a general FME model became adjusted for the particular power unit. Model verification is based on a vast amount of experimental material obtained by commissioning test systems, vibration monitoring systems (VMSs) and vibration noise diagnostic systems (VNDSs).

As it is known, computational models do not identify a root cause of abnormal vibration. FEM models are built on the RP design data, and they represent an operable object. Such models can describe the occurrence on an abnormality (up to a certain extent of development) if some of its settings are adjusted. Equipment failure or malfunction may introduce changes into the structure of an object, which are not considered in the FEM model.

Applying different setting strategies to the computational model, one can get an idea or build a hypothesis on the cause of the change in vibration parameters of the object. But no more than that. It would be incorrect to apply a computational model for automatic diagnostics. In order to verify such a model by experimental frequencies and amplitudes, it is necessary to variate the settings, but there is no guarantee that a single solution exists in the form of a fixed parameters vector. Furthermore, there is no specific rule how to find a solution for this task.

The summary of computational FEM studies performed by several groups of experts can be found in [382]. There is also a description of FEM model shown in Fig. 4.2.

This model has tens of thousands of degrees of freedom. It was simplified sequentially so that the main vibration effects are represented by a smaller number of settings. For this purpose, cladding elements were used instead of rod elements, and the model itself was reduced sequentially with no practical change in the modeling quality. As a result, they managed to reduce the following models:

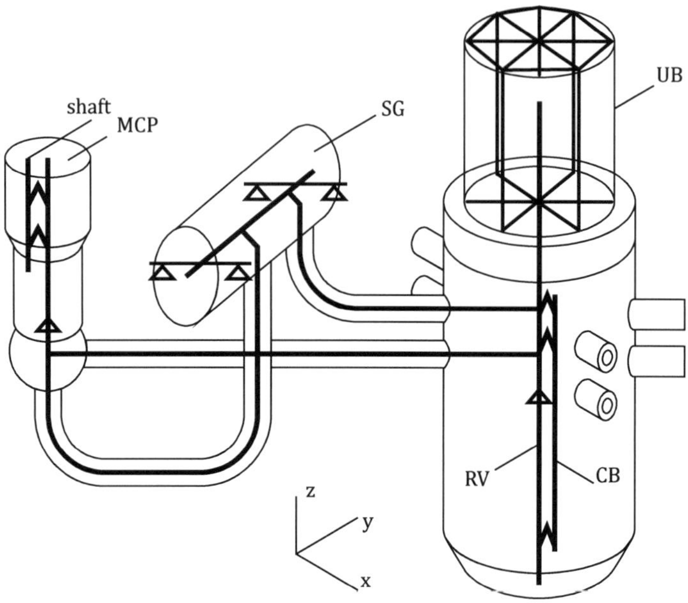

Figure 4.2: Schematic representation of VVER reactor.

- MCP set model — to six mass points and four spring units;
- SG model — to 10 rod elements and four spring units;
- MCC model — to 1,866 degrees of freedom.

Spring units were represented by three-dimensional stiffness matrices.

Figure 4.3 shows the lowest type of MCC oscillations and MCC natural frequencies for a 36 tones computational model. One natural frequency may correspond to several tones of the model. Depending on the place of maxima in vibration displacements spectra, different natural frequencies are assigned to some or to several MCC elements.

Verified FEM models give quite accurate values of natural frequencies of oscillations as a numerical solution of a multidimensional task on eigenvalues and eigenvectors. However, FEM models are very sensitive to the spatial distribution of the field of disturbing forces, and to the description of dissipative forces, i.e. the approximation of FEM model to the experimental MCC characteristics is

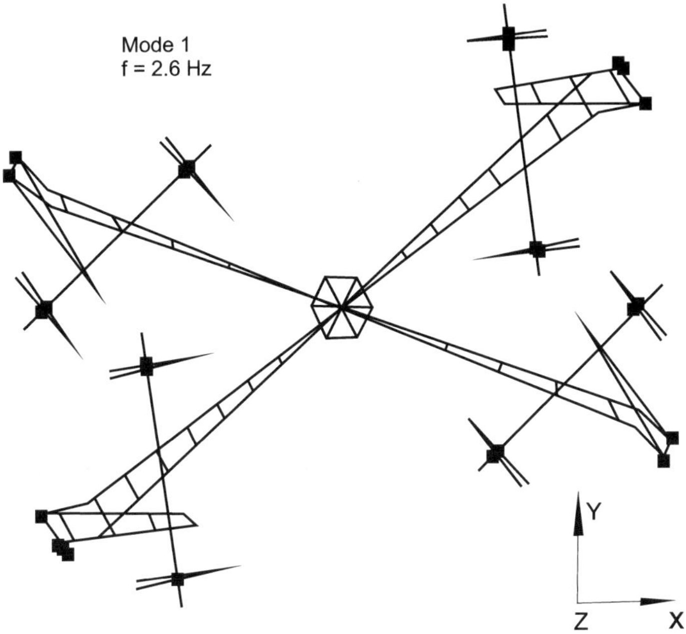

Mode nmb.	1–4	5–6	7–9	10	11–12	13–14	15	16–19	20
Frequen,Hz	2.6	3.5	4.4	4.6	4.7	6.4	7.7	7.9	8.0
Units	MCP	UB	UB	SG	CB	CB	MCP	MCP	MCP

Mode nmb.	21–22	23	24–27	28	29	30	31–34	35	36
Frequen,Hz	2.6	3.5	4.4	4.6	4.7	6.4	7.7	7.9	8.0
Units	MCP SG	MCP SG	MCP	SG	SG	SG	MCP	MCP SG, RV	UB

Figure 4.3: The lowest type of MCC oscillations at 2.6 GHz and the table of natural frequencies of MCC oscillations, where MCP is the natural frequency of MCP oscillations, UB is the natural frequency of oscillations of the reactor upper block, SG is the natural frequency of SG oscillations.

quite adequate in terms of the location of resonances on the frequencies axes, but not in terms of their Q-factors and amplitudes. Further on, analytical, semi-empirical and phenomenological models shall be involved. The latter ones postulate the patterns observed during the experiment in the analytical form.

4.3 Uncertainties of Mathematical Modeling

4.3.1 *External field of generating forces and interpretation of classical concept "natural frequency of oscillations of a structural element"*

It is known that an element of MCC cannot be characterized solely by its own vibration properties regardless of other MCC elements. MCC, being a distributed system of various masses coupled to each other with different stiffnesses, is characterized by natural frequencies of the whole MCC. They are not a "simple" sum of natural frequencies of each individual MCC element. That is why, strictly speaking, it would be wrong to "assign" some experimentally defined individual resonance frequency to a certain MCC element, as it belongs to the whole MCC.

If the root cause of the change of a particular observed frequency or a set of frequencies is the malfunctioning of some particular piece of equipment, then the inverse task is (diagnostics also deals with inverse tasks, when an unknown cause has to be identified based on the known consequence) to find the fact of changes in stiffness and/or damping of supports and/or mass of a particular element. At the same time, spectral characteristics are changing, both at this (faulty) element, and at its nearest operable neighbors. The faulty element, however, will exhibit larger changes in its spectral characteristics compared to the spectrum of unfalty elements. That is why, in the vibration measurement practice, resonance frequencies are assigned to certain MCC elements, though this is incorrect. The described logic, obviously, cannot be applied to global spatially distributed vibration sources, like ASW and oscillations of the whole MCC as one unity.

Adequate full-scope vibration models produce the calculated set of MCC vibration frequencies, indeed, as a number of MCC natural frequencies, not classifying them into the global and the local ones. Apart from frequency, oscillations are also characterized by spatial distribution, or the so-called mode or type of oscillations. Each of these types have the points of maximum and minimum vibration displacements. They may be related to the notion "localization" of oscillations: max vibration displacements are assigned to the nearest

MCC element, and their frequency potentially may be the natural frequency of oscillations of this element.

Some natural frequency oscillations may be assigned to a single quite well-defined unit of equipment, or to one well-defined heat exchange loop. In calculations, it is usually assumed that all four loops are identical, and the so-called loop oscillations become automatically the oscillations of the whole circuit, which has never been observed in practice. Due to the structural differences of the loops, loop oscillations will not be synchronized with each other.

Spatial distribution of MCC oscillation amplitudes may also be obtained by calculations. For this purpose, we need to set the perturbing forces, which are also spatially distributed across the MCC. This is the right part of the motion equation, and the main reason for uncertainty in computational modeling. Four MCP sets, as four sources of MCC perturbations, have polyharmonic (i.e. discrete) vibration spectrum, while the model is supposed to provide a solution in the form of a set of natural frequencies, also a discrete one, which actually do not coincide with the set of perturbation frequencies. For excitation of all kinds of oscillation frequencies, the model has an implicit "white" continuous component of perturbances, which is also a source of uncertainty. Anyway, the model adequacy may be challenged unless it is verified by full-scope *in situ* measurements.

No matter how complex a computational model is, diagnostics engineers use simple categories — frequency: ω_0, amplitude: $H(\omega_0)$ and Q-factor: Q of a particular resonance. In any case, in addition to the verified computational model, one has to know how to switch from the three vibration resonance parameters $(\omega_0, H(\omega_0), Q)$ to the three parameters of an elementary dynamic system of the second order, (M, K, C), where M is mass, K is stiffness, C is DS damping. The same interrelations at the qualitative level, or, which is more preferable, at the level of simple analytical dependencies, shall be used to solve the inverse task: finding a cause for the observed abnormal vibration.

Before that, one has to solve the task of identifying the vibration sources, knowing that, strictly speaking, vibration state of any

MCC element has an impact on the vibration state of any other MCC element. It is acceptable to classify resonance frequencies in the following terms:

- global — local (the whole circuit/a single loop);
- natural — forced.

Chapter 5

MCC Acoustics in Lumped Parameters

5.1 Analytical and Empirical Approaches Applied to Description of MCC Acoustics

The earliest and the most fruitful studies included simplified descriptions of acoustics of elastic media and, particularly, coolant circulation circuits [1–11, 20, 23, 25, 30, 32–34, 55, 56, 99, 104, 106, 107, 240, 250, 253, 254, 255, 321, 332, 341–353]. Most of these studies used analogies in lumped parameters from electrical engineering and/or analogies from theory of transmission lines. Adequate application of such methods enables one to estimate such global parameters of acoustic media as coolant natural frequency, coolant void fraction, sound velocity in the coolant, etc. Analogies in lumped parameters are not applicable to the full-scope numerical modeling of acoustic media, but simplified analytical methods are efficient for building the diagnostic features of coolant thermophysical anomalies.

Acoustics equations use pressure, velocity and density of an elastic medium — $P(x,t)$, $V(x,t)$, $\rho(x,t)$ — as space–time functions. For many practical tasks, it is acceptable to consider that the density of a single-phase medium is constant $\rho(x,t) \equiv \rho_0$. The change of pressure and velocity is understood as elastic vibration of a volume of medium around its equilibrium position when a longitudinal wave is propagating in it. It emerges from some acoustic source of excitation (e.g., for VVER coolant, from MCP set), or in the ideal case it always exists, if the medium does not absorb acoustic energy.

At the same time, the fluid — coolant, for example — may move with a constant velocity V_0. The longitudinal wave, in its turn, propagates with some velocity C, which is usually much higher than the medium transport velocity ($C \gg V_0$). It may propagate in a closed space (a resonator), or in an open or a semi-closed pipe, both in still water and in cases when the fluid is riven with the velocity V_0. A longitudinal wave propagates both upstream and downstream. This principle is used in Doppler ultrasound coolant flowmeters.

Sometimes these variables are somewhat incorrectly called pressure pulsations or velocity pulsations designated with symbols $\delta P(x,t)$, $\delta V(x,t)$. Sometimes, somewhat incorrectly, they are called variations, and the equations are called variational equations for variables:

$$P(x,t) = P_0 + \delta P(x,t)$$
$$V(x,t) = V_0 + \delta V(x,t)$$

Sometimes the functions $\delta P(x,t)$, $\delta V(x,t)$, also somewhat incorrectly, are called fluctuations, and assumptions are made on their random origin and minor range of amplitudes.

Theoretical acoustics, binding the variables $P(x,t)$, $V(x,t)$, $\rho(x,t)$ with equations, provides solutions in the form of standing and traveling pressure and velocity waves of elastic fluid. Moreover, the waves may be small or large in amplitude, random or deterministic, harmonic or non-harmonic. In each case, specific acoustic models are introduced with certain assumptions, however, solutions of acoustic equations would always include standing and traveling acoustic waves.

An acoustic standing wave is not capable of energy propagation along the pipeline axis, and it changes its amplitude not only in space but also in time. Strictly speaking, an acoustic standing wave is a non-stationary phenomenon, but its changes are stable in space. On the contrary, traveling waves transport energy along the way of their propagation and actually do not necessarily have sinusoidal or any other periodical shape.

On the one hand, problems of MCC acoustics may be set up as quite complex theoretical problems of mathematical physics. On the other hand, they can be partially solved at the intuitive and primitive level, based on the analysis of MCC structural features.

When we need to evaluate the frequency and length of a standing wave, the latter approach turns out to be quite productive.

The simplest analysis of boundary conditions of acoustics equations confirms the applicability of a wide-spread method when ASW frequency is obtained in a simplified way and in a rough representation of ASW amplitude distribution. That is, when there are two reflecting walls at the ends of the pipeline, the length of the lowest-type ASW is equal to the length of the pipeline multiplied by two (Fig. 5.1): $\lambda = 2L$, $f = \frac{C}{\lambda}$.

That is, the length of the pipeline hosts a half-wave with nodes at the ends and an antinode in the middle of the pipeline. This is the case with ideally reflecting walls.

A complete ASW description, including not only the values of natural frequencies, but also the spatial amplitude distribution of the standing wave, is provided by analytical equations of mathematical physics, or numerically, but again, via the same equations. In order to solve them, one has to set the spatial distribution of external disturbing forces. When it is unknown a priori, or is set very approximately, the quality of calculations becomes quite doubtful. Further on, the following logic goes: if the simple ways enable us to determine natural frequencies of standing waves, and in order to get their amplitudes through a computational and analytical way we need some *a priori* information which is difficult to find, then why take so much effort for ASW analytical description? This section presents various analytical approaches leading to practical results when describing acoustic standing waves.

The full-scope analytical description of MCC acoustics is impossible in principle due to the complex MCC geometry and, consequently,

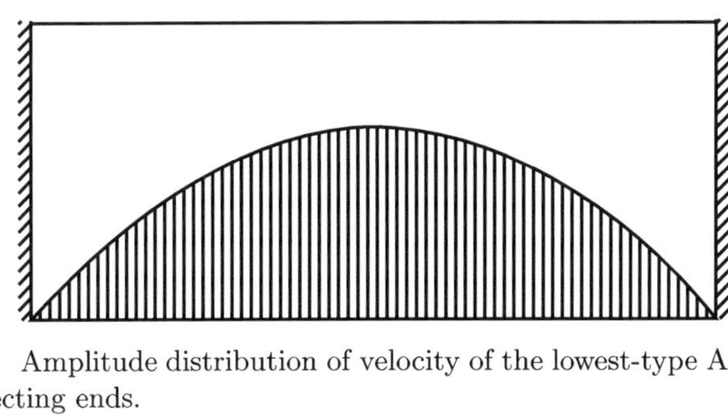

Figure 5.1: Amplitude distribution of velocity of the lowest-type ASW with two ideally reflecting ends.

the complex analytical description of boundary conditions. The second reason why there is no closed analytical solution for acoustics equations is their nonlinearity. Thus, the analytical approach often degenerates into computational schemes implemented via numerical methods. We may try to find as accurate solutions as can be, especially with the existing computer performance allowing for the most complex computational models. However, the quality of the solution will largely depend on the assumptions made when switching from the analytical form to the numerical model, and on the way how the a priori information on external disturbing forces was formalized.

In analytical methods used in acoustics, all main types of mathematical physics equations are applied: elliptic, hyperbolic and parabolic. A variety of mathematical physics equations, accurate nonlinear and approximate linearized systems of hydrodynamics equations are applied.

But before we go to partial differential equations, let us consider the method of direct transfer of an ordinary differential equation of mechanical motion for the problems of acoustics.

5.2 Acoustic Mass, Acoustic Stiffness, Acoustic Damping as Analogs of Motion Equation Parameters, Analogs from Mechanics and Electrical Engineering

The acoustic parameters mentioned in the headline appeared as analogs of mechanical motion parameters. Mass — M, stiffness — K and damping, C, are the three parameters of equation of spring pendulum motion under an external force $f(t)$:

$$M \cdot \frac{d^2x(t)}{dt^2} + C \cdot \frac{dx(t)}{dt} + K \cdot x(t) = f(t) \tag{5.1}$$

Based on the mechanical analogies, the corresponding coefficients preceding the derivatives in such acoustic equation are named, as acoustic mass, acoustic damping and acoustic stiffness. Other terms are also used, unambiguously related to the above, such as acoustic compliance, acoustic elasticity, etc., along with their specific and linear normalized values.

The three mechanical parameters (M, C, K) also have the corresponding three electrical parameters: inductance — L, capacitance — $\mathcal{C}$, active resistance — R, with the flow of current $I(t)$ under voltage $U(t)$:

$$L \cdot \frac{d^2q}{dt^2} + R \cdot \frac{dq}{dt} + \frac{1}{\mathcal{C}} \cdot q = U \tag{5.2}$$

where q is the charge, $\frac{dq}{dt} = I$ is the current.

The corresponding coefficients preceding the derivatives in Eq. (5.2), based on electrical analogies, can also be declared acoustic inductance, acoustic capacitance and acoustic active resistance. The notion of acoustic impedance is also introduced, based on the analogy with electrical impedance, as a complex resistance which includes all three parameters from Eq. (5.1). In practice, in addition to the above mechanical analogs, we still need the analogs from electrical engineering, such as acoustic current, acoustic resistance, acoustic impedance, acoustic capacitance. Now, one and the same acoustic notion may be called by its mechanical or electrical analog, which also leads to ambiguity.

The physical interpretation of acoustic analogs of the parameters (M, C, K) shall again trace their roots to acoustic patterns and to assumptions and constraints in which such analogies are acceptable. Definitions of the listed notions can be found in classical papers on the theory of acoustics. However, in practice, acoustic analogies are very inconsistent, and vibroacoustic measurements are often interpreted ambiguously, if not erroneously.

Probably, the main reason for this is that there are electrical analogs for lumped parameters and electrical analogs for long lines. It is also possible that the acoustic parameters which define the loss of energy of a longitudinal wave (all kinds of viscosity mechanisms of elastic media) represent a range of physical phenomena, unlike the single active resistance in electrical engineering. Consideration of viscosity, which, strictly speaking, leads to equations of nonlinear acoustics, breaches the "direct" analogies with mechanics and electrical engineering.

Another disorienting thing is the nonlinear algebraic dependency between pressure differential ΔP at a certain hydraulic resistance ϵ

and coolant velocity, V_0:

$$\Delta P = \text{const} \cdot \epsilon \cdot V_0^2 \tag{5.3}$$

where V_0 is the coolant velocity, not a phase speed of a longitudinal wave in the elastic medium of the coolant. This is the so-called interrelation of constant components which has nothing to do with the time functions characterizing a longitudinal wave in elastic and viscous medium. Sometimes Eq. (5.3) is erroneously used as the closing equation in a system of differential equations.

Acoustics equations, as partial differential equations, are reduced to ordinary differential equations by various methods, most often by complex harmonic substitution, but in the frequency domain, which is not equivalent to recording in the time domain (5.1). It is impossible to reduce partial differential equations to an ordinary differential equation of motion by introducing some new acoustic concept only to the equation of motion. In acoustic analogies, it is necessary to find new variables, apart from the fundamental parameters, like pressure — $P(x,t)$, velocity — $V(x,t)$, or density — $\rho(x,t)$, of the medium during the propagation of a longitudinal wave introducing new understanding of the parameters (M, C, K).

The terms for the three parameters (M, C, K) taken from mechanics, and — later — the notion of complex resistance taken from electrical engineering, have lead to the unified equations of motion in which triples of parameters are designated in the following unified way:

$$m_M \cdot \ddot{x}(t) + r_M \cdot \dot{x}(t) + \frac{1}{C_M} \cdot x(t) = F(t) \quad \text{in mechanics} \tag{5.4}$$

$$L \cdot \ddot{q}(t) + r_E \cdot \dot{q}(t) + \frac{1}{C_E} \cdot q(t) = U(t) \quad \text{in electrical engineering} \tag{5.5}$$

$$m_a \cdot \ddot{\xi}(t) + r_a \cdot \dot{\xi}(t) + \frac{1}{C_a} \cdot \xi(t) = P(t) \quad \text{in acoustics.} \tag{5.6}$$

Now, instead of stiffness K, for unification, the mechanical equation of motion uses an inverse value — $C_M = \frac{1}{K}$ — compliance.

In acoustics, it has the meaning of acoustic capacitance and/or acoustic compressibility, C_a.

The equation of motion for acoustics will look exactly the same way as two other equations of motion if for the time-dependent variation of current $\dot{\xi}(t)$ we select the so-called acoustic flow rate of the fluid. In fact, this is the fluctuation of volumetric flow rate with the dimension $\left[\frac{\mathrm{m}^3}{\mathrm{s}}\right]$. The flow rate is proportional to the transferred volume — $\mho$ and velocity — $V(t)$. In acoustics, pressure $P(t)$ acts as an external forcing effect. Velocity $V(t)$ and pressure $P(t)$ are fundamental variables which are bound by acoustics equations.

The acoustic variable $\xi(t)$ is called volume displacement. Remember that we are speaking about an elementary volume in which pressure $P(t)$ changes when a longitudinal wave propagates with the velocity $V(t)$. Instead of a material point, whose displacements (oscillations) are designated by x in the mechanical equations of motion, in acoustics, some volume v is displaced, which may compress/expand and, correspondingly, has a certain oscillation speed. The acoustic equation of motion is the equation of displacement, not of a point, but of a volume. The medium of this volume has both inertial and elastic properties. Moreover, a cross-section area is introduced, S, which is displaced by its linear dimension x, resulting in the motion of the mass. In acoustics, an elementary volume can be represented by an elementary bottle with the medium split into two parts:

- the medium in the neck with cross-section S displaced by the value x is an inertial element similar to a weight of a spring pendulum M;
- the medium in the rest of the bottle (hereinafter referred to as the volume) is compressing/expanding due to displacement x, representing an elastic element similar to the spring of a pendulum.

If a mechanical equation is the sum of three forces — inertial, dissipative (frictional) and elastic – an acoustic equation is the sum of three pressures — inertial, viscous (frictional) and elastic. The first pressure is concentrated in the neck of the elementary bottle, and friction is most often modeled also in the neck, while the elastic pressure is concentrated in the volume of the bottle. It must be noted that the shape of the bottle can be different, but the bottle must have a neck with the cross-section S and, moreover, some volume, which completely describes the elastic properties of the medium. An infinite elastic medium cannot be represented as an infinite long ensemble of elementary bottles, it must have some end walls. Besides, the elementary bottle should have a single hole through which an external pressure is connected. At the same time, one part of the medium shall be a pipeline, and another part, a vessel. Although the same medium fills the neck and the volume of the bottle, these two parts correspond to the elements of the oscillating system with different properties: a weight and a spring of a pendulum. In the neck, there is the kinetic energy of mass motion $\rho \cdot S \cdot x$, in the volume of the bottle, there is an elastic element, every time its volume is changed by ΔV, potential energy is accumulated

$$E_{P_a} = \frac{1}{2} \cdot \frac{(\Delta \mho)^2}{C_a}$$

In order to match the form of the acoustics equation with the similar equations of motion in mechanics and electrical engineering, it is necessary to derive the analogs of mass, stiffness and resistance. They will still have the same physical meaning: inertia, compliance, resistance, but the dimensions will differ essentially from the mechanical analogs. As pressure is an external impact in acoustics, in order to switch from mechanical force to pressure, let us normalize the equation of mechanical motion (5.4) by cross-section area

$$\frac{m}{S} \cdot \ddot{x}(t) + \frac{r}{S} \cdot \dot{x}(t) + \frac{1}{C \cdot S} \cdot x(t) = \frac{F(t)}{S} = P(t) \qquad (5.7)$$

and switch from linear units $x(t)$ to volume units $\xi(t) = S \cdot x(t)$.

$$\frac{m}{S^2} \cdot \ddot{\xi}(t) + \frac{r}{S^2} \cdot \dot{\xi}(t) + \frac{1}{C \cdot S^2} \cdot \xi(t) = P(t) \qquad (5.8)$$

Let us set:

- parameter $m_a = \frac{m}{S}$, $\left[\frac{\text{kg}}{\text{m}^4}\right]$ as acoustic mass,
- parameter $C_a = C \cdot S$, $\left[\frac{\text{m}^4 \cdot ts}{\text{kg}}\right] = \left[\frac{\text{m}^5}{N}\right]$ as acoustic flexibility (compliance, capacitance),
- parameter $r_a = \frac{r}{S^2}$, $\left[\frac{\text{kg}}{\text{m}^4 \cdot \text{s}}\right]$ as acoustic resistance.

Then the acoustic equation of motion has the following form:

$$m_a \cdot \ddot{\xi} + r_a \cdot \dot{\xi} + \frac{1}{C_a} \cdot \xi = P$$

The unusual acoustic dimensions of such intuitively comprehensible categories as inertia, compliance and resistance may lead to erroneous conclusions in interpretation of VVER vibroacoustic parameters. For two acoustic parameters, m_a and r_a, normalization $1/S^2$ is introduced, while for the third one, C_a, there is normalization S^2. This is not a mistake, as normalization by $1/S^2$ is applied to stiffness, K, which is an inverse value of compliance $C = 1/K$ and $C_a = C \cdot S^2$. As soon as the dimension of acoustic mass $m_a \left[\frac{\text{kg}}{\text{m}^4}\right]$ is defined, the dimensions of r_a and $1/C_a$ are obtained automatically — by successive division by time in seconds: $r_a \left[\frac{\text{kg}}{\text{m}^4 \cdot \text{s}}\right]$, $1/C_a \left[\frac{\text{kg}}{\text{m}^4 \cdot \text{s}^2}\right]$.

From the three introduced acoustic parameters, further on we can obtain energies and powers defined by various forces. For example, the expression for kinetic energy of a spring pendulum

$$E_K = \frac{m \cdot (x)^2}{2}$$

through formal replacement $m \to m_a$, $x \to \xi$ becomes the expression for kinetic energy of the acoustic medium

$$E_{K_a} = \frac{m_a \cdot (\xi)^2}{2}$$

with the dimension $\left[\frac{\text{kg}}{\text{m}^4} \cdot \frac{\text{m}^6}{\text{s}^2} = \text{N} \cdot \text{m} = \text{J}\right]$. As for the expression for potential energy of a spring pendulum,

$$E_P = \frac{K \cdot x^2}{2}$$

formal replacement $\frac{1}{K} = C \to C_M$, $x \to \xi$ leads to the result

$$E_{P_a} = \frac{\xi^2}{2 \cdot C_a}$$

with equally correct dimension $\left[\text{m}^6 \cdot \frac{\text{N}}{\text{m}^5} = \text{J}\right]$.

For a spring pendulum, displacement of mass and spring were measured by one and the same value x, while in acoustics, mass is displaced within the neck, and volume is changed within the bottle.

Table 5.1 provides the summary of analogies of oscillation parameters of a spring pendulum, a resonant electric circuit and a resonant acoustic medium.

For oscillating mechanical systems, there are a lot of mechanisms for energy dissipation. There is no need to provide their detailed analytical description for diagnostic purposes. For example, a pendulum slows down due to the medium where the pendulum mass is moving, and due to friction in its suspension. Most often, it is enough to consider only the first friction mechanism as the dominating one. For a long dynamic structure, it is difficult to localize the source of dissipation as the loss of energy occurs in different places: in movable bearings, kinematic pairs, fixtures and joints, etc. Sometimes all these mechanisms are integrated into one single parameter — damping. That is, oscillations of a complex mechanism are reduced to the oscillation of a single equivalent pendulum with the three parameters (M, C, K).

Sometimes this is enough to discover a new consumer of oscillation energy, i.e. via abnormal oscillation (abnormal dissipation) to identify abnormal parameters of the mechanism. When precise localization of abnormal damping is needed, distributed small-scale motion models are used.

The same is true for acoustic objects: if MCC is described with one acoustic equation, it is possible to get some global impression of its elastic medium in the form of three integral parameters. Acoustic mass, acoustic damping and acoustic stiffness do not characterize the MCC coolant as a whole, since in different points the coolant has different temperature, density and, maybe, state of matter (vapour phase due to the subcooled boiling). When the circulation loop includes such acoustically complex objects as the core and pressurizer, simplified acoustic schemes are still possible, but within a narrow frequency subrange where a single acoustic effect dominates (e.g. the lowest type ASW). Simplified acoustic schemes are also acceptable for:

- representation of objects with small number of acoustic volumes;
- acoustically homogeneous extensional structures, like pipelines;
- a part of MCC with small number of acoustically active sources, e.g. one loop without a pressurizer, or simply a pressurizer;
- small volumes of MCC in direct proximity to a pressure pulsation sensor.

Table 5.1: Analogies from mechanics, electrical engineering and acoustics.

	Mechanics	Electrical engineering	Acoustics
1	Schematic view 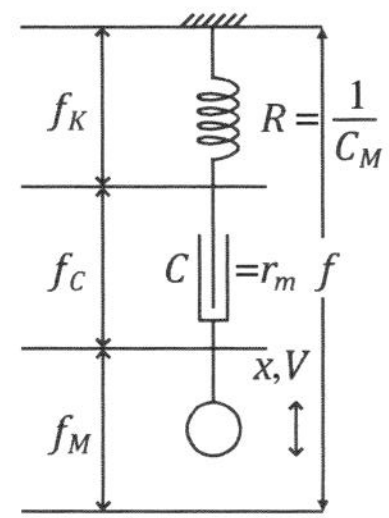	Schematic view 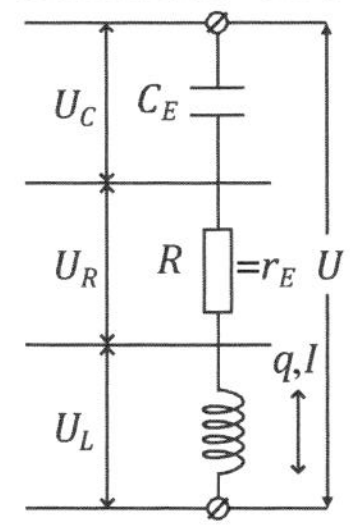	Schematic view 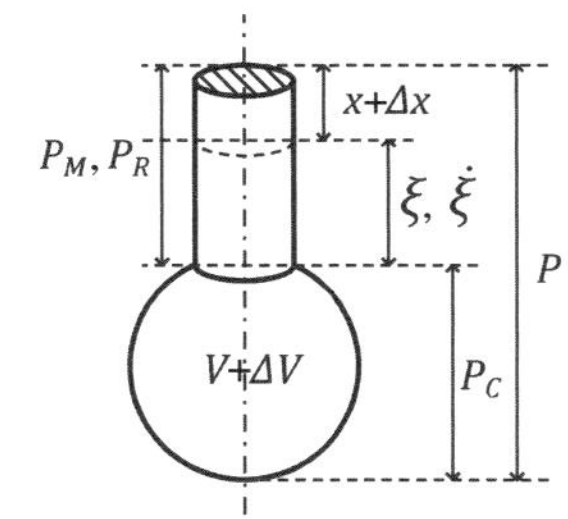
2	Point displacement	Charge	Volume displacement
	$x = \int v\,dt$ [m]	$q = \int I dt$ [C]	$\xi = x \cdot S$ [m^3]
3	Velocity	Current	Volume velocity
	$\dot{x} = \frac{dx}{dt} = v \; \left[\frac{\mathrm{m}}{\mathrm{s}}\right]$	$\dot{q} = \frac{dq}{dt} = I \left[a = \frac{\mathrm{C}}{\mathrm{s}}\right]$	$\dot{\xi} = \dot{x} \cdot S \left[\frac{\mathrm{m}^3}{\mathrm{s}}\right]$
4	Acceleration	Current change rate	Volume acceleration
	$\ddot{x} = \frac{d^2x}{dt^2} = \frac{dv}{dt} \; \left[\frac{\mathrm{m}}{\mathrm{s}^2}\right]$	$\ddot{q} = \frac{dI}{dt} \left[\frac{a}{s}\right]$	$\ddot{\xi} = \ddot{x} \cdot S \left[\frac{\mathrm{m}^3}{\mathrm{s}^2}\right]$

(*Continued*)

Table 5.1: (*Continued*)

	Mechanics	Electrical engineering	Acoustics
5	Force	Voltage	Pressure
	f [N]	U [V]	$P\left[\frac{\mathrm{N}}{\mathrm{m}^2}\right]$
6	Mass	Inductance	Acoustic mass
	m [kg]	$L\left[H=\frac{J}{a^2}\right]$	$m_a=\frac{m}{S^2}\left[\frac{\mathrm{kg}}{\mathrm{m}^4}\right]$
7	Compliance	Capacitance	Acoustic flexibility (acoustic capacitance)
	$C_M=\frac{1}{K}\left[\frac{\mathrm{m}}{\mathrm{N}}=\frac{\mathrm{s}^2}{\mathrm{kg}}\right]$	$C_E\left[F=\frac{a^2\cdot s^2}{J}\right]$	$C_a=C_M\cdot S^2\left[\frac{\mathrm{m}^5}{\mathrm{N}}=\frac{\mathrm{m}^4\cdot\mathrm{s}^2}{\mathrm{kg}}\right]$
8	Damping	Active resistance	Acoustic resistance
	$\not{C}=r_m\left[\frac{\mathrm{kg}}{\mathrm{s}}\right]$	$R=r_e\left[\mathrm{Ohm}=\frac{J}{a^2\cdot s}\right]$	$r_a=\frac{r_m}{S^2}\left[\frac{\mathrm{kg}}{\mathrm{s}\cdot\mathrm{m}^4}=\frac{\mathrm{N}\cdot\mathrm{s}}{\mathrm{m}^5}\right]$
9	Inertial force Newton's second law	Voltage at inductance	Inertial pressure
	$f_m=m\cdot\ddot{x}$ [N]	$U_L=L\cdot\ddot{q}$ [V]	$P_m=m_a\cdot\ddot{\xi}\left[\frac{\mathrm{N}}{\mathrm{m}^2}\right]$
10	Damping force	Voltage at active resistance	Friction (viscosity) pressure
	$f_C=C\cdot v=r_M\cdot v$ [N]	$U_R=R\cdot I=r_e\cdot I$ [V]	$P_a=r_a\cdot\dot{\xi}\left[\frac{\mathrm{N}}{\mathrm{m}^2}\right]$

11	Elastic force (Hooke's law)	Voltage at capacitance	Elastic pressure
	$f_K = K \cdot x = \frac{x}{C_M}$ [N]	$U_C = \frac{q}{C}\left[V = \frac{C}{F}\right]$	$P_C = \frac{\xi}{C_a}\left[\frac{\text{N}}{\text{m}^2}\right]$
12	Inertial resistance	Inductive resistance	Inertial acoustic resistance
	$j\omega m \left[\frac{\text{kg}}{\text{s}}\right]$	$j\omega L \left[\frac{H}{s} = \text{Ohm}\right]$	$j\omega m_a \left[\frac{\text{kg}}{\text{s}\cdot\text{m}^4}\right]$
13	Friction resistance	Active resistance	Viscosity resistance
	$r_M \left[\frac{\text{kg}}{\text{s}}\right]$	R [Ohm]	$r_a \left[\frac{\text{kg}}{\text{s}\cdot\text{m}^4}\right]$
14	Elastic resistance	Capacitive resistance	Elastic acoustic resistance
	$\frac{1}{j\omega C_M}\left[\frac{\text{kg}}{\text{s}}\right]$	$\frac{1}{j\omega C_e}\left[\frac{s}{F} = \text{Ohm}\right]$	$\frac{1}{j\omega C_a}\left[\frac{\text{kg}}{\text{s}\cdot\text{m}^4}\right]$
15	Dynamic damping	Complex resistance	Acoustic impedance
	$z_M = j\omega L + r_M + \frac{1}{j\omega C_M}\left[\frac{\text{kg}}{\text{s}}\right]$	$z_e = j\omega L + r_e + \frac{1}{j\omega C_e}$ [Ohm]	$z_a = j\omega m_a + r_a + \frac{1}{j\omega C_a}\left[\frac{\text{kg}}{\text{s}\cdot\text{m}^4}\right]$
16	Transfer function	Transfer function	Transfer function
	$H_M(f) = \frac{x}{f} = \frac{C_M}{1 - F^2 + 2j\alpha F}\left[\frac{\text{m}}{\text{N}}\right]$	$H_e(f) = \frac{q}{U} = \frac{C_e}{1 - F^2 + 2j\alpha F}$ [F]	$H_a(f) = \frac{\xi}{P} = \frac{C_a}{1 - F^2 + 2j\alpha F}\left[\frac{\text{m}^4\cdot\text{s}}{\text{kg}}\right]$
17	Natural frequency	Natural frequency	Natural frequency
	$\omega_{0_M} = \frac{1}{\sqrt{C_M \cdot m}}$ [Hz]	$\omega_{0_e} = \frac{1}{\sqrt{C_e \cdot L}}$ [Hz]	$\omega_{0_a} = \frac{1}{\sqrt{C_a \cdot m_a}}$ [Hz]

(Continued)

Table 5.1: (*Continued*)

	Mechanics	Electrical engineering	Acoustics
		Q-factor	
18	$Q_M = r_M\sqrt{\frac{m}{C_M}} = \frac{1}{2\alpha}$	$Q_e = \frac{1}{r_e}\sqrt{\frac{L}{C_e}} = \frac{\rho}{R} = \frac{\rho}{r_e}$	$Q_a = \frac{1}{r_a}\sqrt{\frac{m_a}{C_a}}$ Dimensionless
19	Damping factor	Damping factor	Damping factor
	$\sigma_M = \frac{\omega_{0_M}}{2Q_M}$ [Hz]	$\sigma_e = \frac{\omega_{0_e}}{2Q_e}$ [Hz]	$\sigma_a = \frac{\omega_{0_a}}{2Q_a}$ [Hz]
20	Relative damping	Relative damping	Relative damping
	$\alpha = \frac{1}{2Q}$	$\alpha = \frac{1}{Q} = \frac{r_e}{2\rho}$	$\alpha = \frac{1}{2Q}$ – dimensionless
21	Characteristic damping	Wave resistance	Wave acoustic resistance
	$\varphi_{x_M} = \frac{r_M}{2m} = Q \cdot r_M \left[\frac{\text{kg}}{\text{s}}\right]$	$\rho = \sqrt{\frac{L}{C}} = Q \cdot r_a$ [Ohm]	$\rho_a = \sqrt{\frac{m_a}{C_a}} \left[\frac{\text{kg}}{\text{s} \cdot \text{m}^4}\right]$
22	Potential energy	Electrical energy	Potential energy
	$E_{\pi_M} = \frac{x^2}{C_M \cdot 2}$ [J]	$E_{\pi_e} = \frac{q^2}{C_e \cdot 2}$ [J]	$E_{\pi_a} = \frac{(\Delta V)^2}{C_a \cdot 2}$ [J]
23	Kinetic energy	Magnetic energy	Kinetic energy
	$E_{K_M} = \frac{m \cdot V^2}{2}$ [J]	$E_{K_e} = \frac{L \cdot I^2}{2}$ [J]	$E_{K_a} = \frac{m_a \cdot \xi^2}{2}$ [J]

24	Total energy $E_M = E_{\pi_M} + E_{K_M}$ $= \frac{m}{2}[\omega_0^2 \cdot x^2 + V^2]$ [J]	Total energy $E_e = E_{\pi_e} + E_{K_e}$ $= \frac{L}{2}[\omega_0^2 \cdot q^2 + I^2]$ [J]	Total energy $E_a = E_{\pi_a} + E_{K_a}$ $= \frac{m_a}{2}[\omega_0^2 \cdot (\Delta V)^2 + \xi^2]$ [J]
25	Total energy derivative $\frac{dE_M}{dt} = \left(m \cdot \dot{x} \cdot \ddot{x} + \frac{x \cdot \dot{x}}{C_M}\right) \cdot \frac{1}{2}$	Total energy derivative $\frac{dE_e}{dt} = \left(L \cdot \dot{q} \cdot \ddot{q} + \frac{q \cdot \dot{q}}{2 \cdot C_e}\right) \cdot \frac{1}{2}$	Total energy derivative $\frac{dE_a}{dt} = \left(m_a \cdot \dot{\xi} \cdot \ddot{\xi} + \frac{\xi \cdot \dot{\xi}}{C_a}\right) \cdot \frac{1}{2}$
26	Input power $\dot{x} \cdot f \left[\frac{m}{s} \cdot N = W\right]$	Input power $U \cdot I[a \cdot V = \mathrm{W}]$	Input power $\dot{\xi} \cdot P \left[\frac{m^3}{s} \cdot \frac{N}{m^2} = \frac{J}{s} = W\right]$
27	Dissipated power $N_{C_M} = \frac{r_M \cdot V^2}{2}$ [W]	Active power $N_{C_e} = \frac{r_e \cdot I^2}{2}$ [W]	Dissipated power $N_{C_a} = \frac{r_a \cdot \xi^2}{2} \left[\frac{N \cdot m}{s} = W\right]$
28	$\frac{dE_M}{dt} = N_M$ $m \cdot \ddot{x} + r_M \cdot \dot{x} + \frac{x}{C_M} = f$ $f_M + f_r + f_c = f$	$\frac{dE_e}{dt} = N_e$ $L \cdot \ddot{q} + r_e \cdot \dot{q} + \frac{q}{C_e} = U$ $U_L + U_r + U_c = U$	$\frac{dE_a}{dt} = N_a$ $m_a \cdot \ddot{\xi} + r_a \cdot \dot{\xi} + \frac{\xi}{C_a} = P$ $P_m + P_r + P_c = P$

5.3 MCC Elements in Lumped Electrical Analogies — Pressurizer as a Helmholtz Resonator

In the above-mentioned references, attempts were made to represent a heat exchange loop with similarity to an electric circuit consisting of R–C–L elements, e.g. in the form of their serial connection (Fig. 5.2).

Following the coolant flow and transferring the three-dimensional non-homogeneous structures inside reactor equipment to flat pipeline elements (Fig. 5.3), we may also find electrical analogies using a mnemonic rule: thin pipe for inductance, its neighboring pipes with larger diameter for capacitance. In the reactor vessel, the lengths of acoustic fragments are much shorter than the pipeline lengths of the heat exchange loop. Correspondingly, frequencies of ASWs generated by inhomogeneities inside the reactor vessel are much higher than the ones of ASW inside the loop. While the range of ASW frequencies in the VVER loop, with regard to the harmonics of various loop ASWs with the significant amplitudes, may be limited by 50 Hz, frequencies of ASWs inside the reactor vessel are only starting from 50 Hz. Due to the tightly packed design of reactor vessel and the reactor core itself, which contain relatively small part of total coolant volume, low-amplitude ASWs appear inside the reactor vessel. However, it does not mean that we should not take them into

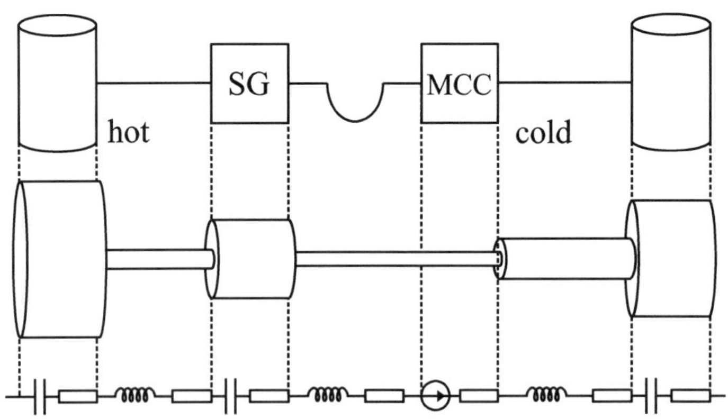

Figure 5.2: Equivalent diagram of a circulation loop as an alternation of pipelines with various cross-sections, and its electrical analogy in lumped parameters.

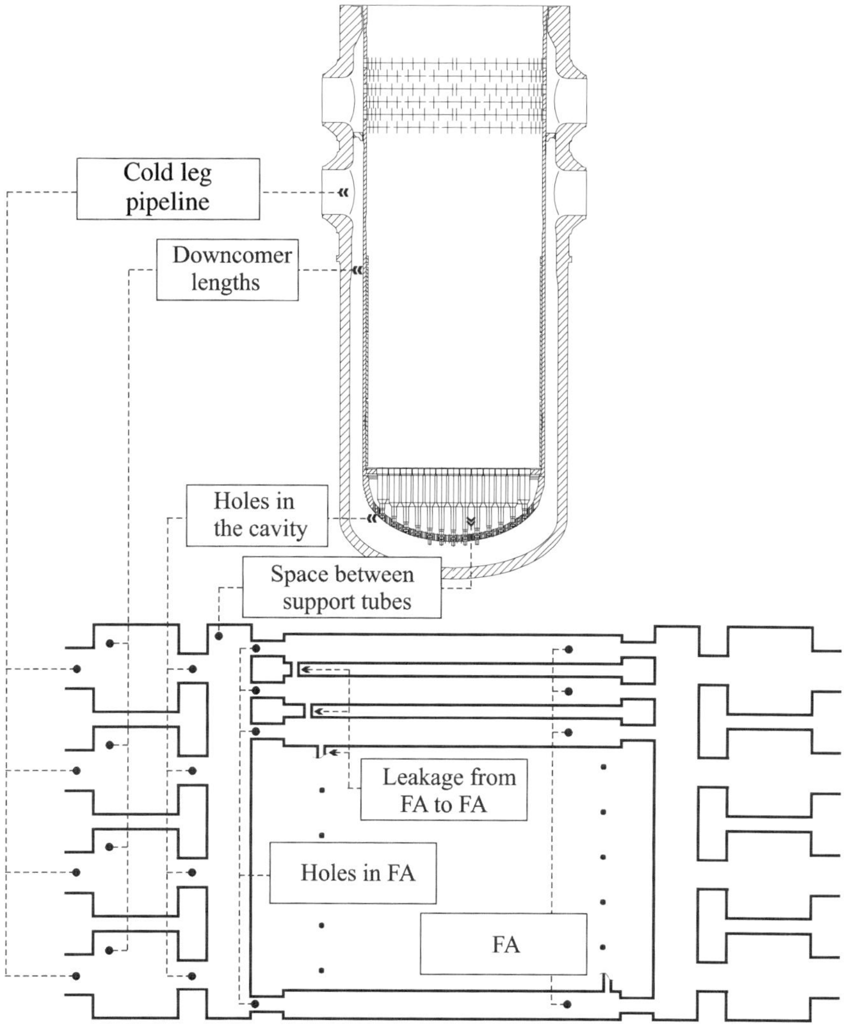

Figure 5.3: Abrupt acoustic inhomogeneities in VVER reactor.

account. The range above 50 Hz includes natural oscillation frequencies of FE parts between two spacer grids, spacer grids themselves, pipe lengths in PTU, etc.

At a very low frequency, such that its wavelength $\lambda = \frac{C}{f}$ definitely exceeds the dimensions of an MCC element, this kind of approach is acceptable. However, electrical analogies in lumped parameters are

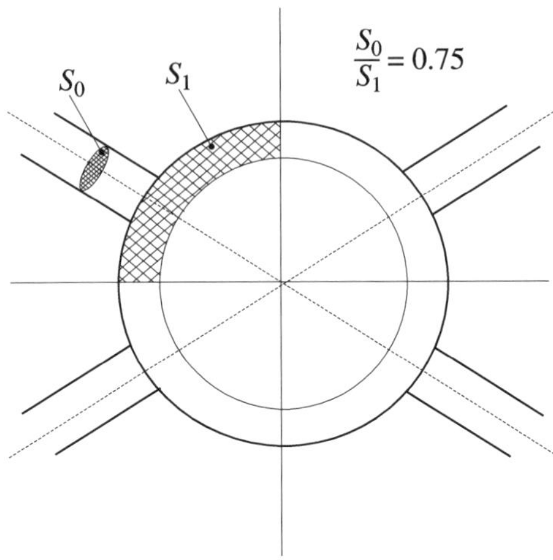

Figure 5.4: Abrupt acoustic inhomogeneity in the nozzle of reactor vessel.

quite approximate, as there is no one-to-one correspondence between electrical elements and elements of the hydraulic circuit. It would be incorrect to claim that a pipeline is an equivalent to an inductance, and a header is an equivalent to a capacitance. A pipeline may have analogies both in the form of inductance with complex resistance m_a, $m_a = \frac{\rho L}{S}$, and in the form of capacitance with complex resistance $\frac{1}{j\omega C_a}$, $C_a = \frac{V}{\rho C^2}$. Strictly speaking, it would be incorrect to go along the MCC and assign electrical equivalents to each and every part of it.

Further on, we will demonstrate that it would be better to approximate MCC by mixed electrical analogies: MCC hot and cold legs may be represented in the form of long electric lines with lumped (point) complex resistance (for locations of pipeline cross-section jumps (Fig. 5.4), bends, branching, etc.). Point analogies are efficient when the acoustic object may be represented as a Helmholtz resonator [9, 20, 30, 32, 34, 37, 56, 99, 107, 147, 216–218, 348, 349, 352]. This is the very elementary acoustic object consisting of a pipeline with length L, cross-section S and volume V (Fig. 5.5).

The natural frequency of medium oscillation in the Helmholtz resonator is described by the formula

$$\omega_0 = C\sqrt{\frac{S}{V \cdot L}}$$

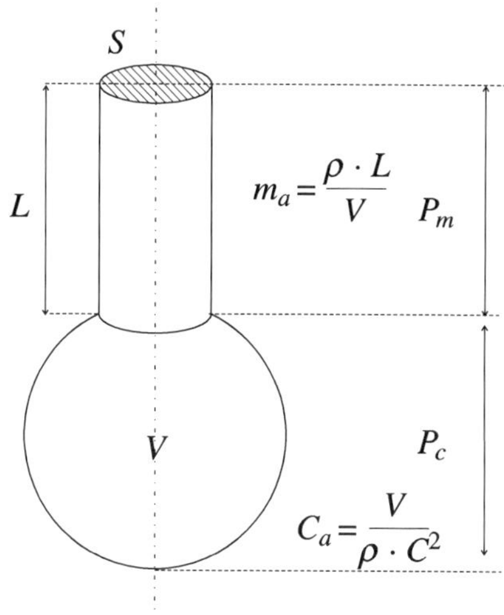

Figure 5.5: Helmholtz resonator.

If we apply the formula for natural oscillation frequency of a spring pendulum as a mechanical analogy:

$$\omega_0 = \sqrt{\frac{K}{M}}$$

we will find that the volume of the bottle, V, in the Helmholtz resonator is equivalent to the pendulum compliance, $\mathbb{C} = \frac{1}{K}$, that is, it plays an elastic role, and the ratio $\frac{\mathrm{L}}{\mathrm{S}}$ is equivalent to the mass of the pendulum weight, M. The acoustic analogy is not obvious. Indeed, in the formula for Helmholtz resonator frequency, there are only geometrical parameters and sound velocity as the single parameter representing an elastic medium. Natural frequency may be derived from the previous section using the notions of acoustic mass and acoustic capacitance, with mass included explicitly in their dimensions:

$$m_a \left[\frac{\mathrm{kg}}{\mathrm{m}^4}\right], \quad C_a \left[\frac{\mathrm{m}^4 \cdot \mathrm{s}^2}{\mathrm{kg}}\right]$$

Using the known parameters $m_a = \frac{\rho L}{S}$, $C_a = \frac{V}{\rho C^2}$ and the resonance condition, we come to the formula for Helmholtz resonator:

$$\omega_0 = \frac{1}{\sqrt{m_a \cdot C_a}} = C \cdot \sqrt{\frac{S}{V \cdot L}} = \omega_{0_a}$$

Medium density, ρ, also included in the expressions for acoustic mass and acoustic compliance, is canceled, with sound velocity remaining, C, as the parameter related to compliance only.

Unlike the one-dimensional pipeline where ASW frequency is defined by the spatial wavelength, in the Helmholtz resonator, natural frequency depends on a one-dimensional parameter pipe length L, a two-dimensional parameter pipe cross-section area S and a three-dimensional parameter volume V, with which the pipeline is connected. The exact shape of the pipe cross-section is not specified (round, square, rectangular, etc.), and the exact type of volume also remains unknown (sphere, cylinder, etc.).

Due to the extended character of the parameters m_a, C_a in the real resonator, the formula $\omega_0 = C\sqrt{\frac{S}{V \cdot L}}$ is inaccurate. Empirical corrections to this formula can be found in some studies [216–218]. For example, the following formula is applied for compliance:

$$C_a = \frac{S}{L + 0,4 \cdot \pi \cdot R},$$

where R is the pipeline radius.

Paradoxically, a Helmholtz resonator has a character of a point object in spite of the fact that it consists of two spatial elements since it is described by the equation of motion of a material point (5.6) without space coordinate. That is, it would be incorrect to say that a field of standing waves exists in it. A Helmholtz resonator has random dimensions, i.e. it may have any values of resonance frequency, but the resonator itself consists of point elements, i.e. it is a point object. And, at the same time, a Helmholtz resonator is an elementary image of an acoustic medium.

Representations of a circulation loop in the form of lumped (point) parameters similar to capacitance and inductance are often erroneously based on a Helmholtz resonator.

Example. Electrical analogies in lumped two-pole parameters.

Let us consider a pipeline system consisting of sequential sectors with smaller and larger cross-sections (Fig. 5.6). A pipe sector with larger diameter will be represented by an equivalent capacitance $\not{C}$, a pipe with a smaller diameter — by an equivalent inductance L. As a result, we shall get a serial connection of L–C-circuits, which can be

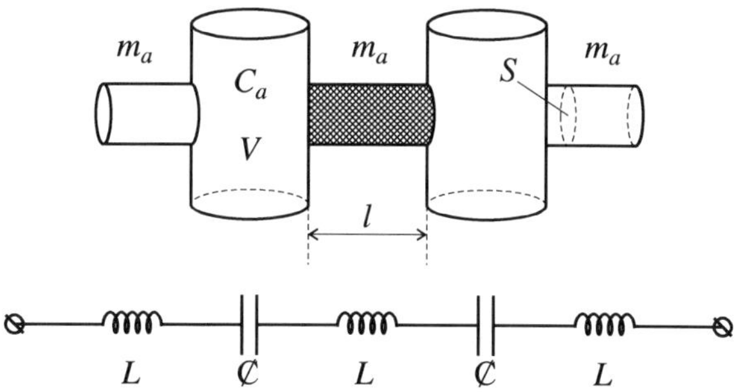

Figure 5.6: Serial pipeline system and its two-pole equivalent circuit.

compressed into a single L–C-circuit of two elements, consequently into a Helmholtz resonator.

Strictly speaking, the initial representation in the form of serial connections of capacitances and inductances is erroneous. This example demonstrates the approximate character of modeling with two-pole lumped elements. Instead of multi-resonance spectral characteristics with various ASW frequencies, we get a single resonance frequency. Two serial neighbouring $L_i - C_i$-connections in Fig. 5.6, strictly speaking, are not a Helmholtz resonator. Only the final equivalent representation, obtained by adding all inductances and inverse capacitances, becomes a Helmholtz resonator. Through capacitance, we model the volume V, which has a single inlet with the area S. In Fig. 5.6, volume V has both the inlet and the outlet.

Sometimes a Helmholtz resonator is incorrectly compared to a half-open pipe (with one end open and another one closed). For example, volume $\mho$ in the Helmholtz resonator is replaced by a reflecting wall, which makes a quarter-wave transformer, or a long line with an open circuit at the end, in which a standing wave emerges with the base frequency $f = \frac{C}{4L}$ and a constant spatial distribution of quarter-wave amplitude. For the long line resonance: $f \sim \frac{1}{L}$; and the resonance frequency of a Helmholtz resonator, keeping all the other parameters unchanged, is $f \sim \frac{1}{\sqrt{L}}$ In this replacement, there is an implicit misconception: a Helmholtz resonator, as a point element, was replaced by a long line, i.e. a spatially extended acoustic medium. Variating the three parameters of a Helmholtz resonator, its resonance frequency may take any values, while the resonance frequency of the long line is limited.

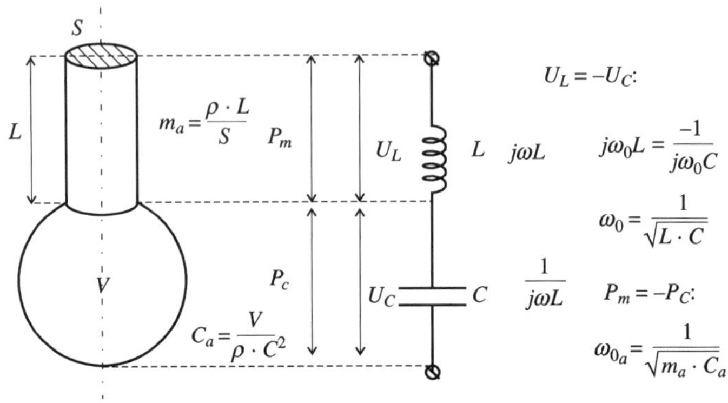

Figure 5.7: Helmholtz resonator and its electrical analogy.

In these essentially different acoustic objects — a long line and a Helmholtz resonator — different physical phenomena take place. A Helmholtz resonator at its natural frequency is an element consisting of two parts, with pressure drops in each part getting equal in amplitude and opposite in sign, thus making the resulting pressure drop equal to zero (Fig. 5.7).

This is equivalent to the resonator turning into a point without acoustic resistance and, thus, without pressure drop on it. In a Helmholtz resonator, at its natural frequency, there is an interchange between potential energy enclosed in volume V and kinetic energy of the medium enclosed in the pipe with length L. During a single period of natural oscillations, energies interchange twice.

Both spatial parts of the Helmholtz resonator are modeled by their point parameters: acoustic compliance C_a and acoustic mass m_a.

Inside the resonator, between any two points located in different parts, the pressure has different signs. With no dissipation, the Helmholtz resonator is a purely harmonic oscillator. It can be represented both as a generator of sinusoidal pressure and a resonance element of the circuit with the external impact filtered by a transfer function shown in Fig. 5.8. It ensures limitless amplitude gain at a single frequency, blocks high frequencies, and in the sub-resonance area we get a transmission band with non-uniform amplification.

Long lines with closed and open ends are one-dimensional not point-like acoustic objects. A resonance in a long line is the act of coincidence of amplitudes of a direct wave and a reflected wave with

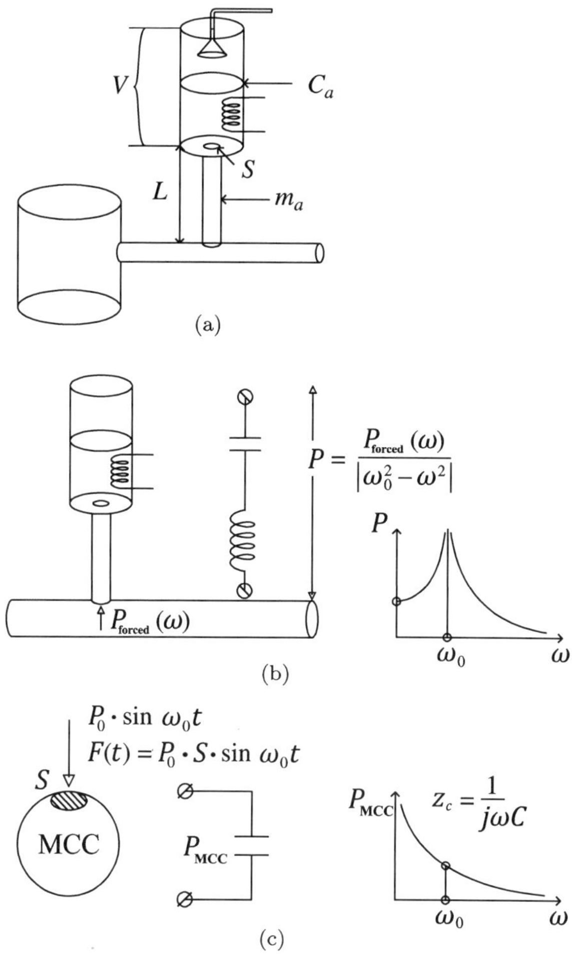

Figure 5.8: Schematic representation of pressurizer acoustic parameters (a), impact on pressurizer from MCC (b), impact on MCC from pressurizer (c).

a certain frequency. Pressure drop in any two points of the quarter-wave line is not equal to zero, and amplitudes have the same sign.

A Helmholtz resonator may approximately represent a pressurizer with geometric characteristics L, and S related to the connecting pipeline, and V — to the volume of the pressurizer vessel (Fig. 5.8). While the pressurizer volume, V, is defined based on the required safety and controllability of the object, as an element which automatically regulates pressure in the MCC, two other parameters, S

and L, may be used to control the natural frequency of the medium oscillation in the pressurizer.

Resonance oscillations of the pressurizer medium are often erroneously taken as a standing wave, and this standing wave is called the lowest-frequency ASW in the MCC. In reality, these are the resonance oscillations of only a part of the MCC medium, namely, the pressurizer volume and the connecting pipeline. These oscillations affect the rest of the MCC medium as an external harmonic disturbing force (Fig. 5.8). This is why it makes no sense to find stationary spatial positions of nodes or antinodes of this wave within the MCC volume. If we switch from a real experimental resonance frequency of the pressurizer of about 0.6 Hz to a characteristic length of ASW half-wave, it will exceed the dimensions of the MCC. These oscillations have a global, co-phased, general-circuit character and have no nodal points within the MCC volume.

Pressurizer natural frequency in VVER is within the range of (0.4–0.8) Hz for all WWER types (WWER-440, WWER-1000, WWER-1200) [364–366]. This makes quite a good choice of pressurizer resonance properties. On the one hand, the range of the pressurizer resonance frequencies covers extremely low values only and does not include natural oscillation frequencies of FA, RI, RP vessel and main equipment of the loops. On the other hand, time constants typical for standard dynamic modes of the RP are incommensurably higher than the time constant of the pressurizer, which is defined as a value inverse to resonance frequency. Broadband transients in the MCC (relatively short-time, or impulse ones), related to the coolant pressure, including emergency transients, are unacceptable. They may cause resonance pressure response in the pressurizer.

So, it is important to know the direction of pressure pulsations (Fig. 5.8). If the MCC has an impact on the pressurizer system, resonance excitation is possible. The transfer function here is similar to the transfer function of a mechanical equation of motion.

If the pressurizer, as an external harmonic generator, has an impact on the MCC, the latter represents a vessel or an elastic medium with a monotonically decreasing frequency dependency. Quite possibly, other physical phenomena, apart from the elastic properties of the MCC coolant medium, will cause some resonance features as additional components to the monotonic frequency characteristic.

The amplitude of pressure fluctuations in the MCC at the pressurizer resonance frequency exceeds all other resonance phenomena in the MCC. Its resonance dominates in the spectra of neutron noise. Neutron flux is modulated by a relatively low pressurizer frequency, which is a global barometric effect of reactivity, sometimes called an element of instability of the core. As a result, we can see periodical fluctuations of the reactor power. Their amplitude could be reduced significantly if we introduce an acoustic absorption element in the pressurizer system. For this purpose, it is enough to cover the inner surface of connecting pipeline to the pressurizer with a material which absorbs acoustic energy.

Actually, the pressurizer volume does not need such a coating as absorption is found in places where kinetic energy of the Helmholtz resonator is accumulated, i.e. in places where the acoustic medium has a non-zero velocity, meaning where the acoustic friction mechanism has place.

Acoustic mass and acoustic compliance of the pressurizer system are changing even in the stationary nominal mode of RP operation due to the processes of automatic pressure regulation. It is ensured automatically via two actions with the pressurizer medium: injection of "cold" water into the pressurizer volume and water heating with electric heaters. Water injection leads to steam condensation, which reduces the steam volume in the pressurizer, and water heating, vice versa, increases it.

Water injection reduces the compliance of the pressurizer volume due to partial reduction of steam volume, and increases the inertial mass in the connecting pipeline due to the motion of the "cold" MCC water towards the pressurizer. When electric heaters are on, the coolant faces an inverse impact: the increasing steam volume, higher compliance and lower inertia due to cold water being removed from the connecting pipeline and replaced by "hot" water with lower density from the pressurizer volume. In both cases, the changes in acoustic mass and acoustic compliance are opposite their effect, which prevents major changes of the natural frequency of the oscillation of the medium in the pressurizer. Nevertheless, the changes of these two factors are incomparable with each other, and the natural frequency of acoustic oscillation of the medium in the pressurizer ω_0 changes following a particular action (either water injection and/or starting electric heaters).

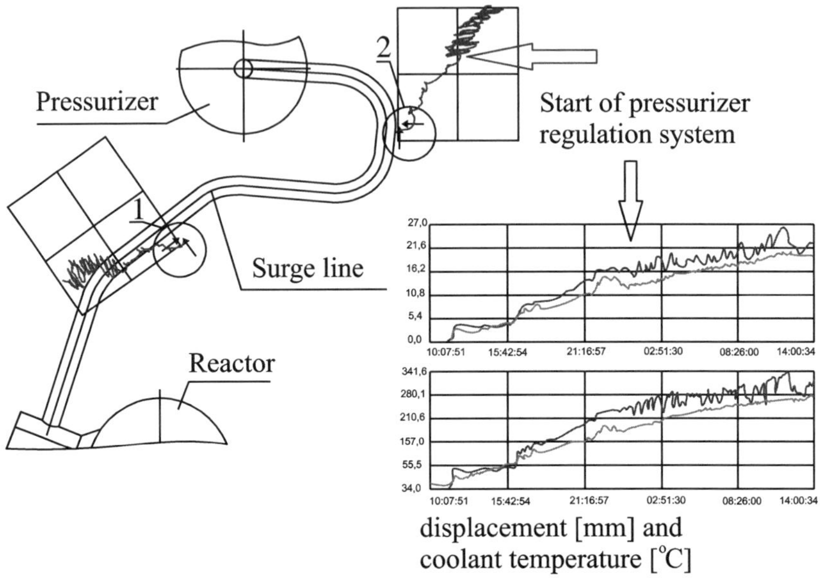

Figure 5.9: Displacement of two points of a connecting pipeline observed for 4 hours during the unit heating up: the upper graph — temperature and displacement in the point near the pressurizer vessel, the lower graph — temperature and displacement in the point near the middle of a surge line.

Thus, the change of the pressurizer resonance frequency is a diagnostic sign which characterizes the operability of the pressurizer system. The automatic pressure regulation in the pressurizer system should not happen too often, otherwise the fatigue damage of MCC pipelines increases. Figure 5.9 shows the displacement of the connecting pipeline when coolant injection and electric heaters are switched on/off. The absolute values of the displacement range are insignificant: about several mm. The number of such cycles is also not too high, not more than several dozens per day. Thus, the monitoring of displacement of the connecting pipeline of the pressurizer system can be used as an indirect method of its diagnostics. Particularly, the following is monitored: thermoelectric heater, valves of injection line, compliance of the regulation system adjustment (i.e. the number of injections).

Such displacements cannot be related to the vibration displacements. They have been observed by the authors via the Pipeline

Displacement Monitoring System (PDMS) in WWER-1000. They may be related to a "slow" response of the structure to the dynamic change of the coolant state in the pressurizer.

At the same time, the pressurizer resonance frequency is another integral diagnostic parameter which also enables us to monitor the parameters of a two-phase coolant in the pressurizer. At the pressurizer resonance frequency, there is some vibration displacement of the pressurizer connecting pipeline, though it is low-frequency, and its amplitude is at least by an order of magnitude lower than the amplitude of its thermal displacement caused by injection of coolant into the pressurizer. A single instrument channel with a relative displacement sensor (RDS) can measure both thermal and vibration displacement continuously and simultaneously.

A low-frequency resonance impact on the circulation loop from the pressurizer may be easily interpreted as some structural bench mark. This is a kind of continuous test input which can assist, through RDS, PPS signals, or through signals of low-frequency accelerometers located in different points of the MCC, in estimation of thermophysical state of the coolant. The impact from the pressurizer itself has a frequency which is defined only by the properties of the pressurizer system.

There is a single point in the MCC — a nozzle of the connecting pipeline of the loop (usually loop number four) — where practically harmonic disturbing pressure pulsations with quite high amplitude appear. This is equivalent to a case when the whole pressurizer system is "cut off" the MCC and replaced by a single sinusoidal impact in the form of pressure pulsations, the point harmonic piston type. The rest of the MCC is replaced with a single large vessel with the volume V_{MCC}, which is equivalent to a single electrical capacitance which is affected by sinusoidal voltage U.

Resistance of capacitance $(jC\omega)^{-1}$, just like the voltage which drops on it, U_C, shall be constant as far as the capacitance value is unchanged. In the acoustic analogy, sound velocity and density, ρ, shall be constant. If the elastic medium of the MCC between some two pressure pulsation sensors PPS1 and PPS2 has decreased its density due to an abnormal coolant heating, i.e. it has increased its compliance, capacitive resistance will decrease. It will decrease significantly if the abnormal heating is accompanied by a vapour phase appearing in the coolant. At the same time, the amplitude of pressure pulsations at the forced oscillations frequency ω_0 in the point of the pressurizer pipeline connection will decrease.

The transfer of pressure inhomogeneities between the two sensors PPS1 and PPS2 will also slow down when the vapour phase appears due to the decreasing sound velocity. The phase between the signals of PPS1 and PPS2 at the frequency ω_0 will increase as the phase is proportional to delay time $\varphi = \tau_3 \cdot \omega_0$. Both the amplitudes of pressure pulsations of PPS1 and PPS2 signals and the phase characteristic PPS1–PPS2 at the frequency ω_0 will be the diagnostic markers of the vapour phase in the MCC.

If we represent the MCC in a more detailed but still simplified way as a serial connection of two complex acoustic resistances z_1, z_2, one of which simulates the core, and the other one — the equivalent steam generator (Fig. 5.10), it would be difficult to assume *a priori*, which character the resulting overall resistance $z = \frac{z_1 \cdot z_2}{z_1 + z_2}$ is going to have. Actually, both complex resistances z_1 and z_2 are supposed to also have capacitive and inductive components. When the vapour phase appears, the core "bridges" the SG with its minor capacitive component. Indeed, a single-phase medium in the SG tubes is rigid in comparison to a compliant medium with the vapour phase. In this case, the information coming with the signals from two PPSs (Fig. 5.10) will be mainly defined by the medium of the core. In case of a single phase, we need to compare hydraulic resistance of the core and all steam generators.

As it was mentioned above, the frequencies of standing waves are also diagnostic indicators, and their frequency decreases when the vapour phase appears. The scale of the change depends on the volumes of the water and vapour phases. In case of a "subcooled" boiling, the vapour phase is negligible, and the change of ASW frequencies is also insignificant.

When modeling an equivalent acoustic MCC diagram, it should be noted that in a closed circulation loop the signals of any two PPSs are equivalent to the signals in a feedback circuit (Fig. 5.10). If one of the acoustic paths is too long, with too much resistance for the disturbance to overcome, this connection (whether forward or backward) may be neglected, and it may be considered that there is only one, short, "bypassing" way for the disturbance to spread. In this case, it is useless to introduce excessive localization or a large

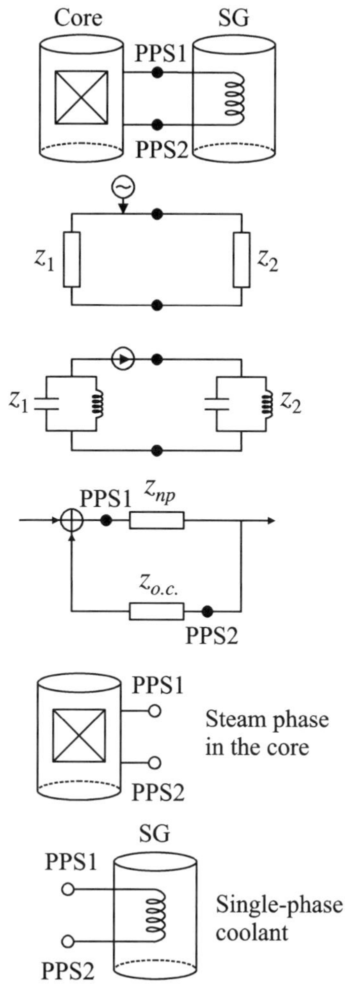

Figure 5.10: MCC representation with a small number of acoustic elements.

number of MCC elements into the model. That is, the acoustic model shall be "bound" to the measurement points which are, for example, PPSs or low-frequency accelerometers.

Thus, if two PPSs are located on a pipeline at a small distance l from each other, this pipe length is equivalent to inertial mass $m_a = \frac{\rho \cdot l}{S}$ with acoustic resistance $j\omega m_a$. There is no dependency on the sound velocity, and the medium is modeled only by its density ρ. As it is known, with a single-phase coolant, the hydraulic resistance of the core is much higher than the hydraulic resistance of the SG. That is why two PPSs installed near the cold and hot SG headers will be sensitive to the change of coolant density in heat exchange tubes (HET) of the SG. In the length of the SG tube of about 11 meters, the coolant density changes significantly due to heat transfer to the secondary circuit and the resulting temperature decrease and coolant density increase at the SG outlet.

A small section of the vast number of SG HETs (approximately 11,000) is "blocked" when intrinsic defects are revealed in the process of operation, and another part of SG HETs looses its heat transfer properties with time. These two operational trends are opposite in direction in terms of the change of SG acoustic resistance. On the one hand, when less HETs are functioning, SG acoustic resistance increases. On the other hand, with the partial loss of SG heat transfer properties, which is equal to "undercooling" of the coolant within the SG HETs volume, the coolant density remains relatively low, and the acoustic resistance decreases. It is also supported by the deposition of coolant impurities from the secondary circuit side. The continuous process of coolant impurities deposition from the primary circuit side initiates two other mechanisms, opposite in direction. On the one hand, due to the growth of impurities thickness on the inner surface of HETs, the coefficient of heat transfer to the secondary circuit also decreases, which leads to the lower coolant density in the primary circuit. On the other hand, the reduced flow area of HETs increases the acoustic mass $m_a = \frac{\rho \cdot l}{S}$, which increases the acoustic resistance again (Fig. 5.11). This approach to HETs monitoring suggested by the authors still needs to be proved in experiments.

Let us demonstrate how the change in character of SG acoustic resistance influences the coolant pressure pulsations at the SG outlet. By replacing a steam generator tube with an equivalent inductance, L_0, assuming that they are uniform we can get the whole SG to be represented in the form of a single equivalent inductance comprising many inductances connected in parallel (Fig. 5.12): $L = \frac{L_0}{N}$, where N is the number of HETs.

If n HETs were blocked in the process of operation, the resulting inductance (acoustic mass) will increase: $L_1 = \frac{L_0}{N-n}$. If HETs' flow area S decreases in the process of operation, the corresponding acoustic mass of the coolant increases: $S \downarrow$, $m_a \uparrow$. This unobvious

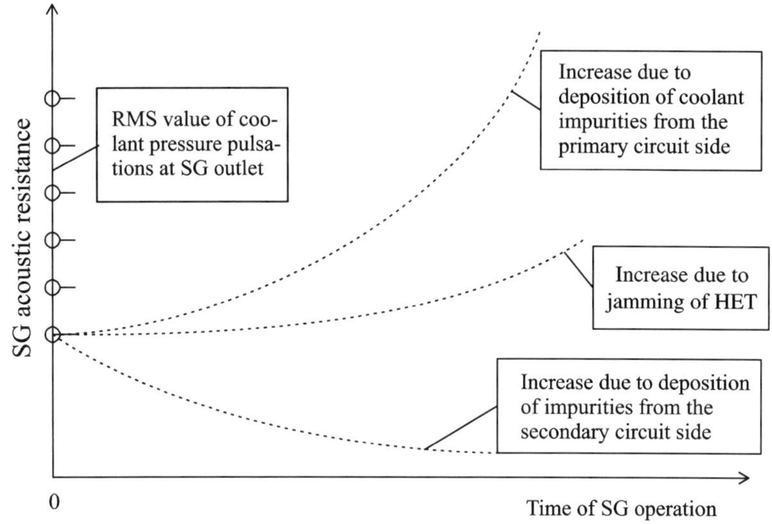

Figure 5.11: Change factors of coolant pressure pulsations at the SG outlet.

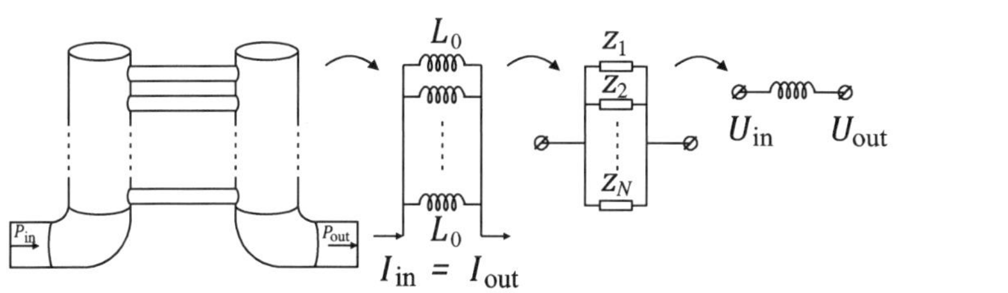

$$\frac{1}{z} = \frac{1}{z_1} + \cdots + \frac{1}{z_N} = \frac{N}{z_0} \rightarrow \frac{1}{j\omega L} = \frac{1}{j\omega L_0} + \cdots + \frac{1}{j\omega L_0} = \frac{N}{j\omega L_0} \rightarrow L = \frac{L_0}{N} \rightarrow U_{out} = U_{in} \cdot j\omega L$$

Figure 5.12: SG in electrical analogy.

result is related to the specific way of defining the acoustic mass — its inverse relation to the flow area $m_a \sim \frac{1}{S}$, while the mechanical mass of water in HETs decreases as far as the layer of deposits from the coolant is growing. Here we will see a quadratic dependency on the thickness of deposits. If the number of the jammed tubes has changed by 1%, SG resistance will also change by 1% approximately: $1 - n/N$, and if the thickness of inner deposits ΔR increases by 1%, the resistance will change by about 2%: $1 - 2\frac{\Delta R}{R}$.

So, pressure pulsations at the SG outlet will increase or decrease in the process of long-term operation $P_{out} \sim j\omega m_a$, which is a diagnostic indicator of a certain factor described above (Fig. 5.11).

The higher the frequency at which the factor reveals itself, the bigger the change observed in pressure pulsations at the SG outlet. The inductive character of SG resistance $j\omega m_a$ means that between the inlet and the outlet coolant pressure pulsations there will always be a 90° phase difference: $j = e^{j\frac{\pi}{2}}$.

A simplified representation of MCC using such electrical equivalents as capacitance and inductance makes it easier to analyze pressure pulsations in the MCC. However, transition to quantitative parameters $m_a = \frac{\rho L}{S}$, $C_a = \frac{V}{\rho C^2}$, and what is even more, to resonance properties of the MCC or its elements, requires a representation of the acoustic object in the form of a Helmholtz resonator.

5.4 Fresnel Equations, Nearly Optical Analogy for Point Acoustic Inhomogeneitics

Suppose a traveling wave is progressing in an elastic medium with density ρ. Let us call it an incident wave, U_{inc}. As it is known, for a traveling wave, an elastic medium is characterized by acoustic resistance $R_1 = P_{inc}/V_{inc}$, which is a real number. Suppose in coordinate $x = 0$ there is a jump of acoustic resistance of the medium, from R_1 to R_2. Such acoustic inhomogeneity causes two more waves: reflected (P_{ref}, V_{ref}) and transmitted (P_{tr}, V_{tr}). The reflected wave moves in inverse direction, and if an incident wave with a unit amplitude is $P_{inc} = e^{jkx}$, the expression for a reflected wave is $P_{ref} = \Gamma \cdot e^{-jkx}$. Obviously, the amplitude of an incident wave is not less than the amplitude of a reflected wave and, thus, $0 \ll |\Gamma| \ll 1$. It is also obvious that the transmitted wave will be the following: $P_{tr} = (1 - \Gamma) \cdot e^{jkx}$. In point $x = 0$, we get a balance:

$$P_{inc}(0) = 1 = P_{ref}(0) + P_{tr}(0) = \Gamma + (1 - \Gamma) \tag{5.9}$$

For the velocities of these three waves, the following ratios are true as an analogy of Ohm's law

$$\begin{cases} P_{inc} = R_1 \cdot V_{inc} \\ P_{ref} = -R_1 \cdot V_{ref} \\ P_{tr} = R_2 \cdot V_{tr} \end{cases}$$

from which we get

$$\frac{V_{\text{ref}}}{V_{\text{inc}}} = -\frac{P_{\text{ref}}}{P_{\text{inc}}} = \frac{R_1 - R_2}{R_1 + R_2} = \Gamma \tag{5.10}$$

$$\frac{V_{\text{tr}}}{V_{\text{inc}}} = \frac{R_1 \cdot P_{\text{ref}}}{R_2 \cdot P_{\text{inc}}} = \frac{2R_1}{R_1 + R_2} = 1 - \Gamma \tag{5.11}$$

Formulas 5.10 and 5.11, based on simplest assumptions, are called the Fresnel equations. These relations between the acoustic resistance of two media R_1, R_2 and the reflection coefficient Γ enable us to interpret the properties of acoustic point inhomogeneities. Going back to the expression for acoustic resistance $R = \rho \cdot C$, we have

$$\Gamma = \frac{\rho_1 \cdot C_1 - \rho_2 \cdot C_2}{\rho_1 \cdot C_1 + \rho_2 \cdot C_2} \tag{5.12}$$

If the second medium is much denser than the first one ($\rho_2 \gg \rho_1$) and the sound velocity in it is higher than in the first one ($C_2 \gg C_1$), then

$$\Gamma \approx -\frac{\rho_2 \cdot C_2}{\rho_2 \cdot C_2} = -1 \tag{5.13}$$

In this case, total reflection takes place, with a velocity phase shift by π. Such pairs may be "gas–metal", "water–metal". This case is also very close to the case of a closed-end of a pipeline (Fig. 5.13).

If, on the contrary, $\rho_1 \gg \rho_2$, $C_1 \gg C_2$, then $\Gamma = 1$. This is equivalent to an open end of a pipeline, also with total reflection, but with phase shift by π for pressure (Fig. 5.14).

The transfer from the narrow pipe to the wide one (Fig. 5.14) in its limit has a narrow pipe with an open end, i.e. with zero acoustic resistance.

This means that at the reflecting wall the velocity has zero amplitude, while pressure amplitude is at its maximum (Fig. 5.15).

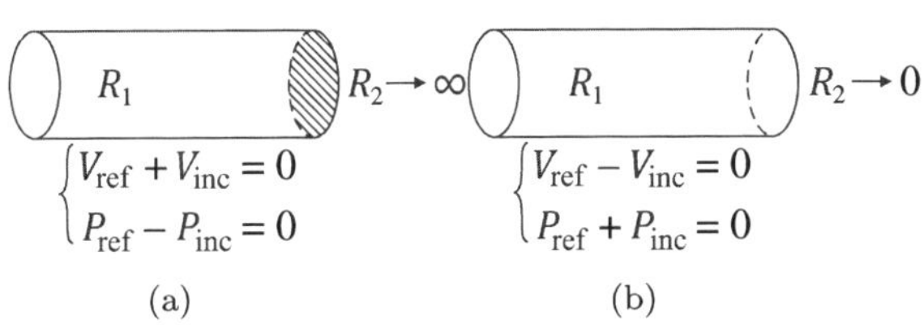

Figure 5.13: Closed end of a pipeline (a) and an open pipeline (b).

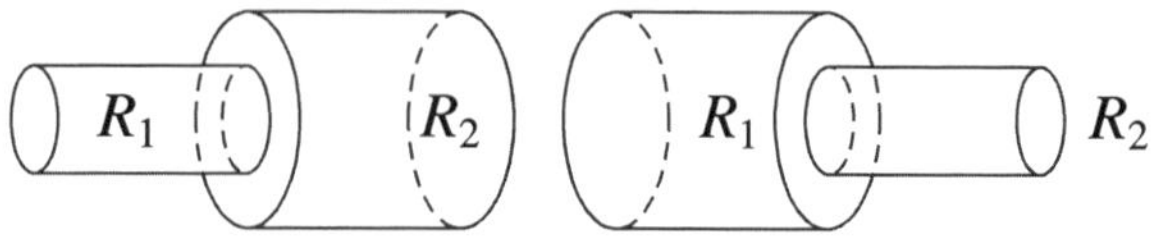

Figure 5.14: Jump of acoustic resistance of a pipeline in case of diameter change.

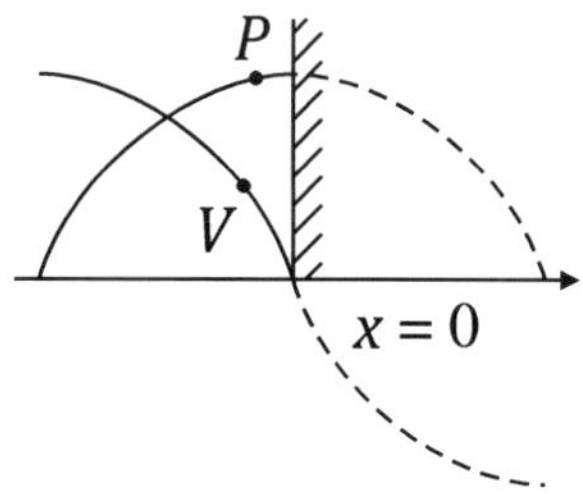

Figure 5.15: Velocity and pressure at an absolutely reflecting wall.

In an electrical analogy to acoustics, the following inversion takes place. Velocity in acoustics is equal to current in electrical engineering $V \leftrightarrow I$, and the current reaches its maximum in case of a short-circuit. Here, in acoustics, if there is a reflecting wall, the velocity is equal to zero. That is, short-circuit in electrical engineering corresponds to an open end of a pipeline in acoustics, which is quite unobvious. And vice versa: open circuit in electrical engineering corresponds to a reflecting wall in acoustics.

Formula 5.12 for an infinite medium did not take into account its specific spatial features, and if the medium is constant, i.e. if the equalities $\rho_1 = \rho_2$, $C_1 = C_2$ are true, it is necessary to obtain the jump of acoustic resistances from purely geometrical considerations. Thus, for two pipelines with different diameters filled with the same medium

$$\frac{R_1}{R_2} = \frac{\xi_1}{\xi_2} = \theta \tag{5.14}$$

(where ξ is the cross-section area of the corresponding pipeline), we get

$$\Gamma = \frac{1-\theta}{1+\theta}$$

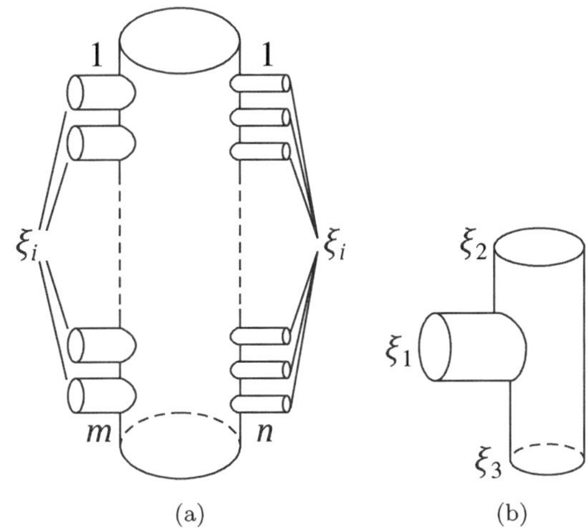

Figure 5.16: Pipeline branching: (a) "m" pcs. of inlet pipelines with cross-section area ξ_i, "n" pcs. of outlet pipelines with cross-section area ξ_j; (b) T-pipe.

From the same simple formula 5.14 an intuitive empirical rule follows which is used by designers of hydraulic circulation loops. In case of pipeline branching, acoustic reflections will be nearly non-existent, if the balancing of flow areas is fulfilled (Fig. 5.16)

$$\sum_{i=1}^{m} \xi_i = \sum_{j=1}^{n} \xi_j$$

In particular, the so called T-pipe where

$$\xi_1 = \xi_2 + \xi_3$$

will hardly form any standing waves. Heat exchange branching of pipelines with the property

$$\xi = \sum \xi_i$$

will also form no standing waves.

When Fresnel equations were derived, it was assumed that the acoustic resistances of the two lines — R_1 and R_2 — were real numbers. In the expression, "$-$" sign meant phase shift by π. If we go on with our generalization and assume that Γ is a complex number due to the fact that acoustic resistances of the lines are actually

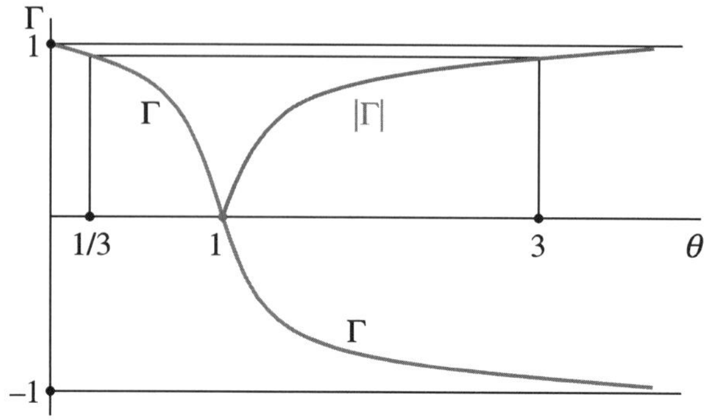

Figure 5.17: Dependency of "Γ" reflection coefficient on the parameter θ.

complex functions of space, we will get a generalization of the Fresnel equations, which is true at the point $x = 0$:

$$\Gamma(x) = \frac{z_1(x) - z_2(x)}{z_1(x) + z_2(x)}$$

Another important property of the formula for reflection coefficient $\Gamma(x)$ is the so-called "reciprocity by modulus". When transferring from a pipeline with a larger cross-section to the one with smaller cross-section, the parameter $\theta > 1$, and the reflection coefficient $\Gamma > 0$. Otherwise, $\theta < 1$, and Γ coefficient is negative (but still not exceeding one modulo). The sign before Γ coefficient characterizes the co-phase or counter-phase reflection from an inhomogeneity. One reflection happens with no change of phase (co-phase), and another, the phase jump (phase shift by "π": $\varphi = e^{-j\pi} = -1$) — is counter-phase. The property of "reciprocity by modulus" means that when a transfer from cross-section $S1$ to cross-section $S2$ takes place, the modulus of the reflection coefficient remains unchanged when a backward transfer happens, from $S2$ to $S1$ (Fig. 5.17). Indeed,

$$|\Gamma| = \left|\frac{1-\theta}{1+\theta}\right| = \left|\frac{1-\frac{1}{\theta}}{1+\frac{1}{\theta}}\right|$$

Chapter 6

Wave Equation

6.1 Features of Wave Equation

This chapter presents wave equation as the basis for the theory of elastic media acoustics [21–67]. It puts emphasis on its solutions in specific applications, depending on the initial and boundary conditions, as well as on the type of external driving force. It should be emphasized that in any case there is a standing wave in the solution of the wave equation.

The wave equation (d'Alembert equation) for a function of two variables — $U(x,t)$ has the following form:

$$\frac{\partial^2 U(x,t)}{\partial t^2} = C^2 \cdot \frac{\partial^2 U(x,t)}{\partial x^2} + f(x,t) \tag{6.1}$$

Equation (6.1) is written in the most elementary form for a one-dimensional spatial coordinate and time function — $U(x,t)$. For multidimensional and, in particular, three-dimensional space (x, y, z), the equation is written using the Laplace operator — $\Delta U(x,y,z,t)$:

$$\frac{\partial^2 U(x,y,z,t)}{\partial t^2} = C^2 \cdot \Delta U(x,y,z,t) + f(x,y,z,t) \tag{6.2}$$

where $\Delta = \frac{\partial^2}{\partial x^2} + \frac{\partial^2}{\partial y^2} + \frac{\partial^2}{\partial z^2}$ — Laplace operator, $f(x,y,z,t)$ is external driving force, C is the velocity of the longitudinal wave propagation in the medium.

As the desired function of coordinates and time — $U(x, y, z, t)$ — different physical fields can act:

$P(x, y, z, t)$ — pressure,
$V(x, y, z, t)$ — velocity,
$\rho(x, y, z, t)$ — density,
$U(x, y, z, t)$ — displacement, etc.

A homogeneous wave equation often is written using the d'Alembert operator $\square$:

$$\square = \Delta - \frac{1}{C^2} \cdot \frac{\partial^2}{\partial t^2} \tag{6.3}$$

Further, both the operator form and the traditional notations of derivatives through indices and strokes and, if possible, in one-dimensional form by coordinates will be applied. For examples, the following formulas

$$\frac{\partial^2 U(x,t)}{\partial t^2} = C^2 \cdot \frac{\partial^2 U(x,t)}{\partial x^2}$$

$$\square U(x,t) = 0$$

$$\frac{\partial^2 U(x,t)}{\partial t^2} = C^2 \cdot \Delta U(x,t)$$

$$U_{tt}(x,t) = C^2 \cdot U_{xx}(x,t)$$

$$U_t''(x,t) = C^2 \cdot U_x''(x,t)$$

are the same homogeneous wave equation.

The important special cases of the wave equation are stationary degenerations:

$\Delta U(x,t) = f(x,t)$ — Poisson's equation and its special case,
$\Delta U(x,t) = 0$ — Laplace equation (homogeneous Poisson's equation).

In these equations, differentiation is performed by a spatial coordinate, and the solution is also obtained as a function of time. The wave equation can be stationary due to an incompressible medium when the velocity of the disturbance propagation $C \to \infty$ and the "non-stationary" component of Eq. (6.3) tend to zero.

Let's consider one particular solution of the homogeneous wave equation (6.1). Let the two-dimensional field of time and coordinates $U(x,t)$ have a sinusoidal shape

$$U(x,t) = A \cdot \sin(\omega t + k \cdot x), \tag{6.4}$$

where $k = \frac{\omega}{C}$ is a wave number.

The parameters ω and k reduce dimensional units of time and space to the dimensionless argument of the sinusoidal function through the velocity or speed of sound C.

Let's differentiate it twice with respect to time and twice to the coordinate:

$$\frac{\partial^2 U}{\partial t^2} = -A \cdot \omega^2 \cdot \sin(\omega t + kx) = -\omega^2 \cdot U \tag{6.5}$$

$$\frac{\partial^2 U}{\partial x^2} = -A \cdot k^2 \cdot \sin(\omega t + kx) = -k^2 \cdot U \tag{6.6}$$

After normalization and addition of Eqs. (6.5) and (6.6), we obtain a wave equation

$$\frac{\partial^2 U(x,t)}{\partial t^2} = C^2 \cdot \frac{\partial^2 U(x,t)}{\partial x^2} \tag{6.7}$$

This was not a derivation of the wave equation, but only a formal technique that establishes differential relations of the coordinates of a sinusoidal two-dimensional field $U(x,t)$. Here is the reciprocal sequence: from a specific explicit form of the particular solution (6.4) — through the equations of motion in the form of ordinary differential equations (6.5) and (6.6) — to the standard form of the wave partial differential equation (6.7). Thus, it is established that if one argument of the sinusoidal function contains two variables — space and time — then separately by the time variable t the motion is subject to the harmonic spatial oscillator equation with a spatial frequency ω and by the spatial variable x is subject to one with a spatial frequency Λ. It is replaced by the inverse value, namely, the wave number $k = \frac{\omega}{C} = \frac{1}{\Lambda}$. Two frequency parameters of oscillations ω and k are connected by the velocity (speed) of sound C. Only the velocity of sound as a single parameter is present in the original wave equation (6.1).

Let's generalize the solution of Eqs. (6.5) and (6.6):

$$\frac{d^2U}{dt^2} + \omega_0^2 \cdot U = 0 \tag{6.8}$$

$$\frac{d^2U}{dx^2} + k_0^2 \cdot U = 0 \tag{6.9}$$

By a comprehensive solution

$$U(x,t) = A \cdot e^{j(\omega_0 t + k_0 x)} \tag{6.10}$$

Any of those equations are called harmonic oscillator equations. They are also called motion equations or Lagrange equations (sometimes Euler–Lagrange equations, sometimes Euler's equations), which, in particular, describe the movement of the spring pendulum $U(x,t)$ with a natural frequency $\omega_0 = \sqrt{\frac{K}{m}}$, where K is the spring rigidity and m is the weight of the pendulum load. The function (6.10) can be considered both a time and a spatial oscillator. When $k_0 = \frac{\omega_0}{C}$ and the equation of motion has the form

$$\frac{d^2U}{dx^2} + \left(\frac{\omega_0}{C}\right)^2 \cdot U = 0$$

then $\frac{\omega_0}{C}$ has a physical meaning of the natural spatial frequency (with a wavelength $\Lambda = \frac{C}{\omega}$). When

$$\frac{d^2U}{dt^2} + \omega_0^2 \cdot U = 0$$

then ω_0 is the time natural frequency.

The wave equation implicitly contains the equations of the spatial and time oscillator. After the elimination from the two equations of the terms with the function $U(x,t)$, we again obtain a wave equation

$$U_{tt}(x,t) - C^2 \cdot U_{xx}(x,t) = 0$$

where there is only one parameter — C — the velocity of the disturbance propagation (the speed of sound). Further, based on the speed of sound in some spatial constraints, we can obtain the spatial wavelength, and in some initial time constraints, the natural frequency of the oscillations is obtained.

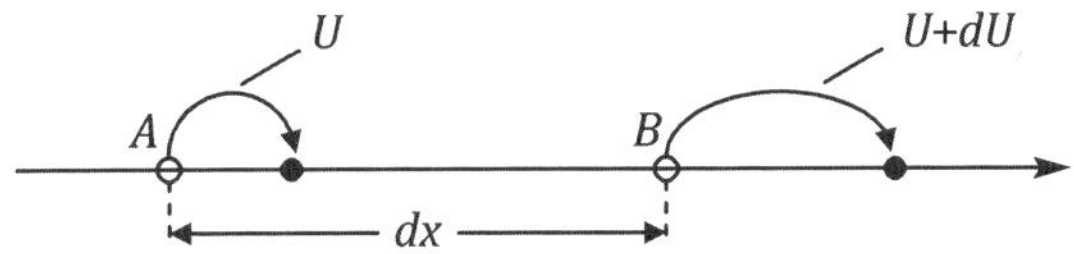

Figure 6.1: Displacements of two points of the elastic medium under the effect of a longitudinal wave.

In physics, the wave equation is obtained from the fundamental laws of mechanics (Newton's laws of motion), electrical engineering (Kirchhoff's laws), the field theory (Maxwell's equations), hydrodynamics (continuity and motion equations). Different types of conservation laws are also applied depending on the specific field $U(x,t)$.

The above formal conclusion of the wave equation can be given a mechanical sense. To do this, let us consider two points of the elastic medium "A" and "B" (Fig. 6.1).

Let the distance between them be dx, and under the influence of a longitudinal wave, the first point "A" moves to the distance U, and the second — "B" — to the distance $U + dU$. The elastic medium was thus deformed (compressed or stretched). This deformation can be described by a coefficient $d = \frac{\partial U}{\partial x}$.

If the longitudinal wave is sinusoidal

$$U = A \cdot \sin(\omega t + kx)$$

and its two variables, namely, time and coordinate are "involved" in one argument, then

$$\frac{dU}{dt} = A \cdot \omega \cdot \cos(\omega t + kx) \tag{6.11}$$

$$\frac{dU}{dx} = A \cdot k \cdot \cos(\omega t + kx) \tag{6.12}$$

and

$$C \cdot \frac{dU}{dx} = \frac{dU}{dt} = d \cdot C \tag{6.13}$$

Physically, this means that the greater is the deformation of the medium $\partial U/\partial x$, the higher is the speed of its oscillations $\partial U/\partial t$. For harmonic oscillations, the deformation is maximum in the equilibrium position and minimal in the position of changing the direction of the oscillations (when the maximum displacement is reached). If we divide Eq. (6.11) by Eq. (6.12), we obtain a differential form of space–time relation

$$\frac{dx}{dt} = \frac{\omega}{k} = C$$

After its integration with the zero initial condition, we obtain the usual path-to-speed relationship in case of uniform motion: $x = t \cdot C$. It is important to note that the path x is overcome by an elastic wave with the speed of sound "C", and not with the speed of the medium movement in case of its transportation along the MCC.

By differentiating the expressions (6.11) and (6.12) once again

$$\frac{\partial^2 U}{\partial t^2} = -A \cdot \omega^2 \cdot \sin(\omega t + kx) \tag{6.14}$$

$$\frac{\partial^2 U}{\partial x^2} = -A \cdot k^2 \cdot \sin(\omega t + kx) \tag{6.15}$$

we come again to the wave equation

$$\frac{\partial^2 U(x,t)}{\partial t^2} = C^2 \cdot \frac{\partial^2 U(x,t)}{\partial x^2}$$

The sinusoidal field is connected not only with the second-order wave equation, but also with partial differential equations of the first order:

$$\frac{\partial U}{\partial x} = A \cdot k \cdot \cos(\omega t + kx)$$

$$\frac{\partial U}{\partial t} = A \cdot \omega \cdot \cos(\omega t + kx)$$

where

$$\frac{1}{C} \cdot \frac{\partial U}{\partial t} = \frac{\partial U}{\partial x}$$

That is, the sinusoidal field has interrelated coordinates both on their velocity change (by $\frac{\partial}{\partial t}\frac{\partial}{\partial x}$) and on their acceleration change (by $\frac{\partial^2}{\partial t^2}$, $\frac{\partial^2}{\partial x^2}$). The wave equation, as the relationship between accelerations, is derived from Newton's second law and/or from the law of conservation of energy, operating via quadratic functions:

$$U^2 \quad \text{and} \quad \left(\frac{\partial U}{\partial x}\right)^2, \left(\frac{\partial U}{\partial t}\right)^2$$

If the solution of the first-order equation

$$\frac{1}{C} \cdot \frac{\partial U}{\partial t} = \frac{\partial U}{\partial x} \tag{6.16}$$

by its argument defines a traveling wave in one direction $U(\omega t + kx)$, for example,

$$U(x,t) = A \cdot \sin(\omega t + kx),$$

then when changing the direction of the wave — $U(\omega t - kx)$, for example,

$$U(x,t) = A \cdot \sin(\omega t - kx),$$

the sign of the left part of the differential equation also changes:

$$-\frac{1}{C} \cdot \frac{\partial U}{\partial t} = \frac{\partial U}{\partial x}$$

Second-order wave equation

$$\frac{1}{C^2} \cdot \frac{\partial^2 U}{\partial t^2} = \frac{\partial^2 U}{\partial x^2} \tag{6.17}$$

is invariant to the direction of motion. Regardless of the sign, its solution will be both the direct wave $U(\omega t + kx)$ and the inverse one $U(\omega t - kx)$.

By directly substituting into a homogeneous wave equation (6.1), one can verify that its particular solution is written as

$$U(x,t) = \frac{1}{2}U_0(kx + \omega t) + \frac{1}{2}U_0(kx - \omega t) \tag{6.18}$$

where, still, $k = \frac{\omega}{C}$ and U_0 is an arbitrary, not necessarily sinusoidal function.

From the solution of (6.18) it follows that the form of initial deflection is distributed in an unchanged form in the opposite directions. If we simplify the task, assuming that the initial deviation has a sinusoidal shape

$$U(x,t) = \frac{1}{2}\sin(kx + \omega t) + \frac{1}{2}\sin(kx - \omega t)$$

then we obtain that the particular solution is a product of two sinusoids

$$U(x,t) = \sin kx \cdot \cos \omega t \tag{6.19}$$

For an infinite elastic medium, it does not matter in what form the sinusoidal shape of wave is given. For example, it can be specified by a cosine, in the form

$$U(x,t) = \frac{1}{2}\cos(kx + \omega t) + \frac{1}{2}\cos(kx - \omega t),$$

and the final solution will be given as the product of sinuses:

$$U(x,t) = \sin kx \cdot \sin \omega t \tag{6.20}$$

Expressions (6.19) and (6.20) are the most primitive ASW model revealing the essence of the ASW phenomenon. No boundary conditions were laid into the original model, that is, the infinite medium is only perturbed by the initial stationary effect. No reflections in a homogeneous infinite medium can arise and, nevertheless, standing waves are formed as a result of addition of two waves running in opposite directions.

The wave equation refers to the hyperbolic type of equations of mathematical physics and can describe vibrations of a string, rod, membrane, vibrations of continuous media (acoustics of volumes filled with liquid or gases), etc. To get the solution of the equation, it is necessary to set the boundary and initial conditions. The boundary conditions determine functions of time at given spatial coordinates. The initial conditions are coordinate functions at a zero point in time. Vibrations of the medium will also occur in the absence of external influence, only due to non-zero initial conditions.

6.2 Particular Solutions of the Harmonic Oscillator Equation

As shown in the previous section, there is a genetic relationship between the wave partial differential equation and ordinary differential equations of motion (harmonic oscillators). Let us consider their particular solutions anticipatorily.

Substituting in the form of complex solution

$$U(x,t) = U(x) \cdot e^{j\omega t}$$

leads the wave equation

$$U_{xx}(x,t) = \frac{1}{C^2} \cdot U_{tt}(x,t)$$

to the ordinary differential equation of motion

$$\frac{d^2U(x)}{\partial x^2} + \left(\frac{\omega}{C}\right)^2 \cdot U(x) = 0$$

which is called the harmonic oscillator equation. Its non-trivial solution $U(x) \not\equiv 0$ is called a harmonic solution.

By directly substituting into Eq. (6.1), it is easy to verify that there are functions among particular solutions:

$$U_1(x) = \sin\frac{\omega}{C}x = \sin kx$$

$$U_2(x) = \cos kx$$

It is obvious that their linear combination

$$U_3(x) = U_1(x) + U_2(x) = \sin kx + \cos kx = A \cdot \sin(kx + \varphi)$$

is also a solution to (6.1). Solutions to (6.1) remain such if they are multiplied by an arbitrary constant (amplitude) and, in particular, by a complex unit. As a result, we will obtain a comprehensive generalization of the solution

$$U_4(x) = \cos kx + j \cdot \sin kx = e^{jkx}$$

As well as

$$U_5(x) = e^{-jkx}$$

Their linear combinations

$$U_6(x) = \frac{1}{2}(U_4(x) + U_5(x)) = \cos kx$$

$$U_7(x) = \frac{1}{2j}(e^{jkx} - e^{-jkx}) = \sin kx$$

When the number k is a pure imaginary $j \cdot k$, we obtain

$$U_8(x) = \sin(jkx) = -\frac{1}{2j}(e^{kx} - e^{-kx}) = j \cdot shkx$$

As well as

$$U_9(x) = e^{kx}$$

$$U_{10}(x) = e^{-kx}$$

$$U_{11}(x) = \frac{1}{2}(e^{kx} + e^{-kx}) = chkx, \quad U_{12}(x) = shkx$$

$$U_{13}(x) = A \cdot e^{kx} + B \cdot e^{-kx}$$

Thus, particular solutions of the harmonic oscillator equation can be written as trigonometric, exponential and hyperbolic functions, as well as their complex generalizations. At the qualitative level, solutions of the elementary equation of the harmonic oscillator represent a wide class of functions, ranging from limited in amplitude periodic functions to exponentially decreasing and exponentially growing in amplitude functions. A natural frequency ω_0 can function as a parameter forming a spatial wave length in the solution $\sin \frac{\omega_0}{C} x$, and as a constant of spatial attenuation/growth in the solutions $e^{-\frac{\omega_0}{C} x}$ and $e^{+\frac{\omega_0}{C} x}$.

The harmonic oscillator represented by a second-order differential equation has a unique solution when two initial conditions are set: a condition for displacement at certain time (without limiting the generality — at the zero time moment) $x(0) = x_0$, and for velocity $v(0) = v_0$. A complex solution, that incorporates all the solutions of the harmonic oscillator discussed above, is:

$$x(t) = C_1 \cdot e^{-j\omega_0 t} + C_2 \cdot e^{j\omega_0 t}$$

Considering the provided initial conditions, it can be written as the sum of two sinusoids

$$x(t) = x_0 \cdot \cos \omega_0 t + \frac{v_0}{\omega_0} \cdot \sin \omega_0 t$$

or as a single one

$$x(t) = \sqrt{x_0^2 + \left(\frac{v_0}{\omega_0}\right)^2} \cdot \sin \left[\omega_0 t + \operatorname{arctg}\left(\frac{x_0 \cdot \omega_0}{v_0}\right)\right]$$

or as sum of two complex exponents

$$x(t) = \left(\frac{x_0}{2} - \frac{v_0}{j2\omega_0}\right) \cdot e^{-j\omega_0 t} + \left(\frac{x_0}{2} + \frac{v_0}{j2\omega_0}\right) \cdot e^{j\omega_0 t}$$

The last notation can be interpreted as the interaction of two harmonic processes — with the "direct" "$+t$" and "reverse" "$-t$" passage of time as two-multidirectional time traveling waves. Two complex harmonic processes can also be interpreted in the language of frequency as two sinusoids: one with positive frequency "$+\omega_0$", and the other — with negative one "$-\omega_0$". Neither negative time, nor negative frequency are natural phenomena, so they are ignored in the interpretation of experiments and complex representation is the mathematical abstraction.

6.3 Types of Complex Harmonic Solutions of Wave Equations

The transition from the partial differential equation to the ordinary differential equation, or even to the algebraic equation, as a rule, is made by means of a complex harmonic substitution or representation of the particular solution of the equation in the form of a complex harmonic function. In addition to the fact that the solution is only found in the class of periodic functions, it is usually assumed that the spatial and temporal dependences of the solution are separated, for example, in the form of a product of two functions of one variable. In the following, we will use several types of harmonic complex substitutions.

The simplest case is when the solution is represented by two complex exponents separately in time and space:

$$U_1(x,t) = U_0 e^{j(\omega t - kx)} \tag{6.21}$$

After substituting (6.21) into the homogeneous wave equation

$$U_{tt} - C^2 \cdot U_{xx} = 0$$

we have

$$-\omega^2 \cdot U_1 + C^2 \cdot k^2 \cdot U_1 = 0$$

or

$$k = \pm\frac{\omega}{C} \tag{6.22}$$

Such a simple representation of the solution (6.22) is valid for any amplitude U_0 and after the connection between the spatial parameter k and the time parameter ω is determined, the homogeneous wave equation does not serve any additional information. Substitution (6.21) makes sense when finding the ratio of the disturbance amplitudes and the response of the elastic medium for non-homogeneous wave equations.

Another type of complex solution is the periodic time component with an arbitrary spatial factor:

$$U_2(x,t) = U(x) \cdot e^{j\omega t} \tag{6.23}$$

Substituting it into the homogeneous wave equation (6.1) will give an ordinary differential equation

$$-\omega^2 \cdot U(x) - C^2 \cdot U_{xx}(x) = 0$$

or

$$\frac{d^2U(x)}{dx^2} + k^2 \cdot U(x) = 0, \tag{6.24}$$

which is a harmonic oscillator with a spatial natural parameter k. Its solution, as shown in Section 1, can be a sinusoid, hyperbolic function, or complex exponent, which extends the class of simple harmonic solutions. The solution of (6.23) is a traveling wave that carries in time the spatial non-harmonic (generally speaking) amplitude distribution $U(x)$.

Similarly, it is easy to write a complex solution of a periodic spatial component with an arbitrary time factor

$$U_3(x,t) = U(t) \cdot e^{jkx} \tag{6.25}$$

After substituting it into the homogeneous wave equation, we obtain the equation of the harmonic oscillator with the time natural parameter ω (the natural frequency)

$$U_{tt} - C^2 \cdot (-k^2) \cdot U = 0$$

or

$$\frac{d^2U(t)}{dt^2} + \omega^2 \cdot U(t) = 0 \tag{6.26}$$

We obtain, in particular, the equation of motion of the spring pendulum where $U(t)$ is its displacement amplitude, and the natural frequency is determined by the mass and stiffness.

6.4 Fundamental Properties of Complex Harmonic Wave Representation

Only, from the analytical form of the harmonic complex function

$$U(x,t) = U_0 \cdot e^{j\varphi} \cdot e^{j(\omega t - kx)} \tag{6.27}$$

where $U_0 \cdot e^{j\varphi}$ is the complex amplitude, one can get the basic parameters of the wave in an elastic medium. Let's introduce the phase of the complex exponent as a linear function of two variables: time t and space x:

$$\theta(x,t) = \omega t - kx + \varphi \tag{6.28}$$

Two variables t, x are reduced to a dimensionless form by the corresponding coefficients ω and k. Frequency ω and wavelength $\Lambda = \frac{1}{k}$ are two fundamental dimensions of a wave propagating in an elastic medium.

Let's differentiate the phase (6.28) with time and space

$$\frac{\partial\theta(x,t)}{\partial t} = \omega, \tag{6.29}$$

$$\frac{\partial\theta(x,t)}{\partial x} = -k \tag{6.30}$$

as a result, we obtain that the parameters ω and k correspond to the rate of change of a phase in time and space. It follows from the partial derivatives (6.29) and (6.30) that

$$\omega \cdot dt + k \cdot dx = 0$$

or

$$\left|\frac{dx}{dt}\right| = \left|-\frac{\omega}{k}\right| = C = V_f \tag{6.31}$$

This expression is the definition of the phase velocity of the wave V_f propagating in an elastic medium. The phase velocity completely determines the velocity of any point of the wave, if it does not change its shape. Determination of the phase velocity (6.31) is only obtained from the analytical form of the linear phase (6.28). If we differentiate the function of the wave itself (6.27),

$$\begin{cases} \dfrac{\partial U(x,t)}{\partial t} = j\omega \cdot U(x,t) \\ \dfrac{\partial U(x,t)}{\partial x} = -jk \cdot U(x,t) \end{cases}$$

we obtain a first-order equation for a traveling wave

$$\frac{\partial U(x,t)}{\partial t} + C \cdot \frac{\partial U(x,t)}{\partial x} = 0 \tag{6.32}$$

where the phase velocity "C" acts as a coefficient in the exponent, which determines the vibrational properties of the solution $U(x,t)$:

$$e^{-j\frac{\omega}{C}(x-C\cdot t)} \tag{6.33}$$

Based on the complex representation (6.27) and (6.28), it is possible to introduce the dispersion relation. Dispersion defined as the dependence of the phase velocity on the frequency,

$$V_f = V_f(\omega),$$

is not quite a correct definition. When the dependence $V_f = \frac{\omega}{C}$ is linear, it corresponds to the absence of distortion of the signal form, that is, the absence of dispersion effect. To distort the signal, the phase velocity must be a nonlinear dependence. It would be more natural if dispersion is defined as a nonlinear dependence of frequency on wavelength

$$\omega = \omega(k) \tag{6.34}$$

or as an inverse function

$$k = k(\omega) \tag{6.35}$$

or as an implicit function

$$D(\omega, k) = 0 \tag{6.36}$$

In these cases, the coordinates of space and time become dependent on each other. Now, if the spatial function "runs" along the time axis, and the dependences (6.34)–(6.36) are nonlinear, it automatically means that the wave changes its shape.

Also, from the complex representation (6.27) and (6.28), the group wave velocity is introduced, V_{gr}, as a derivative of the dispersion relation (6.34):

$$V_{\text{gr}} = \frac{d\omega(k)}{dk} \tag{6.37}$$

Bearing in mind that the phase velocity is defined as $V_f = \frac{\omega}{k}$, we obtain the connection between the group and phase velocity

$$V_{\text{gr}} = \frac{d}{dk}[k \cdot V_f] = V_f + k \cdot \frac{dV_f}{dk} \tag{6.38}$$

that is called Rayleigh formula. It can be written using logarithmic derivatives

$$V_{\text{gr}} = V_f \left\{ 1 + \frac{[\ln V_f]'}{[\ln k]'} \right\} \tag{6.39}$$

The graph of dispersion relation (Fig. 6.2) shows the group and phase velocity through the corresponding tangents of the angles of inclination.

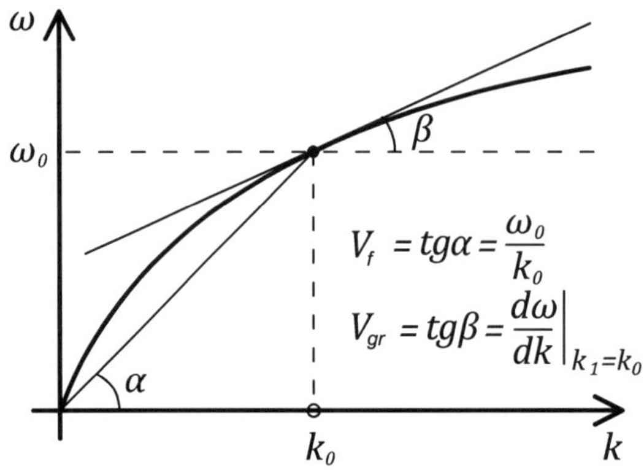

Figure 6.2: Dispersion curve and determination of phase and group velocity.

The phase velocity coincides with the group velocity only when the dispersion relation is linear (which corresponds to the absence of dispersion). It follows from the definition of phase and group velocity (6.31) and (6.37) that they are interconnected by a differential expression

$$\frac{V_{\text{gr}}}{V_f} = \frac{\omega \cdot dk}{k \cdot d\omega}$$

or after integration

$$\omega = \text{const} \cdot (k)^{\frac{V_{\text{gr}}}{V_f}} \tag{6.40}$$

The curves (6.40) with the condition $V_f > V_{\text{gr}}$ are referred to as normal dispersion relations, and the curves with the condition $V_f < V_{\text{gr}}$ are referred to as abnormal dispersion relations (Fig. 6.3). The concepts of phase and group velocity are introduced, thus only the power dependence of the frequency on the wave number is determined (6.40). It is clear that the dependences (6.40) are, although a wide class of functions, not the only kind of dispersion relations observed in real elastic and viscous media. The two velocities introduced, V_f, V_{gr}, are generally insufficient to represent the propagation of an arbitrary wave in an arbitrary elastic medium. There should be infinite number of velocities.

The simplification of the description of the wave propagation by means of two velocities — V_f and V_{gr} — can be illustrated as the

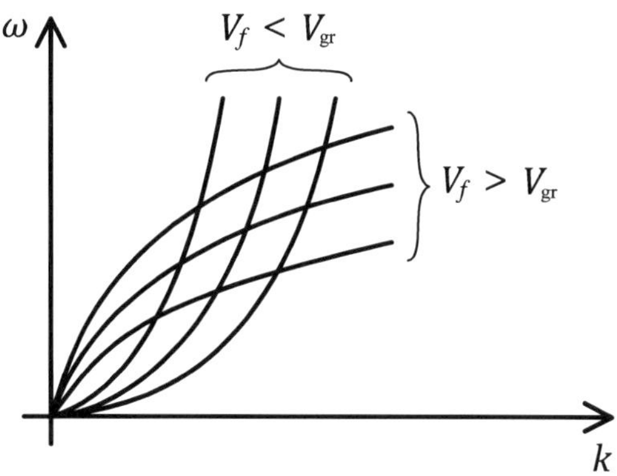

Figure 6.3: Normal and abnormal dispersion functions.

movement of the envelope with the group velocity and the movement of the carrier inside the envelope with the phase velocity.

Approximately, we can assume that for the envelope $A(x, t)$ the same equation of traveling wave (6.32) is true if the phase velocity is replaced by the group velocity

$$\frac{\partial A(x,t)}{\partial t} + V_{\text{gr}} \cdot \frac{\partial A(x,t)}{\partial x} = 0 \tag{6.41}$$

The properties of the group velocity are demonstrated by the example of beating between two frequencies. The most elementary representation of beats is through the difference of two sinusoids, each of which has similar parameters $(\omega_0 - \frac{\Delta\omega}{2}, k_1)$ and $(\omega_0 - \frac{\Delta\omega}{2}, k_2)$ and, respectively, similar phase velocities V_{f1} and V_{f2}. Beats are represented by the following formulas:

$$U(x,t) = \cos\left(\frac{\Delta\omega}{2} \cdot t - \frac{\Delta k}{2} \cdot x\right) \cdot \cos\left(\omega_0 \cdot t - \frac{k_1 + k_2}{2} \cdot x\right)$$

The derivative by which the group velocity is determined is approximated by finite differences:

$$V_{\text{gr}} = \frac{d\omega}{dk} \approx \frac{\omega_1 - \omega_2}{k_1 - k_2}$$

Then for the amplitude of the slowly varying function in the limit, we obtain

$$\frac{\Delta\omega}{\Delta k} \approx V_{\text{gr}},$$

and for a fast oscillating function in the limit, we obtain:

$$\frac{\omega_0}{k_0} \approx V_f$$

Thus, each of the two sinusoids has its own velocity of movement: one envelope moves approximately with the group velocity V_{gr}, and the sine wave of the carrier frequency moves with the phase velocity V_f.

For any real dispersion relations, the wave is not damped by amplitude, since

$$|U(x,t)| = U_0 \cdot |e^{j\varphi} \cdot e^{[\omega(k) \cdot t - kx]}| = U_0 \tag{6.42}$$

If the dispersion relation is complex: $k(\omega) = k_1(\omega) + jk_2(\omega)$, then there is damping determined by the imaginary part of the expression — $k_2(\omega)$:

$$|U(x,t)| = |e^{-k_2(\omega) \cdot x}| \cdot U_0$$

That is, we obtain a spatial damping factor $k_2(\omega)$. If the dispersion relation has the form of an inverse function $\omega(k) = \omega_1(k) + j\omega_2(k)$, then

$$|U(x,t)| = |e^{-\omega_2(k) \cdot t}| \cdot U_0$$

and we obtain the damping factor $\omega_2(k)$.

If a wave stores energy E, then the passage of a viscous medium with decrement ω_2 absorbs the following power:

$$\frac{dE}{dt} = -2 \cdot \omega_2 \cdot E$$

or

$$E \sim e^{-2\omega_2 \cdot t}$$

By how quickly the wave loses its energy we can judge from the absorbing properties of the elastic medium, for example, by the viscosity of the medium.

6.5 Connection of Helmholtz Equation and d'Alembert Equation with Lagrange Ordinary Differential Equation of Motion — Standing Waves in the Solution of a Homogeneous Wave Equation

A homogeneous (without external disturbing force) wave equation is able to describe acoustic standing waves arising from some initial stationary action $U_0(x)$. The solution is a complex harmonic time function $e^{j\omega t}$ with spatially dependent amplitude $U_0(x)$:

$$U(x,t) = U_0(x) \cdot e^{j\omega t}$$

After substitution in the wave homogeneous equation (after double integration with respect to x and to t), we have

$$e^{j\omega t} \cdot \frac{\partial^2 U_0}{\partial x^2} + \frac{1}{C^2} \cdot e^{j\omega t} \cdot U_0 \cdot \omega^2 = 0$$

or

$$\frac{\partial^2 U_0(x,t)}{\partial x^2} + \frac{\omega^2}{C^2} \cdot U_0(x,t) = 0 \tag{6.43}$$

This equation, called the Helmholtz equation, also applies to the class of stationary equations, since the differentiation is only carried out by one spatial coordinate. The Helmholtz equation is the degeneracy of the partial differential equation into an ordinary differential equation, which means, it is a special case of the wave equation, which is valid for the harmonic function. If we do not take into account the time variable, the Helmholtz partial differential equation coincides in shape with the ordinary differential equation of Lagrange motion (or the equation of the spatial harmonic oscillator):

$$\frac{d^2 U(x)}{dx^2} + k^2 \cdot U(x) = 0$$

The Lagrange equation of motion describes the trajectory $U(t)$ in time, and Eq. (6.43) does it in space. However, the analogy of the Helmholtz equation and the Lagrange equation is useful to apply for the so-called "stopped" time, namely, the spatial distribution of the medium disturbance at any time $U(x,t_0)$ is similar to the displacement of an equivalent pendulum in time. The analogy of the natural frequency of oscillations of the spring pendulum with the spatial parameter $\frac{\omega}{C}\left[\frac{1}{\mathrm{m}}\right]$ of the Helmholtz equation is also useful. For example, in neutron physics it is called material constitutive parameter, m_0, associated with the spatial harmonics of neutron waves in a nuclear reactor. The stationary neutron field of the reactor — $F(x)$ is also described by the Helmholtz equation

$$\frac{d^2F(x)}{dx^2} + m_0^2 \cdot F(x) = 0$$

The homogeneous Lagrange equation, or as it is also called the harmonic oscillator equation, has a sinusoidal solution. For example, half of the sine period, as a half wave of an acoustic standing wave from one reflecting wall to another, is the spatial distribution of the amplitude of the acoustic standing wave. A harmonic factor $e^{j\omega t}$, also sinusoidal, is a change in the amplitude of the acoustic standing wave in time.

Now, for each fixed frequency, the distribution of the standing wave amplitudes in space is a harmonic sinusoidal function. We shall show that when the boundary conditions are imposed, the wave equation is satisfied not by the continuum of frequencies, but by the countable series of discrete harmonic frequencies.

In both analogs with the "stopped" time and fixed frequency of the Lagrange equation, the damping $(C \cdot dU/dx)$, which describes the friction losses (for the absorption of the oscillation energy), is missing, thus, in the solution there is no decreasing multiplicant in the envelope amplitude $U(x,t)$. The wave equation in the form (6.1) does not take into account the losses due to the viscosity of the medium, the friction on the pipeline wall, the absorption during reflection from the walls and the radiation of acoustic power into the open space.

To obtain the solution of a second-order equation, it is necessary to set initial conditions in the form of two functions of the coordinates:

- $U(0,x) = U_0(x)$ is the value of desired function at zero point in time;
- $U_t(0,x) = U_1(x)$ is the derivative of the desired function at the same zero time.

The easiest way to illustrate the initial conditions is by the example of string vibrations, the movement of its points $U(x,t)$ also follows wave equation. For the string, its static deviation from the initial equilibrium state $U_0(x)$ is the first initial condition, and the second initial condition $U_1(x)$ is the velocity that was given to the deviated string (the string was not only deviated, but also strummed). If no other effects on the string were made, then for $t > 0$ its undamped oscillations will be observed because the losses of the oscillation energy are not included in the wave equation.

6.6 Standing Wave in the Solution of Homogeneous Wave Equation with a Single Initial Equation, d'Alembert Method

Let's consider the simplest case of excitation of the string vibrations, when it was unhandled, without giving the initial velocity:

$$U(0,x) = U_0(x)$$
$$U_t(0,x) = U_1(x) \equiv 0 \tag{6.44}$$

Only one primary deviation gives two identical in form traveling waves, which added up gave the standing wave. The following expression is taken as a definition of a standing wave $U_{\text{st}}(x,t)$:

$$U_{\text{st}}(x,t) = \frac{1}{2}U_0(x + C \cdot t) + \frac{1}{2}U_0(x - C \cdot t) \tag{6.45}$$

(sometimes the 1/2 coefficient is omitted for easy recording, which does not change the essence of this definition).

The string vibrations are the most significant example of the solution of the wave equation, as these oscillations are visualized. The half-wave of the oscillation amplitude of the first harmonic with two nodes is visualized: one node at one fixed end and another node at the other fixed end. The maximum amplitude of oscillations of the first harmonic is visualized, which is observed exactly in the middle of the string, regardless of where the initial impact was made. Thus, the spatial half-period of oscillations is determined exclusively by the distance between the nodes of the first harmonic or the distance between the two points of fixing the string. The same, but only invisible fluctuations in the medium pressure will occur in the pipeline, closed at both ends.

In case of sinusoidal spatial perturbation, the standing wave has the following properties:

- for each fixed point in time, the standing wave is a spatial sine wave;
- for each fixed point in space, the standing wave is a temporal sine wave;
- two traveling waves composing one standing wave have the same amplitudes and the same frequencies.

The general view of the standing wave (6.45) does not contain any periodic components. They appear in the spectral language. If the function of the initial spatial deviation $U_0(x)$ is converted by Fourier transformation:

$$F[U_0(x)] = \int_{-\infty}^{+\infty} U_0(x) \cdot e^{-j\lambda x} \cdot dx = U(\lambda)$$

where $\lambda \left[\frac{1}{\mathrm{m}}\right]$ is the spatial frequency, the Fourier image of the standing wave will look like:

$$\begin{aligned} F[U_{\mathrm{st}}(x,t)] &= \frac{1}{2}F[U_0(x+Ct)+U_0(x-Ct)] \\ &= \frac{1}{2}U(\lambda)\cdot(e^{jCt}+e^{-jCt}) = U(\lambda)\cdot\cos(C\cdot t) \end{aligned} \tag{6.46}$$

That is, a standing wave in the spectral representation is a certain spatially dependent stationary spectrum $U(\lambda)$ with an amplitude as a sinusoidal function of time, $\cos(C \cdot t)$.

The inverse Fourier transformation on the expression (6.46)

$$F^{-1}F[U_{\mathrm{st}}(x,t)] = F^{-1}[U(\lambda)\cdot\cos(C\cdot t)] = U_0(x)\cos(C\cdot t) \tag{6.47}$$

obviously is not identical to the original expression

$$U_{\mathrm{st}}(x,t) = \frac{1}{2}U_0(x+C\cdot t) + \frac{1}{2}U_0(x-C\cdot t) \tag{6.48}$$

This paradox is explained by the fact that the direct and inverse Fourier transformation on the spatial coordinate x above is executed incorrectly: they treat spatial variables $\pm Ct$ as constants. In other words, a completely incorrect transition to the complex plane is made:

$$F[U_0(x + C \cdot t)] \longrightarrow F[U_0(x)] \cdot e^{jCt}$$
$$F[U_0(x - C \cdot t)] \longrightarrow F[U_0(x)] \cdot e^{-jCt} \tag{6.49}$$

This technique is widely used in the analysis of AC circuits. Mathematical transformations (6.47) and (6.48) are not identical, but the physical meaning does not suffer from such substitution if the mixed argument is only a spatial addend and only a temporary addend.

A wave of arbitrary shape $U_0(x)$ "runs" along the imaginary time axis. If the same wave, only different in the direction, runs towards it, then because of their interference, the imaginary time axis degenerates, and only the real time axis remains. Thus, the attribute of the traveling wave becomes a complex exponent $e^{j\omega t}$. The standing wave is characterized by a real time function (6.47), and the coordinate function of the standing wave $U_0(x)$ can be complex.

Euler Formula

$$e^{jy} = \cos y + j \cdot \sin y$$

or, consequently, the expression

$$\cos y = \frac{1}{2}(e^{jy} + e^{-jy})$$

reflects the fact that the standing wave is the interference of two waves traveling in opposite directions yet identical in amplitude and frequency. If the first harmonic of Fourier transformation of the function $U_0(x)$ corresponds to the function $\sin x$, then the most simplified form of the standing wave is trigonometric:

$$U_{\mathrm{st}} = (\sin x)(\cos C \cdot t)$$

In the following, a mnemonic rule will be applied: if a certain, generally speaking, complex field $U(x,t)$ has a summand of the form $U_0(x) \cdot e^{j\omega t}$, then this summand is a traveling wave, if there is another summand as the product of a spatial function and a sinusoidal real-time function $U_0(x) \cdot \cos \omega t$, then it is a standing wave.

Complex representation of a standing wave in the form of

$$U(x,t) = U_0 \left[e^{j\omega\left(t+\frac{x}{C}\right)} + e^{-j\omega\left(t-\frac{x}{C}\right)} \right] = 2U_0 \cdot e^{j\omega\frac{x}{C}} \cdot \cos\omega t \quad (6.50)$$

($2U_0 \cdot e^{j\omega\frac{x}{C}}$ is a complex spatial function) is also the solution of the wave equation

$$U_{tt} = C^2 \cdot U_{xx}$$

which turns out by a direct substitution.

A traveling wave presented in a complex form

$$U(x,t) = U_0 \cdot e^{j\omega\left(t+\frac{x}{C}\right)} = U_0 \cdot e^{j\omega t} \cdot e^{j\omega\frac{x}{C}}$$

is also a solution to the wave equation.

If the separately standing wave is a particular solution of the wave equation, and the traveling wave is also a particular solution of the wave equation, then their sum is also the solution of the wave equation

$$U(x,t) = U_{\text{st}}(x,t) + U_{\text{tr}}(x,t)$$

where $U_{\text{tr}}(x,t)$ is the designation of a traveling wave.

A simplistic initial condition in the form of a sinusoidal function

$$U_0(x) = \sin x$$

is an idealization and not feasible in practical cases. Indeed, it is impossible to instantly excite the entire infinite elastic medium. Because of the fact that both the waves traveling towards each other, $U_0(kx+\omega t)$ and $U_0(kx-\omega t)$, are defined in the infinite interval $(-\infty,\infty)$, their interference occurs and, as a result, a standing wave. In practice, a perturbation of a certain finite volume of the medium is possible. It is described by so-called localized functions. The localized functions are identically zero everywhere except for the finite segment of their argument (Fig. 6.4).

At the zero moment of time $t=0$, the initial condition is a certain localized function $U_0(x)$, which after the moment $t=0$ is split into two identical waves running in the opposite directions, $U_0(kx+\omega t)$ and $U_0(kx-\omega t)$. The interaction occurs as long as they have a common spatial non-zero intersection.

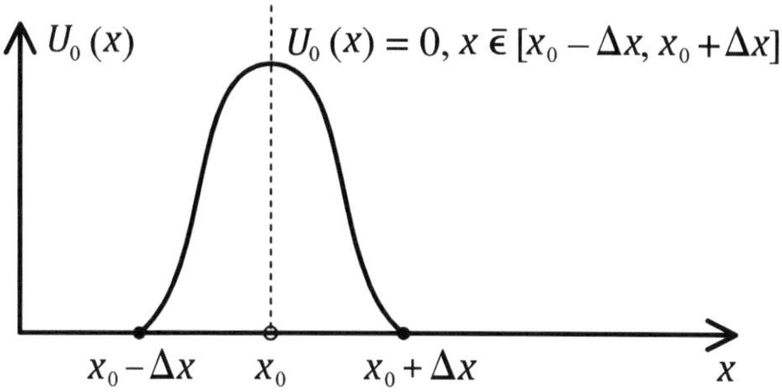

Figure 6.4: Finite spatial component of external influence.

That is, for the initial condition in the form of a localized function, a standing wave in an infinite medium is a short-term phenomenon, it can be attributed to the transient process, during which there is no standing wave.

If, in the case of an infinite initial perturbation, one of the traveling waves has an amplitude less than the amplitude of the other one, then not only is a standing wave formed, but a traveling wave with an uncompensated amplitude remains. For example,

$$\begin{aligned} &A \cdot \sin(x + C \cdot t) + B \cdot \sin(x - C \cdot t) \\ &\quad = 2A \cdot \sin x \cdot \cos C \cdot t + (B - A) \cdot \sin(x - C \cdot t) \\ &\quad = U_{\mathrm{st}}(x, t) + U_{\mathrm{tr}}(x, t) \end{aligned} \tag{6.51}$$

In the general case of initial conditions,

$$U(0, x) = U_0(x) \neq 0$$
$$U_t(0, x) = U_1(x) \neq 0$$

The solution is written like this:

$$\begin{aligned} U(x, t) &= \frac{1}{2}[U_0(x + C \cdot t)] + \frac{1}{2}[U_0(x - C \cdot t)] \\ &\quad + \frac{1}{2 \cdot C} \cdot \int_{x-Ct}^{x+Ct} U_1(\varepsilon) d(\varepsilon) \\ &= \frac{1}{2}[U_0(x + C \cdot t) + \frac{1}{C} \cdot F(x + C \cdot t)] \\ &\quad + \frac{1}{2}\left[U_0(x - C \cdot t) + \frac{1}{C} \cdot F(x - C \cdot t)\right] \end{aligned}$$

$$
\begin{aligned}
&= \theta_1(x + C \cdot t) + \theta_2(x - C \cdot t) \\
&= [\theta_1(x + C \cdot t) + \theta_1(x - C \cdot t)] \\
&\quad + [-\theta_1(x - C \cdot t) + \theta_2(x - C \cdot t)] \\
&= U_{\text{st}}(x, t) + U_{\text{tr}}(x, t)
\end{aligned}
$$

where $F(x)$ is a primitive function $U_1(x)$.

It is called a d'Alembert solution of the wave equation. Now we observe movement in the opposite directions of two different functions, θ_1 and θ_2. Their interaction not only brings about a standing wave but also leaves an uncompensated traveling wave. The solution of the wave equation can be any two functions with "against-running" arguments $\theta_1(x + C \cdot t)$ and $\theta_2(x - C \cdot t)$, therefore, in the solution, the ratio between the standing and traveling waves can be arbitrary.

6.7 Energy Properties of a Standing Wave

Let us show that a standing wave does not propagate energy. Let a certain volume of medium with equivalent mass M and a stiffness K in respect to perturbation be exposed to a standing wave at which the velocity changes according to the law

$$
V_{\text{st}}(x, t) = V(x) \cdot \cos \omega t
$$

where $\omega = C/\Lambda$.

Then the displacement of this volume is expressed by the formula

$$
x_{\text{st}} = \int V_{\text{st}}(x, t) dt = \frac{1}{\omega} \cdot V(x) \cdot \sin \omega t + C
$$

The total energy of this elastic, as a spring pendulum, oscillating volume is written as the sum of the potential and kinetic energies

$$
\begin{aligned}
E = E_p + E_k &= \frac{K}{2} \cdot x_{\text{st}}^2 + \frac{m}{2} \cdot V_{\text{st}}^2 \\
&= \frac{V^2(x)}{2} \cdot \left(\frac{K}{\omega^2} \cdot \sin^2 \omega t + m \cdot \cos^2 \omega t \right) \\
&= \frac{V^2(x)}{2} \cdot m \cdot \left[\frac{\omega_0}{\omega} \cdot \sin^2 \omega t + \cos^2 \omega t \right],
\end{aligned}
$$

$$
\max E_k = \frac{m \cdot V^2(x)}{2}
$$

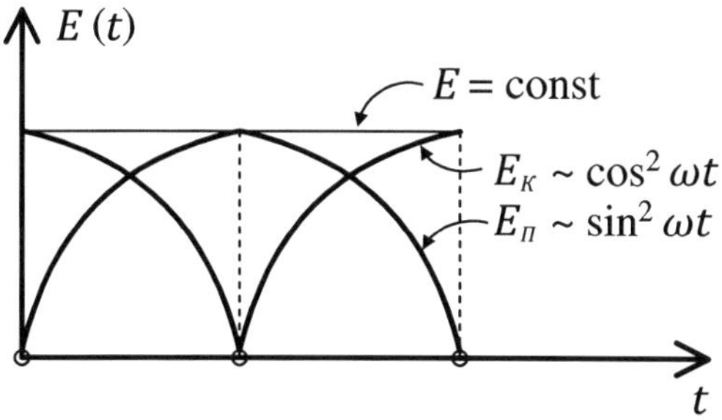

Figure 6.5: Total energy — E, kinetic energy — E_k and potential — E_p is energy of a standing wave.

Thus, the total energy of the standing wave — E at $\omega_0 = \omega$ does not depend on time (Fig. 6.5). This sets the resonance property of the standing wave. The volume element with its mass and stiffness must be in resonance caused by perturbation with the velocity of sound. The standing wave is only observed at one frequency. A standing wave does not carry energy, more precisely, its energy does not spread in space, but this does not mean that the standing wave does not contain energy. It is stationary, constant in time and distributed in space.

$$E(x) = \frac{m \cdot V^2(x)}{2}$$

Moreover, the energy continuously passes from the kinetic to the potential one and back on one time half-cycle of the standing wave

$$\frac{1}{2}T = \frac{\Lambda}{\pi C} \tag{6.52}$$

so that the total energy remains the same.

6.8 Non-homogeneous Wave Equation

6.8.1 *Zero initial conditions*

In the VVER Main Circulation Loop, the main circulation pumps during their operation generate an acoustic field, and it can be expected that the solution will contain both standing waves (the general solution of the homogeneous equation) and waves formed by the MCPs themselves (a particular solution). There is always a standing wave in the general solution of a wave equation. However, in the case of forced oscillations it should be neglected. The acoustic field that exists in the MCC to the moment $t = 0$ or resulting from any initial disturbance is negligible so that its energy parameters can be considered as identically zero compared with the enormous energy that is brought into the MCC by an MCP in operation.

If the initial field is identical to zero, it can be disturbed not only by the stationary initial deviation, as it was done in the previous item, but also by the inclusion at the time $t = 0$ of an external source $f(x, t)$. Further on, for $t > 0$ this source functions continuously.

In a homogeneous case, if the medium does not have an absorbing effect on the oscillations, the initial deviation will be in some way transported and will never disappear. In a non-homogeneous case, even if the medium has an absorbing effect, the flow of energy from the outside as a result of the continuous impact of an external force complements this loss. As a result, the two considered cases of solutions of homogeneous and non-homogeneous wave equations should have some qualitatively similar properties. For example, the presence of a superposition of standing and traveling waves.

Now, let the initial conditions be zero and at the time moment $t = 0$ an external driving source $f(x, t) \neq 0, U_0 \equiv 0, U_1 \equiv 0$ has occurred, which then operates continuously. This is a model of acoustic field excitation in the MCC by the inclusion of the first MCP. This practical problem is formalized by a non-homogeneous wave equation with zero initial conditions:

$$U_{tt} - C^2 \cdot U_{xx} = f(x, t)$$

$$\begin{cases} U(0, x) = U_0 = 0 \\ U_t(0, x) = U_1 = 0 \end{cases} \tag{6.53}$$

Let's consider particular solutions of this equation. Let the elastic medium be affected by a traveling wave with a unit amplitude as an external force:

$$f(x, t) = e^{j(\omega t + kx)}$$

$$|f(x, t)| \equiv 1 \tag{6.54}$$

and the initial conditions are zero

$$U_0(x) \equiv 0, U_1(x) \equiv 0$$

We will look for a particular solution in the form of

$$U(x,t) = A \cdot e^{j(\omega t + kx)} \tag{6.55}$$

By a direct substitution of (6.54) into (6.53), we obtain

$$-A \cdot \omega^2 + k^2 \cdot C^2 = 1$$

Hence, the solution amplitude (6.55) has the form

$$A = \frac{1}{C^2 \cdot (k^2 - \omega^2/C^2)} = \frac{1}{C^2 \cdot (k^2 - k_0^2)} \tag{6.56}$$

Thus, the traveling wave, forcing the oscillations of the elastic medium at each point, generates a new traveling wave with the same frequency ω, but with a spatial amplitude depending on the wavelength (6.56). The elastic medium is a linear filter that passes all the frequencies of the external force, but imposes an amplitude-frequency weight on them (Fig. 6.6).

This filter suppresses high-frequency spatial components, it has a resonance frequency $k = k_0$, at which the amplitude increases infinitely, and a passband at low frequencies. This filter corresponds exactly to the transfer function "force–displacement" for continuous (non-dispersive) oscillations of the spring pendulum.

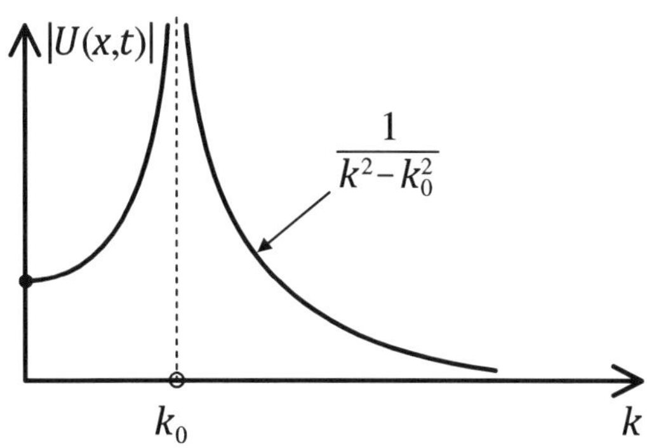

Figure 6.6: Module of the wave equation solution when exposed to a traveling wave with a unit amplitude as an external force.

Now let us imagine that the elastic medium is forced by a standing wave with a unit amplitude

$$f(x,t) = \cos\omega t \cdot \cos kx$$

It is easy to check that we obtain the solution

$$U(x,t) = \frac{1}{C^2 \cdot (k^2 - k_0^2)} \cdot \cos\omega t \cdot \cos kx \qquad (6.57)$$

similar to the previous case. If a standing wave acts as a driving force, then at each point of the elastic medium a standing wave will be observed, but filtered by the transfer function

$$\frac{1}{C^2 \cdot (k^2 - k_0^2)}$$

It is obvious that if the driving force consists of a standing and traveling waves

$$f(x,t) = f_{\text{st}}(x,t) + f_{\text{tr}}(x,t)$$

then at each point of the elastic medium the former and the latter wave will be observed

$$U(x,t) = U_{\text{st}}(x,t) + U_{\text{tr}}(x,t)$$

The external force, a product of the sinusoidal frequency function and the sinusoidal spatial function, is an idealization. It is practically impossible to force almost all points of the infinite medium at the same time.

6.8.2 *General case*

By direct substitution, it is easy to verify that the general solution of the equation (6.53) under zero initial conditions is written as

$$U(x,t) = \frac{1}{2C} \cdot \int_0^t d\tau \left[\int_{x-C(t-\tau)}^{x+C(t-\tau)} f(\xi,\tau) d\xi \right] \qquad (6.58)$$

Here the function $f(x,t)$ is arbitrary, including the localized types. If in the example from the previous section the solution only depended on the initial conditions (U_0, U_1), now it only depends on an external influence $f(x,t)$. If earlier for the initial deviation the arguments ran in the opposite directions "$x - Ct$, $x + Ct$", now such arguments are observed in the integral transformation over the external force "$x - C(t - \tau)$; $x + C(t - \tau)$:"

$$\int_{x-C(t-\tau)}^{x+C(t-\tau)} f(\xi, \tau) d\xi \tag{6.59}$$

Let the external influence again be a spatially distributed harmonic function, but with non-unit spatial amplitude:

$$f(x, t) = f_0(x) \cdot e^{jC_1\tau} \tag{6.60}$$

After substitution of (6.60) into (6.59), we have

$$\begin{aligned} U(x, t) &= \frac{1}{2C} \cdot \int_0^t e^{jC_1\tau} \left[\int_{x-C(t-\tau)}^{x+C(t-\tau)} f_0(x) dx \right] d\tau \\ &= \frac{1}{2C} \cdot \int_0^t e^{jC_1\tau} \{ F_0[x + C(t - \tau)] \\ &\quad - F_0[x - C(t - \tau)] \} d\tau \end{aligned} \tag{6.61}$$

where $F_0(x)$ is a primitive function of $f_0(x)$:

$$F_0(x) = \int f_0(x) dx$$

We make a complex generalization under the assumption that $F_0(x)$ is the real co-factor of the complex function

$$F_0(x, t) = F_0(x \pm C \cdot t) = F_0(x) \cdot e^{\pm jC\tau} \tag{6.62}$$

And, in particular:

$$F_0(x) = F_0(x, 0) = F_0(x) \cdot e^{j0 \cdot \tau}$$

Then the solution has the form

$$\begin{aligned} U(x, t) &= \frac{1}{2C} \cdot F_0(x) \cdot \int_0^t e^{jC_1\tau} [e^{jC(t-\tau)} - e^{-jC(t-\tau)}] d\tau \\ &= \frac{2j}{2C} \cdot F_0(x) \cdot \int_0^t e^{jC_1\tau} \cdot \sin[C(t - \tau)] d\tau \\ &= j \cdot F_0(x) \cdot \frac{1}{C^2 - C_1^2} \cdot e^{jC_1\tau} + F_0(x) \cdot \frac{1}{C^2 - C_1^2} \cdot \sin(Ct - \psi) \\ &= U_{\text{tr}}(x, t) - U_{\text{st}}(x, t) \end{aligned} \tag{6.63}$$

Thus, the solution is a linear combination of a traveling wave with the velocity of the external influence C_1 and a standing wave. Each summand has the same weight function analogous to the frequency function (6.56) with a resonance property.

If the external influence is spatially localized, then its primitive for any practical cases is also a localized function. The elastic medium integrates spatial perturbations of the external force and transports the traveling wave with velocity C_1. The same medium in the form of a primitive also determines the spatial form of a standing wave. The nodes of the standing wave are determined by the properties of the medium, and the spatial form of the driving force determines its amplitude.

So, non-zero initial conditions in the absence of an external driving force generate both a standing wave and a traveling wave. Zero initial conditions under the influence of an external driving force give the same qualitative result: both standing and traveling waves.

If the operating MCP is represented as δ-shaped, that is, a spatially point harmonic source with frequency $\omega = \omega_{\text{MCPs}}$:

$$f(x, t) = A \cdot \delta(x - x_0) \cdot e^{j\omega t}$$

then the solution will be much easier:

$$U(x, t) = \frac{A}{C \cdot \omega} \cdot \sin \frac{\omega t}{2} \cdot e^{j\frac{\omega t}{2}}$$

In this almost unrealizable case, the spatial dependence disappears, and the perturbed medium is a single traveling wave. The external influence in the form of δ-functions infinitely localized in one spatial point does not generate standing waves. Continuously generated δ-functions at any time do not intersect in space, and therefore do not interact with each other, which means there is no interference, and therefore no standing waves.

On the contrary, continuously generated localized functions running into the opposite directions interact with each other on the overlapping domains of their non-zero definition.

In applied problems, a standing wave occurs, as a rule, in the presence of reflecting surfaces. These are the walls at the boundaries of the acoustic path or reflecting non-homogeneities in the direction of movement of the sound wave, for example, such as an abrupt change of the acoustic path cross-section. As shown above, the possibility of the formation of standing waves is a fundamental property of any elastic medium where waves propagate. Even an elastic acoustic medium without boundaries, excited in a certain finite volume or even in the entire infinite volume, will be a carrier of a superposition of two types of waves. For a one-dimensional case, two traveling waves are multidirectional. Depending on how the acoustic medium is excited, interference can result in:

- only one standing wave,
- a standing wave in combination with an uncompensated wave running in one direction.

For an infinite elastic homogeneous medium, the concepts of "frequency" or "standing wave length" are coupled with the velocity of sound C. The elastic medium does not have natural frequency, but has its own characteristic — the velocity of sound C, and it is the carrier (conductor) of the external perturbation (no natural) frequencies. For example, perturbation of the MCC coolant by circulation pumps has the property of a standing wave with the characteristic C, as an attribute of the circulation loops and a traveling wave forced rotational frequency of MCP.

Four MCPs, synchronized in respect to the rotor frequency in a single power grid, create a field of traveling and standing waves presented above. However, because the MCC is a closed loop, two waves running in the opposite directions at the reverse frequency will meet and begin to interact. As a result, there will be another standing wave at the rotational MCPs frequency. At the rotational frequency, the signals of any two pressure pulsation sensors are either in phase or out of phase. When the synchronicity of four MCP rotations is lost whether due to poor-quality power supply or due to the significant difference in the motor slips of four rotors, these phases will significantly depart from common phase or opposite phase. The mode of standing waves in the circulation circuit is not most favorable from the point of view of excitation of equipment vibrations. In addition to the fact that the standing wave has a double amplitude in comparison to the corresponding traveling wave, it has completely defined spatial points in which the amplitude of the oscillations has its maxima.

In these cross-sections of the pipeline, the maximum scale of its vibration deformations takes place. A traveling wave, in contrast, does not generate a stationary field of vibration load.

When the functions of the initial deviation or external influence are arbitrary, the standing wave, generally speaking, is a non–harmonic function $(U(x - Ct) + U(x + Ct))$ and the frequency or wavelength are incorrect terminologies. Such frequencies, generally speaking, form a continuum ensemble. Frequency representation is no more effective here, better to use time square presentation. According to it, the initial perturbation (and/or perturbation by an external force) "splits" into the opposite directions and interacts where there are common spatial parts. So, if the perturbation is infinite in space (for example, sinusoidal one), i.e. it was possible to generate it in the entire medium at the same moment, the total spatial part of the interaction of two traveling waves is infinite, and an infinitely extended standing wave is formed on it.

Closed circulation loops, such as the VVER MCC, can be regarded as an infinite elastic medium. It has both own ASW defined by the loop geometry and the similarity of ASW at the rotational MCPs frequency — ω_{MCPs}. The easiest way to show the effect of the formation of similarity of a forced ASW is the case of one functioning MCP, when the forced oscillations of the coolant pressure at the rotational frequency will be coherent, by definition. These pressure oscillations are longitudinal waves running in the opposite directions, and due to the closed circulation loop they certainly will meet and interfering they form a standing wave. Its spatial period is determined not by the geometry of the reflecting planes, but by a well-known formula:

$$\Lambda = \frac{C}{4f_{\text{MCPs}}}$$

where f_{MCPs} is the rotational MCPs frequency. Taking a very rough (estimated) velocity of sound in water at 100% power level of VVER-1200 for the value of 800 m/s, we obtain that the spatial wavelength of the ASW similarity will be about 12 m. Consequently, in the MCC pipelines vertical planes will appear with the maximum amplitude of vibration stresses at the rotational MCPs' frequency. We have to estimate the positions of these planes from the installation point of

MCPs in the loop in both directions along the pipeline. There will be no more than four such planes in one loop.

6.9 Standing Waves in the Solution of a Wave Equation with Boundary Conditions — Fourier Method

Let us return to the homogeneous wave equation to obtain its analytical solution not for an infinite elastic medium, but with boundary conditions. The analogy with the reflecting walls of the pipeline is illustrated by the oscillations of the fixed string, which are represented by the same wave equation. If the string is fixed in two points spaced apart by a distance L, we get the boundary conditions in their most elementary form: $U(0,t) = 0$, $U(L,t) = 0$.

If the solution of the wave equation is represented as a product of the coordinate function and the time function

$$U(x,t) = X(x) \cdot T(t) \tag{6.64}$$

and we substitute it into a homogeneous wave equation, then we obtain

$$C^2 \cdot \frac{\partial^2 X(x)}{\partial x^2} \cdot T(t) + X(x) \cdot \frac{\partial^2 T(t)}{\partial t^2} = 0 \tag{6.65}$$

or

$$\frac{X_{xx}}{X} = -\frac{1}{C^2} \cdot \frac{T_{tt}}{T} \tag{6.66}$$

In the expression (6.66), where on the left-hand side there is only the function of coordinates, and on the right is only the function of time, equality is possible if the left and right parts are equal to the same constant, which we denote as $(-\lambda^2)$:

$$\frac{X_{xx}}{X} = -\lambda^2$$

$$\frac{1}{C^2} \cdot \frac{T_{tt}}{T} = \lambda^2$$

As a result, one wave partial differential equation was transformed into two ordinary Lagrange equations and into the harmonic oscillator equations: the first is fully space-dependent the second is fully time-dependent. In the spatial equation the parameter λ determines the reciprocate of the natural frequency (wavelength), while in the time equation the reciprocate of the parameter $\frac{\lambda}{C}$ is the natural frequency of oscillations. It is important that these two natural properties of standing waves are proportional to each other through the velocity C. For acoustic standing waves, C is the velocity of sound, and if it is known, you can determine both the ASW frequency and its wavelength. For the string

$$C = \sqrt{\frac{N}{\rho_0}}$$

where N is a tensile force along the string, ρ_0 is the linear density of the string.

For longitudinal oscillations of a thin rod

$$C = \sqrt{\frac{E}{\rho}}$$

where E is the modulus of elasticity of the rod material and ρ is its density.

The non-trivial solution of the Lagrange equation is obviously a harmonic or sinusoidal function:

$$X(x) = C_1 \cdot \cos \lambda x + C_2 \cdot \sin \lambda x$$

For elementary boundary conditions

$$X(0) = 0 \to C_1 = 0; X(L) = 0 \to \sin \lambda x = 0$$

we obtain that the frequency of the sinusoid must satisfy the condition

$$\lambda_n = \frac{\pi \cdot n}{L}, \quad n = 1, 2, 3, \ldots.$$

A discrete infinite series of frequencies λ_n are called eigenvalues and there are infinite number of particular solutions in the form of harmonic functions

$$X_n(x) = A_n \cdot \sin \frac{\pi \cdot n}{L} \cdot x$$

which are called eigenfunctions. Thus, boundary conditions transformed the continuum of frequencies associated with the velocity

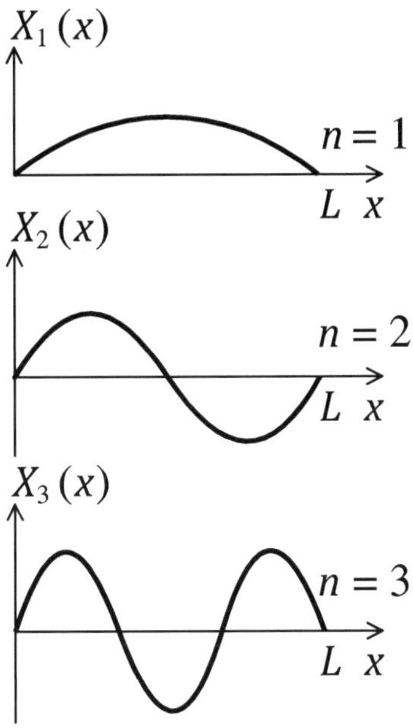

Figure 6.7: The first three spatial harmonics (eigenfunctions) of the standing wave in the segment $[0, L]$.

parameter to the discrete series of frequencies (wavelengths) — λ_n. Now, only those sinusoidals can exist as solution in the segment $[0, L]$ which have integral value of half-waves (Fig. 6.7).

An equation with a time variable

$$T''(t) + C^2 \cdot \lambda^2 \cdot T(t) = 0$$

has also a harmonic solution

$$T_n(t) = a_n \cdot \cos \frac{\pi \cdot C \cdot n}{L} \cdot t + b_n \cdot \sin \frac{\pi \cdot C \cdot n}{L} \cdot t, \tag{6.67}$$

But the amplitudes a_n and b_n already depend on the initial conditions $U_0(x)$, $U_1(x)$:

$$a_n = \frac{2}{L} \cdot \int_0^n U_0(x) \cdot \sin \frac{\pi \cdot n \cdot x}{L} \cdot dx$$

$$b_n = \frac{2}{\pi \cdot C \cdot n} \cdot \int_0^n U_1(x) \cdot \sin \frac{\pi \cdot n \cdot x}{L} \cdot dx \tag{6.68}$$

The method of obtaining the solution of the wave equation is called the Fourier method. As a result, the general view of the solution is written as a series

$$U(x,t) = \sum_{n=1}^{\infty} \left(a_n \cdot \cos \frac{\pi \cdot C \cdot n \cdot t}{L} + b_n \cdot \sin \frac{\pi \cdot C \cdot n \cdot t}{L} \right) \cdot \sin \frac{\pi \cdot n \cdot x}{L} \tag{6.69}$$

The solution of the wave equation in the form of d'Alembert equation

$$U_{\text{st}}(x,t) = \frac{1}{2} U_0(x + C \cdot t) + \frac{1}{2} U_0(x - C \cdot t)$$

in contrast to the solution in the form of Fourier equation (6.69) does not give an explicit form of the function, but only demonstrates that two traveling waves running in the opposite directions can be distinguished in the solution. In the Fourier solution, which involves additional conditions as the boundary ones, it is demonstrated that the solution can be represented as a product of periodic functions (in particular, two sinusoids) — spatial and temporal.

In the form of d'Alembert equation, only initial conditions were used, that is, the medium was infinite. If the boundaries from which multiple reflections occur are introduced, the result will take the solution in the form of Fourier equations. It is represented as trigonometric functions, that is, in an explicit form. If the solution in the form of the d'Alembert equation was written using implicit functions of the initial conditions $U_0(x)$, $U_1(x)$ or an implicit function of external influence $f(x,t)$, then the simultaneous assignment of non-zero initial conditions and some boundary conditions lead to the solution in the form of trigonometric Fourier series. Moreover, the boundary conditions are only included in the spatial argument of standing wave harmonics — $\sin \frac{\pi \cdot n \cdot x}{L}$, and the initial conditions are only included in the coefficients (amplitudes) a_n, b_n in front of the time functions $\cos \frac{\pi \cdot C \cdot n \cdot t}{L}$ $\sin \frac{\pi \cdot C \cdot n \cdot t}{L}$.

These properties of the solution of the wave equation by the Fourier method explain the attractiveness of the empirical approach of qualitative description of standing waves: it is enough to determine the reflecting walls at the ends of the pipeline to depict all the harmonics of standing waves at a qualitative level (without the amplitude values).

As a special case, let's consider an important practical example. Let the string fixed from two sides ($U(0,t) = 0$; $U(L,t) = 0$) be only driven by one initial deviation: $U_0(x) \not\equiv 0$; $U_1(x) \equiv 0$. As follows from the solution (6.69), its subsequent motion will be described by the formula:

$$U(x,t) = \sum_{n=1}^{\infty} a_n \cdot \cos\frac{\pi \cdot C \cdot n \cdot t}{L} \cdot \sin\frac{\pi \cdot n \cdot x}{L} \tag{6.70}$$

where

$$a_n = \frac{2}{L} \cdot \int_0^L U_0(x) \cdot \sin\frac{\pi \cdot n \cdot x}{L} dx \tag{6.71}$$

This is a description of the superposition of an infinite number of standing wave harmonics. That is, a string fixed on both sides is a mechanical device decomposing the initial deviation U_0 into a series of spatial harmonics $\sin\frac{\pi \cdot x}{L}$, $\sin\frac{2\pi \cdot x}{L}$, $\sin\frac{3\pi \cdot x}{L}, \ldots$, whose amplitudes are the coefficients of Fourier transformation of the function $U_0(x)$.

Let's make the problem simpler: let the initial deviation be applied in the middle of the string and is δ-shaped. Then the series (7.60) will only consist of odd harmonics ($n = 1, 3, 5, \ldots$):

$$a_n = \frac{2}{L} \cdot \begin{cases} 1, n = 1 \\ 0, n = 2 \\ -1, n = 3 \\ 0, n = 4 \\ \ldots \end{cases} \tag{6.72}$$

That is, those that have an extremum in the middle of the segment $\left(x = \frac{L}{2}\right)$:

$$U(x,t) = \frac{2}{L} \cdot \left(\cos\frac{\pi \cdot C \cdot t}{L} \cdot \sin\frac{\pi \cdot x}{L} - \cos\frac{3\pi \cdot C \cdot t}{L} \cdot \sin\frac{3\pi \cdot x}{L} + \cos\frac{5\pi \cdot C \cdot t}{L} \cdot \sin\frac{5\pi \cdot x}{L} - \cdots\right) \tag{6.73}$$

Figure 6.8 shows the change in time of the first harmonic $\cos\frac{\pi \cdot C \cdot t}{L} \cdot \sin\frac{\pi \cdot x}{L}$. Any other initial deviation $U_0(x)$, symmetrical with respect to the middle of the string, will have the same quality

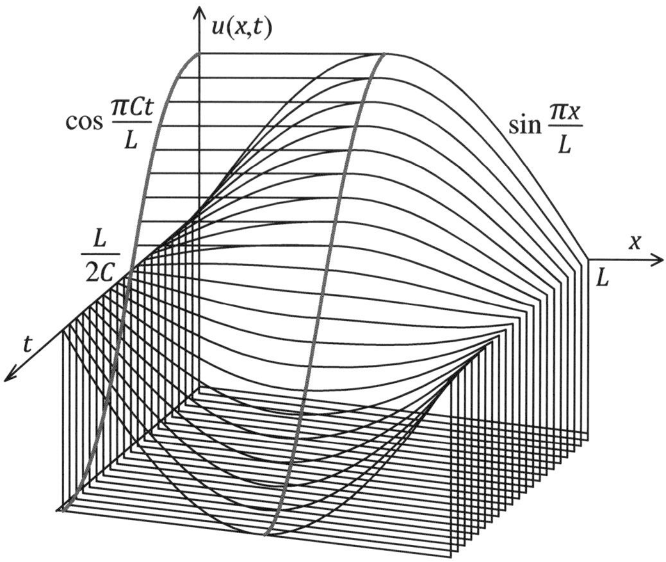

Figure 6.8: The first harmonic of the standing wave as a function of time and coordinate.

as in Fig. 6.8 as defined by the formula (6.73). The difference will be that the amplitude of the harmonic will decrease with its increasing number. From the type of the initial condition $U_0(x)$, we can make a preliminary conclusion about the form of the final solution. As is known, the Fourier transformation of odd and even functions will only consist, respectively, of odd and even harmonics. If a string fixed on both sides is perturbed exactly in the middle, so that an even function (symmetric with respect to the middle of the string) is obtained, the resulting string oscillations will only consist of odd harmonics ($n = 1, 3, 5, \ldots$). Only these harmonics have extremum in the middle of the string and even harmonics have a knot in the middle of the string, that is, they are odd functions. If even harmonics are not observed in practice, then most likely, the field excitation is produced near the middle of the segment $[0, L]$.

No matter in which point of the string the deviation has been produced from the equilibrium, the solution will always include the first harmonic. Indeed, under such conditions, the initial function is unimodal, which means, it is a function with a single maximum inside the segment $[0, L]$ and with zero values at its ends. Any unimodal

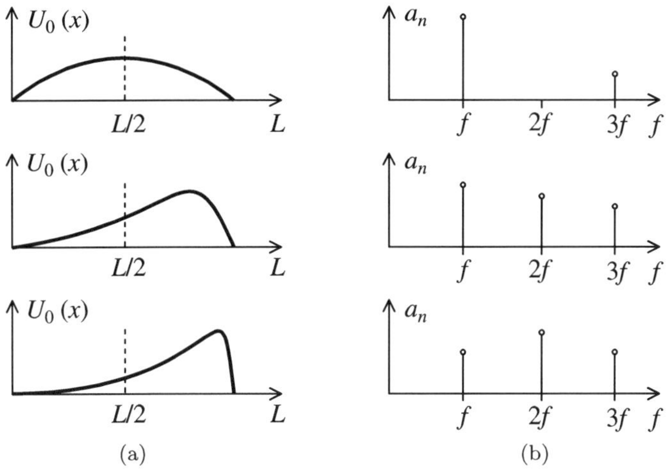

Figure 6.9: The shape of the initial deviation is $U_0(x)$ (a) and harmonic amplitudes are a_n (b).

function in the Fourier transformation has the first harmonic with non-zero amplitude, and it is higher if the maximum is closer to the middle of the segment $[0, L]$ (Fig. 6.9).

In the VVER-1200 design excitation of an acoustic wave is produced by the MCPs located exactly in the point $x = \frac{3L}{4}$, where L is the loop length. It will be shown in the following that the length of the loop fits into half-period of the lowest node of the loop ASW1p. Moreover, at the ends of the half-wave there are reflecting walls (the reactor cavity surfaces), so all the ASW in these boundary conditions must correspond to the idling, when the pressure is maximum at the ends of the pipeline. Here, the analogy with string is only valid for velocity and not for pressure.

Let us illustrate via a specific example that at such boundary conditions only those ASW1p harmonics appear whose number is the power of two (Fig. 6.10). The conditions for excitation of the first ASW1p harmonic are quite favorable: The MCPs are located where the ASW1p amplitude is large enough (see the top left graph in Fig. 6.10). If the MCPs are placed exactly in the middle of the loop, then the first ASW1p harmonic would not exist. This was to happen to the second harmonic, having zero ASW2p amplitude at the place of MCPs' installation. On the one hand, the second harmonic can exist in these boundary conditions, but, on the other hand, it is

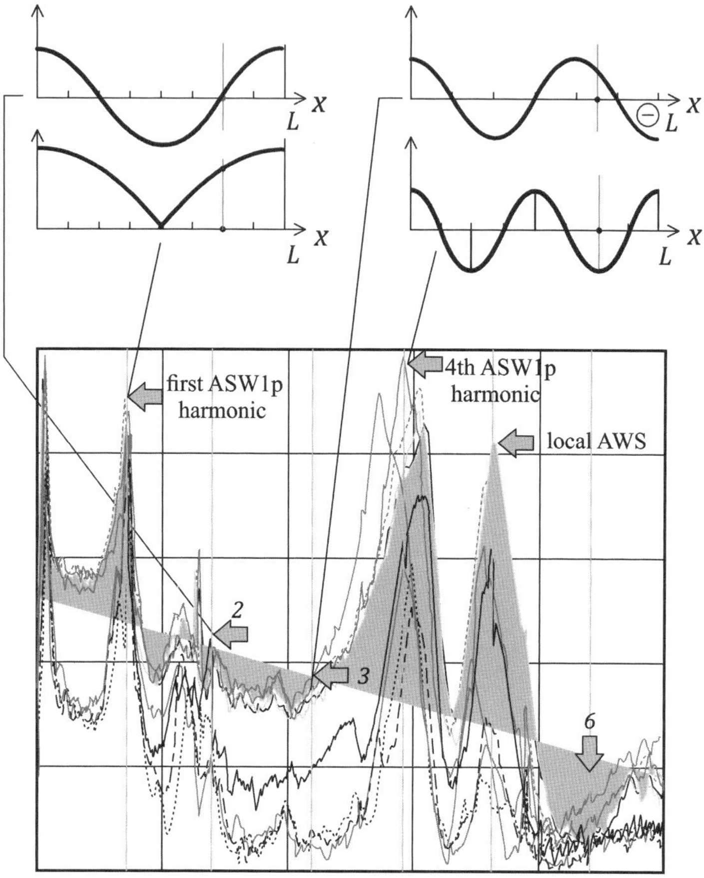

Figure 6.10: Amplitude of Spatial Harmonics from the first to the fourth (see the graphs). The vertical line marks the location of the MCPs installation. There are only the first, second and fourth harmonics. The second harmonic is in the wave node, the third and sixth harmonics cannot exist.

practically not excited because of the MCPs' location. Third ASW3p harmonic cannot survive with two closed ends of the pipeline: maximum pressure at one end always corresponds to the minimum pressure at the other end.

The most favorable conditions for excitation took place for the fourth harmonic. The MCPs installation is exactly located in the maximum amplitude of modulus of ASW4p. In the experimental data, the ASW4p amplitude even exceeds the ASW1p amplitude.

The fifth and sixth harmonics cannot exist in these boundary conditions. In addition, at the frequency near the fifth harmonic there is a local ASW with other boundary conditions.

The next practically essential case occurs when the string fixed from two sides is only influenced by external force $f(x,t)$, and the initial conditions are zero — $U_0(x) \equiv 0$, $U_1(x) \equiv 0$:

$$\begin{cases} U_{tt} = C^2 \cdot U_{xx} + f(x,t) \\ U_0(x) \equiv 0, U_1(x) \equiv 0 \\ 0 \leq x \leq L \end{cases}$$

Let's find a particular solution in the form:

$$U(x,t) = a^2 \cdot f(x,t)$$

$$f(x,t) = \varphi(x) \cdot \psi(t)$$

Here, external influence is a product of the coordinate function and the time function, and a is a constant. This means that a field follows the external influence with an accuracy up to a constant. After substitution of the particular solution to the initial equation, we obtain

$$\psi_{tt}(t) \cdot \varphi(x) - C^2 \cdot \varphi_{xx}(x) \cdot \psi(t) = a^2 \cdot \varphi(x) \cdot \psi(t)$$

$$\frac{\psi_{tt}}{\psi} - C^2 \cdot \frac{\varphi_{xx}}{\varphi} = a^2$$

The time function $\frac{\psi_{tt}}{\psi}$ and the spatial coordinate function $\frac{\varphi_{xx}}{\varphi}$ will sum to a constant only if each function is equal to a constant:

$$\psi_{tt} = \psi \cdot \text{const}_1$$

$$\varphi_{xx} = \varphi \cdot \text{const}_2$$

Again, we come to two ordinary Lagrange differential equations, each characterized by its own frequency. This is the so-called Sturm–Liouville problem, the solution to which is found in the form of a series

$$U(x,t) = \sum_{n=1}^{\infty} U_n(t) \cdot X_n(t)$$

where $X_n(t) = \sin \frac{\pi n x}{L} = \sin \lambda_n \cdot x$ — eigenfunctions, $\lambda_n = \frac{\pi n}{L}$ — eigenvalues.

No matter how complex the function $U(x,t)$ is, qualitatively the Sturm–Liouville problem has a solution similar to the solution of a homogeneous wave equation with a single initial condition ($U_0(x) \not\equiv 0$; $U_1(x) \equiv 0$). The difference is that the function continuously influencing on the medium $f(x,t)$ generates traveling waves forming some standing wave field and uncompensated traveling wave field where $U_n(t)$ is not just any function but only those satisfying the Lagrange equation of motion.

$$\frac{d^2U_n(t)}{dt^2} + C^2 \cdot \lambda_n \cdot U_n = f_n(t)$$

$$U_n(0) = 0$$

$$\frac{dU_n}{dt}(0) = 0$$

where $f_n(t)$ is not just any function but only those satisfying the condition

$$f(x,t) = \sum_{n=1}^{\infty} f_n(t) \cdot X_n(x)$$

6.10 Features of Solutions to Wave Equation

Table 6.1 summarizes the solutions to the wave equation discussed above.

Case 1 is demonstrated via an example: at the moment of time $t = 0$ the string was deviated from balance — $U_0(x) \not\equiv 0$:

$$U(x,t) = \frac{1}{2}U_0(x + C \cdot t) + \frac{1}{2}U_0(x - C \cdot t) = U_{\text{st}}(x,t)$$

As a result, we have a "pure" standing wave, the spatial shape of which is determined by the initial deviation.

Case 2 is demonstrated via an example: at the moment of time $t = 0$ the string has deviated balance, $U_0(x) \neq 0$, and given a speed $U_1(x) \neq 0$. As a result, we have a superposition of a standing and traveling waves:

$$U(x,t) = \frac{1}{2}U_0(x + C \cdot t) + \frac{1}{2}U_0(x - C \cdot t) + \frac{1}{2C} \cdot \int_{x-Ct}^{x+Ct} U_1(\tau)d\tau$$
$$= \theta_1(x + C \cdot t) + \theta_2(x - C \cdot t) = U_{\text{st}}(x,t) + U_{\text{tr}}(x,t)$$

Table 6.1: Properties of solutions to wave equation in various initial and boundary conditions and presence/absence of external influence.

Case	1	2	3	4	5	6	7
Initial conditions							
$U_0(x,0)$	$U_0 \neq 0$	$U_0 \neq 0$	$U_0 = 0$	$U_0 = 0$	$U_0 \neq 0$	$U_0 \neq 0$	$U_0 \neq 0$
$U_1(x,0)$	$U_1 = 0$	$U_1 \neq 0$	$U_1 = 0$	$U_1 = 0$	$U_1 \neq 0$	$U_1 = 0$	$U_1 \neq 0$
Boundary conditions							
$U(0,t)$	—	—	—	$U(0,t) = 0$	$U(0,t) = 0$	$U(0,t) = 0$	$U(0,t) = 0$
$U(L,t)$				$U(L,t) = 0$	$U(L,t) = 0$	$U(L,t) = 0$	$U(L,t) = 0$
Forcing influence							
$f(x,t)$	—	—	$f(x,t) \neq 0$	$f(x,t) \neq 0$	—	$f(x,t) \neq 0$	$f(x,t) \neq 0$
Solution presentation	"Pure" standing wave U_{st}	Standing and running wave $U_{\text{st}} + U_{\text{tr}}$	Standing and running wave $U_{\text{st}} + U_{\text{tr}}$	Harmonics of standing waves $\sum_n U_{\text{st}}^{(n)}$	Harmonics of standing waves and a traveling wave $\sum_n U_{\text{st}}^{(n)}$	Harmonics of standing waves and a traveling wave $\sum_n U_{\text{st}}^{(n)} + U_{\text{tr}}$	Harmonics of standing waves and a traveling wave $\sum_n U_{\text{st}}^{(n)} + U_{\text{tr}}$

Case 3 is demonstrated via an example: at the moment of time $t = 0$ on the identically zero boundless field $U(x,t)$, an external force has an effect $f(x,t) \neq 0$, affecting continuously. As a result, we have a superposition of a standing and traveling waves:

$$U(x,t) = \frac{1}{2C} \cdot \int_0^t d\tau \int_{x-C(t-\tau)}^{x+C(t-\tau)} f(\xi,\tau)d\xi = U_{\text{st}}(x,t) + U_{t2}(x,t)$$

Case 4 is demonstrated via an example: at the moment of time $t = 0$, on the identically zero field in the boundary conditions $0 \leq x \leq L$, an external force has an effect $f(x,t) \neq 0$, which further on operates continuously. As a result, we only have a superposition of standing wave harmonics:

$$U(x,t) = \sum_{n=1}^{\infty} U_n(t) \cdot \sin\frac{\pi n x}{L}$$

Case 5 is demonstrated via an example: at the moment of time $t = 0$ on the identically zero field in the boundary conditions $0 \leq x \leq L$, only the initial deviation has an effect at zero external force $f(x,t) \equiv 0$. As a result, we only have a superposition of standing wave harmonics:

$$U(x,t) = \sum_{n=1}^{\infty} \left(a_n \cdot \cos\frac{\pi C n t}{L} + b_n \cdot \sin\frac{\pi C n t}{L} \right) \cdot \sin\frac{\pi n x}{L} = \sum_{n=1}^{\infty} U_{\text{st}}^{(n)}(x,t)$$

Case 6 is a generalization of Cases 1 and 3, that is, the string was deviated, $U_0 \not\equiv 0$, and simultaneously an external influence was included, $f(x,t) \not\equiv 0$. The solution is a superposition of two functions:

$$U(x,t) = v(x,t) + w(x,t) = U_{\text{st}}(x,t) + U_{\text{tr}}(x,t)$$

where $v(x,t)$ — matches the solution for Case 1, $w(x,t)$ — matches the solution for Case 3.

Case 7 is a general case of non-zero initial conditions, boundary conditions and non-zero external influences. There is no explicit analytical solution.

Thus, if only the initial deviation is present either with boundary conditions or without them, it generates a "pure" field of standing waves. If only external influence on the medium is present with fixed boundaries, it also generates a "pure" field of standing waves. In all others, there are superpositions of standing and traveling waves.

Chapter 7

Equations of Fluid and Gas Dynamics

7.1 Standing Waves in Solution of Fluid and Gas Dynamics Equations

A coupled, complete system of fluid and gas dynamics equations [15, 20, 23, 30–34, 37–39, 47–56, 255, 319, 332] connects pressure $P(x,t)$, velocity $V(x,t)$, density $\rho(x,t)$ and energy $E(x,t)$ of the coolant flow by the following equations:

$$
\begin{aligned}
&\frac{\partial \rho(x,t)}{\partial t} + \frac{\partial}{\partial x}[\rho(x,t)\cdot V(x,t)] = 0 && \text{continuity equation}\\
&\frac{\partial}{\partial t}[\rho(x,t)\cdot V(x,t)] \\
&\quad + \frac{\partial}{\partial x}[\rho(x,t)\cdot V(x,t)] = 0 && \text{motion equation}\\
&\frac{\partial E(x,t)}{\partial t} + V(x,t)\cdot\frac{\partial E(x,t)}{\partial x} \\
&\quad = -\frac{E(x,t)}{\rho(x,t)}\cdot\frac{\partial V(x,t)}{\partial x} && \text{energy equation}\\
&F[P(x,t),\rho(x,t),E(x,t)] = 0 && \text{state equation}
\end{aligned}
\tag{7.1}
$$

The equations of the system (7.1) are nonlinear; after linearization, considering the medium density as a constant and homogeneous, the first two equations are changed as follows:

$$\begin{cases} \dfrac{\partial P(x,t)}{\partial t} + \rho \cdot C^2 \cdot \dfrac{\partial V(x,t)}{\partial x} = 0 & \text{continuity equation} \\ \rho \cdot \dfrac{\partial V(x,t)}{\partial t} + \dfrac{\partial P(x,t)}{\partial x} = 0 & \text{motion equation (Euler)} \end{cases} \tag{7.2}$$

where C is the sound velocity.

These two equations connecting two functions of two variables represent fundamental laws. The continuity equation resolves itself into the law of conservation of mass. The motion equation is the representation of Newton's second law. Indeed, the summand $\rho \cdot \frac{\partial V}{\partial t}$ is a product of acceleration by a unit mass, and the second summand $\frac{\partial P(x,t)}{\partial x}$ of the equation is proportional to the force arising as a reaction to the change in volume, i.e. acceleration $\frac{\partial V}{\partial t}$ is reported to some volume element by mass m, which causes a Newtonian inertial force $F_\rho = m \cdot \frac{\partial V}{\partial t}$. The Newtonian inertial force is balanced by the reaction force $F_p = v \cdot \frac{\partial P}{\partial x}$ due to compression (expansion) of volume: $F_\rho = F_p$. For the one-dimensional hypothetical case:

$$F_p = v \cdot \frac{\partial P}{\partial x} = v \cdot \frac{\partial F}{\xi \cdot \partial x} = L \cdot \frac{\partial F}{\partial x}, \quad v = \xi \cdot L.$$

Two first-order equations (7.2) can be easily converted to the wave equation (second-order) through differentiation by the coordinate $\left(\frac{\partial}{\partial x}\right)$ of the first equation and through differentiation by time $\left(\frac{\partial}{\partial t}\right)$ of the second equation, and then by subtracting one from the other. As a result, we obtain the wave equation for velocity:

$$\rho \cdot C^2 \cdot \frac{\partial^2 V(x,t)}{\partial x^2} = \rho \cdot \frac{\partial^2 V(x,t)}{\partial t^2}$$

or

$$V_{xx} = \frac{1}{C^2} \cdot V_{tt}$$

The same system of Eqs. (7.2) is used to obtain the wave equation for pressure by means of similar actions:

$$P_{xx} = \frac{1}{C^2} \cdot P_{tt}$$

Compression (expansion) of some volume of the elastic medium causes an excess (lack) of pressure in this volume, resulting in the change of this pressure in time by the sound velocity. Equations (7.2) are also recorded in a symmetrical form by introducing parameter $\beta = \frac{1}{\rho C^2}$, called a compressibility:

$$\begin{cases} \beta \cdot \dfrac{\partial P}{\partial t} + \dfrac{\partial V}{\partial x} = 0 \\ \rho \cdot \dfrac{\partial V}{\partial t} + \dfrac{\partial P}{\partial x} = 0 \end{cases} \tag{7.3}$$

Now the medium is characterized by two parameters: density ρ and compressibility β. One parameter — ρ — defines inertial properties of the medium, the other — β — defines elastic properties. It may seem to be analogous with a spring pendulum equation. However, it is just a seeming analogy since the equations of the spring pendulum motion will never lead to the fluid and gas dynamics equations. However, the corresponding parameters can be compared at a qualitative level in order to emphasize the difference between these two oscillatory systems.

For an acoustic medium, parameters ρ, β are interconnected by the sound velocity, and the compressibility makes sense for sensitivity of the normalized density to the pressure change: $\beta = \frac{1}{\rho} \cdot \frac{d\rho}{dP}$. That is, the inertial parameter — ρ and the elastic parameter — β — of the medium include the medium density ρ. In case of the spring pendulum, the inertial parameter M (mass) and the elastic parameter K (stiffness) of the spring were independent and their ratio

$$\frac{M}{K} = \frac{1}{\omega_0^2}$$

defined the natural frequency ω_0 of oscillations of the spring pendulum. In acoustics, similar relation $\beta\rho = \frac{1}{C^2}$ provides the only parameter of the elastic medium — the sound velocity C as a characteristic, similar to the natural frequency ω_0.

When for the spring pendulum with the following motion equation:

$$M \cdot \frac{d^2x(t)}{dt^2} + K \cdot x(t) = 0$$

the natural frequency is $\omega_0 = \sqrt{\frac{K}{M}}$, then for the wave equation

$$V_{xx} = \frac{1}{C^2} \cdot V_{tt}$$

or

$$V_{xx} = \rho \cdot \beta \cdot V_{tt}$$

the sound velocity $C = \frac{1}{\sqrt{\rho \cdot \beta}}$ is the same medium characteristic as ω_0. Indeed, by comparing the mechanical and acoustic parameters, we will obtain the following:

$$\begin{gathered} \rho\left[\frac{\text{kg}}{\text{m}^3}\right] \leftrightarrow M[\text{kg}] \\ \frac{1}{\beta}\left[\frac{\text{N}}{\text{m}^2}\right] \leftrightarrow K\left[\frac{\text{N}}{\text{m}}\right] \\ C\left[\frac{\text{m}}{\text{s}}\right] \leftrightarrow \omega_0\left[\frac{1}{\text{s}}\right] \end{gathered} \tag{7.4}$$

However, the complete analogy of the mechanical pendulum and the elastic medium does not hold, as the motion equation in the form of a standard differential equation is not similar to the wave partial differential equation. Note that acoustic parameters (7.4) have the property of geometric normalization either in three or two or one dimension, which is a kind of specific property to be further developed in other models.

7.2 Acoustic Resistance

The solution of the wave equation can be both pressure $P(x, t)$ and velocity $V(x, t)$. Generally speaking, each of these solutions refers to a superposition of standing and traveling waves. Pressure and velocity are physically interrelated space–time functions. Considering them separately as independent solutions of the corresponding equation may lead to erroneous results.

The traveling pressure wave can be presented both as a real space–time function

$$P(x,t) = P(\omega t - kx) \tag{7.5}$$

and as a complex space–time function

$$P(x,t) = P_0 e^{j(\omega t - kx)} \tag{7.6}$$

Expressions for the traveling velocity waves can be written in the same representations:

$$\begin{aligned} V(x,t) &= V(\omega t - kx) \\ V(x,t) &= V_0 e^{j(\omega t - kx)} \end{aligned} \tag{7.7}$$

When the traveling pressure wave $P(x,t)$ and the traveling velocity wave $V(x,t)$ propagate in an infinite medium (i.e. without reflections), they are in phase, i.e. their space–time arguments coincide, and their relation — the constant

$$z = \frac{P(x,t)}{V(x,t)} = \frac{P_0}{V_0} = \text{const} \tag{7.8}$$

is called the acoustic resistance. It is a real number here. In this case, z is called an active resistance referring to the electrical analogy with an active electrical resistance, where no delay of one process $P(x,t)$ occurs in relation to another process $V(x,t)$. With the active resistance of the medium, the velocity and pressure are proportional to each other:

$$\begin{gathered} P(x,t) = \pm z \cdot V(x,t) \\ \left[\frac{\text{N}}{\text{m}^2}\right] \left[\frac{\text{kg}}{\text{c} \cdot \text{m}^2}\right] \cdot \left[\frac{\text{m}}{\text{c}}\right] \end{gathered} \tag{7.9}$$

and a symbol $\pm$ considers the direction of wave movement.

Acoustic resistance $z = \rho \cdot C$ is a characteristic of the medium and can be estimated from its density — ρ — and sound velocity in it — C. The traveling wave in the real representation relates to pressure traveling wave by the same expression as the Ohm's law connects current and voltage:

$$V(\omega t + kx) = \pm\frac{1}{\rho C} \cdot P(\omega t + kx) \tag{7.10}$$

However, there is one discrepancy with the electrical analogy. According to the latter, the active resistance irrevocably absorbs energy, but for the traveling waves under consideration, this constant is the wave resistance.

The existence of a single traveling wave and consequently phase coincidence between velocity and pressure is just an idealization. There are no mediums in nature, which give rise to solely traveling waves only. Even an infinite elastic medium generates standing waves, which in turn are a superposition of two waves traveling in opposite directions. Their interference causes other interrelations between pressure and velocity. Now their phase relationships (delay–advance) are not in-phase consequently, there is one understanding of resistances for the traveling waves, and another one is for standing waves.

If from the fundamental equations of fluid and gas dynamics two wave equations (continuity and motion equations) follow separately for velocity and pressure, this means that all conclusions on solution of the wave equation, including conclusions on inevitability of the standing waves, are the fundamental properties of the elastic medium. Then the velocity and pressure are represented by the following formulas as solutions of wave equations in the d'Alembert form:

$$V(x,t) = F_1(x - Ct) - F_2(x + Ct) \tag{7.11}$$

$$P(x,t) = \rho \cdot C \cdot F_1(x - Ct) + \rho \cdot C \cdot F_2(x + Ct) \tag{7.12}$$

Only due to the sign "–" in the velocity formula the acoustic resistance for the standing waves $z = P/V$ becomes a function of the coordinate

$$z = \rho \cdot C \cdot \frac{1 + G(x,t)}{1 - G(x,t)}$$

where $G(x,t) = \frac{\mathrm{F}_2(x+Ct)}{\mathrm{F}_1(x-Ct)}$ serves as a reflection coefficient.

While making a complex transition:

$$F_2(x + Ct) \to F_2(x) \cdot e^{jCt} \tag{7.13}$$

$$F_1(x - Ct) \to F_1(x) \cdot e^{-jCt} \tag{7.14}$$

we obtain

$$G(x,t) = \frac{F_2(x)}{F_1(x)} \cdot e^{j2Ct} \tag{7.15}$$

or for the modulus

$$|G(x,t)| = G(x) = \frac{F_2(x)}{F_1(x)}$$

Let us consider what kind of phase relationships can take pressure and velocity connected with Eq. (7.2), i.e. being bound by the fundamental equations of fluid and gas dynamics.

We consider the idealized case when pressure is a traveling wave

$$P(x,t) = P_0(x) \cdot e^{j\omega t} \tag{7.16}$$

and velocity is also a traveling wave, but with a different phase — φ:

$$V(x,t) = V_0(x) \cdot e^{j(\omega t + \varphi)} \tag{7.17}$$

Now let's ask ourselves: what phase φ is acceptable for velocity within the fluid and gas dynamics? Substituting the traveling waves into Eq. (7.3) the symmetrical system becomes simplified

$$\begin{cases} \dfrac{dP_0(x)}{dx} = -j \cdot \omega \cdot \rho \cdot e^{j\varphi} \cdot V_0(x) \\ \dfrac{dV_0(x)}{dx} = -j \cdot \omega \cdot \beta \cdot e^{-j\varphi} \cdot P_0(x) \end{cases} \tag{7.18}$$

After dividing the first equation by the second, we will obtain the following:

$$\frac{dP_0}{dV_0} = \frac{\rho}{\beta} \cdot e^{j2\varphi} \cdot \frac{V_0}{P_0} \tag{7.19}$$

or

$$P_0 \cdot dP_0 = \frac{\rho}{\beta} \cdot e^{j2\varphi} \cdot V_0 \cdot dV$$

After integrating with the zero initial condition, we have:

$$\frac{P_0^2}{2} = \frac{\rho}{\beta} \cdot e^{j2\varphi} \cdot \frac{V_0^2}{2}$$

or

$$\frac{P_0}{V_0} = \pm\sqrt{\frac{\rho}{\beta}} \cdot e^{j\varphi} = \pm\rho \cdot C \cdot e^{j\varphi} = z = z_0 \cdot e^{j\varphi}$$

where $z_0 = \rho \cdot C$ is the active wave resistance for the in-phase traveling waves.

The acoustic resistance of the phase became a complex number as there is *a priori* phase change between pressure and velocity. Its modulus has not changed, $|z| = \rho \cdot C$, i.e. it coincides with a real case, and the non-zero phase component means that the velocity starts to fall behind the pressure. This is a case of the so-called "pure" delay without attenuation. If the delay is a quarter of the traveling wave period, then $\varphi = \frac{\pi}{2}$ and acoustic resistance becomes a purely imaginary value:

$$z = \rho \cdot C \cdot e^{j\frac{\pi}{2}} = j \cdot \rho \cdot C = j \cdot z_0 \tag{7.20}$$

In electrical analogy, elastic acoustic medium is equivalent to capacitive resistance $(j \cdot \omega \cdot C)^{-1}$ or inductance with resistance $\cdot\, \omega \cdot L$. Thus, the phase change between the traveling waves of pressure and velocity of 90° is possible when the acoustic resistance is complex.

Now, let us consider a "pure" field of the standing wave, when both pressure

$$P(x,t) = \sin kx \cdot \cos \omega t \tag{7.21}$$

and velocity

$$V(x,t) = \frac{1}{\rho \cdot C} \cdot \sin(kx + \varphi) \cdot \cos \omega t \tag{7.22}$$

are the standing waves without additional traveling waves. The pressure amplitude here is the unit amplitude, and velocity amplitude $\frac{1}{\rho \cdot C}$ is coordinated with that. Now, let's ask ourselves: what phase is acceptable between the pressure and the velocity in accordance with the fluid and gas dynamics equations for standing waves? To answer this question, let us substitute hydromechanical expression (7.21) and (7.22) into a symmetrical equation system (7.3) of fluid and gas dynamics and we obtain a system of trigonometric equations:

$$\begin{cases} -\beta\omega + \dfrac{k}{\rho \cdot C} \cdot \dfrac{\cos(kx+\varphi)}{\sin kx} \cdot ctg\ \omega t = 0 \\ -\dfrac{\omega}{C} + k \cdot \dfrac{\cos kx}{\sin(kx+\varphi)} \cdot ctg\ \omega t = 0 \end{cases} \tag{7.23}$$

which leads to a single equation

$$\frac{\sin 2(kx+\varphi)}{\sin 2kx} = \beta \cdot \rho \cdot C^2 = 1$$

or

$$tg\ \varphi = ctg\ 2kx = \mp tg\left(\pm\frac{\pi}{2} - 2kx\right) \tag{7.24}$$

Thus, it is allowed to have a shift between pressure and velocity to only one quarter of the spatial period of the standing wave or to $\varphi = \pm\frac{\pi}{2}$ in "pure" standing waves mode. Such medium has acoustic resistance depending on the spatial variables:

$$\begin{cases} P(x,t) = \sin kx \cdot \cos\omega t \\ V(x,t) = \pm\dfrac{1}{\rho \cdot C} \cdot \cos kx \cdot \cos\omega t \end{cases} \tag{7.25}$$

$$z(x) = \frac{P(x,t)}{V(x,t)} = \rho \cdot C \cdot tgkx \tag{7.26}$$

This is a tangential (purely nonlinear) spatial dependence which changes resistance in the infinite limits with periodic discontinuity. For the low frequencies and/or short distances $k \cdot x = \frac{\omega}{C}x \ll 1$, we have a linear dependence of the acoustic resistance on the distance and frequency

$$z(x) = \rho \cdot C \cdot tgkx \simeq \rho \cdot C \cdot \frac{\omega}{C} \cdot x = \rho \cdot \omega \cdot x \tag{7.27}$$

and

$$z(x) = \rho \cdot C \cdot tgkx = \begin{cases} \sim 0; & \omega, x \sim 0, \\ \rho \cdot \omega \cdot x; & \dfrac{\omega x}{C} \ll 1 \\ \rho \cdot C, & \text{at } x = \dfrac{\pi C}{4\omega} \\ \infty, & \text{at } x = \dfrac{\pi C}{2\omega} \end{cases} \tag{7.28}$$

At very short distances, the medium does not actually resist the standing wave. While the distance is increasing, the resistance is proportional to the density ρ and frequency ω. At a distance $x = \frac{\pi C}{4\omega}$, it exactly coincides with the acoustic wave resistance for the traveling wave — $\rho \cdot C$. At $x_\infty = \frac{\pi C}{2\omega}$, it is infinitely large, i.e. the velocity is zero.

For instance, for parameters close to TN VVER-1200 parameters in nominal conditions $C \approx 0.8\,\text{km/s}$, $f = 7\,\text{Hz}$, we get $x_\infty = 28.5\,\text{m}$, which is roughly equal to the quarter of the spatial period of the standing wave.

Two extreme cases have been considered above:

- pressure and velocity are traveling waves;
- pressure and velocity are standing waves.

In this case, the acoustic resistance changes from a fixed number to unlimited value, z can be a complex number, and a function of the spatial coordinate. It turns out that in practice, when there is a combination of the traveling and standing waves of pressure and velocity, the acoustic resistance can have any values.

In electrical engineering, the resistance is always a real positive number for the direct current and voltage. The same will apply to constant pressure P_0 and velocity V_0, which are not the subject of acoustics of the elastic media. The resistance will be capacitive or inductive (i.e. a purely imaginary complex number) for alternating current and voltage, when there is no energy loss, i.e. no active resistance. At this stage, the analogy of acoustics and electrical engineering ends in terms of concentrated (point) resistances, as resistance in acoustics becomes a function of spatial coordinates. The analogy will be restored and can be continued if we move from the lumped electrical engineering parameter model to distributed parameter model, to the theory of the so-called long lines.

7.3 Power Characteristics of Acoustic Waves

Solution of the wave equation for pressure

$$P_{xx}(x,t) = \frac{1}{C^2} \cdot P_{tt}(x,t)$$

written in the following form:

$$P(x,t) = A \cdot e^{j(kx-\omega t)} \tag{7.29}$$

is called a plane wave.

Exactly the same wave equation is true for velocity $V(x,t)$

$$V_{xx}(x,t) = \frac{1}{C^2} \cdot V_{tt}(x,t) \tag{7.30}$$

and its complex harmonic solution will have the same exponent but different amplitude:

$$V(x,t) = \frac{A}{\rho \cdot C} \cdot e^{j(kx-\omega t)} \tag{7.31}$$

We have a traveling wave both for pressure and velocity. They are in-phase with each other. The sound intensity — I — as a power characteristic can be estimated as the product of $P(x,t) \cdot V(x,t)$. As a rule the intensity modulus is estimated averaging for a period:

$$\begin{aligned} |I| &= |\langle \operatorname{Re} P(x,t) \cdot \operatorname{Re} V(x,t) \rangle| \\ &= \frac{|A|^2}{\rho \cdot C} \cdot \langle \cos^2(kx - \omega x) \rangle = \frac{|A|^2}{2\rho \cdot C} \end{aligned} \tag{7.32}$$

where Re is the real part, $\langle \ldots \rangle$ is the averaging operation for the period, I is sound intensity (current power density).

The system of Eq. (7.3), consisting of the motion equation and continuity equation, is the law of energy conservation in an implicit form. To obtain an explicit form, we will multiply the first equation by the velocity $V(x,t)$, and the second equation — by the pressure $P(x,t)$; then let us sum up:

$$V \cdot \left(\rho \cdot \frac{\partial V}{\partial t} + \frac{\partial P}{\partial x}\right) = 0$$

$$P \cdot \left(\beta \cdot \frac{\partial P}{\partial t} + \frac{\partial V}{\partial x}\right) = 0 \tag{7.33}$$

$$\rho \cdot V \cdot \frac{\partial V}{\partial t} + \beta \cdot P \cdot \frac{\partial P}{\partial t} + V \cdot \frac{\partial P}{\partial x} + P \cdot \frac{\partial V}{\partial x} = 0$$

where parameter β — compressibility — defines the elastic properties of the medium: $\beta = \frac{1}{\rho C^2}$, with dimension $\left[\frac{\mathrm{m}^2}{\mathrm{N}}\right]$, which is inverse of dimension of pressure.

After converting Eq. (7.33)

$$\frac{\partial}{\partial t}\left(\frac{1}{2} \cdot \rho \cdot V^2 + \frac{1}{2} \cdot \beta \cdot P^2\right) + \frac{\partial}{\partial x}(P \cdot V) = 0 \tag{7.34}$$

the summands $\frac{1}{2} \cdot \rho \cdot V^2 = E_k^0$ and $\frac{1}{2} \cdot \beta \cdot P^2 = E_n^0$ are, respectively, the specific kinetic and specific potential energy per volume unit of the medium, in which the perturbation propagates. Their sum $E^0 = E_k^0 + E_n^0$ is the total specific energy of the medium with dimension $[\mathrm{J/m^3}]$.

The product $P(x,t) \cdot V(x,t) = I(x,t)$ is the current power density with dimension $\left[\frac{\mathrm{N}}{\mathrm{m \cdot s}}\right]$. Thus, as follows from (7.34), the differential law of conservation of the fluid energy will be expressed as follows:

$$\frac{\partial E^0}{\partial t} = -\frac{\partial I}{\partial x}$$

That is, the change of the specific energy of the medium in time is equal to the change of current power density in space.

For the traveling wave, the velocity V and pressure P are proportional to each other via the two parameters of the medium (ρ, β) or (ρ, C)

$$V(x,t) = \frac{1}{\rho \cdot C} \cdot P(x,t) = \sqrt{\frac{\beta}{\rho}} \cdot P(x,t)$$

the current power density is

$$I(x,t) = V(x,t) \cdot P(x,t) = \frac{1}{\rho \cdot C} \cdot P^2(x,t) = \rho \cdot C \cdot V^2(x,t)$$

or

$$I(x,t) = 2 \cdot C \cdot E_n^0(x,t)$$

The power flux density in the traveling wave is proportional to both specific kinetic energy and specific potential energy, which are equal to each other, and therefore

$$I(x,t) = C \cdot E^0(x,t)$$

The traveling wave affects each dS cross-section perpendicular to the direction of propagation of perturbation with force due to sound pressure:

$$dF = P(x,t) \cdot dS$$

which increases the power of the sound pressure

$$dQ = V(x,t) \cdot dF = V(x,t) \cdot P(x,t) \cdot dS = I(x,t) \cdot dS = C \cdot E^0 \cdot dS$$

When we say that the traveling wave carries energy, we mean that the sound pressure in any coordinates "x" can produce work.

For the standing wave which is a result of the superposition of two traveling waves

$$[P_1(\omega t - kx), V_1(\omega t - kx)]$$

and

$$[P_2(\omega t + kx), V_2(\omega t + kx)]$$

the resulting pressure is sum, but the velocities are subtracted, since these two waves have different directions

$$\begin{aligned} P(x,t) &= P_1(\omega t - kx) + P_2(\omega t + kx) \\ V(x,t) &= V_1(\omega t - kx) - V_2(\omega t + kx) \\ &= \frac{1}{\rho \cdot C}[P_1(\omega t - kx) - P_2(\omega t + kx)] \end{aligned}$$

Then the current power density of the standing wave

$$I = P \cdot V = \frac{1}{\rho \cdot C}[P_1^2(\omega t - kx) - P_2^2(\omega t + kx)] = I_1 - I_2 \equiv 0$$

since the traveling waves have similar parameters. This is the meaning of the saying that the standing wave does not carry energy.

Even in the simplest case, when writing the solutions of wave equations using the d'Alembert method

$$\begin{cases} P(x,t) = P_1(\omega t - kx) + P_2(\omega t + kx) \\ V(x,t) = V_1(\omega t - kx) - V_2(\omega t + kx) \end{cases}$$

there is too little information on their phase relations, although the sign "–" in the formula for the velocity terms, which appeared due to the opposite direction of the two traveling waves, introduces difference into the pressure and velocity representations. To provide more clarity, it is necessary to turn to the pure frequency language in the phase relations of velocity and pressure.

The wave equation, which is valid both for pressure and velocity, does not provide their phase relationship, if these equations are considered in separate ways. In that case, it produces always an in-phase solution

$$\begin{cases} P(x,t) = P_0(x) \cdot e^{j(\omega t - kx)} \\ V(x,t) = V_0(x) \cdot e^{j(\omega t - kx)} \end{cases}$$

which is valid only for the traveling wave.

But the wave equation is also valid for the standing wave with a constant phase shift between the pressure and velocity:

$$P(x,t) = 2P_0(x) \cdot \cos kx \cdot \cos \omega t$$

$$V(x,t) = 2V_0(x) \cdot \cos kx \cdot \cos \omega t$$

Generally, this means that the spatial functions $P_0(x)$ and $V_0(x)$ are complex functions. Their interrelation is introduced through the complex acoustic resistance by analogy with Ohm's law in electrical engineering

$$z(x) = \frac{P_0(x)}{V_0(x)}$$

Let us recall that the complex resistance for the traveling wave is a real number $z_0 = \frac{P_0}{V_0}$.

Chapter 8

Long Lines as an Analogy to the Acoustic Medium

8.1 Standing Wave in Solution of the Telegraph Equation System

This chapter is devoted to a brief presentation of that part of the long line theory which has a formal similarity in its main equations to the equation of elastic media. However, the information that follows will be enough to build competent analogies between acoustics and long lines for reactor applications. This approach refers to a summary of those cases when the standard "point" electrical analogies in the centralized parameters have not been valid. For a deeper understanding of this issue, we recommend studying not only the classics of electrical engineering [14], but also the theory of ultra-high-frequency transmission lines [12, 13].

The telegraph equations have the form of the system of two partial differential equations

$$\begin{cases} L \cdot \dfrac{\partial I(x,t)}{\partial t} + \dfrac{\partial U(x,t)}{\partial x} = 0 \\ \not{C} \cdot \dfrac{\partial U(x,t)}{\partial t} + \dfrac{\partial I(x,t)}{\partial x} = 0 \end{cases} \tag{8.1}$$

connecting current $I(x,t)$ with voltage $U(x,t)$ in a long line, defined by two parameters:

- $\not{C}$ — linear (specific) capacitance $\left[\frac{F}{m}\right]$,
- L — linear (specific) inductance $\left[\frac{henry}{m}\right]$.

The form of the telegraph equation system is similar to the form of fluid and gas dynamics equation system:

$$\begin{cases} \rho \cdot \dfrac{\partial V(x,t)}{\partial t} + \dfrac{\partial P(x,t)}{\partial x} = 0 \\ \beta \cdot \dfrac{\partial P(x,t)}{\partial t} + \dfrac{\partial V(x,t)}{\partial x} = 0 \end{cases} \tag{8.2}$$

when

- voltage $U(x,t)$ is replaced with pressure $P(x,t)$;
- current $I(x,t)$ is replaced with velocity $V(x,t)$;
- specific capacitance $\not{C}$ is replaced with compressibility β;
- specific inductance L is replaced with density ρ.

Here, we observe the first discrepancy: parameters $L, \not{C}$ are distributed (linear) in long lines; in acoustics the density ρ and compressibility β are point-to-point parameters, defining the entire acoustic medium in a global manner. Direct analogies in terms of $L \Longleftrightarrow \rho$ and $\not{C} \Longleftrightarrow \beta$ parameters are not obvious at all. But correspondence of relations of these parameters

$$\mathcal{R} = \sqrt{\frac{L}{\not{C}}} \Longleftrightarrow z_0 = \rho \cdot C = \sqrt{\frac{\rho}{\beta}}$$

(or correspondence of the long line wave resistance — $\mathcal{R}$ — to acoustic wave resistance $z_0 = \rho \cdot C$), when the line dimension of inductance and capacitance is reduced into two-point global parameters $\mathcal{R}$ and z_0, makes it on the contrary quite obvious.

The natural frequency of the long line element $\omega_0 = \frac{1}{\sqrt{L \cdot \not{C}}}$, with a linear dimension [Hz · m], instead of the usual dimension [Hz] is also similar to the sound velocity – $C = \frac{1}{\sqrt{\rho \cdot \beta}}$ since both parameters have the same dimension $[\mathrm{Hz} \cdot m] = \left[\frac{\mathrm{m}}{\mathrm{s}}\right]$. Consequently, the following analogies are true:

$$\omega_0 = \frac{1}{\sqrt{L \cdot \not{C}}} \Longleftrightarrow C = \frac{1}{\sqrt{\rho \cdot \beta}},$$

$$\mathcal{R} = \sqrt{\frac{L}{\not{C}}} \Longleftrightarrow z_0 = \rho \cdot C$$

In addition, the products $\frac{1}{\sqrt{L\cdot ¢}}$ rather than the absolute parameters of the linear inductance L and linear capacitance ¢ serve as parameters in solutions for current and voltage. Indeed, by differentiating the equations of systems (8.1 and 8.2) in terms of time and space, they are transformed into the wave equations with parameters ω_0 and C:

$$\begin{cases} \dfrac{\partial^2 U}{\partial x^2} - \dfrac{1}{\omega_0^2}\cdot\dfrac{\partial^2 U}{\partial t^2} = 0 \\ \dfrac{\partial^2 I}{\partial x^2} - \dfrac{1}{\omega_0^2}\cdot\dfrac{\partial^2 I}{\partial t^2} = 0 \end{cases}$$

$$\begin{cases} \dfrac{\partial^2 P}{\partial x^2} - \dfrac{1}{C^2}\cdot\dfrac{\partial^2 P}{\partial t^2} = 0 \\ \dfrac{\partial^2 V}{\partial x^2} - \dfrac{1}{C^2}\cdot\dfrac{\partial^2 V}{\partial t^2} = 0 \end{cases}$$

For the complex harmonic wave

$$P(x,t) = e^{j\omega t}\cdot P(x)$$

$$V(x,t) = e^{j\omega t}\cdot V(x)$$

the acoustic system takes the form of harmonic oscillators

$$\frac{\partial^2 P(x)}{\partial x^2} - \gamma^2\cdot P(x) = 0$$

$$\frac{\partial^2 V(x)}{\partial x^2} - \gamma^2\cdot V(x) = 0$$

with a wave number (or a propagation constant)

$$\gamma = \sqrt{j\omega\rho\cdot j\omega\beta} = j\frac{\omega}{C} = jk\left[\frac{1}{m}\right]$$

by particular solutions

$$P(x) = P_0 e^{-jkx}$$

$$V(x) = V_0 e^{-jkx}$$

and by global solutions

$$P(x,t) = P_0 e^{j\omega t}\cdot e^{-jkx}$$

$$V(x,t) = V_0 e^{j\omega t}\cdot e^{-jkx}$$

Parameter $\gamma = \sqrt{j\omega L \cdot j\omega \cancel{C}} = j\frac{\omega}{\omega_0} = jk[\frac{1}{\mathrm{m}}]$ and solutions will be similar for the long line. In both cases, the medium parameters ρ, β (and parameters of the long line $L, \cancel{C}$) take part in the solutions in the form of the product. The analogy is valid for both relations and products:

$$\omega_0 = \frac{1}{\sqrt{L \cdot \cancel{C}}} \iff C = \frac{1}{\sqrt{\rho \cdot \beta}},$$

Mistakes are very frequent in applications due to the use of the standard dimension of the natural frequency for the long lines instead of the linear natural frequency and to improper analogies of the linear parameters with point parameters.

Direct analogies of specific electrical and global acoustic parameters are impossible by definition. However, the global acoustic parameters can be represented by specific parameters: the density can be represented by acoustic mass m_a since $\rho = m_a \cdot l\,s$ of the dimension $\left(\left[\frac{\mathrm{kg}}{\mathrm{m}^4}\right] \cdot [\mathrm{m}]\right)$, and compressibility can be represented by acoustic compressibility as $\beta = \frac{C_a}{\mho}s$ of $\left(\left[\frac{\mathrm{m}^5}{\mathrm{N}}\right] \cdot (\mathrm{m}^{-3}\right)$ dimension, where l is a linear element, and $\mho$ is a volume element:

$$L\left[\frac{\mathrm{henry}}{\mathrm{m}}\right] \iff \rho = m_a \cdot l \left[\frac{\mathrm{kg}}{\mathrm{m}^4}\right] \cdot [\mathrm{m}]$$

$$\cancel{C}\left[\frac{F}{\mathrm{m}}\right] \iff \beta = \frac{C_a}{\mho}\left[\frac{\mathrm{m}^5}{\mathrm{N}}\right] \cdot [\mathrm{m}^{-3}] \text{ or } \frac{1}{\cancel{C}} \iff \frac{1}{\beta} = \frac{\mho}{C_a}$$

That is, the density, being a global parameter, is a product of specific parameters — the acoustic mass — m_a, by length of the acoustic object and the stiffness is a specific parameter — $\frac{1}{C_a}$, which shall be multiplied by the volume of the acoustic object. If the element of the acoustic volume can be represented as a complex of acoustic mass at micro-length l and acoustic capacity in micro-volume $\mho$, then the electrical engineering analogy seems to implicitly exist after all. This logic corresponds to the reverse transition from the distributed long line to the Helmholtz resonator in the lumped parameters.

A gross error is made when the natural frequency of the long line $\omega_0 = \frac{1}{\sqrt{L \cdot \cancel{C}}}$ is matched with the natural frequency of the acoustic medium in the form $\omega_a = \frac{1}{\sqrt{m_a \cdot \cancel{C}_a}}$. Let us recall that the natural frequency of the Helmholtz resonator is derived from the latter formula. There we had to spatially divide acoustic medium into inertial and elastic parts. The long lines do not impose such a requirement. It seems that on the same length l there exist both parts. When substituting the expressions $m_a = \frac{\rho L}{S}$, $C_a = \frac{V}{\rho C^2}$ to the natural frequency formula $\omega_a = \frac{1}{\sqrt{m_a \cdot C_a}}$, we obtain $\omega_a = \frac{C}{l}$ [Hz]. The frequency ω_a was obtained in absolute units even though the frequency $\omega_0 = \frac{1}{\sqrt{L \cdot \cent}} - \text{B}$ was obtained in linear units [Hz $\cdot$ m]. This lead to an incorrect analogy exactly to a constant

$$\omega_0 = \frac{1}{\sqrt{L \cdot \cent}} \iff \omega_a = \frac{C}{l} \quad \text{instead of}$$

$$\omega_0 = \frac{1}{\sqrt{L \cdot \cent}} \iff C = \frac{1}{\sqrt{\rho \cdot \beta}}$$

Pair of variables — $V(x,t)$ and $P(x,t)$ — a field of velocity and pressure, as a result of propagation of a longitudinal wave in an elastic acoustic medium can be changed if the acoustic medium is enclosed in a pipeline with constant cross-section $S(\text{m}^2)$. In such conditions, the velocity can be replaced by the flow rate $G(x,t) = S \cdot V(x,t)$, and the pressure field can be replaced by the force field: $F(x,t) = S \cdot P(x,t)$. The system of Eqs. (8.2) would remain unchanged:

$$\begin{cases} \rho \cdot \dfrac{\partial G(x,t)}{\partial t} + \dfrac{\partial F(x,t)}{\partial x} = 0 \\ \beta \cdot \dfrac{\partial F(x,t)}{\partial t} + \dfrac{\partial G(x,t)}{\partial x} = 0 \end{cases}$$

Let us note that $G(x,t)$ is the flow rate of the longitudinal wave and not a flow rate of a moving coolant, as well as the force $F(x,t)$ emerges due to wave propagation through the elastic medium, and it is not an external disturbing force. Here lies the reason for another analogy error, when one tries to describe the coolant environment pumping with flow rate $G(x,t)$ by acoustic parameters.

Let us return to the properties of long lines: as shown above, by differentiating by ≪x≫ and by ≪t≫ of the system equations (8.1), we obtain a wave equation for the voltage

$$U_{xx} = \frac{1}{\omega_0^2} \cdot U_{tt} \tag{8.3}$$

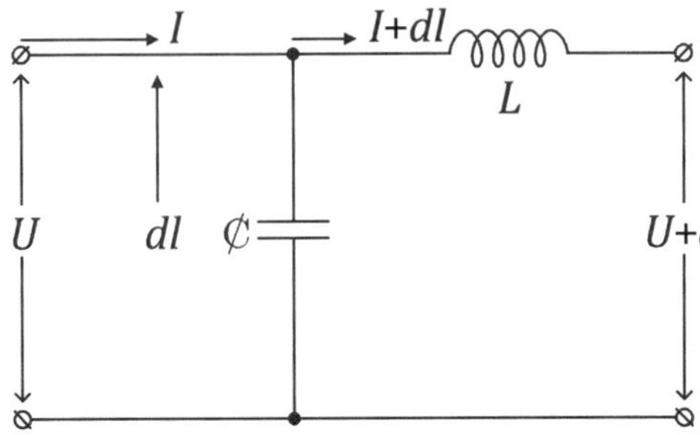

Figure 8.1: A long line element of dimension dx, as a four-pole (two-port) circuit.

and a wave equation for the current

$$I_{xx} = \frac{1}{\omega_0^2} \cdot I_{tt} \tag{8.4}$$

It is difficult to use these second-order equations to make an electrical equivalent. However, the line element can be represented without losses by a simple four-pole (two-port) circuit, consisting of a linear capacitance and a linear inductance (Fig. 8.1).

The input voltage of the four-pole (two-port) circuit $U(x,t)$ is converted on the elementary length dx into the output voltage $U(x,t) + dU(x,t)$ through the voltage drop on the inductance:

$$-\partial U = \frac{\partial I}{\partial t} \cdot L \cdot dx \tag{8.5}$$

and change of the output current $I(x,t) + dI(x,t)$ is determined by the current in capacitance

$$-\partial I = \frac{\partial U}{\partial t} \cdot C \cdot dx \tag{8.6}$$

The combined representation of voltage on inductance and current through capacitance will provide a system of telegraph equations:

$$\begin{cases} L \cdot \dfrac{\partial I}{\partial t} + \dfrac{\partial U}{\partial x} = 0 \\ \not{C} \cdot \dfrac{\partial U}{\partial t} + \dfrac{\partial I}{\partial x} = 0 \end{cases} \tag{8.7}$$

The long electrical line can be imagined as two long wires, which are located so close to each other that they have distributed capacitance and a distributed inductance at the same time due to electromagnetic properties of metal conductors. In this case, the acoustic analogy is not obvious. In the point (lumped parameter) case, one electrical element cannot be both capacitance and inductance at the same time. They must be spatially separated. Then the Helmholtz resonator is the acoustic analogy. Ultra-high-frequency waveguides, which are round or rectangular shaped tubes, can be represented as long lines [12, 13]. They are the most suitable for the analogy of the pipelines filled with acoustic media.

Let us again recur to complex harmonic variables

$$\begin{aligned} \tilde{U}(x,t) &= e^{j\omega t} \cdot U(x), \\ \tilde{I}(x,t) &= e^{j\omega t} \cdot I(x) \end{aligned} \tag{8.8}$$

Then equations (8.7) are reduced to the system

$$\begin{cases} e^{j\omega t} \cdot \dfrac{\partial U}{\partial x} = -L(j,\omega) \cdot e^{j\omega t} \cdot I \\ e^{j\omega t} \cdot \dfrac{\partial I}{\partial x} = -\text{¢}(j,\omega) \cdot e^{j\omega t} \cdot U \end{cases} \tag{8.9}$$

or

$$\begin{cases} \dfrac{\partial U}{\partial x} = -j\omega L \cdot I \\ \dfrac{\partial I}{\partial x} = -j\omega \text{¢} \cdot U \end{cases} \tag{8.10}$$

Now let's divide Eq. (8.10) by the right and left sides

$$\frac{\partial U}{\partial I} = \frac{L}{\text{¢}} \cdot \frac{I}{U} \tag{8.11}$$

or

$$U \cdot dU = R^2 \cdot I \cdot dI \tag{8.12}$$

where $\mathcal{R} = \sqrt{\frac{L}{\text{¢}}}$ is wave resistance.

The latter equation is Ohm's law in a differential form. After integrating with a zero initial condition, we obtain Ohm's law as the usual algebraic expression:

$$U^2 = R^2 \cdot I^2 + \text{const} \quad \text{or} \quad U = R \cdot I \tag{8.13}$$

The fundamental Ohm's law is neither a consequence nor a solution of wave equations (8.3) and (8.4). Being a relation between current and voltage, it is used whenever the system is reduced to two wave equations individually for voltage and current.

Each of the wave equations (8.3) and (8.4) has only one unknown function: either $U(x,t)$, or $I(x,t)$, while the original system of Eqs. (8.1) has a relation $U(x,t) \leftrightarrow I(x,t)$.

As it is shown above, the standing and traveling waves occur in the medium described by the wave equation. For the one-dimensional spatial case, this can be represented in other physical categories: at any point of the elastic medium there are two multidirectional traveling waves at the same time — the so-called incident and reflected traveling waves. In the long line, both the current $I(x,t)$ and voltage $U(x,t)$ can be represented in any point as a relevant sum of the incident and reflected wave. In case of complex harmonic solutions, we have:

$$\begin{aligned} U(x,t) &= e^{j\omega t} \cdot (U_{\text{fall}} \cdot e^{jkx} + U_{\text{refl}} \cdot e^{-jkx}) \\ I(x,t) &= e^{j\omega t} \cdot (I_{\text{fall}} \cdot e^{jkx} - I_{\text{refl}} \cdot e^{-jkx}) \end{aligned} \tag{8.14}$$

where $k = \frac{\omega}{C}$ is a wave number.

The voltage $U(x,t)$, as a superposition of the incident and reflected wave, is a standing wave and uncompensated traveling wave in any point of the long line. This conclusion is also true for the current $I(x,t)$.

For simplicity, we will omit the multiplier as a complex exponent $e^{j\omega t}$. The voltage normalized to the amplitude of the incident wave in the long line is represented by the following formula:

$$\begin{aligned} \tilde{U}(x,t) = \frac{U(x,t)}{U_{\text{fall}}} &= e^{jkx} + G \cdot e^{-jkx} \\ &= 2 \cdot G \cdot \cos kx + (1 - G) \cdot e^{jkx} \end{aligned} \tag{8.15}$$

where $0 \leq |G| \leq 1$ and G are the reflection coefficients, which is a complex number.

According to the formula (8.15), in the absence of reflection – $|G| = 0$ – or if the load fully absorbs the wave energy (and absorption in the long line itself is zero)

$$\tilde{U} = e^{jkx}$$

That is, the long line contains only one traveling wave, which is the incident wave. There is no attenuation of the voltage amplitude in the line, since $|\tilde{U}| \equiv 1$. However, there is s phase shift of the output voltage with respect to the input voltage.

The expression $\tilde{U} = e^{jkx} = e^{j\varphi}$ is a so-called "pure" delay with phase-shift φ, and the ideal long line is that according to electrical engineering, called delay line. These lines are literally defined by the delay time — τ_3. If we recall that the spatial coordinate $\frac{1}{k}$ is overcome by a sound wave at the sound velocity — $k = \frac{\omega}{C}$, then

$$k \cdot x = 2\pi \cdot \frac{x}{C} \cdot f = 2\pi \cdot \tau_3 \cdot f$$

is a linear function of frequency, and

$$\tilde{U} = e^{jkx} = e^{j2\pi \cdot \tau_S \cdot f} \tag{8.16}$$

where $\tau_3 = \frac{x}{C}$ is a long-time delay time, proportional to the long line length — x.

Looking ahead, we note that the non-zero attenuation, introduced by the non-ideal long line actively absorbing the energy of the longitudinal wave, establishes the so-called dispersion effect. This effect, based on the fact that the phase ceases to be a linear function of frequency and the form of the input voltage at $x = 0$, will be distorted till the long line output at $x = l$. Phase nonlinearity between input and output of the long line is a sign of non-zero active attenuation of the long line.

Only when there is a full reflection $|G| = 1$

$$\tilde{U} = 2 \cdot \cos kx$$

there is a standing wave of double amplitude.

In practice, $0 < |G| < 1$ we have a superposition of the standing wave and non-compensated incident traveling wave

$$\tilde{U} = 2 \cdot G \cdot \cos kx + (1 - G) \cdot e^{jkx}$$

This expression can also be written in terms of the reflected wave

$$\tilde{U} = 2 \cdot \cos kx - (1 - G) \cdot e^{-jkx}$$

Both current $I(x,t)$ and voltage $U(x,t)$ in the long line will be represented without losses by the standing wave, which is similar to the standing wave of pressure and velocity in the acoustic medium.

8.2 Long Lines with Losses

For a line with losses (Fig. 8.2), due to non-ideal isolation between conductors (linear conductivity of isolation $G \neq 0$) and zero ohmic resistance of conductors (linear resistance of the line $R \neq 0$), the telegraph equations are extended by the so-called active terms G, R.

By analogy with Eqs. (8.5) and (8.6), it is easy to write the following equations for increments:

$$\begin{cases} -dU = I \cdot R \cdot dx + \dfrac{\partial I}{\partial t} \cdot L \cdot dx \\ -dI = U \cdot G \cdot dx + \dfrac{\partial U}{\partial t} \cdot C \cdot dx \end{cases} \tag{8.17}$$

The energy irreversibly dissipated in the non-ideal line is proportional to the linear conductance and the linear resistance:

$$E_a = R \cdot I^2 + G \cdot U^2$$

This energy is determined only by the active elements, so it is called the active energy (in this case, it is the energy absorbed by the line), and the energy stored in the reactive elements (capacitance and inductance) is called the reactive (passive) energy.

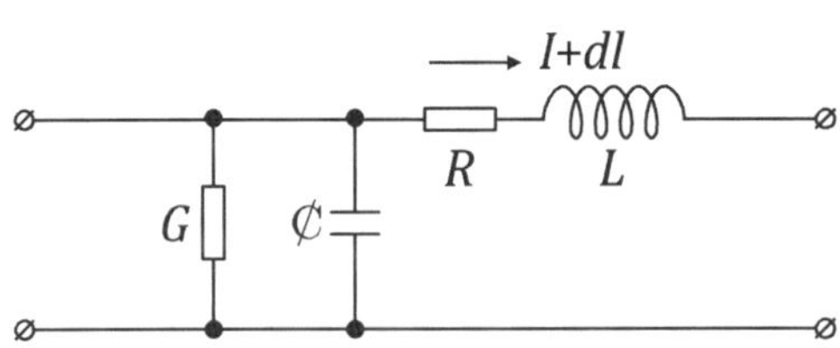

Figure 8.2: Long line with active losses.

The telegraph equations will take the following form:

$$\begin{cases} \dfrac{\partial U}{\partial x} = -L \cdot \dfrac{\partial I}{\partial t} - R \cdot I \\ \dfrac{\partial I}{\partial x} = -\cent \cdot \dfrac{\partial U}{\partial t} - G \cdot U \end{cases} \tag{8.18}$$

and after differentiating we will obtain extended wave equations:

$$\begin{cases} \dfrac{\partial^2 U}{\partial x^2} = L \cdot \cent \cdot \dfrac{\partial^2 U}{\partial t^2} + (R \cdot \cent + G \cdot L)\dfrac{\partial U}{\partial t} + G \cdot R \cdot U \\ \dfrac{\partial^2 I}{\partial x^2} = L \cdot \cent \cdot \dfrac{\partial^2 I}{\partial t^2} + (R \cdot \cent + G \cdot L)\dfrac{\partial I}{\partial t} + G \cdot R \cdot I \end{cases} \tag{8.19}$$

The extension consists of two last terms describing the energy losses. They are represented by only two parameters of loss — G, R — which have a well-defined physical meaning, i.e. losses in the isolation and losses in the conductor. In the acoustics, there are much more mechanisms of energy losses of the longitudinal wave, so the analogies with only two mechanisms of losses in the long lines are not always true. In acoustics, one every time has to find analogues of the linear ohmic resistance R and the linear isolation conductivity G, but this may not be satisfactory.

The "acoustics–electrical engineering" analogy is not the only one, there is also the "acoustics–mechanics" analogy at the qualitative level. The acoustic volume element can be presented as a four-pole (two-port) circuit, electrical components of which are replaced with mechanical components of the spring pendulum (Fig. 8.3):

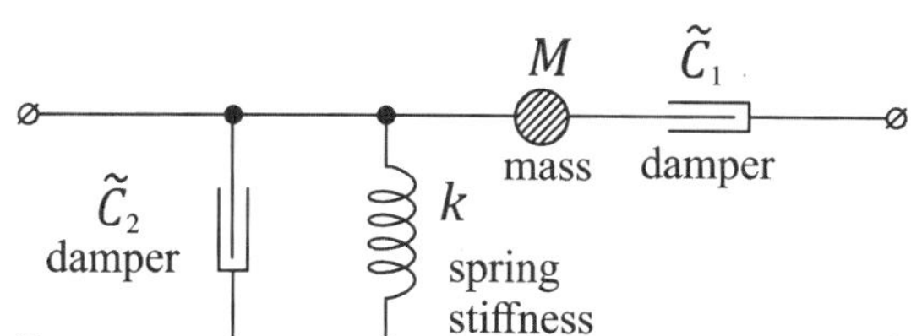

Figure 8.3: "Acoustics–mechanics" analogy.

where matches are applied as follows:

- mass $m \leftrightarrow$ inductance $L \leftrightarrow$ density ρ;
- stiffness $k \leftrightarrow$, capacitance $₵ \leftrightarrow$ compressibility β;
- damping $\leftrightarrow$ resistance $\leftrightarrow$ (conductance) friction (viscosity).

The entire acoustic environment at the micro level seems to consist of an infinite number of these mechanical four-pole (two-port) circuits–pendulums. This qualitative correspondence between acoustics and mechanics does not allow replacing acoustic parameters with the relevant mechanical parameters in the corresponding equations of mechanical motion. This analogy rather illustrates the genetic relationship of the mechanical motion equations and the acoustics wave equation.

Let us continue to examine the properties of long lines. For harmonic complex variables, similarly to equations of the ideal long line

$$\ddot{U}(x,t) = e^{j\omega t} \cdot U(x),$$
$$\tilde{I}(x,t) = e^{j\omega t} \cdot I(x)$$

the telegraph equations take the form (Fig. 8.4) of the harmonic spatial oscillators

$$\begin{cases} \dfrac{\partial^2 U}{\partial x^2} = -z_1 \cdot \dfrac{\partial I}{\partial x} = z_1 \cdot Y_1 \cdot U = \gamma^2 \cdot U \\ \dfrac{\partial^2 I}{\partial x^2} = -Y_1 \cdot \dfrac{\partial U}{\partial x} = z_1 \cdot Y_1 \cdot I = \gamma^2 \cdot I \end{cases} \tag{8.20}$$

where the parameter γ has an analogy of the natural frequency in the Lagrange equation of motion but it is called the propagation

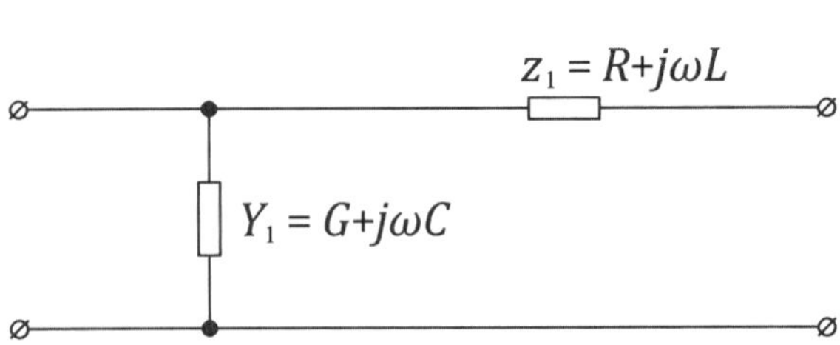

Figure 8.4: Long line with complex impedance.

coefficient (constant) and refers to the "I/O" relationship parameter of the long line.

Now the impedance of the long line defined by Ohm's law is a complex number

$$z = \sqrt{\frac{z_1}{Y_1}} = \sqrt{\frac{R + j\omega L}{G + j\omega \text{₵}}}, \text{ [Ohm]}$$

where $z_1 = R + j\omega L$, $\left[\frac{\text{Ohm}}{\text{m}}\right]$ is a complex linear resistance of conductors, $Y_1 = G + j\omega \text{₵}$ is a complex linear resistance of isolation.

Resistance z_1 and conductance Y_1, which are the factors in the expression for the propagation coefficient γ, are also present in the complex resistance expression, but in the form of a fraction: $z = \frac{z_1}{Y_1}$.

Two complex functions of frequency $z(\omega)$ and $\gamma(\omega)$, depending on all four parameters of the long line, become coefficients, constants or parameters only in special cases. In this regard, their names do not accurately define their physical essence. The complex resistance coincides with the wave resistance at $R = 0$, $G = 0$ and it is a real number $\mathcal{R} = \sqrt{\frac{L}{\text{₵}}}$.

Theoretically, a wave propagation coefficient

$$\gamma = \sqrt{z_1 \cdot Y_1} = \sqrt{(R + j\omega L) \cdot (G + j\omega \text{₵})} \tag{8.21}$$

at $R = 0$, $G = 0$ is a purely imaginary number: $\gamma = j \cdot \frac{\omega}{\omega_0} = j \cdot F$, $\left[\frac{1}{\text{m}}\right]$ is a normalized linear frequency.

The wave equations (8.20) both for the current and for the voltage set the same law of propagation via the same propagation coefficient — γ. Ohm's law connects these fields to each other through the complex resistance $z(x)$:

$$U(x, t) = z(x) \cdot I(x, t) \tag{8.22}$$

The telegraph equations in the form (8.20) were reduced to two Lagrange equations. Their general solutions are written as a sum of the exponents

$$\begin{cases} U = B_U \cdot e^{\gamma x} + A_U \cdot e^{-\gamma x} = U_{\text{refl}} - U_{\text{fall}} \\ I = B_I \cdot e^{\gamma x} + A_I \cdot e^{-\gamma x} = I_{\text{refl}} + I_{\text{fall}} \end{cases} \tag{8.23}$$

As a complex number, we will represent the propagation coefficient γ, in a form of $\gamma = \xi + j \cdot \eta$, where the real part ξ is a damping ratio, and an imaginary part η is a phase constant. Two traveling waves are the general solutions of the telegraph equation. They are marked as a reflected wave U_{refl} and an incident wave U_{fall}.

Ohm's law is valid both for the incident wave and for the reflected wave, so when multiplying the second equation in the system (8.23) by the complex resistance z and taking intro account directions of the incident and reflected currents, we obtain the following system equation from (8.23):

$$\begin{cases} U = B_U \cdot e^{\gamma x} + A_U \cdot e^{-\gamma x} \\ z \cdot I = B_U \cdot e^{\gamma x} - A_U \cdot e^{-\gamma x} \end{cases} \tag{8.24}$$

In the zero coordinate at $x = 0$, we have – $\begin{bmatrix} U_0 \\ I_0 \end{bmatrix}$ — an initial vector (an input perturbance) which allows to convert to the matrix hyperbolic functions:

$$\begin{cases} U = U_0 \cdot \text{ch}(\gamma x) + I_0 \cdot z \cdot \text{sh}(\gamma x) \\ I = I_0 \cdot \text{ch}(\gamma x) + \dfrac{U_0}{z} \cdot \text{sh}(\gamma x) \end{cases}$$

This system corresponds to the matrix equation

$$\begin{bmatrix} \text{ch}(\gamma x) & z \cdot \text{sh}(\gamma x) \\ \dfrac{1}{z} \cdot \text{sh}(\gamma x) & \text{ch}(\gamma x) \end{bmatrix} \times \begin{bmatrix} U_0 \\ I_0 \end{bmatrix} = \begin{bmatrix} U \\ I \end{bmatrix} \tag{8.25}$$

with determinant $det = 0$

The long line (in our case acoustic medium) corresponds to the matrix which is defined by two complex numbers, z_1, Y_1 or four real parameters: $R, G, L, \text{₵}$. The matrix equation (8.25) provides a simple method of recalculating the input perturbation

$$\begin{bmatrix} U_0(x,t) \\ I_0(x,t) \end{bmatrix} \quad \text{into the output perturbation} \quad \begin{bmatrix} U(x,t) \\ I(x,t) \end{bmatrix}.$$

In the ideal case with no losses ($R = 0, G = 0$)

$$\gamma = j\omega\sqrt{LC} = j\frac{\omega}{\omega_0}, \quad z = \sqrt{\frac{L}{\cancel{C}}} = \mathcal{R}, \quad \xi = 0, \quad \eta = \frac{\omega}{\omega_0} = F,$$

The hyperbolic functions are converted to the relevant trigonometric functions

$$\begin{cases} U = U_0 \cdot \cos \eta x + jI_0 \cdot R \cdot \sin \eta x \\ I = I_0 \cdot \cos \eta x + j\dfrac{U_0}{R} \cdot \sin \eta x \end{cases} \tag{8.26}$$

That is, the standing waves occur regardless of whether there is a decay ($\xi \neq 0$) or not ($\xi = 0$).

The complex resistance ceases to be a frequency-dependent function when we have the line with no losses

$$z = \sqrt{\frac{L}{C}} = \mathcal{R}$$

or when a condition is true

$$z = \sqrt{\frac{L \cdot \left(\frac{R}{L} + j\omega\right)}{\cancel{C} \cdot \left(\frac{G}{\cancel{C}} + j\omega\right)}} = \sqrt{\frac{L}{\cancel{C}}}, \quad \text{at } \frac{R}{L} = \frac{G}{\cancel{C}}$$

That is, when the ratio of losses is equal to the square of the wave resistance

$$\frac{G}{\cancel{C}} = \frac{R}{L}; \quad \frac{L}{C} = \mathcal{R}^2$$

This means that the phase relation between current and voltage will be constant with exponentially decreasing amplitudes. The long line with decay and a property $\frac{R}{L} = \frac{G}{\cancel{C}}$ is called a line with no distortion.

Let us consider a practically significant case of minor losses (R, G — small values), the acoustic analogue of which is the VVER water coolant with low viscosity. The lower the losses in the active elements R and G, the more the time constants $\tau_L = \frac{L}{R}$, $\tau_{\cancel{C}} = \frac{\cancel{C}}{G}$, defining the decay time of amplitudes of the long line current and voltage. Taking into account in this case the values $\frac{1}{\tau_L}$ and $\frac{1}{\tau_{\not{C}}}$ are small, we obtain an approximate expression for the complex resistance

$$z = \sqrt{\frac{R + j\omega L}{G + j\omega \not{C}}} = \sqrt{\frac{L}{\not{C}}} \cdot \sqrt{\frac{\frac{1}{\tau_L} + j\omega}{\frac{1}{\tau_{\not{C}}} + j\omega}} = R \cdot \sqrt{\frac{1 + \frac{1}{j\omega\tau_L}}{1 + \frac{1}{j\omega\tau_{\not{C}}}}}$$

$$\simeq R \cdot \sqrt{1 + \frac{1}{j\omega} \cdot \left(\frac{1}{\tau_L} - \frac{1}{\tau_{\not{C}}}\right)}$$

$$\approx R \cdot \left[1 + \frac{1}{j2\omega} \cdot \left(\frac{1}{\tau_L} - \frac{1}{\tau_{\not{C}}}\right)\right], \quad \omega\tau \gg 1 \tag{8.27}$$

By selecting parameters $L, \not{C}, R, G$, one can achieve $\tau_L = \tau_{\not{C}}$ and $z \approx \mathcal{R}$, but the attenuation will always be zero.

The propagation constant will be changed accordingly

$$\gamma = \sqrt{(R + j\omega L)(G + j\omega \not{C})}$$

$$= \sqrt{L\not{C}} \cdot \sqrt{-\omega^2 \left(1 + \frac{1}{j\omega\tau_L}\right)\left(1 + \frac{1}{j\omega\tau_{\not{C}}}\right)}$$

$$\approx j\frac{\omega}{\omega_0} + \frac{1}{2\omega_0\left(\frac{1}{\tau_L} + \frac{1}{\tau_{\not{C}}}\right)} \tag{8.28}$$

Therefore, the small losses add a purely imaginary component to the wave resistance. This component is called a reactive component and it can be either a capacitance or an inductance. Thus, the presence of losses (the presence of viscosity in acoustics) will always introduce a phase change between voltage (pressure) and current (velocity). If the propagation constant was a purely imaginary value $\gamma = j\frac{\omega}{\omega_0}$ in an ideal case and this resulted only in delay without any change of the amplitude $e^{j\omega\left(t - \frac{x}{\omega_0}\right)}$, then now the presence of a small attenuation causes decrease of the amplitude in $e^{-\left[\frac{1}{2\omega_0}\left(\frac{1}{\tau_L} + \frac{1}{\tau_{\not{C}}}\right)\right] \cdot x}$ times when adding a purely real summand $\frac{1}{2\omega_0}\left(\frac{1}{\tau_L} + \frac{1}{\tau_{\not{C}}}\right)$ compared to the magnitude of the input current or voltage. This attenuation at low losses is almost independent from the frequency.

Thus, we obtained a practical result of the study of a long line with small losses, which is similar to the acoustic medium with low viscosity: the dispersion effect will be weak as well as the own decrease of the longitudinal wave amplitude in space. Small losses are equivalent to the small purely reactive additive to the wave line impedance — R. The additive has capacitive characteristics.

$$\frac{1}{j\omega \cdot \frac{2\tau_{\cent} \cdot \tau_L}{(\tau_{\cent} - \tau_L) \cdot R}}$$

At the same time, $(R - L)$-component competes with $(G - \cent)$-component and at $\tau_{\cent} = \tau_L$ the reactive component disappears. This is the case of the line without phase distortion: $\mathcal{R} = P$.

Similarly, we obtain an approximate formula for the propagation coefficient

$$\begin{aligned}\gamma &= \sqrt{(R + j\omega L)(G + j\omega \cent)} \\ &= j\omega\sqrt{L \cdot \cent} \cdot \sqrt{\left(1 + \frac{1}{j\omega\tau_L}\right)\left(1 + \frac{1}{j\omega\tau_{\cent}}\right)} \\ &\approx j\frac{\omega}{\omega_0} - \frac{1}{2\omega_0} \cdot \left(\frac{1}{\tau_L} + \frac{1}{\tau_{\cent}}\right)\end{aligned}$$

Small losses are equivalent to the low active resistance added to the imaginary constant propagation $j\frac{\omega}{\omega_0}$. It causes frequency-independent attenuation between the input and output of the long line with the following magnitude:

$$U_{\text{output}}(x, t) = e^{-\frac{1}{(2\omega_0)\left(\frac{1}{\tau_L} + \frac{1}{\tau_{\cent}}\right)} \cdot x} \cdot U_{\text{input}}(x, t)$$

In the first approximation, the linear phase dependence is replaced by a constant phase $\varphi_0 = \frac{1}{2\omega_0} \cdot \left(\frac{1}{\tau_L} + \frac{1}{\tau_{\cent}}\right)$, and its slope $\frac{\omega}{\omega_0}$ remains unchanged. The interaction of components $R - L$ and components $G - \cent$ does not compensate either:

$$\varphi(\omega) = j\frac{\omega}{\omega_0} - \varphi_0 = jk - \varphi_0$$

8.3 Properties of Complex Resistance and Propagation Coefficient of the Long Line

Complex resistance of the long line

$$z = \sqrt{\frac{R + j\omega L}{G + j\omega \cent}} = \alpha + j\beta \tag{8.29}$$

and the constant (coefficient) of propagation of the long line

$$\gamma = \sqrt{(R + j\omega L)(G + j\omega \cent)} = \xi + j\eta \tag{8.30}$$

are the functions of four parameters of the long line R, L, G, $\cent$. They define essentially different interrelations. The propagation coefficient as a power of the exponential function or exponent of exponential function appears in solutions of the wave equation, linking the input and output of the long line:

$$U_{\text{output}}(x,t) = U_{\text{input}}(x,t) \cdot e^{-\gamma x} \tag{8.31}$$

and the complex resistance links current and voltage of the long line as

$$U(x,t) = z(x) \cdot I(x,t) \tag{8.32}$$

If the imaginary component η determines the phase relations between the input and output of the long line, or, in other words, relation of the output signal delay to the input signal, the parameter β determines the relation between current $I(x,t)$ (velocity $V(x,t)$) and voltage $U(x,t)$ (pressure $P(x,t)$). There are no causal restrictions here: the current can get ahead of the voltage or lag behind it. In both cases, the "delay" concept is used, which confuses the interpretation of the long line properties.

The expression for the complex resistance (8.29) is equivalent to the system of equations:

$$\begin{cases} \alpha^2 - \beta^2 = \mathcal{R}^2 + \dfrac{P^2 - \mathcal{R}^2}{1 + (\omega\tau_{\cent})^2} \\ 2\alpha\beta = \omega\tau_{\cent} \cdot \dfrac{P^2 - \mathcal{R}^2}{1 + (\omega\tau_{\cent})^2} \end{cases} \tag{8.33}$$

with notations having a fundamental role in the theory of long lines:

$$\mathcal{R}^2 = \frac{L}{\cent}, \quad P^2 = \frac{R}{G}, \quad \tau_{\cent} = \frac{\cent}{G}, \quad \tau_L = \frac{L}{R}, \quad \omega_0 = \frac{1}{\sqrt{L \cdot \cent}} \tag{8.34}$$

where $\mathcal{R}$ is wave resistance, P are generalized active losses, $\tau_{\cent}$ is a time constant of capacitive component, τ_L are time constants of inductive component, $\tau_{\cent}$ and τ_L define the inertia of the capacitive and inductive components, respectively. The bigger the time constants $\tau_{\cent}$, τ_L, the slower the energy accumulation (losses) on the corresponding reactive elements (L or $\cent$). The time constants are linked through the wave resistance and loss parameter

$$\frac{\tau_L}{\tau_{\cent}} = \left(\frac{\mathcal{R}}{P}\right)^2$$

Two biquadratic equations (8.33) have the following solutions:

$$2\alpha^2 = \mathcal{R}^2 + \frac{P^2 - \mathcal{R}^2}{1 + (\omega\tau_{\cent})^2} \pm \sqrt{\frac{P^4 + \mathcal{R}^4 \cdot (\omega\tau_{\cent})^2}{1 + (\omega\tau_{\cent})^2}}$$

$$2\beta^2 = -\mathcal{R}^2 - \frac{P^2 - \mathcal{R}^2}{1 + (\omega\tau_{\cent})^2} \pm \sqrt{\frac{P^4 + \mathcal{R}^4 \cdot (\omega\tau_{\cent})^2}{1 + (\omega\tau_{\cent})^2}}$$

Let us consider the typical partial solutions. At zero frequency (DC mode), the capacitance is a circuit break and the inductance has zero resistance, therefore the long line circuit can be simplified from the form of Fig. 8.2 to the form shown in Fig. 8.5.

For direct current in the line ($\omega = 0$), there are only ohmic losses and no energy is accumulated. Based on Eqs. (8.33) at $\omega = 0$,

$$\alpha = P, \quad \beta = 0$$

i.e. the voltage decreases exponentially, and no phase progression occurs:

$$U(x,t) = e^{-P \cdot x} \cdot I(x,t)$$

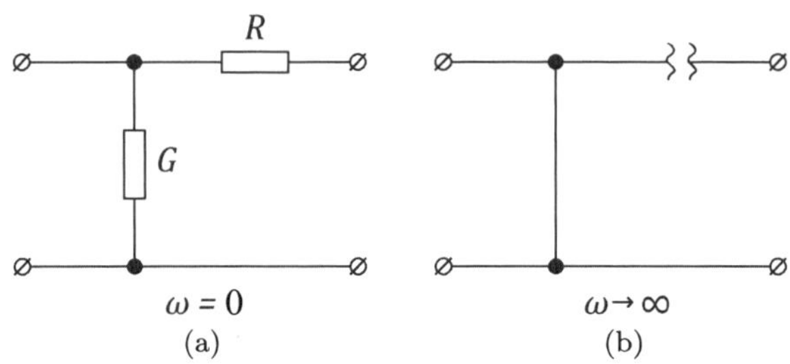

Figure 8.5: Long line circuit for zero frequency (a) and infinitely high frequency (b).

At an infinitely high frequency $\omega \to \infty$, the capacitance is a zero resistance (see Fig. 8.5) so the long line becomes short-circuited:

$$\alpha = 0, \quad \beta = -j \cdot \mathcal{R}$$

In two cases

$$P = \mathcal{R}$$

and

$$\tau_{\text{¢}} = \frac{\text{¢}}{R} \to \infty$$

the phase progression between $U(x,t)$ and $I(x,t)$ does not depend on frequency. The complex resistance will be a purely real number (Figs. 8.6 and 8.7), equal to

$$\alpha = \mathcal{R} \quad \text{at } P = \mathcal{R}$$
$$\alpha = P \quad \text{at } \tau_{\text{¢}} = 0$$

Ohmic losses of the long line

- decrease with a frequency, if $P > R$;
- increase with a frequency, if $\mathcal{R} > P$;
- remain unchanged at $P = \mathcal{R}$.

Very high values of the capacitance time constant $\tau_{\text{¢}}$ (very good isolation at very high capacitance) and $\omega \gg 0$ makes the parameters α, β almost independent from the frequency $\alpha \approx \beta \approx \mathcal{R}$.

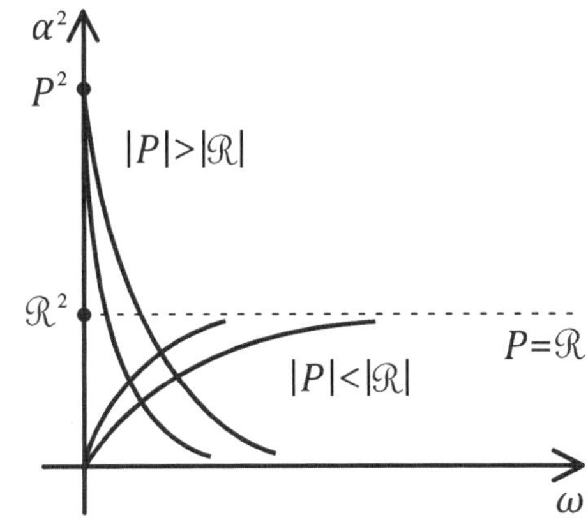

Figure 8.6: Attenuation in the long line depending on frequency.

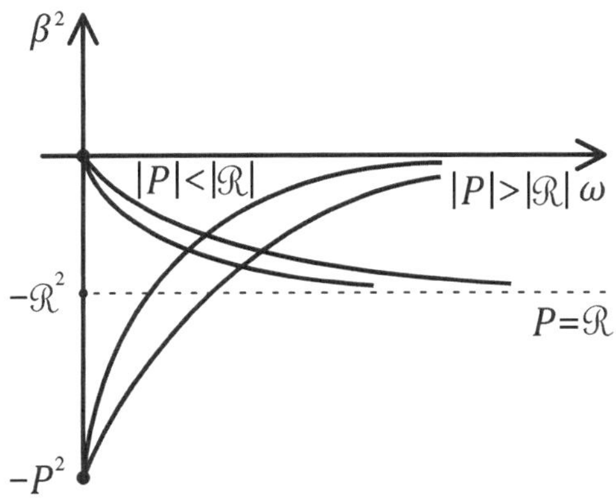

Figure 8.7: Dependence of imaginary component of the complex resistance from frequency.

At a very low capacitance time constant (very poor isolation and very low capacitance), there is also almost no dependence on frequency

$$\alpha \approx \beta \approx P = \sqrt{\frac{R}{G}}, \quad \omega \sim 0$$

Thus, the line with property $\tau_{\not{C}} \gg 0$ tends to properties of the line without losses $R = 0$, $G = 0$ and properties of the line without distortion $\frac{R}{G} = \frac{L}{\not{C}}$.

For the propagation coefficient

$$\gamma = \sqrt{(R + j\omega L)(G + j\omega ¢)} = \xi + j\eta \tag{8.35}$$

we also obtain a system of equations

$$\begin{cases} \xi^4 - A^2 \cdot \xi^2 - \dfrac{1}{4} \cdot B^4 = 0 \\ \eta^4 + A^2 \cdot \eta^2 - \dfrac{1}{4} \cdot B^4 = 0 \end{cases}$$

with solutions

$$2\xi^2 = \theta^2 - F^2 \pm \sqrt{A^4 + B^4}$$
$$2\eta^2 = -\theta^2 + F^2 \pm \sqrt{A^4 + B^4}$$

where

$$A^2 = R \cdot G - F^2;$$
$$B^2 = F \cdot \left(R \cdot G + \frac{R}{R}\right);$$
$$F = \frac{\omega}{\omega_0};$$
$$\omega_0 ¢ = \frac{1}{\sqrt{L \cdot ¢}};$$

$\theta^2 = R \cdot G$ are dimensionless generalized active losses.

Table 8.1 contains the characteristic parameter values at two extreme frequencies for the complex resistance and propagation coefficient.

Another characteristic frequency is obtained in case of equality of the real and imaginary parts of the propagation coefficient $\xi = \eta$:

$$\xi + j\eta = \xi(1 + j) = \sqrt{2} \cdot \xi \cdot e^{j\frac{\pi}{4}}$$

This means that there is a frequency at which the phase progression between the input and output of the long line is 45°, and it

Table 8.1: Characteristic values for complex resistance and propagation coefficient at two extreme frequencies.

	$\omega = 0$	$\omega \to \infty$
Equivalent circuit	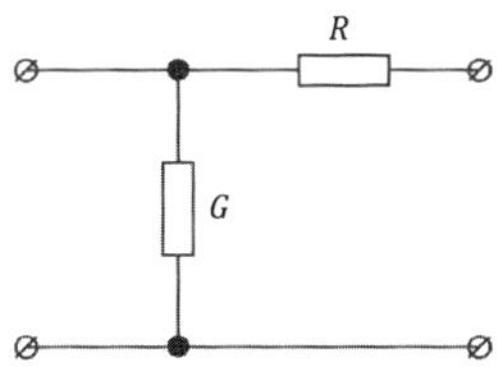	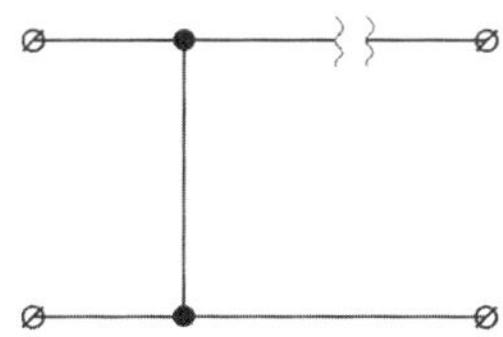
$\gamma = \xi + j\eta$	$\gamma = \sqrt{R \cdot G}$	Uncertain
$z = \alpha + j\beta$	$z = \sqrt{\frac{R}{G}}$	$z = j\mathcal{R}$

occurs only under condition that

$$R \cdot G = \omega^2 \cdot L \cdot \not{C}$$

or

$$\omega_1 = \frac{1}{\sqrt{\tau_L \cdot \tau_{\not{C}}}}$$

When the frequency ω_1 with a 45° phase is obtained, we also receive information on the time constants τ_L and $\tau_{\not{C}}$ or absorption parameters of the wave energy.

The characteristic values of components of the propagation coefficient and complex resistance are summarized in Table 8.2.

8.4 Absorption of Electrical Energy in Long Line and Acoustic Energy in the Pipeline Fluid

An extension of the wave equation for the long line

$$U_{xx} = L \cdot \not{C} \cdot U_{tt} + U_t(L \cdot G + R \cdot \not{C}) + R \cdot G \cdot U \tag{8.36}$$

can be reduced to a standard differential equation not only by a complex substitution, but also by a real substitution of the type

$$U(x,t) = e^{-\frac{1}{2} \cdot \frac{LG + R\not{C}}{L\not{C}} \cdot t} \cdot V(x) \tag{8.37}$$

Table 8.2: Characteristic values of components of the propagation coefficient and complex resistance.

	$z = \alpha + j\beta$	$\gamma = \xi + j\eta$
Line without losses with phase progression 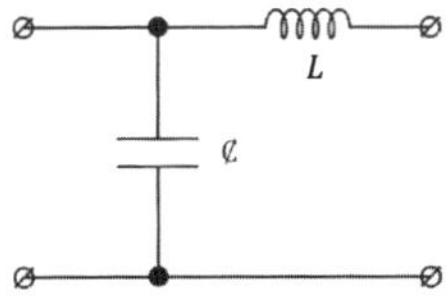 	$R = 0, G = 0,$ $\alpha = 0$ $\beta^2 = -\mathcal{R}^2$ $z = 0 \pm j \cdot \mathcal{R}$ $U = \pm j \cdot \mathcal{R} \cdot I$	$\xi = 0$ $\eta = \dfrac{\omega}{\omega_0} = F$ $\gamma = 0 \pm j \cdot F$ $U_{\text{output}} = e^{\pm jFx} \cdot U_{\text{input}}$
Ohmic line without phase progression 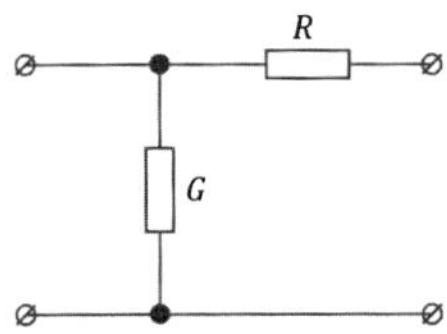 	$L = 0, \not{C} = 0,$ $\beta = 0$ $\alpha^2 = P^2 = \dfrac{R}{G}$ $z = \sqrt{\dfrac{R}{G}} + j \cdot$ $U = \pm P \cdot I$	$\eta = 0$ $\xi^2 = R \cdot G$ $\gamma = \sqrt{R \cdot G} + j \cdot 0$ $U_{\text{output}} =$ $e^{\pm\sqrt{R \cdot G} \cdot x} \cdot U_{\text{input}}$
Idling 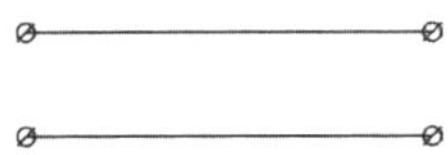	$\alpha = 0, \beta = 0$ $R = 0, G = 0,$ $L = 0, \not{C} = 0$ $z \equiv 0$	$\xi = 0, \beta = 0$ $\gamma \equiv 0$ $U_{\text{output}} = U_{\text{input}}$
Short circuit 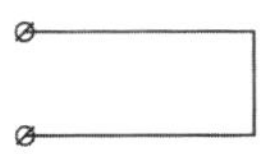	$G \to \infty$ $\alpha = 0$ $z = 0$ $U = 0$	Not defined
Line with ohmic losses but without phase distortion between $U \leftrightarrow z$ 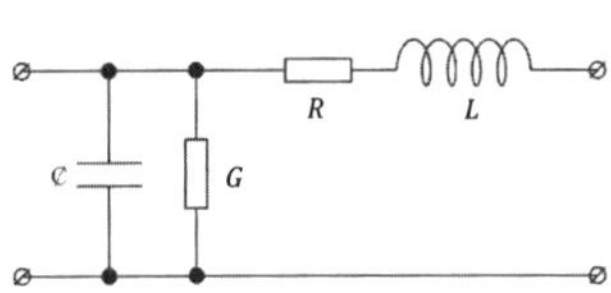 	$\dfrac{R}{G} = \dfrac{L}{\not{C}};$ $P^2 = R^2$ $\alpha = \mathcal{R} = \dfrac{L}{\not{C}}$ $\beta = 0$ $z = \mathcal{R} + j \cdot 0$ $U = \mathcal{R} \cdot I$	$\xi \neq 0$ $\eta \neq 0$

In this solution, the amplitude decreases in time, and in the complex harmonic solution

$$U(x,t) = U_0 \cdot e^{j[\omega t + k(\omega)\cdot x]}$$

we assigned a periodic component for time and an exponentially decreasing function for space variable.

Let us reduce the original Eq. (8.36) to the equation with time coefficients

$$\omega_0^2 \cdot U_{xx} = U_{tt} + \left(\frac{1}{\tau_L} + \frac{1}{\tau_C}\right) \cdot U_t + \frac{1}{\tau_L \cdot \tau_C} \cdot U \tag{8.38}$$

where $\omega_0 = \frac{1}{\sqrt{LC}}$, [Hz · m] is the linear natural frequency of the long line, $\tau_L = \frac{L}{R}$, [s] is the time constant of the serial inductive circuit, $\tau_C = \frac{₵}{G}$, [s] is the time constant of the parallel capacitive circuit.

Parameters τ_L, τ_C are the time constant of loss of the energy, stored in inductance, when R is a serial resistance to it, and the time constant of energies, stored in capacitance, when conductivity G is linked to it in parallel correspondingly.

In an idealized long line without losses

$$R = 0, \quad G = 0, \quad \frac{1}{\tau_L}, \quad \frac{1}{\tau_C} \to 0$$

and

$$\omega_0^2 \cdot U_{xx} = U_{tt},$$

the time constants τ_L, $\tau_L \to \infty$ and energy, stored in induction and capacitance, will never be lost.

After new notation, the solution (8.37) will take the form

$$U(x,t) = e^{-\frac{1}{2}\left(\frac{1}{\tau_L}+\frac{1}{\tau_C}\right)\cdot t} \cdot V(x) \tag{8.39}$$

Substitution of (8.39) into Eq. (8.38) leads to an expression of a spatial harmonic oscillator

$$\omega_0^2 \cdot V_{xx}(x) = -\frac{1}{4}\left(\frac{1}{\tau_C} - \frac{1}{\tau_L}\right)^2 \cdot V(x) \tag{8.40}$$

or

$$V_{xx} + \left[\frac{1}{2\omega_0}\left(\frac{1}{\tau_C} - \frac{1}{\tau_L}\right)\right]^2 \cdot V = 0 \tag{8.41}$$

with partial solution

$$V(x) = \sin\left\{\left[\frac{1}{2\omega_0}\left(\frac{1}{\tau_C} - \frac{1}{\tau_L}\right)\right] \cdot x\right\}, \quad \tau_C \neq \tau_L \tag{8.42}$$

Substitution of (8.37) is chosen in a way to exclude the term with the first derivative from Eq. (8.36). This term is a friction force or a term describing absorption of energy. By its decreasing exponential factor, the substitution already considers the absorption of energy, and the time coefficients of the harmonic oscillator form the argument of the sinusoidal function.

When the time constants $\tau_C = \tau_L$ coincide, we obtain a well-known condition

$$\frac{L}{\not{C}} = \frac{R}{G} \quad \text{or} \quad \rho = \sqrt{\frac{R}{G}}$$

which means that the wave resistance of the long line coincides with a set of loss parameters. In this case, the harmonic oscillator equation degenerates into the equation

$$\frac{d^2V}{dx^2} = 0 \tag{8.43}$$

with a linear solution with regard to spatial coordinate

$$V(x) = C_1 \cdot x + C_2 \tag{8.44}$$

If the harmonic oscillator is considered as a balance of two forces, then, when one of the force becomes zero, the second unbalanced force causes an unlimited increase of space amplitude.

The final partial solution of the original equation is represented by the following formula:

$$U(x,t) = \begin{cases} (U_0 \cdot e^{-\frac{1}{2}\left(\frac{1}{\tau_C}+\frac{1}{\tau_L}\right)\cdot t} \cdot \sin\left[\left(\frac{1}{\tau_C} - \frac{1}{\tau_L}\right) \cdot \frac{1}{2\omega_0}\right] \cdot x, & \tau_L \neq \tau_C \\ (U_0 \cdot e^{-\frac{1}{2}\left(\frac{1}{\tau_C}+\frac{1}{\tau_L}\right)\cdot t} \cdot (C_1 \cdot x + C_2), & \tau_L = \tau_C \end{cases} \tag{8.45}$$

Thus, the solution is either an exponentially decreasing sine wave, or an aperiodic process with linear spatial growth limited by the same exponential time function.

This partial solution provides an ability to interpret a long line without phase distortion at the energy level. In this line, four parameters $(L, \not{C}, R, G)$ are chosen so that the time τ_L of decrease of potential energy, stored in inductance, coincides with the time τ_C of decrease of kinetic energy, stored in capacitance.

Spatial undulation of the longitudinal wave amplitude exists only since $\tau_L \neq \tau_C$. When the capacitance discharge and inductance discharge occur at different characteristic times, the continuous exchange of potential energy (in inductance) and kinetic energy (in capacitance) becomes unequal. For example, if the inductive circuit gives its potential energy fast, and the capacitive circuit is charged slowly, then it is the inductive circuit that will determine the spatial period of energy exchange: $\sim (\frac{1}{\tau_L} - \frac{1}{\tau_C}) \sim \frac{1}{\tau_L}$.

The spatial periodicity of a damped wave in a viscous medium is similar to a standing wave. Both have extrema fixed in space (nodes and antinodes). The difference is that the amplitude of the standing wave is constant. It can be assumed that the viscous medium is a distributed absorbing inhomogeneity with reflective properties. If two viscosity mechanisms are consistent $(\tau_L = \tau_C)$, then the reflective properties disappear.

The long lines without distortions loaded on the wave resistance are characterized by the absence of reflected waves, which means there are no standing waves either. Two energy absorption mechanisms in each point of the long line retain its consistency with the wave resistance $R = \sqrt{\frac{R}{G}}$ by means of resistance $R \neq 0$ and conductivity $G \neq 0$ continuously synchronizing the decrease of potential and kinetic energy. Otherwise, failure to fulfill the condition $(\tau_L = \tau_C)$ is perceived by the incident wave as an inhomogeneity of the acoustic path, from which there is reflection and, accordingly, standing waves appear.

Let us now consider the complex set up

$$U(t, x) = U_0 \cdot e^{j[\omega t + k(\omega) \cdot x]}$$

for Eq. (8.36), which leads to reduction to

$$-\omega_0^2 \cdot k^2(\omega) = -\omega^2 + \left(\frac{1}{\tau_C} + \frac{1}{\tau_L}\right) j\omega + \frac{1}{\tau_C \cdot \tau_L} \tag{8.46}$$

If there are no phase distortions in the line ($\tau_C = \tau_L = \tau_0$), we have

$$\omega_0^2 \cdot k^2(\omega) = \left(\omega - \frac{j}{\tau_0}\right)^2 \tag{8.47}$$

or

$$k(\omega) = \pm\frac{\omega}{\omega_0} \mp \frac{j}{\omega_0 \cdot \tau_0} \tag{8.48}$$

The dispersion relation $k(\omega)$ will not become zero at any frequency, more specifically, such frequency takes place on an imaginary axis $\omega = \frac{j}{\tau_0}$. In the absence of absorption $\left(\frac{1}{\tau_C} = 0, \frac{1}{\tau_L} = 0\right)$, the dispersion relation refers to linear frequency function $k(\omega) = \frac{\omega}{\omega_0}$, and $U(x,t) = U_0 \cdot e^{j\omega\left(1-\frac{x}{x_0}\right)}$ is a traveling wave.

At $\frac{1}{\tau_0} \neq 0$, we obtain a simple solution

$$\begin{aligned} U(x,t) &= U_0 \cdot e^{j\omega\left(t-\frac{x}{\omega_0}\right)} \cdot e^{-\frac{x}{\omega_0 \cdot \tau_0}} \\ &= U_0 \cdot e^{j\omega\left(t-\frac{x}{\omega_0}\right)} \cdot e^{-\frac{R}{\mathcal{R}}\cdot x} = U_0 \cdot e^{j\omega\left(t-\frac{x}{\omega_0}\right)} \cdot e^{-R\cdot G\cdot x} \end{aligned} \tag{8.49}$$

which consists of harmonic space–time function

$$e^{j\omega\left(t-\frac{x}{\omega_0}\right)}$$

and exponential spatial decay

$$e^{-\mathcal{R}\cdot G\cdot x} \quad \text{or} \quad e^{-\frac{R}{\mathcal{R}}\cdot x}$$

In this case, the linearity of the phase in time is not distorted, but it has a spatial shift $\varphi(\omega,t) = \omega \cdot t + \varphi_0(\omega,x)$, where $\varphi_0(\omega,x) = \frac{\omega}{\omega_0}$.

In the general case ($\tau_L \neq \tau_C$), the frequency-dependent factors are obtained both for amplitude attenuation and phase progression:

$$k(\omega) = \frac{1}{\omega_0} \cdot \sqrt{\left(\omega^2 - \frac{1}{\tau_C \cdot \tau_L}\right) + j\omega\left(\frac{1}{\tau_C} + \frac{1}{\tau_L}\right)} \tag{8.50}$$

Let us emphasize that the presence of only linear active elements in linear long lines lead to the nonlinear frequency dependence (8.50). The nonlinearity of dispersion relation is often associated only with the presence of nonlinear elements in the long lines. We could see that linear active elements can also cause nonlinear dispersion relations.

If we assume that the energy absorption in the long line is low $\left(\frac{1}{\tau_C}, \frac{1}{\tau_L} \sim 0\right)$, we obtain an approximate dispersion relation

$$k(\omega) \simeq \frac{\omega}{\omega_0} + \frac{j}{2\omega_0}\left(\frac{1}{\tau_C} + \frac{1}{\tau_L}\right) \tag{8.51}$$

In this case, the linearity is approximately preserved. The phase transfer remains approximately a linear function in space and time

$$\sim e^{j\omega\left(t-\frac{x}{\omega_0}\right)}$$

and the spatial attenuation of the amplitude does not depend on the frequency

$$\sim e^{-\frac{j}{2\omega_0}\cdot\left(\frac{1}{\tau_C}+\frac{1}{\tau_L}\right)}$$

The time constants τ_C, τ_L are included in the solution in symmetrical manner, in the sense that they are interchangeable (replacing one with the other does not change the solution). Although the expression represents two mechanisms of energy absorption due to losses in conductor — $R \neq 0$ and due to losses in isolator — $G \neq 0$, they are equivalent in form.

If we keep only one energy loss mechanism, i.e. to ignore some losses, for example, $\frac{1}{\tau_C} \to 0$, $\frac{1}{\tau_L} = \frac{1}{\tau_0} \neq 0$, the dispersion relation will be the following:

$$k(\omega) \simeq \frac{\omega}{\omega_0} + \frac{j}{2\omega_0 \cdot \tau_0}$$

Again, the amplitude decreases exponentially in space variable, and the phase is a linear space–time function

$$U(x,t) \simeq U_0 \cdot e^{-\frac{1}{2}\cdot\frac{R}{\mathcal{R}}\cdot x} \cdot e^{j\omega\left(t-\frac{x}{\omega_0}\right)}$$

Let us recall that the parameters $\omega_0[\mathrm{Hz}\cdot\mathrm{m}]$, $R\left[\frac{\mathrm{Ohm}}{\mathrm{m}}\right]$, $G\left[\frac{1}{\mathrm{Ohm}\cdot\mathrm{m}}\right]$ are linear parameters, and the wave resistance $\mathcal{R}$ describes the entire line (it is not a linear characteristic).

8.5 Methods for Reducing the Standing Wave Amplitude

8.5.1 *Consensus*

The acoustic standing waves as the main source of VVER equipment vibrations are negative phenomena. It would be tempting to reduce their amplitude by controlling the energy absorption mechanisms (including the coolant viscosity mechanism). For illustration, suppose one absorption mechanism is the coolant friction on the pipeline walls, and the other mechanism is the shearing viscosity in the coolant volume. The friction mechanism is obviously related to velocity of the longitudinal wave, and hence to its kinetic energy. The viscosity mechanism is related to potential energy.

If the time constants of these two phenomena are in consensus ($\tau_L = \tau_C$), the acoustic perturbance will pass through the pipeline medium without phase distortions, i.e. it will not "stretch" in time and will proportionally decrease in amplitude without changing its form. In frequency language, this means that when passing through the pipeline medium all spectral components of the longitudinal wave proportionally reduce their amplitudes solely because of the two effects of energy absorption, and not due to reflection effects. If in the end of its path the longitudinal wave also met the active load with magnitude equal to the wave resistance, then again, no reflection occurs. As a result, no standing waves are generated in the pipeline with such walls, such viscous medium and such load. This approach with the load consistent with the wave resistance of the long line had long been used in the communication theory to fight the "echo effect". For VVER Main Circulation Loop, experimental bench studies shall be performed to select the mechanisms for consensus energy absorption of acoustic waves.

8.5.2 *Absorption at reflection*

No standing wave will occur if the reflecting surface has absolute absorbing properties. The VVER Main Circulation Circuit contains very few surfaces with such properties and all of them are related to generation of ASW1loop, ASW1vessel and their harmonics.

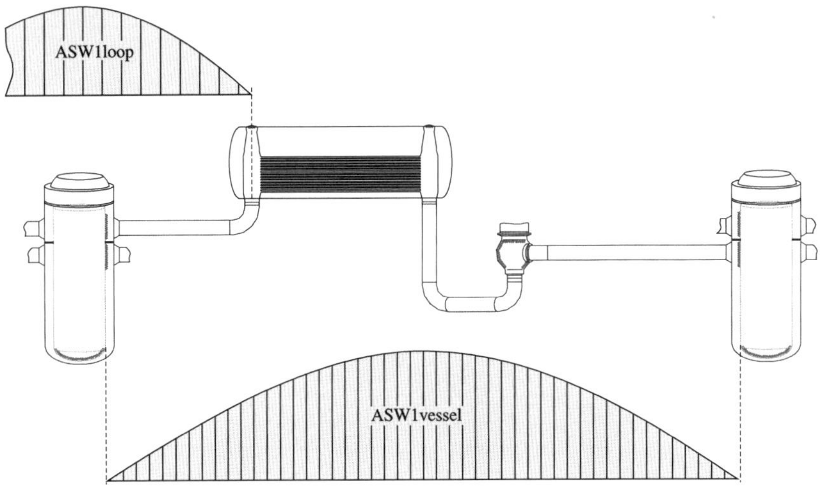

Figure 8.8: Areas (surfaces) of sound absorption in VVER coolant circulation loop to reduce reflection coefficient.

The surfaces to be provided with such absorption properties are marked with thick lines in Fig. 8.8:

- lids of SG headers;
- surface of reactor core barrel visible from the cold and hot nozzles;
- the outer surface of reactor core barrel elliptic bottom;
- the inner surface of the MCP volute.

The absorption of acoustic energy by a special material which covers components of MCC walls shall increase the coolant enthalpy, and therefore increase the reactor plant power. All surfaces of the reactor core barrel contacting the coolant exchange heat with the coolant, so they will automatically increase its enthalpy. In other cases, when the absorbing material covers the lids of headers and the inner surface of MCP, it is necessary to review the isolation structure in corresponding places.

8.5.3 *Flow-around*

The idea of this method is to reduce the reflection coefficient on the acoustic inhomogeneity. It is well known that the bending of the

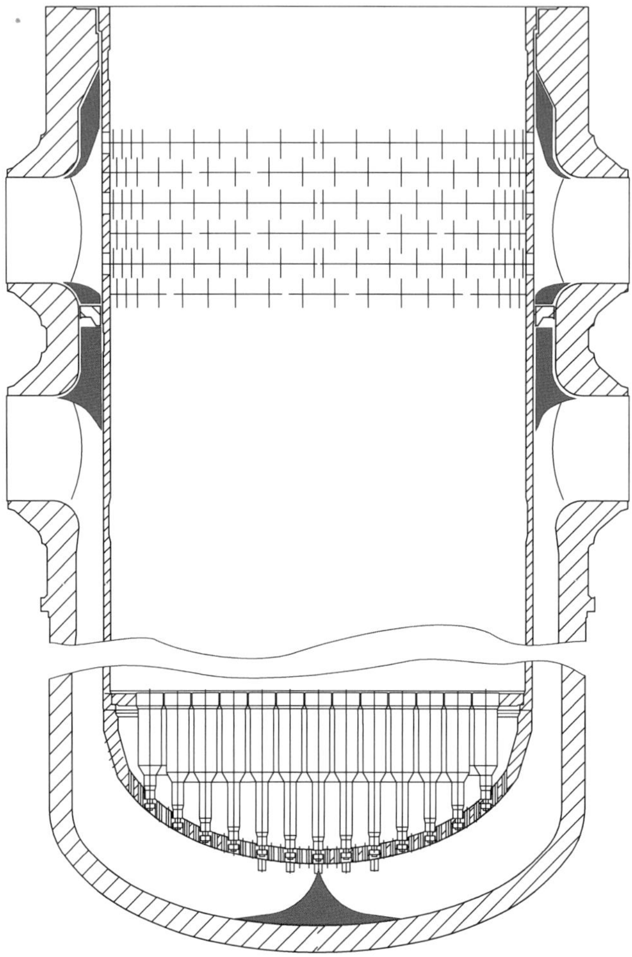

Figure 8.9: Reduction of reflection coefficient in nozzles and on the bottom of reactor plant vessel by flow-around.

pipeline with quite large radius is not practically an inhomogeneity of the acoustic path. The incident longitudinal waves are not reflected from the bended walls of the pipeline. If the "cold" nozzles of the reactor plant vessel, where the coolant flow is diverted, are "smoothed" along the inner surfaces to the maximum extent (Fig. 8.9), then there will be less reflections. Hence, the ASW1loop amplitude with its harmonics, dominating in power in the heat exchange loop, will decrease. In the "hot" nozzle of the reactor plant vessel, which is an outlet component operating for the pipeline, one has to apply an adequate construction of horn (corniculate) type to reduce the reflections also.

The deviation of the coolant flow to the reactor core barrel elliptical bottom also implies a high acoustic resistance. It can be decreased due to the cone-like roundabout island fixed to the bottom of the reactor vessel (Fig. 8.9). It is this local volume that has been upgraded in the VVER-1200 reactor plant, so that its hydraulic resistance is lesser compared to VVER-1000. For safety reasons, the reactor core barrel almost touches the inner surface of the reactor vessel on the vertical VVER-1200 symmetry axis and this reduces the acoustic resistance in the coolant flow deviation.

8.6 Wave Equation with Linear Parameters

Generalized wave equation, which considers losses of internal friction due to viscosity of the medium

$$M \cdot \frac{\partial^2 V}{\partial t^2} + R \cdot \frac{\partial V}{\partial t} - K \cdot \frac{\partial^2 V}{\partial x^2} = 0 \tag{8.52}$$

can be interpreted with linear parameters instead of parameters of the point acoustic analogs m_a, r_a, $\frac{1}{C_a}$.

For the pipeline with cross-section S, we will consider the linear mass $\left[\frac{\text{kg}}{\text{m}}\right]$ as the mass $M = \rho \cdot S$; the linear friction $\left[\frac{\text{kg}}{\text{s}}\right]$ is considered as a friction coefficient $R = r \cdot S$; the linear elasticity coefficient $[\text{kg} \cdot \text{m}]$ is considered as an elasticity coefficient $K = \varkappa \cdot S$. Each term of Eq. (8.52) implies a force. Compared to the motion equation, which is a balance of pressure by means of acoustic parameters m_a, r_a, C_a, the generalized wave equation (8.52) by means of linear parameters M, R, K is a balance of forces. The linear parameters (M, R, K) should not be confused with acoustic parameters m_a, r_a, C_a.

The representation of the wave equation in a form (8.52) is in exact analogy with the long line with losses on ohmic resistance — R; the isolation here is perfect without any losses (Fig. 8.10):

$$L \cdot \frac{\partial^2 i}{\partial t^2} + R \cdot \frac{\partial i}{\partial t} - \frac{1}{\text{¢}} \cdot \frac{\partial^2 i}{\partial x^2} = 0$$

Let us recall that all parameters in Fig. 8.10 with regard to electrical engineering analogy are linear, i.e. they are normalized per length

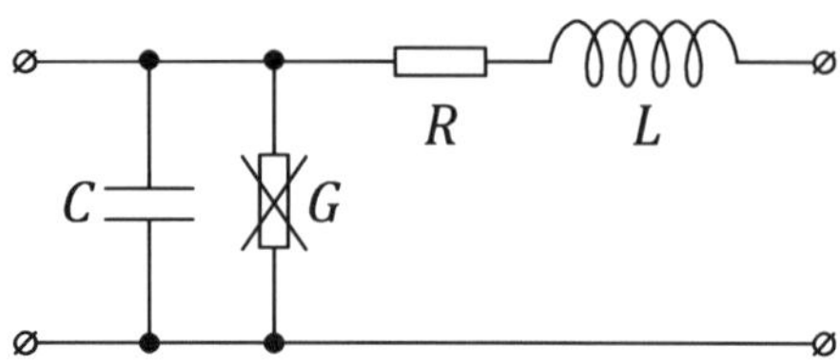

Figure 8.10: Long line with Ohmic losses and perfect isolation.

unit: L[henry/m], R[Ohm/m], $₵$[farad/m]. Compared to the general case, when electrical losses are due to two parameters: ohmic resistance of conductor — R and conductivity of isolation G, the long line here has a perfect isolation, i.e. $G = 0$, and there is a single parameter of losses — the ohmic resistance R. This means that the viscosity coefficient is similar to the ohmic resistance $\eta \leftrightarrow R$, and the kinematic viscosity is similar to specific electrical resistance.

If we normalize Eq. (8.52) to the cross-sectional area of the pipeline — S, we obtain the wave equation in the normalized specific parameters:

$$\rho \cdot \frac{\partial^2 V}{\partial t^2} + r \cdot \frac{\partial V}{\partial t} = \varkappa \cdot \frac{\partial^2 V}{\partial x^2}$$

Now we will normalize this equation once again, this time by the density:

$$\frac{\partial^2 V}{\partial t^2} + \mu_1 \cdot \frac{\partial V}{\partial t} = C^2 \cdot \frac{\partial^2 V}{\partial x^2}$$

This equation can be solved by substitution

$$V(x,t) = V_0 \cdot e^{j(\omega t + K(\omega)\cdot x)} = V_0 \cdot e^{j\omega t} \cdot e^{-\alpha x} \cdot e^{-j\beta x} \tag{8.53}$$

As a result, we obtain an algebraic equation

$$(j\omega)^2 + \mu_1(j\omega) = -C^2 \cdot K^2(\omega),$$

where

$$K^2(\omega) = \frac{\omega^2}{C^2}\left[1 - j\frac{\mu_1}{\omega}\right];$$

$$K(\omega) \simeq K\sqrt{1 - j\frac{\mu_1}{\omega}} \simeq \frac{\omega}{C} - j\frac{\mu_1}{2C}$$

Thus, we confirmed the conclusion of the previous section on low viscosity, but now it has a factual analytical form $\mu_1 = \frac{r}{\rho}$. The medium in the pipeline at low viscosities does not practically distort the linear phase dependence; it introduces an amplitude attenuation with an exponent $-\mu_1$, which does not depend on frequency:

$$V(x,t) = V_0 \cdot e^{j\omega t} \cdot e^{j\frac{\omega}{C}x} \cdot e^{-\frac{\mu_1}{2C}x}$$

8.7 Complex Resistance Equivalent to Long Line with Point Element

The long line in a certain input and/or output section of the acoustic path can be similarly replaced with a complex resistance. For example, the section of the long line with a closed end can be replaced by capacitance. However, representation by capacitance will be an approximation for it as well, valid for short lengths. As it has been shown above, the complex resistance of such a long line

$$z = \frac{\rho \cdot C \cdot S}{j \cdot tg\left(\frac{\omega l}{C}\right)}$$

is represented by capacitive resistance only approximately

$$z \approx \frac{\rho \cdot C \cdot S}{j \cdot \frac{\omega l}{C}} = \frac{1}{j \cdot \omega \cdot \frac{V}{\rho \cdot C^2 \cdot S^2}}; \quad ₵ = \frac{V}{\rho \cdot C^2 \cdot S^2}$$

In contrast to it, a lumped capacitance connected to the long line can be represented as capacitance loaded with equivalent complex resistance. If some volume has an inlet opening as well as an outlet opening, its electrical equivalent is a parallel "$Z_n - C$"-circuit (Fig. 8.11).

Any acoustic inhomogeneity z_n can be described by a dimensionless I/O transfer function — H_ω. To do this, it is necessary to load the inlet z_1 and outlet z_2 by complex resistances, thereby replacing the sections of the inlet and outlet pipeline and to consider the equivalent circuit in Fig. 8.12.

For example, for a capacitive inhomogeneity, the transfer function (or coefficient of transmission from inlet to outlet) will have a form with characteristics of a low pass filter (Fig. 8.13).

The Helmholtz resonator loaded from both the inlet and outlet side has a transfer function of the band-pass filter (Fig. 8.14).

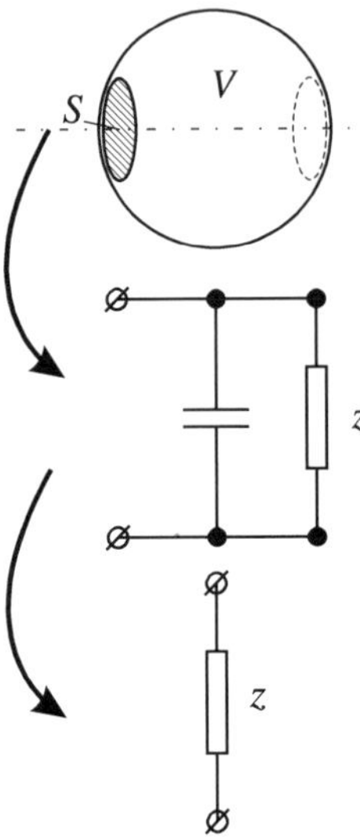

$$\frac{1}{z} = \frac{1}{z_v} + \frac{1}{z_n} = \frac{1}{\dfrac{1}{-j \cdot \omega \cdot C}} + \frac{1}{z_n} = -j \cdot \frac{\omega \cdot V}{\rho \cdot C^2 \cdot S^2} + \frac{1}{z_n}$$

Figure 8.11: Volume with inlet and outlet openings and its equivalent circuit.

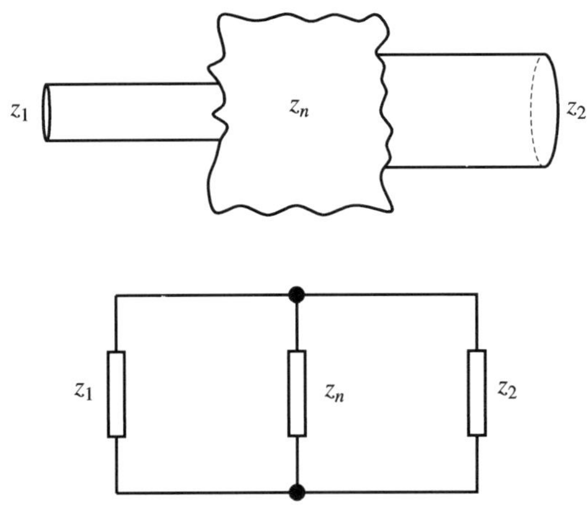

$$|H(\omega)| = \left| \frac{\dfrac{1}{z_n}}{\dfrac{1}{z_1} + \dfrac{1}{z_2} + \dfrac{1}{z_n}} \right|$$

Figure 8.12: Inhomogeneity represented by resistance z_n and loaded on inlet and outlet complex resistances z_1, z_2.

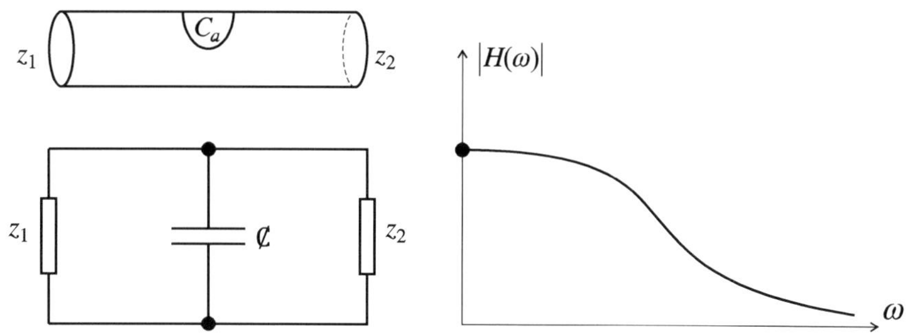

$$|H(\omega)| = \left| \frac{\frac{1}{j\cdot\omega\cdot\not{C}}}{\frac{1}{z_1}+\frac{1}{z_2}+\frac{1}{j\cdot\omega\cdot\not{C}}} \right|$$

Figure 8.13: Transfer function of capacitive inhomogeneity.

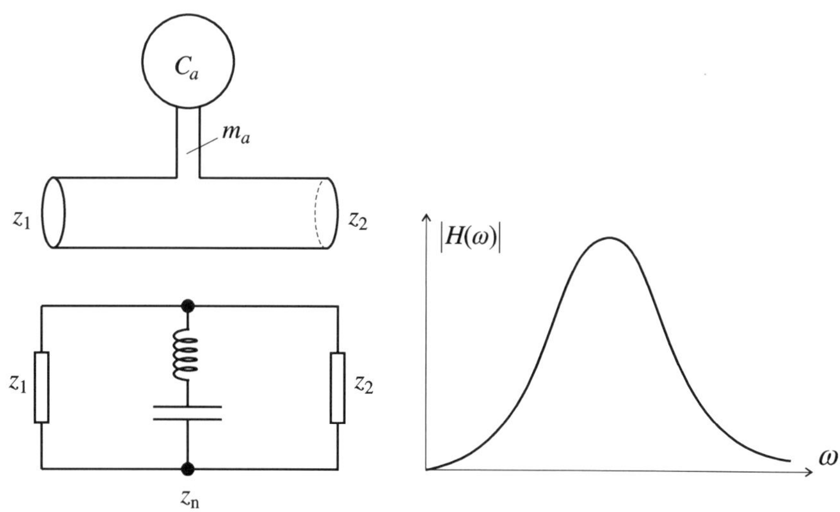

$$|H(\omega)| = \left| \frac{\frac{1}{j\cdot\omega\cdot m_a}+j\cdot\omega\cdot C_a}{\frac{1}{z_1}+\frac{1}{z_2}+\frac{1}{j\cdot\omega\cdot m_a}+j\cdot\omega\cdot C_a} \right|$$

Figure 8.14: Transfer function of resonance inhomogeneity.

8.8 Four Pole (Two-Port) Circuit as Electrical Analogy of the Acoustic Path Element

Long lines considered in the previous sections were represented by four-pole (two-port) circuits with evenly spaced reactive L – $\not{C}$-elements, having measured values in linear units, $\left[\frac{\mathrm{H}}{\mathrm{m}}\right]$ and $\left[\frac{\mathrm{F}}{\mathrm{m}}\right]$, respectively. But L, $\not{C}$ which are the elements of the four-pole (two-port) circuit, equivalent to a certain pipeline element of the acoustic path, can be represented in the lumped classical parameters with measurement units [H] and [F].

The alternation of sections of the pipelines in Fig. 8.15 has been mentioned above as an equivalent two-pole circuit. It can also be represented as a four-pole (two-port) circuit (Fig. 8.15), which coincides with the electrical diagram of the long line.

There is also confusion here: the same type of electrical diagram in Fig. 8.15 is also an analogue of the long line and four-pole (two-port) circuit in the lumped parameters. A distinction is made

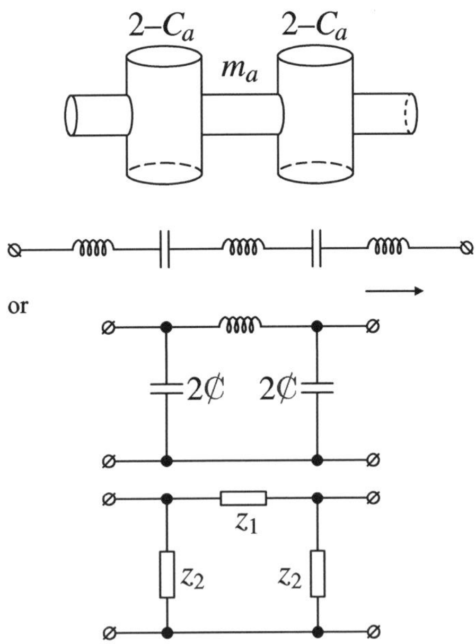

Figure 8.15: Pipeline system and its equivalent circuits by means of two-pole circuit and four-pole (two-port) circuit.

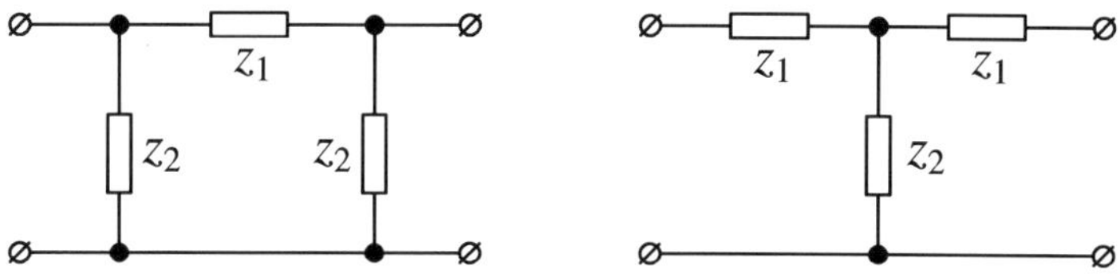

Figure 8.16: U-shaped and T-shaped elementary four pole circuit.

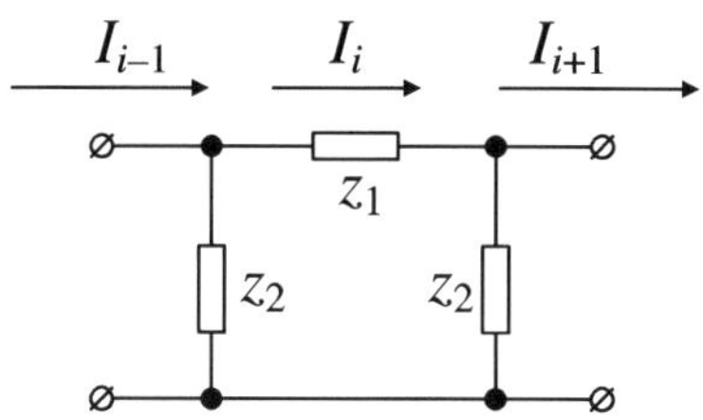

Figure 8.17: Currents in U-shaped four-pole (two-port) elementary component.

between U-shaped and T-shaped elementary four-pole (two-port) circuits (Fig. 8.16).

Knowing the properties of the elementary four-pole (two-port) circuits we can obtain properties of their various combinations, in particular, cascade (serial) connections.

Let us consider the U-shaped connection of four-pole (two-port) circuits. For one elementary component (Fig. 8.17), it is easy to obtain the relationship of three currents under Kirchhoff's law:

$$z_2(I_{i-1} - I_i) - z_1 \cdot I_i = z_2(I_i - I_{i-1})$$

or

$$z_2\left(\frac{I_{i-1}}{I_i} - 1\right) - z_1 = z_2\left(1 - \frac{I_{i+1}}{I_i}\right)$$

The ratio of successive "input-output" currents is a constant number (due to identity of U-shaped components), associated with propagation constant γ by expression

$$\frac{I_i}{I_{i-1}} = \frac{I_{i+1}}{I_i} = \cdots = e^{-\gamma}$$

By substituting this exponent into the formula of the original equation, we obtain

$$z_2(e^{\gamma} - 1) - z_1 = z_2(1 - e^{-\gamma})$$

or

$$\mathrm{ch}\gamma = 1 + \frac{z_1}{2 \cdot z_2}$$

Since there are no losses in the four-pole (two-port) circuit, the $\gamma = j \cdot \theta$ is a purely imaginary number, and the expression is simplified as follows:

$$\mathrm{ch}(j \cdot \theta) = \cos\theta = 1 + \frac{z_1}{2 \cdot z_2}$$

which causes non-strict inequalities

$$-1 \leq 1 + \frac{z_1}{2 \cdot z_2} < +1$$

The right-hand equality

$$1 + \frac{z_1}{2 \cdot z_2} = 1$$

is achieved at $\frac{z_1}{z_2} = 0$; the left-hand equality

$$1 + \frac{z_1}{2 \cdot z_2} = -1$$

is achieved at $\frac{z_1}{z_2} = -4$.

If L is inductance and $z_1 = j\omega L$, and $\not{C}$ is capacitance and $z_2 = \frac{1}{j\omega \not{C}}$, the boundary conditions

$$\begin{cases} \dfrac{z_1}{z_2} = 0 \\ \dfrac{z_1}{z_2} = -4 \end{cases}$$

correspond to the frequency range

$$\left[\omega_1 = 0;\ \omega_2 = \frac{2}{\sqrt{L \cdot \not{C}}}\right]$$

That is, the cascade connection of the U-shaped four-pole (two-port) circuits is a low-pass filter with a transparency band

$$\left[0, \frac{2}{\sqrt{L \cdot ₵}}\right]$$

If the inductance and capacitance are reversed, it is easy to demonstrate that such a cascade connection is a high-pass filter with a stop band coinciding with transparency band in case of LPF.

If we return to the analogies in the lumped parameters,

$$L \leftrightarrow m_a = \frac{\rho \cdot l}{S}$$

$$₵ \leftrightarrow C_a = \frac{V}{\rho \cdot C^2}$$

the cutoff frequency

$$\omega^2 = \frac{2}{\sqrt{m_a \cdot C_a}} = 2 \cdot C \cdot \sqrt{\frac{S}{l \cdot V}}$$

will be equal to doubled frequency of the Helmholtz resonator. This does not mean that the cascade connection of elementary components is equivalent to one Helmholtz resonator. It is known that the latter is an idealized resonance circuit, and not a low-pass filter or a high-pass filter.

The linear four-pole (two-port) circuit is specified by four parameters $a_{11}, a_{12}, a_{21}, a_{22}$ or by a matrix $\|a\| 2 \times 2$ (Fig. 8.18):

$$a = \begin{Vmatrix} a_{11} & a_{12} \\ a_{21} & a_{22} \end{Vmatrix}$$

$$\begin{cases} I_1 = a_{11} \cdot I_2 + a_{12} \cdot U_2 \\ U_1 = a_{21} \cdot I_2 + a_{22} \cdot U_2 \end{cases}$$

The matrix equation

$$\begin{bmatrix} I_1 \\ U_1 \end{bmatrix} = \|a\| \cdot \begin{bmatrix} I_2 \\ U_2 \end{bmatrix}$$

has a property

$$\det a = |a| = a_{11} \cdot a_{22} - a_{12} \cdot a_{21} = 1$$

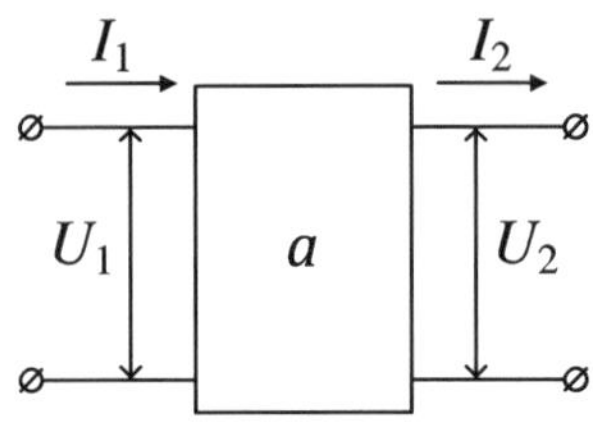

Figure 8.18: Linear four-pole (two-port) circuit.

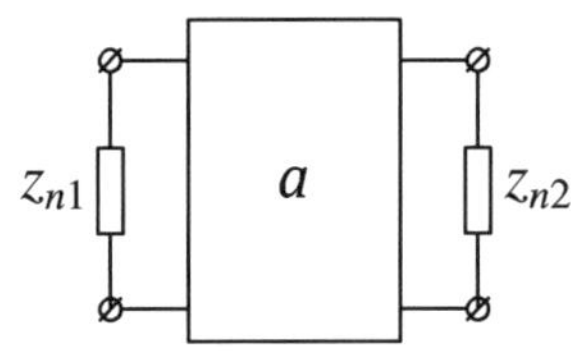

Figure 8.19: Loading of four-pole (two-port) circuit at inlet and outlet.

The four-pole (two-port) circuit is called symmetric when $a_{11} = a_{22}$. If the four-pole (two-port) circuit is loaded by complex resistance z_{H_1} at the inlet side, and loaded for another complex resistance z_{H_2} at the outlet side (Fig. 8.19), we shall call it "agreed upon at inlet and outlet", if

$$z_{H_1} = \sqrt{\frac{a_{21} \cdot a_{22}}{a_{12} \cdot a_{11}}}; \quad z_{H_2} = \sqrt{\frac{a_{21} \cdot a_{11}}{a_{12} \cdot a_{22}}}$$

For symmetric four-pole (two-port) circuit $a_{11} = a_{22}$, $z_{H_1} = z_{H_2}$. Consistency conditions of the four-pole (two-port) circuit are obviously the conditions of the absence of reflected waves at both outlet and inlet.

The long line described by the system

$$\begin{cases} U_1 = U_2 \cdot \mathrm{ch}(\gamma \cdot L) + I_2 \cdot \mathcal{R} \cdot \mathrm{sh}(\gamma \cdot L) \\ I_1 = \dfrac{U_2}{\mathcal{R}} \cdot \mathrm{sh}(\gamma \cdot L) + I_2 \cdot \mathrm{ch}(\gamma \cdot L), \end{cases}$$

has a property of linearity and symmetry, i.e. it shall be consistent with the same wave resistance

$$z_{H_1} = z_{H_2} = \sqrt{\frac{a_{21} \cdot a_{11}}{a_{12} \cdot a_{22}}} = \sqrt{\frac{a_{21}}{a_{12}}} = \mathcal{R}$$

To describe the linear symmetrical four-pole (two-port) circuits, there is no need for four parameters $a_{11}, a_{12}, a_{21}, a_{22}$, two parameters are enough:

$\mathcal{R} = \sqrt{\frac{a_{21}}{a_{12}}}$ is wave resistance and γ is propagation constant.

$$\mathcal{R} = \mathrm{sh}(\gamma \cdot L) = a_{12}$$

Example. The steam generator header can be represented as a cascade connection of U-shaped elementary four-pole (two-port) circuits (Fig. 8.20), which are a prototype of horizontal layers of SG tubes, the header is loaded for an infinitely high resistance — on the header lid. Being an idling based on resistance, it is an inconsistent acoustic load, causing reflections of the incident wave, and hence the presence of the standing waves. In contrast to the idealized long line, where inductances are in serial branches of the four-pole (two-port) circuit, and capacitances are in parallel branches, the equivalent circuit of the steam generator header has an inverse arrangement $L-C$ of the elements. The volume between adjacent holes of the header (Fig. 8.21) with elasticity C_a acts as capacitance.

The coolant mass on one tube m_a acts as inductance. Thus, the steam generator header together with heat exchange tubes are the

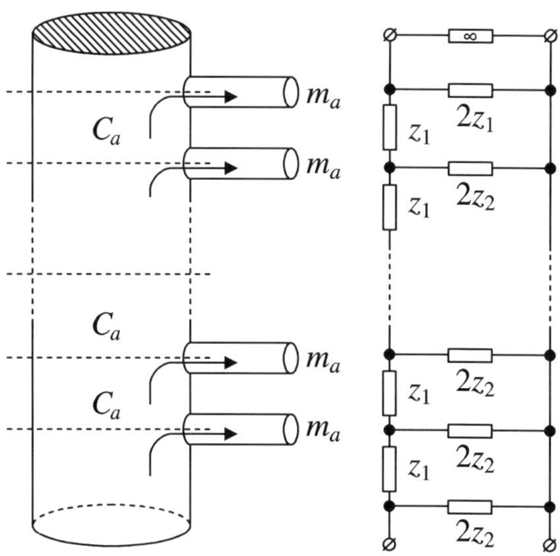

Figure 8.20: Equivalent representation of SG header with heat exchange tubes.

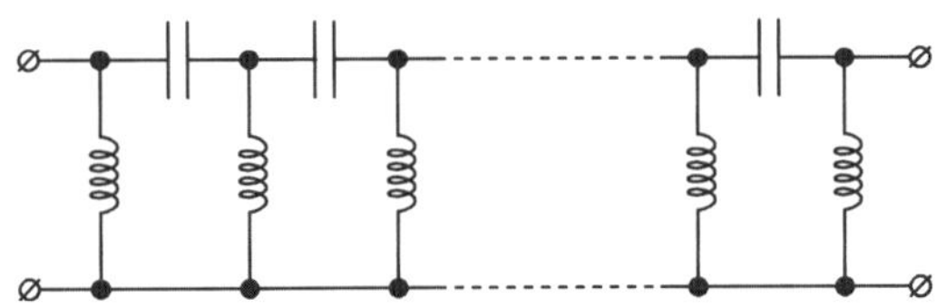

Figure 8.21: Inertial and elastic elements of SG header volumes together with the heat exchange tubes appear in inverse order in equivalent electronic scheme.

high-frequency filter with a stop band $\left[0, \frac{2}{\sqrt{m_a \cdot C_a}} = \omega_c\right]$. The high-pass filter cutoff frequency ω_c is a very small value, which is less than any frequency of ASW within VVER circulation circuit.

If you reduce the equivalent capacitance C, for example, by reducing the distance between adjacent layers of heat exchange tubes or by reducing the SG header diameter, or if you reduce the equivalent inductance L by shortening the length of the heat exchange tubes or by reducing their radius, the frequency of the high-pass filter cutoff — ω_c — will increase. As a result, the frequencies of the lower ASW types will fall into the high-pass filter stop band. The steam generator will become the acoustic barrier, which the lower ASW types cannot cross. Their absence will significantly improve the vibration state of the heat exchange loop equipment. However, most of the above geometric changes of SG worsen intensity of heat exchange between the primary and secondary circuit of the reactor plant.

Chapter 9

Long Lines with Lumped Complex Resistance

9.1 Inhomogeneities Generating a "Pure" Standing Wave

This chapter consistently formalizes the acoustic medium concentrated in the VVER MCC heat exchange loop in accordance with the theory of long lines [12–14].

Closed and open ends of the pipeline are extreme cases of inhomogeneity of the acoustic path. In both cases, there emerges a "pure" standing wave mode (Fig. 9.1).

The pipeline resistance is a purely imaginary value and as a spatial function varies infinitely $(-\infty j, +\infty j)$. In any of its section the resistance

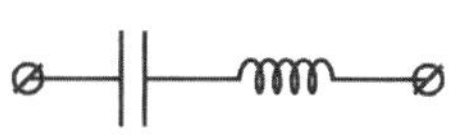

is either capacitive or inductive, or zero (that corresponds to the resonance in a serial L–C circuit), or infinite (that

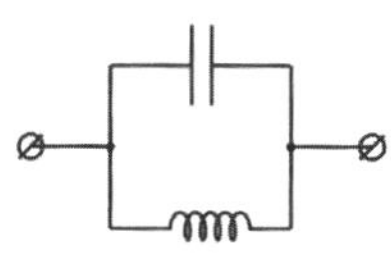

corresponds to the resonance in a parallel L–C circuit).

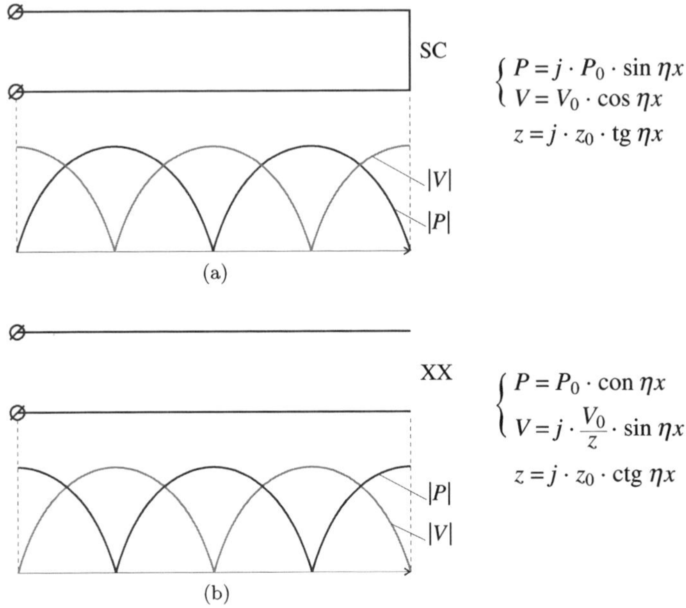

Figure 9.1: Standing waves of pressure (e.g. the second harmonic) and speed in the acoustic pipeline for two extreme inhomogeneity cases at the ends: (a) short circuit (an open pipeline end) and (b) open circuit (closed pipeline end).

In any of these cases, the line does not absorb energy and the reflection of waves from the ends does not contribute to energy absorption, as well.

A short-circuited line is equivalent to a line loaded on a serial circuit in resonance (Fig. 9.2), if the length of the line exactly fits an integer number of half-periods.

It is possible to choose such inductance that by connecting it to the end of the short-circuited line $[0, L_1]$, the distributions $P_1(x)$, $V_1(x)$ would not differ from distributions along the short-circuited line on the length $[0, L]$. It is also possible to choose an appropriate capacitance on the length $[0, L_2]$ (Fig. 9.3).

Such an equivalent replacement to the capacitance and the inductance is only valid for one frequency. Total reflection occurs from any purely reactive (purely imaginary) load–capacitance or inductance, as well as from a serial or parallel resonant L–C circuit.

A purely active load R does not cause standing waves when it is equal to the wave resistance: $\mathcal{R} = R$. In the acoustic path, there is

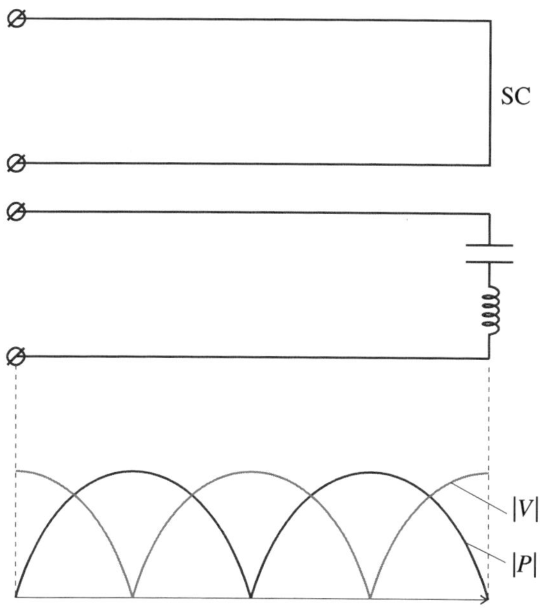

Figure 9.2: Length of a short-circuited long line exactly fitting an integer number of half-periods.

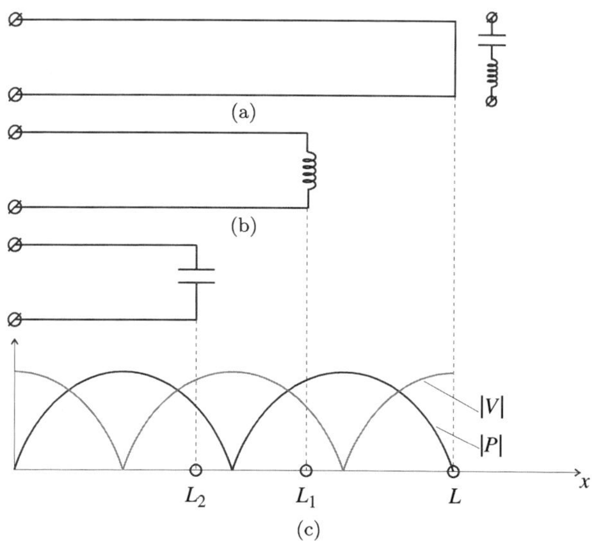

Figure 9.3: (a) An integer number of half-periods on the line length, (b) equivalent inductance with a non-integer number of half-periods and (c) equivalent capacitance with a non-integer number of half-periods.

only an incident-travelling wave, which is completely absorbed (dissipated) on the active load. In this case,

$$U = U_0 \cdot \cos \eta x + j\frac{U_0}{R} \cdot \mathcal{R} \cdot \sin \eta x = U_0 \cdot e^{j\eta x}$$
$$I = I_0 \cdot \cos \eta x + j\frac{I_0}{\mathcal{R}} \cdot R \cdot \sin \eta x = I_0 \cdot e^{j\eta x} \tag{9.1}$$

and

$$|U| = U_0 \cdot \sqrt{\cos^2 \eta x + \left(\frac{R}{R}\right)^2 \cdot \sin^2 \eta x} = U_0$$

A non-ideal line with losses also does not generate a standing wave if, first, the ratio is satisfied,

$$\frac{R}{G} = \frac{L}{C} = \mathcal{R}^2,$$

and, secondly, it is loaded on the resistance equal to the wave one: $R_{\mathrm{H}} = \mathcal{R}$.

Example. As shown in the previous sections, the ASW frequency is easily found along the length of the pipeline — l, closed on both ends. We have shown above that any acoustic inhomogeneity in the pipeline causes a reflected wave, which in interaction with the incident wave forms an ASW. Its frequency is also determined by the distance between two adjacent acoustic inhomogeneities. We will demonstrate the practical value of this approach on the example of the VVER MCC.

Let's choose a passive (without acoustic sources) MCC segment in the form of a "hot" pipeline, connected to the "hot" SG channel head. Its length is measured from the hot nozzle of the reactor vessel to the lid of the hot steam generator channel head.

The L-shaped transition from the pipeline to the SG channel head is not an acoustic heterogeneity. The wall of reactor core barrel located near the reactor vessel nozzle is the reflecting surface.

As a rough approximation, this end of the hot pipeline can be considered closed (from the side of the reactor vessel). The same is the opposite end of the pipeline in the form of the top-lid of the steam generator channel head (Fig. 9.4). Thus, the pipeline in question will be considered closed from two ends. The ASW frequency arising at

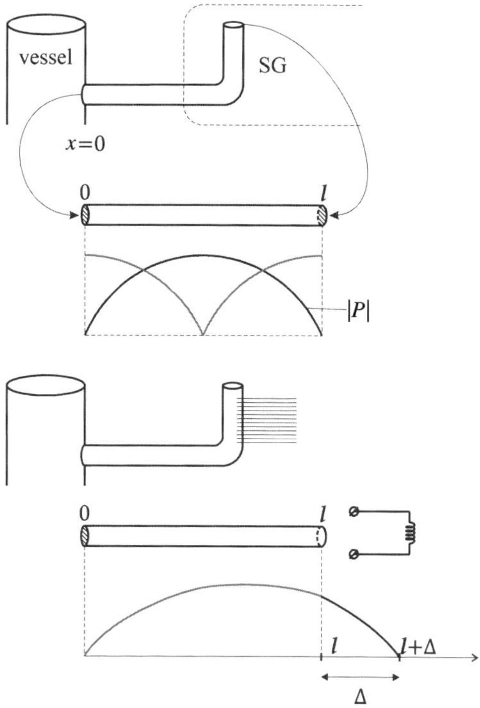

Figure 9.4: Lengthening of wavelength when connecting a reactive load.

this pipeline segment (on this long line) $f = \frac{C}{2l}$ at a sound speed of about $C \sim 800\,\text{m/s}$ will be around 30 Hz. Looking ahead, we note that such an ASW is actually observed in the VVER MCC. Until now, the impact on this ASW of the complex resistance in the form of heat exchanging tubes (HETs) has been neglected.

Now we take into account the HETs of the steam generator manifested as acoustic inhomogeneity in the form of a purely reactance (inductance) load L with complex resistance $z = j\omega L$. Purely reactive load, as well as open circuit, causes a "pure" standing wave. Such a load at $x = 0$ cross-section is converted according to the formula

$$z_0 = \mathcal{R} \cdot \frac{z + j \cdot \mathcal{R} \cdot \operatorname{tg} \eta l}{\mathcal{R} + j \cdot z \cdot \operatorname{tg} \eta l} = \mathcal{R} \cdot \frac{j\omega L + j \cdot \mathcal{R} \cdot \operatorname{tg} \eta l}{\mathcal{R} - \omega \cdot L \cdot \operatorname{tg} \eta l}$$
$$= j\mathcal{R} \cdot \frac{\omega\tau + \operatorname{tg} \eta l}{1 - \omega\tau \cdot \operatorname{tg} \eta l} \tag{9.2}$$

where $\tau = \frac{L}{\mathcal{R}}$ – is a time constant of a loaded long line.

If $\tau = 0$, we have an open circuit, and if $\tau \longrightarrow \infty$, we have a short circuit. If with an idle end (open circuit) in the point $x = l$ a node was observed in pressure distribution, then now in this point there will be some non-zero pressure. In the opposite point $x = 0$, the pressure is still zero.

The connection of inductance which has not been taken into account is equivalent to "stretching" the existing ASW in the direction to the right with a fixed point at the left $x = 0$. In the cross-section at $x = 0$, having idling (open circuit) the resistance is

$$z_0 = j\mathcal{R} \cdot \operatorname{tg} \eta l$$

and the pressure distribution along the pipeline is sinusoidal: $\sin\left(\pi \frac{x}{l}\right)$.

Connecting inductance to the long line will again cause the mode of a ≪pure≫ standing wave but with another effective wavelength $-(l + \Delta) \cdot \eta$, so now its resistance can be represented as

$$\operatorname{tg}[(l + \Delta) \cdot \eta] = \frac{\omega\tau + \operatorname{tg} \eta l}{1 - \omega\tau \cdot \operatorname{tg} \eta l}$$

or

$$\frac{\operatorname{tg}[(l + \Delta) \cdot \eta] - \operatorname{tg} \eta l}{1 + \operatorname{tg}(\eta l) \cdot \operatorname{tg}[(l + \Delta) \cdot \eta]} = \omega\tau$$

or

$$\operatorname{tg}(\Delta \cdot \eta) = \omega\tau$$

Inductive load is equivalent to an extension of the line by an amount

$$\Delta = \frac{1}{\eta} \cdot \operatorname{arc\,tg} \omega\tau$$

For small frequencies — ω — and/or small inductances — L — and/or large wave resistances of a long line — $\mathcal{R}$ — we obtain a linear dependence

$$\Delta = \frac{\omega \cdot L}{\eta \cdot \mathcal{R}}$$

that is why inductance is often called a delay line.

The lengthening Δ is not limitless: when $\omega\tau \to \infty$

$$\Delta = \frac{\pi}{2 \cdot \eta}$$

it reaches its limit and only depends on the propagation constant η.

For capacitive load $z = \frac{1}{j\omega\cancel{C}}$, the line elongation also occurs. From the same formula (9.2) it is easy to obtain

$$\mathrm{tg}(\Delta \cdot \eta) = \frac{1}{\omega \cdot \tau_{\cancel{C}}}, \quad \Delta = \frac{1}{\gamma} \cdot \mathrm{arc\,tg}\left(\frac{1}{\omega \cdot \tau}\right), \quad \tau = \mathcal{R} \cdot C$$

Unlike inductance, capacitance cannot be called a delay line (see Fig. 9.5). With frequency growth ω and/or time constant growth τ_c, the elongation decreases to zero: $\Delta \to 0$.

A standing wave taking place in the hot leg of VVER MCC having frequency close to 30 Hz MCC carries information on the value of equivalent inductance manifested as acoustic resistance of heat exchanging tubes. Its value can change during operation due to the following:

- closing heat exchanging tubes;
- deposits of impurities on the inner surfaces of the HET on the primary circuit side;
- deposits of impurities on the outer surfaces of the HET on the secondary circuit side.

The left end of the hot leg (or the reactor nozzle) can be, of course, very roughly described by open circuit (infinite resistance). However,

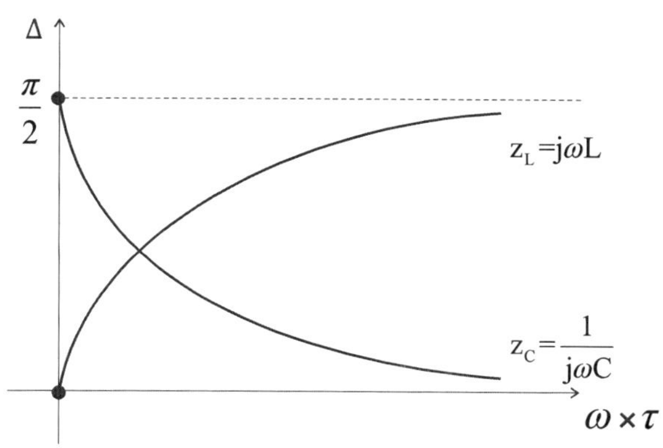

Figure 9.5: Lengthening of wavelength ASW – Δ in connecting a reactive load.

unlike its right end, this finite complex resistance is constant during operational life, and that opens the way for HET diagnostics based on changes of the frequency of the above ASW.

As a diagnostic feature, we can take parameter

$$\Delta \cong l + \frac{\omega\tau}{\eta} = l + \frac{\omega_{\text{ASW}}}{\tau} \cdot \frac{L}{\mathcal{R}} = \text{const1} + \text{const2} \cdot f_{\text{ASW}}$$

Internal HET surfaces with ideal heat transfer have the maximum inductance — L. As the deposition of impurities on the inner HET surfaces grows, the inductance decreases. On the contrary, inductance increases with decreasing HET heat transfer due to the growth of impurity deposits on their outer surfaces. The change of acoustic properties of the SG tubes in the process of a long-term RP operation leading to a change in the highest type ASW frequency is the authors' proprietary methodology.

9.2 Inhomogeneities Causing Mixed Acoustic Field

Let us assume that the acoustic properties of all MCC heat exchange loops are identical to each other. Following sequentially along the coolant flow path, the total loop from reactor via SG till the reactor core back we face the following sudden changes in the cross-section of the flow (Fig. 9.6):

(1) in the volume of the "hot" outlet nozzle of the reactor vessel;
(2) in the "hot" leg of SG channel header;

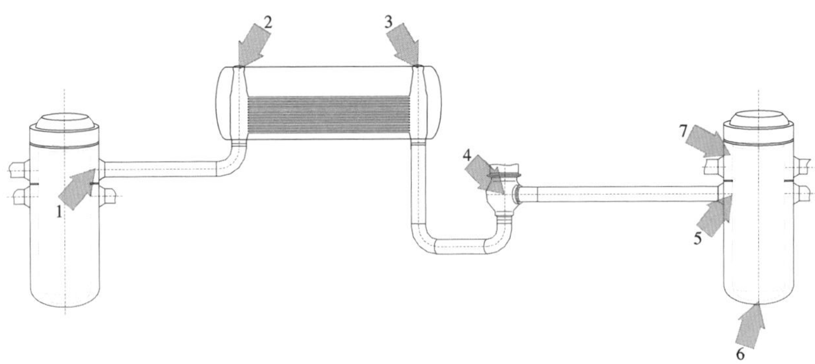

Figure 9.6: Seven acoustic inhomogeneities of a heat exchanging loop of VVER-1200 and VVER-1000.

(3) in the "cold" leg SG channel header;
(4) in MCPs;
(5) in the volume of the "cold" inlet nozzle of the reactor vessel;
(6) in the volume between the reactor vessel bottom and the bottom of the reactor core barrel with a center on the vertical axis of the reactor;
(7) in the reactor volume.

Between the first and fifth inhomogeneities there is the reactor with the highest acoustic resistance in the MCC (Fig. 9.7). It is difficult to imagine the reactor as one equivalent pipe section of MCC, as its acoustic inhomogeneities are volumetric in nature. They occupy a significant cross-sectional area and they are very variable vertically.

Nevertheless, the overall dimensions of the reactor are much smaller than the wavelength of the lowest type ASW, and the inhomogeneities with the largest reflection are formalized by the plane. Therefore, in the roughest constructions of the lowest loop type ASW1p, a reactor can be considered an infinitely large acoustic resistance, similar to idling (open circuit) in electrical engineering.

Using the idea of active resistances we arrive at MCC design values of pressure drops on different components of MCC (constant drops, not pressure fluctuations), see Fig. 9.7. If the constant pressure created by one MCP is taken as 100%, then its largest part is dropped on the VVER-1200 reactor (−73%), while only −21% on the SG and −6% on the pipelines. The volume, including the areas of the coolant flow deflections from the "cold" loop pipelines until the annular downcomer region of the reactor, is presented in the form of an equivalent circuit shown in Fig. 5.3. These volumes are replaced by a reflecting plane with inhomogeneity determined by the ratio of cross-sectional areas S_0/S_1 (Figs. 5.4 and 9.8).

There are subsequent turns of the coolant flow on the bottom of the reactor vessel: on the bottom of the reactor core barrel and the slotted holes at the fuel assembly inlet, which are the largest inhomogeneity of the path with the largest acoustic resistance. These basic inhomogeneities, first of all, are located close to each other and for rough estimates of the ASW parameters they can be represented as one inhomogeneity with the largest total reflection coefficient. Second step: spatially they are approximated by a plane. Further, vertically up both the core and the protective tube unit (PTU) make

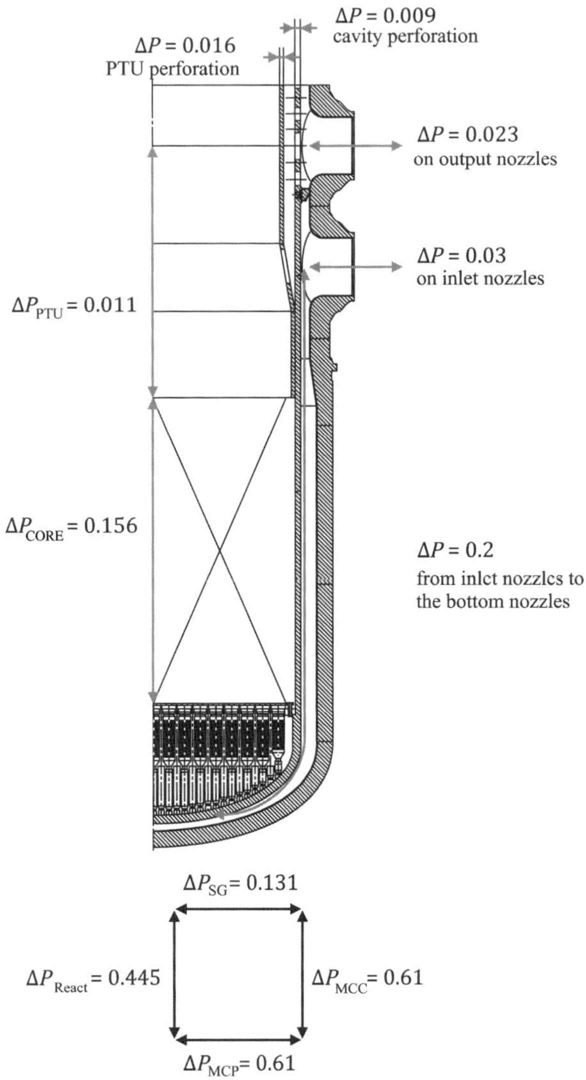

Figure 9.7: Design values of pressure drops [MPa] in the main sections of the coolant flow in the reactor.

insignificant attenuation in acoustic waves. Moreover, the outlet of the reactor, the perforation of the PTU shell and the perforation reactor core barrel also do not have a significant impact on the pressure drop compared with the entrance to the core.

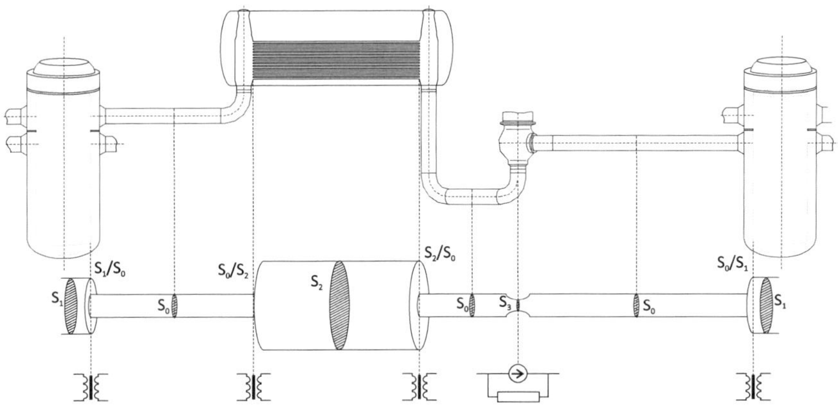

Figure 9.8: Jumps of the MCC flow section and their electrical analogies (transformers).

In practice, total reflection never happens, therefore the modulus of the reflection coefficient is always less than one: $0 < |G| < 1$. That's why simulation of such inhomogeneity using a purely reactive load (capacitive or inductive) is, generally speaking, groundless. With a reactive load there is always a total reflection and $|G| \equiv 1$, however with a phase shift $-\varphi$. The non-zero phase shift that occurs during reflection does not affect the reflectance coefficient because $|e^{j\varphi}| \equiv 1$.

In the simplest representations, an ASW is formed not by the volumes listed above but by the planes perpendicular to the direction of propagation of the longitudinal wave, in which there is a jump of acoustic resistance. The value of the acoustic resistance of the given jump in the roughest approximation, when it is only represented by a real number, is determined by Fresnel equations or by ratios of the effective areas of the flow sections. The total area of flow cross-sections of VVER-1200 steam generator tubing is bigger than the cross-section of the DN850 pipeline: $S_2/S_0 = 2.56$, as well as a quarter of the area of the annular downcomer region of the reactor $S_1/S_0 = 1.74$.

It is also incorrect to introduce active resistance into the acoustic inhomogeneity to make the reflectance coefficient less than one. In electrical analogy, a step-down transformer is most suitable for modeling the jump of the pipeline flow cross-section (Fig. 9.8).

Between adjacent transformers, the medium is modeled by a long line rather than a capacitance or inductance in the lumped parameters (as was done in the previous sections), and it is believed that a mixed field of standing and traveling waves will be formed in it. Moreover, the ASW frequency will be determined by the distance between adjacent transformers, and the ratio between the traveling wave and the ASW will be determined by the transformation coefficient.

If the incident pressure wave P_{inc} corresponds to the input voltage of the transformer U_{in}, and the output voltage of the transformer U_{out} corresponds to the transmitted wave $P_{\text{past}} = 1 - P_{\text{refl}}$, the transformation ratio $|K| = \frac{U_{\text{out}}}{U_{\text{in}}}$ is associated with the reflection coefficient $|G|$ by the formula

$$|K| = \frac{|P_{\text{inc}} - P_{\text{refl}}|}{P_{\text{inc}}} = 1 - |G| \tag{9.3}$$

In any transition from a small diameter to a larger one and, vice versa, from a larger diameter to a smaller one, the transformation ratio will be less than one $|K| < 1$. This does not mean that as the longitudinal wave passes the acoustic inhomogeneity, the energy is absorbed from it. In an ideal transformer, as well as on reactive elements (capacitance and inductance), there is no loss of energy. An ideal transformer is also considered as a reactive component, as a kind of inductance. A transformer, as two coupled inductances, represents some equivalent inductive load. However, its peculiarity is that the energy of the longitudinal wave is not reflected totally on such inductance. Part of the energy overcomes the inhomogeneity which appears here in the form of a transformer and propagates further to the next inhomogeneity. The wave is reflected from the reactive components. Reactive components are capable of accumulating reactive energy, which, unlike active energy, is not lost forever, and can be transferred to another reactive component. The resulting standing waves do not carry energy along the long line. The long line and reactive components accumulate together some constant reactive energy sharing it in dynamic equilibrium and moving from the reactive component to the other and back.

Various combinations of pairs of inhomogeneities cause $\frac{n^2-n}{2}$ of frequencies of standing waves, where n is the number of inhomogeneities (Fig. 9.9). The amplitude of each ASW will thus depend on all the reflection coefficients of all the inhomogeneities and on the distance to the acoustic source. The resulting spatial distribution of

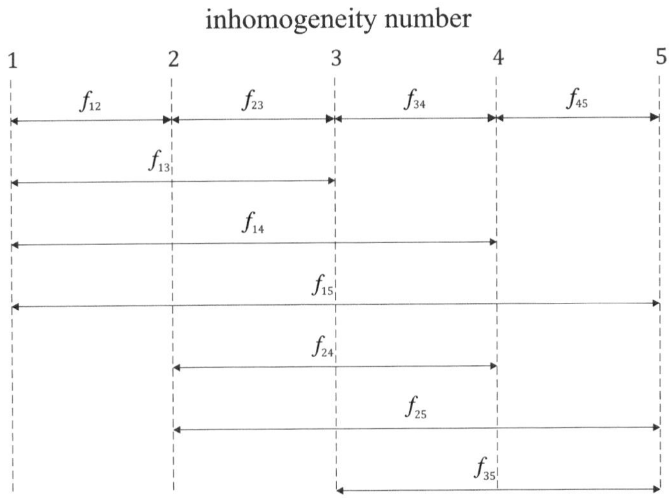

Figure 9.9: Various natural ASW frequencies arising with $\ll n \gg$ inhomogeneities. In this case when $n = 5$, 10 natural frequencies are generated, f_{ij}, where $i, j = 1, \ldots, 5$, $i \neq j$.

the ASW amplitudes is obtained as the sum of an infinite number of incident and reflected traveling waves of the coolant pressure.

The amplification coefficient of standing wave increases linearly with the growth of the ratio $K = \frac{S_1}{S_2}$ (Fig. 9.10) and can reach arbitrarily large values as the heterogeneity increases. If the inhomogeneities on the left and on the right are not identical to each other and have reflection coefficients G_1, G_2, the amplification of the amplitude will be expressed by the formula

$$A \cdot \frac{1 + |G_1|}{1 - |G_1| \cdot |G_2|} \tag{9.4}$$

The volume of the medium between two adjacent inhomogeneities of the pipeline can be understood as a resonator, the frequency of which is determined by the distance between the inhomogeneities L. The more the reflection coefficients from the inhomogeneities, the greater the excited ASW amplitudes. For a group of resonators not connected directly each to the other but separated by heterogeneities, contributing with additional attenuation, the (9.4) type formula will get much more complicated. Multiple reflections from many obstacles are better to model numerically, but first one has to analytically

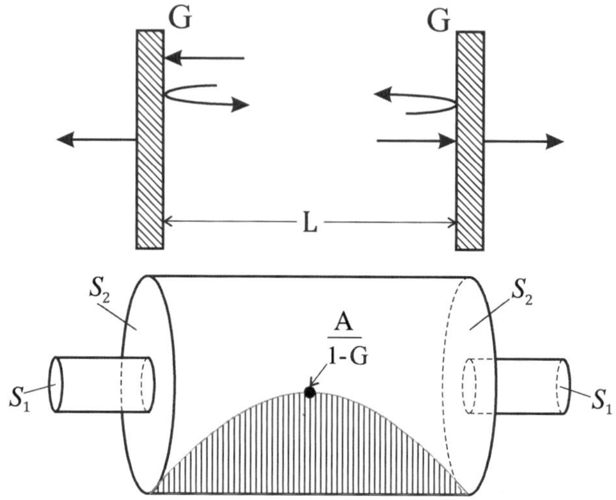

Figure 9.10: Amplitude of a standing wave as a function of the reflection coefficient of inhomogeneity: $\frac{A}{1-G} = \frac{A}{2} \cdot (K+1)$; $K = \frac{S_1}{S_2}$, $|G| = \frac{|K-1|}{K+1}$.

assess "$n-1$" the ASW frequencies and amplitudes arising on adjacent inhomogeneities where n is the number of inhomogeneities.

9.3 ASW Volumetric Acoustic Inhomogeneities and Q-Factors of Their Resonances

When replacing the volumetric inhomogeneities with flat or even point ones, the interrelation between the ASW wavelength and the maximum overall dimension of the inhomogeneity is used. For example, the wavelength of lower ASW types which is commensurate with the loop length, while the diameter of the DN850 pipeline is negligible, can be ignored in a rough approximation.

However, three-dimensional acoustic inhomogeneities create a continuum set of incident and reflected waves or, simply put, a continuum set of incident and reflecting wave paths. As a result, there is a "smearing" of the ASW central frequency. It is well known that there are several physical sources of smearing the ASW resonance, for example, the flow turbulence or the viscosity of the acoustic medium also widen ASW resonance. It is necessary to add a structural design factor, which takes place due to the non-zero radius of the pipeline.

It can shift the location of the ASW central frequency, thus introducing a systematic error in the calculation of its wavelength or the speed of sound based on the ASW frequency and wavelength. The higher the ASW type, greater the error.

The train of ASW harmonics in pipelines having non-zero radii is not exactly harmonics in frequency. The second ASW harmonic is not twice as large as the first in frequency but slightly less than two, the third one is slightly less than three, etc., and the difference increases with the growing number of ASW.

We demonstrate the non-harmonicity of an ASW series on a simple example in case of a non-zero diameter of the pipeline equal to D and its length equal to L. In Fig. 9.11, the shortest and longest path of the incident and reflected wave are shown.

The ratio of the extreme frequencies for the first ASW harmonic will be

$$\frac{f_1^*}{f_1} \approx 1 - \frac{D^2}{2L^2}$$

The value of the difference $\frac{D^2}{2L^2}$ does not necessarily mean a shift of the central ASW frequency. Such a frequency of the spectral component will only be present close to the central ASW frequency. This spectral component will "smear" the ASW resonance, affecting its final Q-factor, but if there are more such spectral components than others, then, indeed, we get a systematic error in estimation of the central frequency. In this case, it is associated with the shortest path of the "round trip" of an acoustic wave that is the minimum ASW wavelength corresponding to the zero diameter of the pipeline.

For the second harmonic, the difference between the maximum path and the minimum path is greater than for the first one (see Fig. 9.11):

$$\frac{f_2^*}{f_2} \approx 1 - 2\frac{D^2}{L^2}$$

For the nth harmonic:

$$\frac{f_n^*}{f_n} \approx 1 - n\frac{D^2}{L^2}$$

The value $\frac{D^2}{L^2}$ is negligible for the first harmonic of ASW1p and is 0.022%, if the MCC is replaced by a straight-line pipeline.

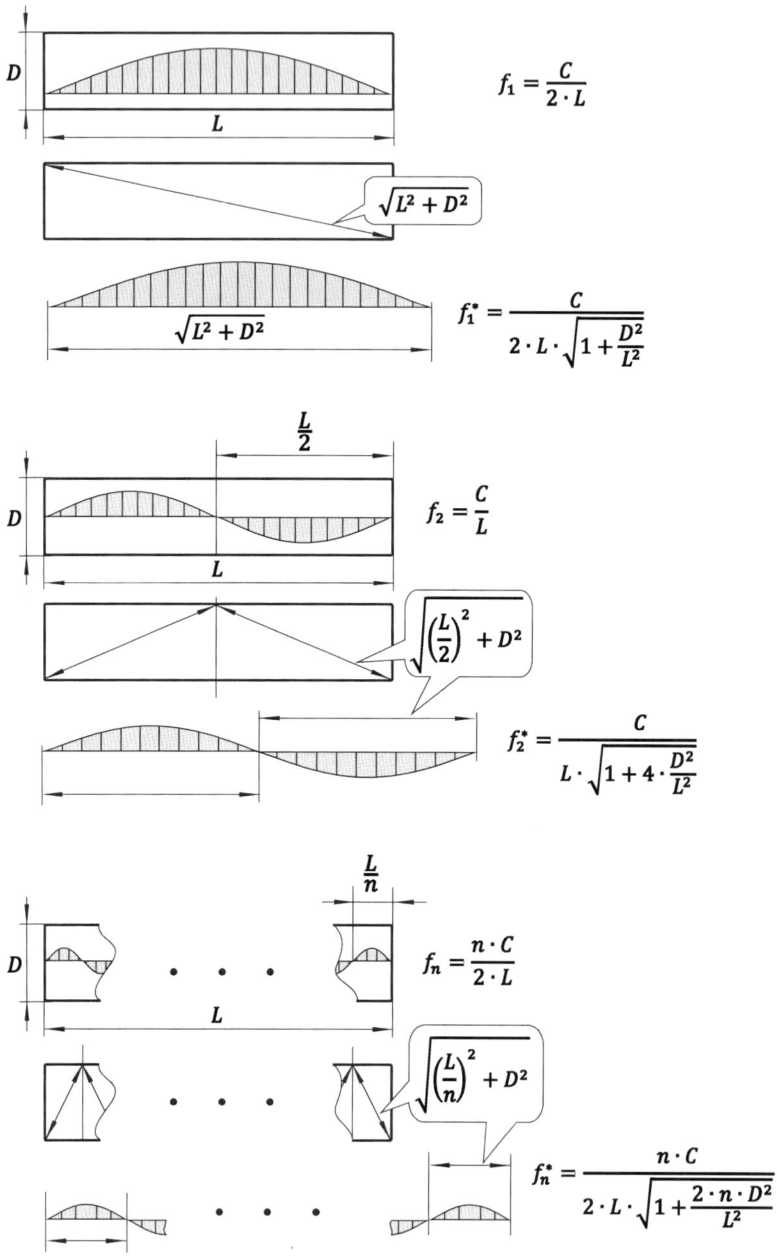

Figure 9.11: Lengthening of the wave of the first and second harmonics due to the non-zero diameter of the pipeline.

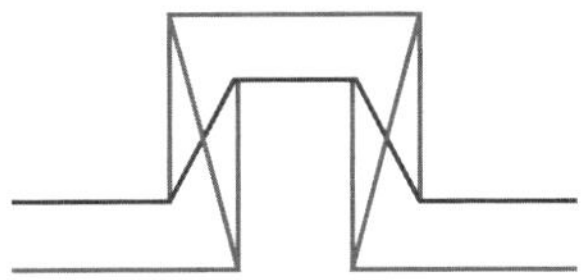

Figure 9.12: Simplified representation of the U-shaped pipe section with the shortest and longest acoustic wave paths.

But the presence of a U-shaped MCC tube fragment significantly increases the difference between the shortest and the longest path (see Fig. 9.12).

More “smearing” of the ASW resonances shall be made by a volumetric MCC inhomogeneity. These include the volume of SG headers and the volume in the vicinity of the bottom of the in-core cavity.

The central ASW frequency (F) is associated with some constant speed of sound (C) in the medium and a constant distance, which has been covered first by the incident and then by the reflected wave: $F = \lambda \cdot C$. That is, the idea of the trajectory based on the wavelength λ is purely one-dimensional and the one based on parameters F, C is purely a single-point one. If the invariance of the sound speed is based on the assumption of the invariance of all the above parameters of the medium, the invariance of the distance is associated with the invariance of the ASW half-wave length between the two acoustic inhomogeneities. In three-dimensional structures of acoustic paths between the two simplest inhomogeneities, for example, in the form of the end walls, there is always a continuum ensemble of different “round paths” of incident and reflected waves, which widen the ASW resonances.

We assume that for simple acoustic paths broadening of the ASW resonance does not lead to a shift in its central frequency and it is associated with the minimum possible distance between the two inhomogeneities. This hypothesis implies that the maximum velocity of the disturbance motion along the elastic medium is located on the minimum path (similarly as the electric current flows along the path of the least resistance).

For a circular cross-section pipeline that is a two-dimensional structure, the one-dimensional ASW trajectory will match its axis of symmetry. Moreover, this will be the trajectory of the minimum length. It is equidistant from the walls of the pipeline; therefore, generalizing this trajectory to all two-dimensional cases, we can introduce the concept of a "mean acoustic trajectory" — MAT. Thus, MAT shall be represented based on the two-dimensional nature of the acoustic path, as a kind of middle line, equidistant from the upper and lower boundaries of the equivalent pipeline in any of its cross-sections. Such inhomogeneity, as a jump in cross-section or a bend of the pipeline, can be easily converted into a flat two-dimensional equivalent. As a result, we will have an equivalent pipeline with a variable cross-section, in which it is easy to draw MAT. However, already in a single bend of the pipeline there is a deviation of the minimum ASW path from the axial line of the pipeline, which leads to an error in calculating the ASW length. Additional difficulties arise on purely three-dimensional inhomogeneities such as branching. For example, the SG channel header together with eleven thousands of HET is a three-dimensional object that requires a complex computational justification in transition to an equivalent inhomogeneous two-dimensional pipeline.

If a single equivalent line, namely, MAT, replaces all possible ways of elastic wave propagation, then calculation of the height of the ASW trajectory in the SG channel header is a separate task. Considering that all HETs are identical to each other and in each of their layers there is the same number of HETs, we will draw MAT in the middle between the upper and lower HET layers. SG tubes are located at about 2.3 m of the channel header height, and above them is the channel header lid, which is a reflecting plane. Apparently, the SG headers will contribute the greatest uncertainty into assessment of the ASW frequency due to the MAT uncertainty.

On the shell-shaped surface with holes, namely, on the bottom of the reactor core barrel with inlet holes for the coolant flow and on cylindrical surfaces of core barrel and the PTU shell-shaped surfaces with outlet holes, we can draw the MAT along their geometric center of symmetry. For example, for the bottom of the reactor core barrel, it will be drawn on the vertical axis of the reactor; for the bended surfaces of PTU and the core barrel – on the axis of symmetry of the "hot leg" pipeline. Thus, it should be noticed that the ASW length calculated by the natural assumptions on the MAT can introduce a systematic error into assessment of the ASW central frequency and, hence, into the speed of sound.

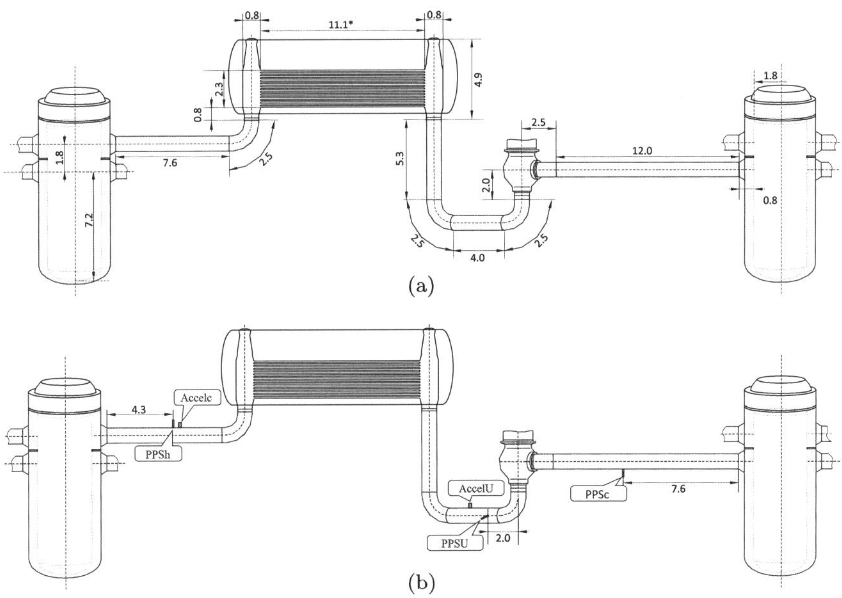

Figure 9.13: Scan of the VVER-1200 MCC heat exchange loop with geometric dimensions (a) and the layout of pressure sensors (marked by Russian Capital letters) and accelerometers (marked (Akc.) (b)).

Each inhomogeneity of the acoustic path has its own individual features. First of all, it is the reflection coefficient, which is a complex function of frequency. In the reflection of the incident wave, not only does its amplitude decrease at the given frequency, but there is also a phase jump.

A two-dimensional representation of the VVER-1200 heat exchange loop (Unit No. of Novovoronezh NPP-2) in Fig. 9.13 with the actual geometric dimensions is, however, its simplified representation. In particular, SG channel headers are shown spaced by the average length of heat exchange tubes of 11.1 m, although they are actually close to each other. Figure 9.13 also shows the layout of sensors on VVER-1200 MCC.

9.4 Branching

So far, branching of pipelines has been interpreted as an abrupt change in the properties of the pipeline, its geometric parameters

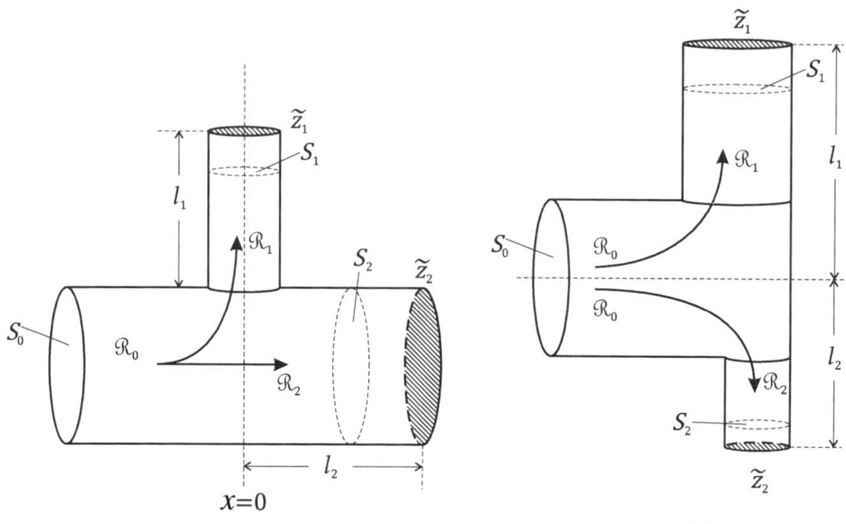

Figure 9.14: Branching as connecting a leg to the main pipeline (a), and as a tee (b).

at the point $x = 0$. It was concluded there is the so-called balance of the flow cross-sections before and after the point $x = 0$. This is a very rough approximation, not taking into account the loading of each of the pipelines. In this point approximation, it was assumed by default that the pipelines have semi-infinite geometry, that is, they are not loaded by any resistance, and because of their infinite length, they have a resistance equal to the wave (that is, active, non-complex) resistance.

Branching can be considered as a pipeline connecting a leg to the main pipeline, and as a tee (Fig. 9.14), which in this approximation is insignificant.

The local geometry of the branch is not considered, and we also neglect the associated changes in the acoustic properties of the medium due to the flow deflections. It is supposed that the resulting local perturbations of the medium are a spatial transition process that dies out at $|x| \gg 0$ and does not affect the steady-state characteristics of the longitudinal waves. In the first case, the main circulation pipeline DN850 acts as the main pipeline, and numerous pipelines of additional circulation circuits mounted in each loop of the VVER reactor plant act as connecting legs. So, for example, Fig. 9.15

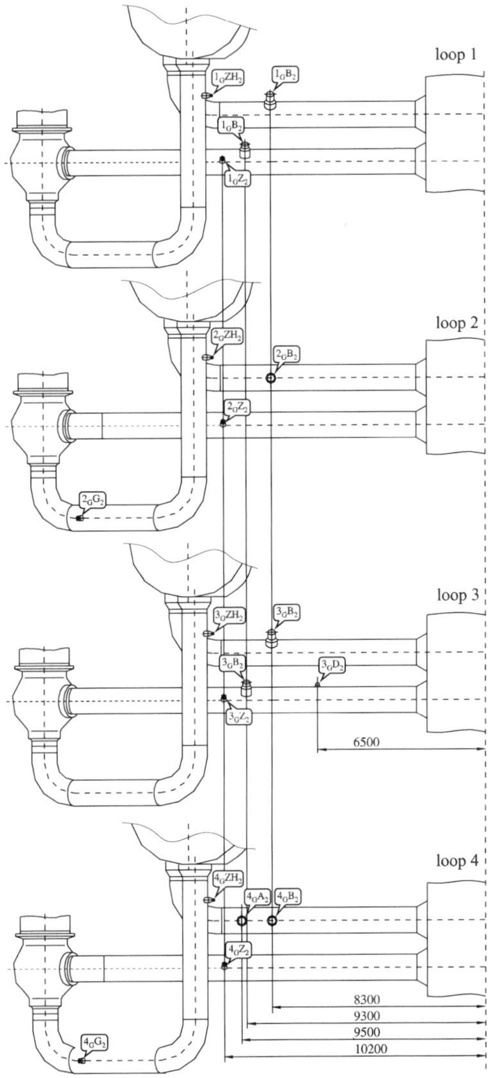

Type	Description	Size
1_GB_2	From ECCS active part	351 × 36
1_GZH_2	Coolant makeup	133 × 14
1_GZ_2	Coolant Blowout	133 × 14

Type	Description	Size
2_GB_2	to Scheduled Primary Cooldown System	351 × 36
2_GG_2	to Scheduled Primary Cooldown System	351 × 36
2_GZH_2	Coolant makeup	133 × 14
2_GZ_2	Coolant Blowout	133 × 14

Type	Description	Size
3_GB_2	From ECCS active part	351 × 36
3_GD_2	Pressurizer Injection Tube	133 × 14
3_GZH_2	Coolant makeup	133 × 14
3_GZ_2	Coolant Blowout	133 × 14

Type	Description	Size
4_GA_2	Connecting Pipeline of Pressurizer	426 × 40
4_GB_2	to Scheduled Primary Cooldown System	351 × 36
4_GG_2	to Scheduled Primary Cooldown System	351 × 36
4_GZH_2	Coolant makeup	133 × 14
4_GZ_2	Coolant Blowout	133 × 14

Figure 9.15: Nozzles of additional circulation circuits for each loop.

shows four circulation loops of the VVER-1200 reactor with nozzles for additional pipelines, and here not every pipeline is specified but only those which have nominal diameter more than 100 mm.

Additional pipelines are, on the one hand, acoustic complex resistances for the main pipeline, and on the other hand, the additional pipeline develops its own mixture of traveling and standing waves. On the one hand, each additional pipeline affects the acoustics of both this heat exchange loop and the whole MCC. On the other hand, as the primary circuit item, the additional pipeline itself needs to take into account its own acoustic properties. The vibration state of the additional pipeline is largely determined by its own ASW, and they, in turn, are determined by the length of the pipeline and its end loading.

The abnormal vibration state of the additional pipeline can provoke leaks of the primary.

Circuit, and can lead to abnormal vibrations of the main equipment and MCC pipelines.

Each heat exchange loop is unique in its composition of additional pipelines (Fig. 9.15), which makes the ASW picture unique for each loop. As a long line, an additional pipeline, theoretically, can contribute any resistance into the MCC, including infinitely large. This means that at some frequency, it can affect as a blocking filter, that is, to play the role of an acoustic damper. For short pipelines, such effects are not observed in practice, since their own ASWs lie in the high-frequency region outside the operating frequency range. Particular attention should be paid to the pressurizer system (Figs. 9.15 and 9.16), in which the connecting pipe has both the maximum length and the maximum diameter in comparison with other additional pipelines. In this sense, the fourth loop, which houses the nozzle of the pressurizer connecting pipeline, always has specific acoustic properties and also a special vibration state of the main equipment.

Some additional pipelines are part of the actuators for controlling parameters of the reactor plant via opening/closing of the relevant valves. This leads to discrete changing in frequencies of ASW of additional pipelines also changing their vibration state.

Also, due to their fairly long connecting legs having also large diameter, it is necessary to consider the pipelines of the passive part of the Emergency Core Cooling System attached to the reactor vessel and designed to connect the ECCS tanks to the reactor (see Fig. 9.16). Two ECCS pipelines are designed to supply boric acid solution to the lower part of the reactor, the other two — to supply boric acid solution to the upper part of the reactor. They have

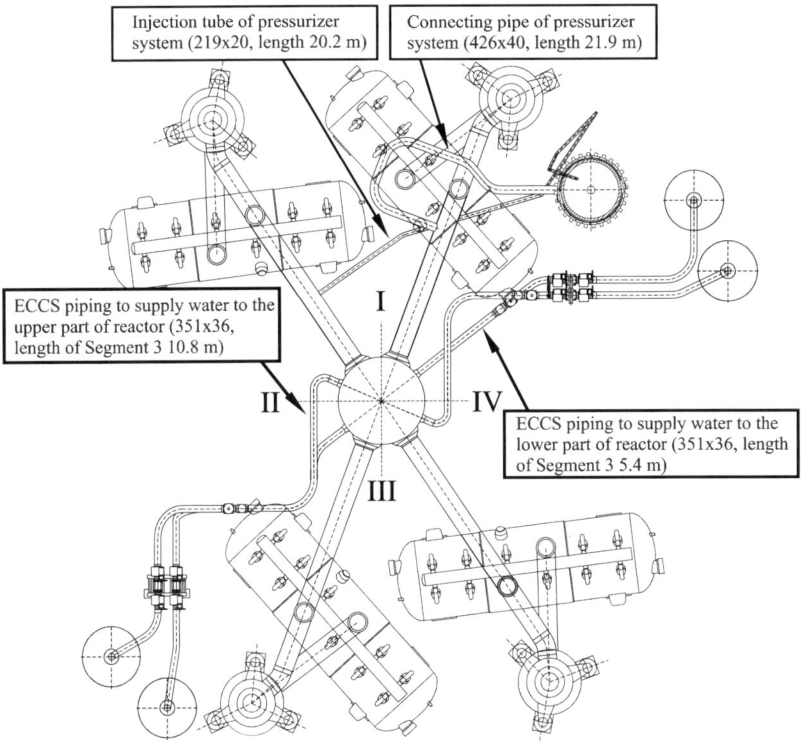

Figure 9.16: Pressurizer piping and piping of the ECCS passive part.

different lengths and thus determine different frequencies impacting on the reactor. The length of the ECCS piping will increase dramatically because of opened stop valves in case of an accident. Therefore, we need justification that such acoustic impact will not cause anomalous resonant responses of the circuit.

The listed problems of acoustic effects of additional pipelines require special painstaking research, beyond the scope of this monograph. In what follows, we will discuss the computing tools for the above research.

Let the main coolant pipeline have a wave resistance R_0, and the connecting leg — $\mathcal{R}_1$, and let them be loaded onto the relevant complex resistances $\tilde{z}_0$ and $\tilde{z}_1$ (Fig. 9.17). To use the electrical analogy it is necessary, in first turn, to bring these two loads into the flow cross section $x = 0$, and in the second turn correctly adjust the given loads right in the flow section $x = 0$. The reduction to the input resistance

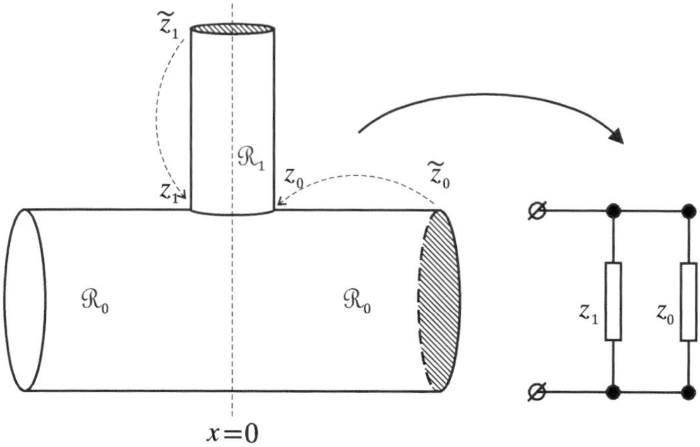

Figure 9.17: The trunk pipeline and the connecting leg are loaded onto complex resistances and their equivalent diagram.

(or to the resistance in the flow section $x = 0$) is made according to the previously derived formula for a long line without attenuation:

$$z = \mathcal{R} \cdot \frac{\tilde{z} + j \cdot \mathcal{R} \cdot \operatorname{tg} \eta l}{\mathcal{R} + j \cdot \tilde{z} \cdot \operatorname{tg} \eta l} \tag{9.5}$$

In the general case

$$z = \mathcal{R} \cdot \frac{\tilde{z} + \mathcal{R} \cdot \operatorname{th}(Kl)}{\mathcal{R} + \tilde{z} \cdot \operatorname{th}(Kl)}$$

where $K = \alpha + j\eta$ is a complex wave number.

In the flow cross-section at $x = 0$, the equivalent diagram is a parallel connection of two complex resistances z_0, z_1. Here we often meet an erroneous equivalent representation in the form of a serial connection of these resistances, when it is believed that the current consistently overcomes first one resistance, then another. This is the insidiousness of electrical analogies. In fact, it is the pressure (or voltage in the electrical analogy) that is common to both pipelines, and the velocity (current) is divided into two parts to two tubes reciprocally of their resistances z_0, z_1, (Fig. 9.17) which correspond to the parallel connection.

Loading of the pipeline is determined by any upcoming inhomogeneity, in particular, by pipeline valves. For example, closing and safety valves in their closed state correspond to closed end of the

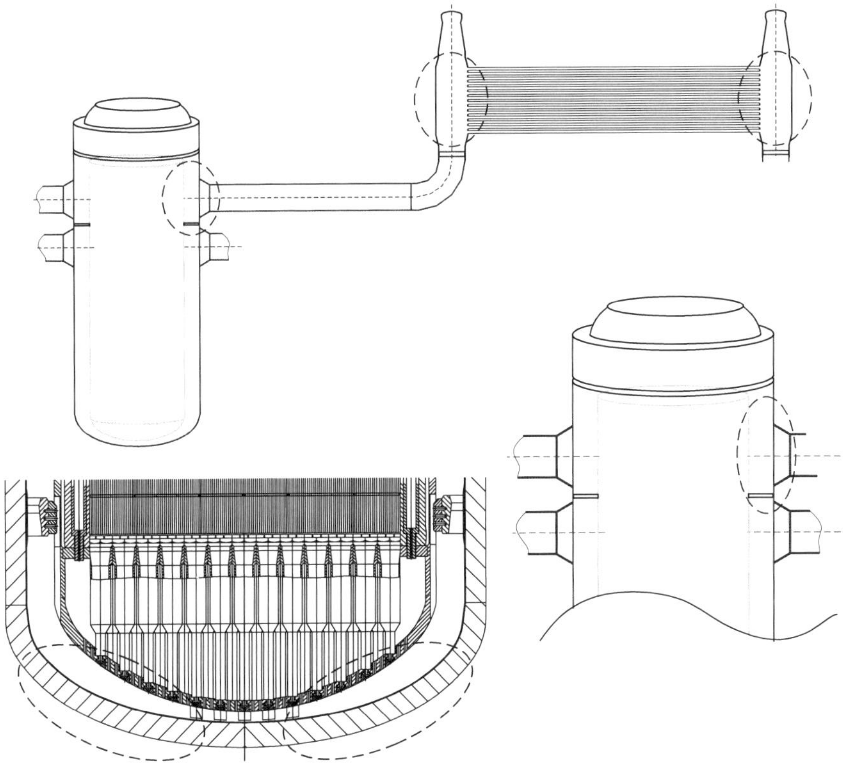

Figure 9.18: MCC tee connectors.

pipeline and can be simulated by idling (open) circuit in the electrical analogy. Control valves used for partial overlap of the flow cross section have changing complex resistance.

One can consider as tees and branches in the VVER circulation circuit: eight volumes of SG channel headers, eight nozzles of the reactor vessels and the bottom of the reactor core barrel (see Fig. 9.18).

Table 9.1 summarizes parameters of various tees. Using those, it is possible to make different calculations for additional pipelines of heat exchange loops of MCC. Moreover, for simplicity, in all cases it is assumed that one branch of the tee is always the main pipeline and the fluid in all of the branches is the same.

If in the flow cross section at zero of the tee of one of the branches the assigned resistance is zero, then this means that the perturbation (longitudinal wave) will not propagate further than

Table 9.1: Complex resistances of tees and branching.

No. item	Tee parameter	Resistance z_0, z_1, estimated for the flow section $x = 0$	Complex resistance of tee
	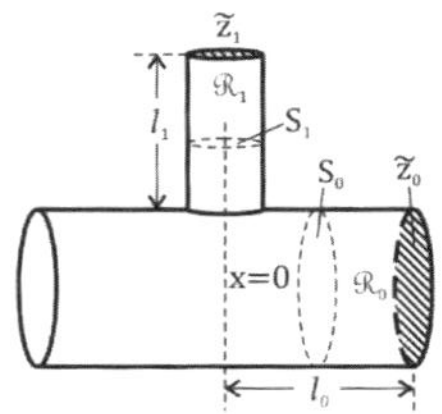 $\mathcal{R}_1 = S_1 \cdot \rho_1 \cdot C_1$ $\mathcal{R}_0 = S_0 \cdot \rho_0 \cdot C_0$	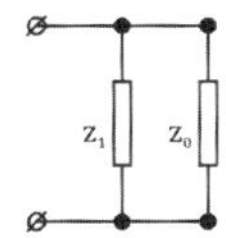 $z_0 = \mathcal{R}_0 \cdot \frac{\tilde{z}_0 + \mathcal{R}_0 \cdot \mathrm{th}(K \cdot l_0)}{\mathcal{R}_0 + \tilde{z}_0 \cdot \mathrm{th}(K \cdot l_0)}$ $z_1 = \mathcal{R}_1 \cdot \frac{\tilde{z}_1 + \mathcal{R}_1 \cdot \mathrm{th}(K \cdot l_1)}{\mathcal{R}_1 + \tilde{z}_1 \cdot \mathrm{th}(K \cdot l_1)}$	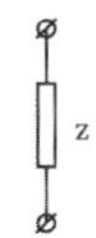 $z = \frac{z_0 \cdot z_1}{z_0 + z_1}$
1	Semi-infite pipelines (main and connecting leg)		
	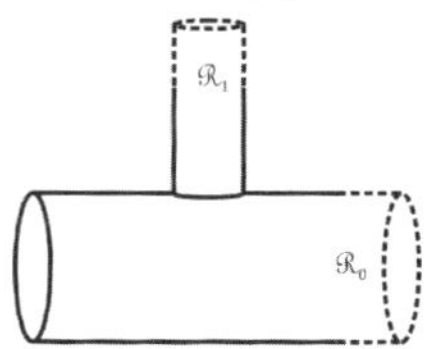	$z_0 = \mathcal{R}_0$ $z_1 = \mathcal{R}_1$	$z = \frac{\mathcal{R}_0 \cdot \mathcal{R}_1}{\mathcal{R}_0 + \mathcal{R}_1}$ $G = \frac{S_0 - (S_0 + S_1)}{S_0 + S_0 + S_1} = \frac{1}{1 + 2 \cdot \theta'}$ $\theta = \frac{S_0}{S_1}$ $z = \frac{1}{2}\mathcal{R}_0$ The tee is equivalent to an inconsistent (with reflection)
1.1	$\mathcal{R} = \mathcal{R}_0(S_1 = S_0)$	$z_0 = \mathcal{R}_0$	$G = \frac{R_0 - 2R_0}{3R_0} = -\frac{1}{3}$ load 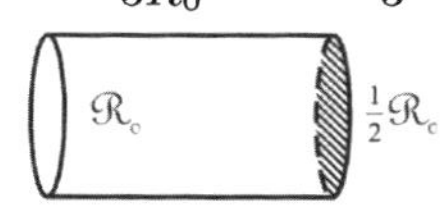

2	Semi-infinite main pipeline		
	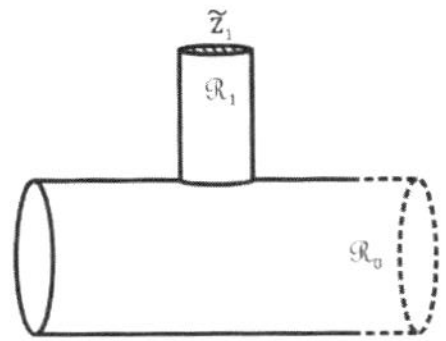	$z_0 = \mathcal{R}_0$ $z_1 = \mathcal{R}_1 \cdot \dfrac{\tilde{z}_1 + \mathcal{R}_1 \cdot \text{th}(K \cdot l_1)}{\mathcal{R}_1 + \tilde{z}_1 \cdot \text{th}(K \cdot l_1)}$	$z = \mathcal{R}_0 \cdot \dfrac{z_1}{\mathcal{R}_0 + z_1}$
2.1	$\tilde{z}_1 = 0$ — an open end of the connecting leg		
	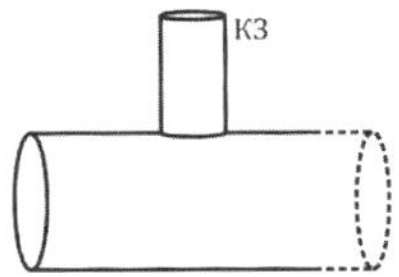	$z_1 = \mathcal{R}_1 \cdot \text{th}(K \cdot l_1)$	$z = \mathcal{R}_0 \cdot \dfrac{\mathcal{R}_1 \cdot \text{th}(K \cdot l_1)}{\mathcal{R}_1 + \mathcal{R}_1 \cdot \text{th}(K \cdot l_1)}$
2.1.1	$l_1 = \dfrac{C}{2f}$ — the length of the connecting leg is equal to half the wave: $K \cdot l_1 = \pi$		
		$z_1 = 0$	$z = 0$ From the tee we get "Short cut Circuit", i.e. the main pipeline with an open end
2.1.2	$l_1 = \dfrac{C}{4f}$ — the length of the connecting leg is a quarter of the wavelength $K \cdot l_1 = \dfrac{\pi}{2}$		
		$z_1 \to \infty$	$z = \mathcal{R}_0$ The connecting leg does not affect the main pipeline

(*Continued*)

Table 9.1: (*Continued*)

No. item	Tee parameter	Resistance z_0, z_1, estimated for the flow section $x = 0$	Complex resistance of tee
2.2	$\tilde{z}_1 \to \infty$ — a closed end of the connecting leg		
	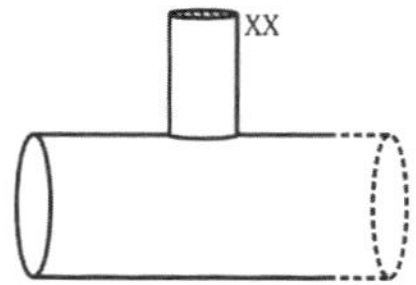	$z_1 = \dfrac{\mathcal{R}_1}{\text{th}(K \cdot l_1)}$	$z = \mathcal{R}_0 \cdot \dfrac{\mathcal{R}_1}{\mathcal{R}_1 + \mathcal{R}_0 \cdot \text{th}(K \cdot l_1)}$
2.2.1	$l_1 = \dfrac{C}{2f}$ — the length of the connecting leg is equal to half the wave: $K \cdot l_1 = \pi$		
		$z_1 \to \infty$	$z = \mathcal{R}_0$ The connecting leg dot not affect the main pipeline
2.2.2	$l_1 = \dfrac{C}{4f}$ — the length of the connecting leg is equal to half the wave: $K \cdot l_1 = \dfrac{\pi}{2}$		
		$z_1 = 0$	$z = 0$ From the tee we get "Short Cut Circuit", the main pipeline
3	Semi-infinite main pipeline		
	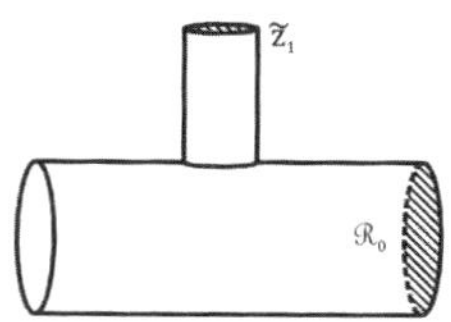 $\tilde{z}_0 \to \infty$	$z_0 = \dfrac{\mathcal{R}_0}{\text{th}(K \cdot l_0)}$ $z_1 = \mathcal{R}_1 \cdot \dfrac{\tilde{z}_1 + \mathcal{R}_1 \cdot \text{th}(K \cdot l_1)}{\mathcal{R}_1 + \tilde{z}_1 \cdot \text{th}(K \cdot l_1)}$	$z = \dfrac{z_0 \cdot z_1}{z_0 + z_1} = z_1 \cdot \dfrac{\mathcal{R}_0}{\mathcal{R}_0 + Z_1 \cdot \text{th}(K \cdot l_0)}$

3.1 $l_0 = \dfrac{C}{2f}$ — the length of the main pipeline to the end is half the wavelength: $K \cdot l_1 = \pi$

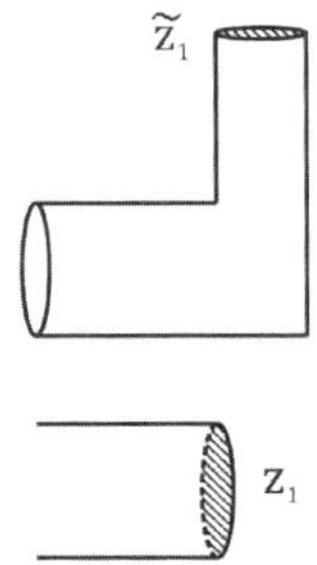

$z_0 \to \infty$

$z = z_1$
The tee is equivalent to the complex resistance z_1

3.1.1 $l_1 = \dfrac{C}{2f}$ — the length of the connecting leg is equal to half the wavelength:

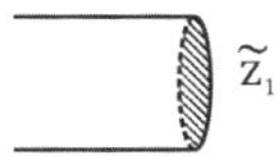

$z_1 = \tilde{z}_1$
$z_0 \to \infty$

$z = \tilde{z}_1$
The tee is equivalent to the complex loading of the connecting leg

3.1.2 $l_1 = \dfrac{C}{4f}$ — the length of the connecting leg is equal to a quarter of the wavelength

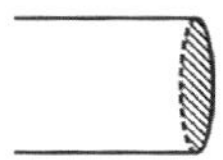

$z \to \infty$

$z_1 \to \infty$
$z_0 \to \infty$

$z \to \infty$
The tee is equivalent to a closed pipeline

(Continued)

Table 9.1: (*Continued*)

No. item	Tee parameter	Resistance z_0, z_1, estimated for the flow section $x = 0$	Complex resistance of tee
3.2	$l_0 = \frac{C}{4f}$ is the length of the main pipeline to the end is equal to a quarter of the wavelength: $K \cdot l_0 = \frac{\pi}{2}$		
3.2.1	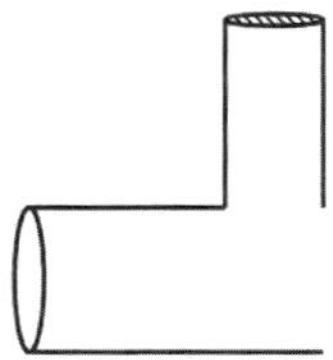	$z_0 = 0$	$z = 0$ The tee is equivalent to an open pipeline
4	Open main pipeline 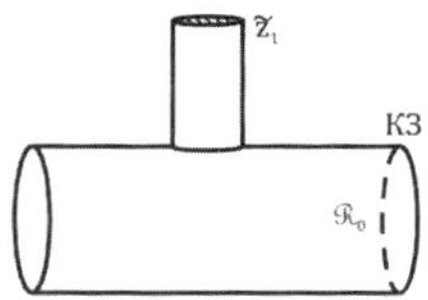 $\widetilde{z}_0 \to 0$	$z_0 = \mathcal{R}_0 \cdot \mathrm{th}(K \cdot l_0)$ $z_1 = \mathcal{R}_1 \cdot \frac{\tilde{z}_1 + \mathcal{R}_1 \cdot \mathrm{th}(K \cdot l_1)}{\mathcal{R}_1 + \tilde{z}_1 \cdot \mathrm{th}(K \cdot l_1)}$	$z = \frac{z_0 \cdot z_1}{z_0 + z_1}$ $= z_1 \cdot \frac{\mathcal{R}_0 \cdot \mathrm{th}(K \cdot l_0)}{z_1 + \mathcal{R}_0 \cdot \mathrm{th}(K \cdot l_0)}$
4.1	$l_0 = \frac{C}{2f}$ — is the length of the main pipeline: $K \cdot l_0 = \pi$		
	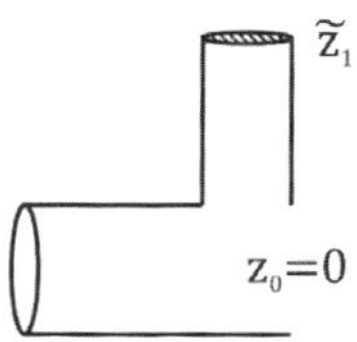	$z_0 = 0$	$z = 0$ The tee is equivalent to an open pipeline 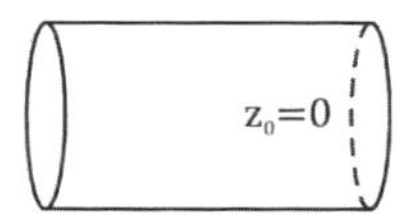

4.2	$l_0 = \frac{C}{4f}$ — is the length of the main pipeline: $K \cdot l_0 = \frac{\pi}{2}$		
	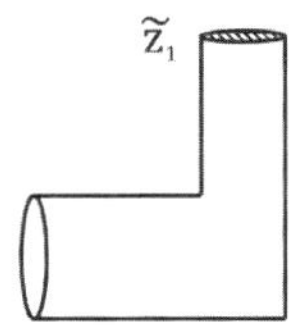	$z_0 \to \infty$	$z = z_1$ The tee is equivalent to the complex resistance z_1 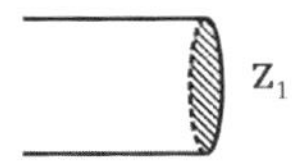
5	Both the pipelines are open (SC)		
	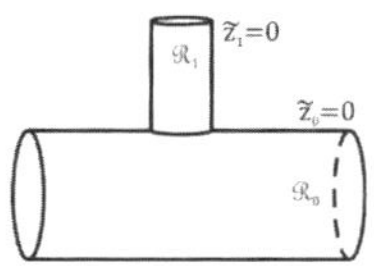	$z_0 = \mathcal{R}_0 \cdot \mathrm{th}(K \cdot l_0)$ $z_1 = \mathcal{R}_1 \cdot \mathrm{th}(K \cdot l_1)$	
5.1	The length of the both pipelines is equal to half the wavelength: $K \cdot l_0 = K \cdot l_1 = \pi$		
	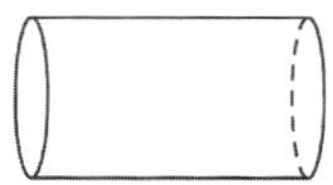 $z = 0$	$z_0 = 0$ $z_1 = 0$	$z = 0$ The tee is equivalent to an open pipeline
5.2	The length of the both pipelines is equal to a quarter wave: $K \cdot l_0 = K \cdot l_1 = \frac{\pi}{2}$		
	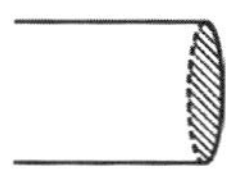 $z \to \infty$	$z_0 \to \infty$ $z_1 \to \infty$	$z \to \infty$ The tee is equivalent to an closed pipeline

the coordinate $x = 0$. This is an important case of a band-stop filter, blocking the spectral component and its harmonics, the frequencies of which are determined by the length of a particular branch of the tee. Particular attention should be paid to the so-called quarter-wave and half-wave lengths of pipelines, which are called $\frac{\lambda}{4}$-transformers and $\frac{\lambda}{2}$-transformers, respectively. The quarter-wave transformer has thc property of inversion: the zero resistance of its load it translates into an infinitely large resistance in the flow section $x = 0$, and vice versa: idle (open) port is translated into a short cut. Half-wave transformers translate themselves into themselves, that is, at a certain frequency, the resistance in the load is transferred to the coordinate without changing $x = 0$.

For example, if a connecting leg of a pipeline with a length of $\frac{\lambda}{4}$ is loaded on a closed valve, it is equivalent to a quarter-wave line loaded at idle (open) circuit and in the flow section $x = 0$ we get a short cut circuit. That is, a wave with frequency $f = \frac{C}{\lambda}$ will not propagate further than such a connecting leg.

Imagine that for the length of the heat exchange loop with $L = 54\,\text{m}$ (as in VVER-1200), the lower ASW type with a frequency of $f = 7.0\,\text{Hz}$ dominates (as in VVER-1200) at a certain coolant temperature. If in the position of the maximum of the ASW (in the middle of the loop) we attach an additional closed end pipeline with a length of 27 m, then this ASW disappears. For this ASW this point $\frac{L}{2}$ will be a short circuit with zero pressure amplitude, completely blocking the propagation of the wave back and forth. Such branching will generate a half wavelength ASW (at a doubled frequency $f = 14\,\text{Hz}$). The same pipeline for the dual frequency is a half-wave transformer, which does not introduce any distortion in the dual frequency ASW. And so on: $\frac{\lambda}{4}$ — the transformer will have the property of inversion, and $\frac{\lambda}{2}$ — the transformer will make no distortions.

Each auxiliary pipeline of heat exchanging loops of the VVER RP is a kind of harmonic filter (a so-called comb filter): It blocks some equidistant frequencies, other equidistant frequencies remain unaffected. Auxiliary pipelines have a length much smaller than the length of the heat exchange loop. Therefore, their harmonic series have high frequency, and they start with a high frequency, often beyond the operating frequency range.

Not only is the length of the auxiliary pipeline important, but also its flow cross-section. In contrast to the mechanical resistance, the acoustic impedance can be normalized by the size of the flow cross-section. An auxiliary pipeline with a small flow cross-section in comparison with that of the main pipeline has a small impact on it. This auxiliary pipeline is a large acoustic resistance and in parallel connection with the equivalent resistance of the main pipeline changes it slightly.

9.5 A Set of Long Lines as Equivalent to VVER Circulation Loop in Electrical Analogies

One of the methods of acoustic calculation of a complex pipeline system is its representation as a set of several long line segments loaded on a point complex resistance. Previously, in such a complex system, all those places (planes perpendicular to the axis of the pipeline) are allocated, where there is a sudden change in the wave resistance and, in particular, a jump in the flow cross-section.

As it has been shown in the previous section, due to the fact that branching can be reduced to a long line with a point and a single complex resistance z_0, a complex pipeline system can be reduced to a serial (cascade) connection of a number of long lines with different wave resistances $\mathcal{R}_k$ and lengths l_k and a single point complex resistance z_0 (Fig. 9.19).

Further, when moving from right to left, the only complex load is sequentially transferred from one section with a jump in resistance to another, neighboring section with another jump in resistance. Repeatedly applying the formula (9.5) at different wave resistances and pipeline lengths, we have

$$\begin{aligned} z_{k+1} &= \mathcal{R}_k \cdot \frac{z_k + j\mathcal{R}_k \cdot \mathrm{tg}(\gamma \cdot l_k)}{\mathcal{R}_k + jz_k \cdot \mathrm{tg}(\gamma \cdot l_k)} \\ &= \mathcal{R}_k \cdot \frac{\mathcal{R}_{k-1} \cdot \frac{z_{k-1}+j\mathcal{R}_{k-1} \cdot \mathrm{tg}(\gamma \cdot l_{k-1})}{\mathcal{R}_{k-1}+jz_{k-1} \cdot \mathrm{tg}(\gamma \cdot l_{k-1})}}{\mathcal{R}_k + j\mathcal{R}_{k-1} \cdot \mathrm{tg}(\gamma \cdot l_k) \cdot \frac{z_{k-1}+j\mathcal{R}_{k-1} \cdot \mathrm{tg}(\gamma \cdot l_{k-1})}{\mathcal{R}_{k-1}+jz_{k-1} \cdot \mathrm{tg}(\gamma \cdot l_{k-1})}} = \cdots \end{aligned} \tag{9.6}$$

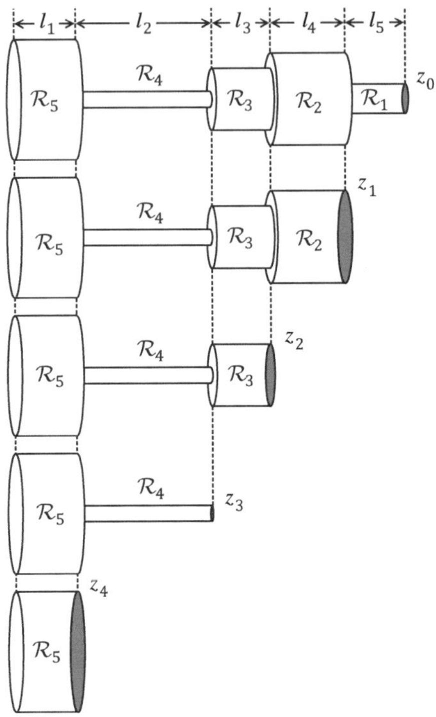

Figure 9.19: Serial connection of long lines with different wave resistances $\mathcal{R}_1, \ldots, \mathcal{R}_5$ and a single complex load z_0. Reduction to the simplest equivalent complex resistance.

where $k = 1, 2, 3, \ldots, N$ is the number of the long lines with a wave resistance of $\mathcal{R}_k$. As a result, we come to a bulky multi-compound fraction. It is simplified in particular cases. For example, in short pipelines $\mathrm{tg}(\gamma \cdot l_k) \approx \gamma \cdot l_k$, where $l_k \ll$ of a quarter of the wavelength $\frac{C}{4 \cdot f}$.

For MCC, the half-wave length of the lowest type of ASW is commensurate with the length of the loop, so when considering any MCC item, including a large one, a point approximation is valid at these frequencies. Even at the length of the steam generator, the assumption of practical invariance of the amplitude of the lowest type of ASW is true. For higher types of ASW, point models should be introduced individually by measuring the MCC item dimensions by a quarter of the spatial ASW wavelength. For a fairly short line, such that its length is much less than a quarter of the wavelength of the lower ASW, we have:

$$z_{k+1} = \mathcal{R}_k \cdot \frac{z_k + j\mathcal{R}_k \cdot \mathrm{tg}(\gamma \cdot l_k)}{\mathcal{R}_k + jz_k \cdot \mathrm{tg}(\gamma \cdot l_k)} \simeq \mathcal{R}_k \cdot \frac{z_k + j\mathcal{R}_k \cdot \gamma \cdot l_k}{\mathcal{R}_k + jz_k \cdot \gamma \cdot l_k}$$
$$= \mathcal{R}_k \cdot \frac{\theta_k + j\gamma \cdot l_k}{1 + j\theta_k \cdot \gamma \cdot l_k}$$

where $\theta_k = \frac{z_k}{\mathcal{R}_k}$ is a reduced load impedance (dimensionless).

Thus, the following simplifications are acceptable:

$z_{k+1} \simeq$

- z_k, when $l_k \sim 0$ a very short line does not affect the transition $z_k \to z_{k+1}$;
- $z_k + j\mathcal{R}_k \cdot \gamma \cdot l_k$, when $\theta_k \cdot \gamma \cdot l_k \ll 1$ a very short line with a large wave resistance $\mathcal{R}_k \gg z_k$ is equivalent to $(R - L)$ of the circuit

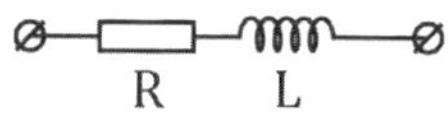

- $j\mathcal{R}_k \cdot \gamma \cdot l_k$, when $z_k \to 0$ (short-cut circuit or open end of the pipeline) the line is equivalent to inductance

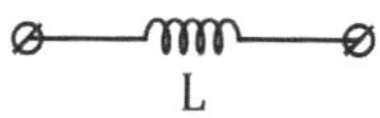

- $\dfrac{\mathcal{R}_k^2}{z_k} + \dfrac{\mathcal{R}_k}{jz_k \cdot \gamma \cdot l_k}$, when $z_k \gg \mathcal{R}_k$ a large load on a short line is equivalent to "$R - C$" circuit

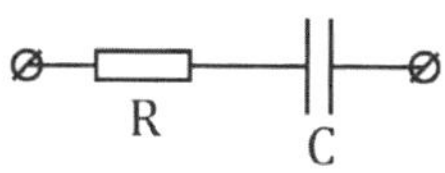

- $\dfrac{\mathcal{R}_k}{j \cdot \gamma \cdot l_k}$, when $z_k \to \infty$ (open circuit or a closed end of the pipeline) the line is equivalent to the capacitance

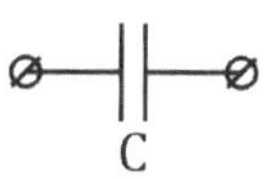

- $\mathcal{R}_k = z_k$, with a purely active load $\mathcal{R}$, equal to the wave resistance the line is equivalent to this purely active load.

The calculation will be easier if we switch from resistance z to the propagation constant γ. The resistance connecting the current (velocity) to the voltage (pressure) leads to two-pole equivalent circuits.

The input $\begin{bmatrix} P_k \\ V_k \end{bmatrix}$ and output $\begin{bmatrix} P_{k+1} \\ V_{k+1} \end{bmatrix}$ two-dimension vectors bound by a square matrix (2×2), lead to quadrupoles:

$$\begin{bmatrix} P_k \\ V_k \end{bmatrix} (2 \times 2) \begin{bmatrix} P_{k+1} \\ V_{k+1} \end{bmatrix}; \tag{9.7}$$

$$\begin{bmatrix} P_{k+1} \\ V_{k+1} \end{bmatrix} = \begin{bmatrix} \cos \gamma l_k & j\mathcal{R}_k \sin \gamma l_k \\ \dfrac{j \cdot \sin \gamma l_k}{\mathcal{R}_k} & \cos \gamma l_k \end{bmatrix} \begin{bmatrix} P_k \\ V_k \end{bmatrix}$$

The matrix in the expression (9.7) only consists of the long line parameters $\mathcal{R}_k$, l_k, γ. The resulting recursive matrix formula (9.7) is easy to implement numerically, thus obtaining a matrix product

$$\begin{bmatrix} P_{k+1} \\ V_{k+1} \end{bmatrix} = \left\{ \prod_{k=1}^{N} \begin{bmatrix} \cos \gamma l_k & j\mathcal{R}_k \sin \gamma l_k \\ \dfrac{j \cdot \sin \gamma l_k}{\mathcal{R}_k} & \cos \gamma l_k \end{bmatrix} \right\} \begin{bmatrix} P_1 \\ V_1 \end{bmatrix}$$

That is, the pressure and velocity in the first section will be consistently reduced through a cascade of pipelines of different flow sections with different wave resistances by successive multiplication of the matrices by the pressure and velocity in the "$k+1$" flow section.

For the first flow cross-section, it is better to choose a pressure node or an antinode, or a place with an active acoustic source (which means that in this place again there is a pressure antinode). Open or closed ends of the pipeline, as well as the section of the pipeline with jumps in acoustic resistance are suitable, too.

Chapter 10

Elastic Dissipative Medium. Attenuation of Acoustic Waves

10.1 Wave Equation with Complex Characteristic Parameter

Multiple generalizations of a wave equation and a motion equation applied in order to consider the actual dissipation properties of elastic media which have been described in classical references [6–18, 23, 30, 32–39, 42–56] do not represent a complete and finalized study. There are still many cases with limitations for their use, there are identified cases when the models do not match the experimental data [106, 107, 147, 174, 208, 212, 216–218, 251–256, 332, 381, 382, 384]. In what follows, we describe the mathematical derivation of each of the wide-spread application methods for consideration of energy losses in dissipative acoustic media.

Let the harmonic wave

$$U(x,t) = U_0 \cdot e^{j(\omega t - kx)} = U_0(x,t) \tag{10.1}$$

which is a solution to the homogeneous wave equation

$$U_{tt}(x,t) - C^2 \cdot U_{xx}(x,t) = 0 \tag{10.2}$$

for an idealized elastic medium with the characteristic "C", be a subject to an exponential spatial attenuation with the attenuation parameter "α" $[\frac{1}{\mathrm{m}}]$. Now let us ask ourselves: "If we have a solution with an attenuated factor

$$U_\alpha(x,t) = U_0 \cdot e^{-\alpha \cdot x} \cdot e^{j(\omega t - kx)} = e^{-\alpha \cdot x} \cdot U_0(t) \tag{10.3}$$

what kind of elastic medium will correspond to it, if it is still described by a homogeneous wave equation

$$U_{tt}(x,t) - C_\alpha^2 \cdot U_{xx}(x,t) = 0 \tag{10.4}$$

but with another characteristic parameter "C_α"?

Let us present the solution (10.1) in another form

$$U(x,t) = U_0 \cdot e^{j[\omega t - (k - j\alpha) \cdot x]} = U_0 \cdot e^{j(\omega t - k_\alpha x)} \tag{10.5}$$

which means that the wave number has become a complex one,

$$k_\alpha = k - j \cdot \alpha$$

and its imaginary part α describes the exponential attenuation of the wave amplitude.

Complex coefficients appear in differential equations when a harmonic complex solution is inserted

$$U(x,t) = U_0 \cdot e^{j(\omega t - k_\alpha x)}$$

as the result of differentiation with respect to time (coefficient $j\omega$) or differentiation with respect to space (coefficient jk). Coefficients before the first derivative with respect to time (before the velocity) are most often related to energy losses, e.g. due to internal friction or oscillations damping in mechanical systems. It is well known that friction coefficient is the proportionality factor in front of velocity in the equation of motion of a dissipative mechanical system. The coefficient before the first derivative with respect to space is often interpreted as deformation ratio. Repeated differentiation of expression (10.5) brings the coefficients before the second derivatives ($-\omega^2$ and $-k^2$) back to the real domain. Using these coefficients before the second derivatives, one can obtain kinetic and potential energy. If we go on with further differentiation, for the third derivatives we will again get complex coefficients, and they will also be related to the loss of energy of the longitudinal wave, etc.

The solution in the form (10.5) has a real amplitude U_0, and the attenuation requirement has degenerated into a complex exponential factor in (10.5) and complex sound velocity "C_α".

The wave number implicitly includes a linear frequency dependence

$$k = \frac{\omega}{C} = \frac{2\pi}{C} \cdot f \tag{10.6}$$

When we consider a single fixed frequency ω_0, the wave number corresponds to a constant spatial phase shift

$$\varphi_0 = k \cdot x = \frac{\omega_0}{C} \cdot x \tag{10.7}$$

If we consider some frequency range, i.e. the signal is not sinusoidal, then phase progression $\varphi(\omega)$ must change linearly with the frequency (10.6), so that the signal could retain its shape when transported through an elastic medium:

$$\varphi(\omega) = \frac{\omega x}{C} = 2\pi f \tau_3, \quad \tau_3 = \frac{x}{C} \tag{10.8}$$

Let us go formally to sound velocity, which is the only characteristic parameter of the medium:

$$\begin{aligned} C_\alpha &= \frac{\omega}{k_\alpha} = \frac{\omega}{k - j_\alpha} = \frac{C}{1 - j\frac{\alpha \cdot C}{\omega}} \\ &= \frac{C}{1 + \left(\frac{\alpha \cdot C}{\omega}\right)^2} + j \cdot \frac{C}{\left[1 + \left(\frac{\alpha \cdot C}{\omega}\right)^2\right] \cdot \omega} \end{aligned} \tag{10.9}$$

Now, sound velocity has also become a complex number. Thus, wave equation also remains in force for the media which absorb the energy of a travelling acoustic wave, assuming that sound velocity is a complex number.

Compared to an idealized case, the absolute value of the sound velocity has decreased:

$$|C_\alpha| = \frac{C}{\sqrt{1 + \left(\frac{\alpha}{k}\right)^2}} < C \tag{10.10}$$

At the intuitive level, it is clear that the medium with the energy absorption property (internal friction) also has lower sound velocity compared to an idealized medium where waves are never attenuated. From (10.10) we can get a criterion for the attenuation smallness:

$$\frac{\alpha}{k} \ll 1 \quad \text{or} \quad \alpha \ll \frac{\omega}{C} \tag{10.11}$$

For MCC in VVER, the lowest-order ASWs have spatial half-period commensurate with the loop length $\frac{\lambda}{2} \sim 50\,\text{m}$, that is, $k = \frac{1}{\lambda} \sim 001 \left[\frac{1}{\text{m}}\right]$. Then the attenuation smallness criterion for VVER coolant and the lowest-order ASWs will be the following inequality:

$$\alpha \ll 001 \left[\frac{1}{\text{m}}\right] \tag{10.12}$$

For an idealized medium, the solution of a wave equation has a phase φ as linear function of frequency

$$U(x,t) = U_0 \cdot e^{j(\omega t - kx)} = U_0 \cdot e^{j(\omega t - \varphi)} \tag{10.13}$$

where

$$\varphi = k \cdot x = 2\pi\tau_3 \cdot f$$

with slope coefficient τ_3, which has the meaning of time delay with respect to zero point $x = 0$. For a medium with attenuation, phase becomes a complex number $\varphi = k_\alpha \cdot x$. Now, the concept of time delay $\tau_{3_\alpha} = \frac{x}{C_\alpha}$ is, strictly speaking, incorrect. It does not concern the case under consideration, as the real part of dispersion relation is constant: $k_\alpha = k - j_\alpha$.

Generally, the dispersion relation $k(\omega) = k_1(\omega) + jk_2(\omega)$ will also affect the concept of time delay. Generalization of the wave equation has lead to the situation when, instead of the "old" solution with constant amplitude and "pure" delay,

$$U(x,t) = U_0 \cdot e^{j\omega(t-\tau_3)}$$

the new solution for the medium with dissipation $U_\alpha(x,t)$ is transformed into the same form

$$\begin{aligned} U_\alpha(x,t) &= U_0 \cdot e^{j\omega\left(t-\frac{x}{C_\alpha}\right)} = U_0 \cdot e^{j\omega\left[t-\frac{x}{C}\left(1-j\frac{\alpha}{k}\right)\right]} \\ &= U_0 \cdot e^{j\omega\left(t-\frac{x}{C}\right)} \cdot e^{-\omega\cdot\frac{x}{C}\cdot\frac{\alpha}{k}} \end{aligned} \tag{10.14}$$

but with decreasing amplitude. The delay was not affected, but it has a contribution to attenuation of the amplitude.

$$U_\alpha(x,t) = U_0(x,t) \cdot e^{-\omega \cdot \tau_3 \cdot \frac{\alpha}{k}}$$

The exponential spatial weight $e^{-\omega \cdot \tau_3 \cdot \frac{\alpha}{k}}$, which appeared here, cannot be interpreted as frequency-dependent, since $\omega \cdot \tau_3 \cdot \frac{\alpha}{k} = \alpha \cdot x$ and

$$e^{-\omega \cdot \tau_3 \cdot \frac{\alpha}{k}} = e^{-\alpha \cdot x}$$

$$U_\alpha(x,t) = U_0(x,t)e^{-\alpha \cdot x} \tag{10.15}$$

In dispersion relations, implicit frequency dependences, like $k = \frac{\omega}{C}$, as well as complex frequency dependences which require the real and imaginary parts to be considered separately, lead to the errors in interpretation of vibroacoustic measurements.

The particular solution of the wave equation which leads to the standing wave

$$U(x,t) = e^{j(\omega t - kx)} + e^{-j(\omega t - kx)} = 2 \cdot \cos(\omega t - kx) \tag{10.16}$$

in case of a dissipative medium will have an attenuating spatial dependence

$$U_\alpha(x,t) = 2 \cdot e^{-\alpha \cdot x} \cdot \cos(\omega t - kx) \tag{10.17}$$

That is, we still have a standing wave with the stationary location of nulls and antinodes in space, but with a monotonic decrease of amplitude.

When an elastic medium is described by the wave equation (10.1), it gains the property of exponential spatial attenuation of amplitude, this leads to the replacement of wave number k with the complex wave number $k_\alpha = k - j \cdot \alpha$, but not a complex frequency-dependent function $k(\omega)$.

Complex generalization of the wave equation (10.5), where both the sought function $U_\alpha(x,t)$ is the complex exponent, and coefficient C_α — sound velocity — is a complex number, seems to be quite an unusual physical postulate. Complex transformations are usually used as intermediate steps. In a complex space, they facilitate the solution procedure, and then, by an inverse complex transformation, the solution is brought back to the real form. In this case, analytical solutions for the running and standing waves have been obtained in the real functions (10.15), and (10.17) with no intermediate complex transformations, just from a single complex generalized homogeneous wave equation.

10.2 Dispersion Relation — Complex Function of Frequency

The practical physical mechanisms of elastic media dissipation are much more complicated than the above formal introduction of an attenuating solution, and this leads to the wave number becoming a complex function of frequency:

$$k(\omega) = k_1(\omega) + j \cdot k_2(\omega) \tag{10.18}$$

In such cases, we say that the elastic medium has the dispersion property (not to be confused with another kind of dispersion, or "variance", which is the parameter of a probabilistic spread of a random value around its average value). Dispersion of an elastic medium is by default a dependence of phase velocity of the wave on frequency. It is characterized by a dispersion relation $k(\omega)$ — a complex function of frequency, which is, in essence, a complex generalization of the wave number — k, where $k_1(\omega)$ is the real part, which defines the new phase velocity as the function of frequency, and $k_2(\omega)$ is the imaginary part, which defines the law of wave amplitude attenuation as the function of frequency.

$$\begin{aligned} U(x,t) &= U_0 \cdot e^{j[\omega t + k(\omega)\cdot x]} = U_0 \cdot e^{j[\omega t + k_1(\omega)\cdot x]} \cdot e^{-k_2(\omega)\cdot x} \\ &= U_\varphi(x,t) \cdot U_\alpha(x,t) \end{aligned} \tag{10.19}$$

Now, different parts of the signal are transported with different velocities, or, more precisely, each frequency component of the solution $U(x,t)$ travels with its own velocity. In an idealized medium, the spatial shape of the travelling wave was not distorted, and we could speak of the same time delay for all frequencies, while the medium with absorption blurs the wave shape as different frequency components arrive to the point x with different velocities, starting from one and the same point $x = 0$. In the above example, $k_\alpha = k - j \cdot \alpha$ and α only reduced the wave amplitude, and now, the dispersion relation also "stretches" the signal in time-space.

Sound velocity in an idealized medium $C = \frac{\omega}{k}$ coincides with the phase velocity of the travelling wave $C = V_{ph}$, which is by default the velocity of motion of some point of a travelling wave with fixed phase, e.g. the velocity of any extremum of the travelling wave. Now, due to the dispersion relation, the notion of phase velocity loses its original meaning, just like time delay would also loose its original meaning.

While for a non-absorbing medium the harmonic solution was a monochromatic wave with the complex amplitude e^{-jkx}:

$$U(x,t) = U_0 \cdot e^{-jkx} \cdot e^{j\omega t} \tag{10.20}$$

and exponential factors in the form of linear frequency dependences

$$j\omega t + j\omega \frac{x}{C} \tag{10.21}$$

then if there is a complex dispersion relation

$$k(\omega) = k_1(\omega) + j \cdot k_2(\omega) \tag{10.22}$$

the solution will be modified to

$$U(x,t) = U_0 \cdot e^{j[\omega t + k_1(\omega)\cdot x]} \cdot e^{-k_2(\omega)\cdot x} \tag{10.23}$$

The real part $k_1(\omega)$ now plays the role of a frequency-dependent phase shift in the component

$$U_{ph} = e^{j\omega t + jk_1(\omega)\cdot x} \tag{10.24}$$

while the imaginary part $k_2(\omega)$ now plays the role of a frequency-dependent attenuation coefficient (not just a simple expression $e^{-\alpha\cdot x}$, as in the previous case). Remember that if an elastic medium does not absorb the energy of travelling waves, then

$$k_1(\omega) = k = \frac{\omega}{C}, \quad k_2(\omega) = 0$$

The impact of dissipation may also be presented as an additional transformation of the idealized medium solution

$$U(x,t) = U_0 \cdot e^{j\omega t + jkx} = U_0(x,t)$$

The solution for a medium with dissipation

$$U_\alpha(x,t) = U_0(x,t) \cdot e^{j\omega t + jk(\omega)\cdot x}$$

may also be transformed in the following way

$$\begin{aligned} U_\alpha(x,t) &= U_0(x,t) \cdot e^{j[k_1(\omega)+jk_2(\omega)]\cdot x - jkx} \\ &= U_0(x,t) \cdot e^{-k_2(\omega)\cdot x} \cdot e^{-jkx\left[1-\frac{k_1(\omega)}{k}\right]} \end{aligned} \tag{10.25}$$

The solution modulus will be subject to attenuation

$$|U_\alpha(x,t)| = |U_0(x,t)| \cdot e^{-k_2(\omega)\cdot x} \tag{10.26}$$

and its phase changes are contained in the multiplier

$$e^{-jkx\left[1-\frac{k_1(\omega)}{k}\right]} \tag{10.27}$$

In many papers, a dispersion relation refers not only to the expression

$$k(\omega) = k_1(\omega) + j \cdot k_2(\omega)$$

but also to the normalized expression

$$\frac{k(\omega)}{k} = \frac{C}{\omega}[k_1(\omega) + j \cdot k_2(\omega)]$$

which causes ambiguity in the frequency representation $k(\omega)$, as $k = \frac{\omega}{C}$ is the linear function of frequency.

10.3 Natural Oscillations of Elastic Dissipative Medium

In Chapter 5, we introduced the acoustic mass, acoustic stiffness and acoustic damping through an ordinary differential equation of motion. As a result, we got one of the possible acoustic interpretations of the parameters (MK). For this purpose, an acoustic medium was represented in the form of an elementary volume limited by walls so that one part of the volume is inertial and the other part is elastic. Obviously, such a model is far from being universal. In an ideal case, an elastic medium may be infinite. It would be more natural to introduce acoustic parameters via a wave equation. In order to do this, let us take advantage of the fact that sound velocity in a wave equation is the only characteristic parameter of the medium, equal to a single natural frequency of a spring pendulum $(C \leftrightarrow \omega_0)$:

$$\frac{\partial^2 U(x,t)}{\partial t^2} - C^2 \cdot \frac{\partial^2 U(x,t)}{\partial x^2} = 0 \tag{10.28}$$

Let us introduce some acoustic mass and some acoustic coefficient of elasticity through the analogy of a motion equation with a wave equation

$$M \cdot \frac{d^2x}{dt^2} + K \cdot x = 0 \Longleftrightarrow M_a \cdot \frac{\partial^2 U}{\partial t^2} - K_a \cdot \frac{\partial^2 U}{\partial x^2} = 0$$

where natural frequency $\omega_0 = \sqrt{\frac{K}{M}}$ and $C = \sqrt{\frac{K_a}{M_a}}$ is the sound velocity.

The wave equation (10.28) is similar to the equation of a mechanical harmonic oscillator

$$\frac{d^2x}{dt^2} + \omega_0^2 \cdot x = 0 \tag{10.29}$$

but it is not the same.

Now, let us add another component to the wave equation — the one which is proportionate to friction coefficient R_a and velocity $\frac{\partial U}{\partial t}$:

$$M_a \cdot \frac{\partial^2 U}{\partial t^2} + R_a \cdot \frac{\partial U}{\partial t} - K_a \cdot \frac{\partial^2 U}{\partial x^2} = 0 \tag{10.30}$$

When there is no energy coming from the outside, the amplitude of natural oscillations of a dissipative medium will be decreasing in time. Note that although this equation is similar to a dissipative equation of motion

$$M \cdot \frac{d^2x}{dt^2} + C \cdot \frac{dx}{dt} + K \cdot x = 0 \tag{10.31}$$

Eq. (10.30) is still a partial differential equation.

Equation (10.31) may also be put down in the normalized form, thus reducing the number of parameters from three (M, C, K) to two normalized (σ, ω_0):

$$\frac{d^2x}{dt^2} + 2\sigma \cdot \frac{dx}{dt} + \omega_0 \cdot x = 0 \tag{10.32}$$

where $\sigma = \frac{C}{2M}$ is the damping coefficient.

Particular solution (10.32) is presented in a simple form of a damped sinusoid.

$$x(t) = e^{-\sigma t} \cdot \cos(\omega_0 t + \varphi)$$

One non-normalized parameter — σ describes attenuation (damping properties of an oscillating system), another one — ω_0 — describes oscillation properties. Both parameters (σ, ω_0) have the dimension of frequency [Hz], although it is better to present the attenuation parameter in the dimensions of time:

$$\tau_c = \frac{1}{2\sigma}$$

One may suppose that the similarity of Eqs. (10.30) and (10.31) will lead to solving Eq. (10.30) solely in the form of a damped sinusoid.

Let us assume again that the solution (10.30) is a complex harmonic function

$$U(x,t) = U_0 \cdot e^{j\omega t} \cdot e^{jk(\omega)\cdot x} \tag{10.33}$$

The solution of Eq. (10.33) is different from a trivial case of a medium without absorption

$$U(x,t) = U_0 \cdot e^{j\omega t} \cdot e^{-jkx} \tag{10.34}$$

that is not a wave number $k = \frac{\omega}{c}$, but a complex function of frequency, i.e. a dispersion relation $k(\omega) = k_1(\omega) + j \cdot k_2(\omega)$.

Then, the original equation reduces to an algebraic equation

$$-\omega^2 + j\omega\frac{R_a}{M_a} + \frac{K_a}{M_a} \cdot k^2(\omega) = 0, \quad \frac{K_a}{M_a} = C^2 \tag{10.35}$$

with a complex solution

$$k^2(\omega) = \frac{1}{C^2}\left(\omega^2 - j\omega\frac{R_a}{M_a}\right) = \left(\frac{\omega}{c}\right)^2\left(1 - j \cdot \frac{1}{\omega\tau_R}\right) \tag{10.36}$$

where $\tau_R = \frac{M_a}{R_a}$ is the time constant of dissipation by mass. The higher the time constant, the lesser the dissipation in the medium.

Taking into account that $k(\omega)$ is a complex function

$$\frac{k(\omega)}{K} = \sqrt{1 - j \cdot \frac{1}{\omega\tau_R}} = a + jb \tag{10.37}$$

we get

$$\begin{cases} a^2 - b^2 = 1 \\ 2ab = -\dfrac{1}{\omega\tau_R} \end{cases} \tag{10.38}$$

and

$$a = \frac{1}{\sqrt{2}} \cdot \sqrt{1 \pm \sqrt{1 + \frac{1}{\omega^2 \cdot \tau_R^2}}} \tag{10.39}$$

$$b = \frac{1}{\sqrt{2}} \cdot \sqrt{-1 \pm \sqrt{1 + \frac{1}{\omega^2 \cdot \tau_R^2}}} \tag{10.40}$$

For small attenuations and/or big time constants ($\omega\tau_R \gg 1$), we have the following approximate formulas:

$$a \simeq \begin{cases} \pm\left(1 \pm \dfrac{1}{8\omega^2 \cdot \tau_R^2}\right) \\ \pm\dfrac{j}{2\omega\tau_R} \end{cases} \tag{10.41}$$

$$b \simeq \begin{cases} \pm j\left(1 \pm \dfrac{1}{8\omega^2 \cdot \tau_R^2}\right) \\ \pm\dfrac{j}{2\omega\tau_R} \end{cases} \tag{10.42}$$

It turns out that the real and imaginary parts of the dispersion relation are equal in modulus, and there are actually two dispersion relations:

$$|k(\omega)| = \sqrt{2} \cdot \begin{cases} \dfrac{\omega}{C} + \dfrac{1}{8 \cdot C \cdot \omega \cdot \tau_R^2} \\ \dfrac{1}{2 \cdot C \cdot \tau_R} \end{cases} \tag{10.43}$$

The two dispersion relations correspond to the existence of two solutions (two waves). Regarding the first solution, as far as the

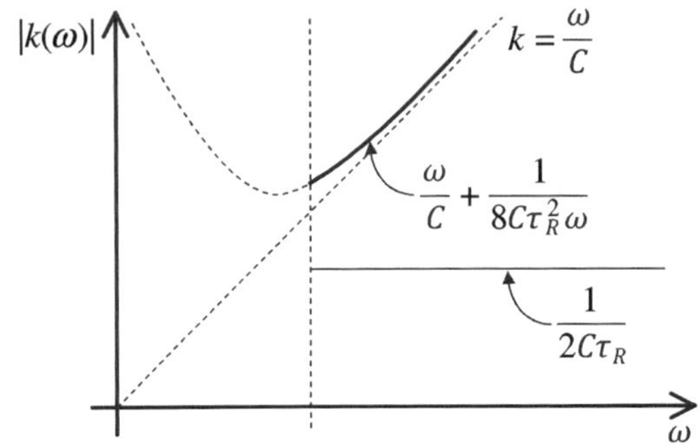

Figure 10.1: Dispersion relation moduli with minor attenuation and/or high frequencies.

frequency is growing, attenuation has less impact on phase shift and amplitude, approaching to an ideal case of a non-dispersive medium. As for the second solution, for relatively high frequencies there is frequency-independent attenuation of amplitude and a constant phase shift (Fig. 10.1).

If we simplify further, bearing in mind the smallness of the attenuation parameter, various combinations of "±" signs, appearing due to the presence of two radicals in expressions (10.39) and (10.40), lead us to three approximate formulas

$$a + jb \simeq \begin{cases} \pm\dfrac{j}{\omega \cdot \tau_R} \\ \pm\left(1 + \dfrac{j}{2 \cdot \omega \cdot \tau_R}\right) \\ \pm\left(1 - \dfrac{j}{2 \cdot \omega \cdot \tau_R}\right) \end{cases} \tag{10.44}$$

Approximation means here that we neglected $\frac{1}{8 \cdot \omega \cdot \tau_R^2}$ before $\frac{1}{2 \cdot \omega \cdot \tau_R}$:

$$k(\omega) \simeq \begin{cases} \dfrac{j}{C \cdot \tau_R} = \dfrac{j \cdot R_a}{C \cdot M_a} \\ \pm\left(\dfrac{\omega}{C} + j\dfrac{R_a}{2 \cdot C \cdot M_a}\right) \\ \pm\left(\dfrac{\omega}{C} - j\dfrac{R_a}{2 \cdot C \cdot M_a}\right) \end{cases} \tag{10.45}$$

The frequency-dependent complex dispersion relation for small viscosities again, just as in the case considered in Section 10.1, has been reduced to complex numbers and linear frequency dependences. The number of solutions reduces to one if we apply the causality principle, according to which the effect cannot occur earlier than the cause (or, if we consider delay only and ignore advance):

$$k(\omega) \simeq \frac{\omega}{C} - j\frac{R_a}{2 \cdot C \cdot M_a} \tag{10.46}$$

If we compare this with the result $k(\omega) = \frac{\omega}{C} - j\alpha$ obtained in the previous section, the only achievement we have is that the attenuation parameter α has got an interpretation in terms of the three parameters M_a, R_a, C.

It is quite obvious that the spatial attenuation is proportional to the acoustic friction coefficient R_a, but it is hardly obvious that it is inversely proportional to acoustic mass M_a and sound velocity C: $\alpha \sim \frac{1}{M_a}, \frac{1}{C}$.

Now, the parameter α may be represented in the following form: $\alpha = \frac{\sigma}{c}\left[\frac{1}{\mathrm{m}}\right]$. This parameter α is exactly similar to the parameter which in mechanics is called relative normalized damping:

$$\alpha_{\mathrm{M}} = \frac{\sigma}{\omega_0} = \frac{C \cdot \sqrt{M}}{2M \cdot \sqrt{K}} = \frac{C}{2\sqrt{K \cdot M}} = \frac{C}{C_{\mathrm{kp}}}$$

In mechanics, the motion equation

$$x_{tt} + 2\sigma x_t + \omega_0^2 \cdot x = 0$$

after a complex substitution $x(t) = x_0 \cdot e^{j\omega t}$

$$(j\omega)^2 + j2\sigma \cdot \omega + \omega_0^2 = 0$$

reduces to an algebraic equation for normalized frequency $F = \frac{\omega}{\omega_0}$:

$$F^2 - j2\alpha_{\mathrm{M}} \cdot F - 1 = 0$$

with the solution

$$F = j\alpha_{\mathrm{M}} \pm \sqrt{1 - \alpha_{\mathrm{M}}^2}$$

or

$$\omega = j \cdot \sigma \pm \sqrt{\omega_0^2 - \sigma^2}$$

Displacement

$$x(t) = x_\sigma \cdot e^{-\sigma t} \cdot e^{\pm j\sqrt{\omega_0^2-\sigma^2}}$$

will attenuate with the factor σ and oscillate with the argument $\sqrt{\omega_0^2 - \sigma^2}$. Thus, there are following analogies between:

- $\omega_0 \leftrightarrow C$ — natural oscillation frequency of a spring pendulum and sound velocity in the elastic medium;
- $\alpha_{\mathrm{M}} = \frac{\sigma}{\omega_0} \leftrightarrow \alpha = \frac{\sigma}{C}$ — relative normalized damping (dimensionless) and normalized damping in acoustics $\left[\frac{1}{\mathrm{m}}\right]$.

Example. The coolant within the MCC is a slightly viscous medium with a very low attenuation parameter $\alpha \ll 0,01\left[\frac{1}{\mathrm{m}}\right]$, that is why the limitation from the previous section is applicable $\omega \cdot \tau_R \gg 1$, meaning that the expression for dispersion relation (10.46) is also true. For the signals from two PPSs located within the pipeline at a short distance L, the attenuation will be the following:

$$e^{-\frac{R_a}{2C \cdot M_a} \cdot L} \tag{10.47}$$

The phase will be nearly linear, close to the idealized dependence

$$\varphi(\omega) \simeq \frac{2\pi f}{C} \cdot L = 2\pi\tau_3 \cdot f$$

where $\tau_3 = \frac{L}{C}$ is the time delay.

Thus, if the signals from two PPSs exhibits an acoustic effect, for example, an ASW resonance, then from the signal attenuation we may take a set of parameters $\frac{R_a}{C \cdot M_a}$, and from the phase we may estimate sound velocity with systematic underestimation C.

A very strong attenuation, as another extreme case, is also interesting in application to a slightly viscous VVER coolant. The criterion $\omega\tau_R$ may be too small for low frequencies and with large values of τ_R. For low frequencies, when $\omega \cdot \tau_R \ll 1$ from the expression (10.46) we have

$$k(\omega) \simeq \frac{\omega}{C}\left[\frac{(1-j)}{\sqrt{\omega \cdot \tau_R}}\right] = \frac{\sqrt{\omega}}{C} \cdot \sqrt{\frac{R_a}{M_a}} \cdot (1-j) = \frac{\sqrt{2\sigma}}{C} \cdot \sqrt{\omega} \cdot (1-j) \tag{10.48}$$

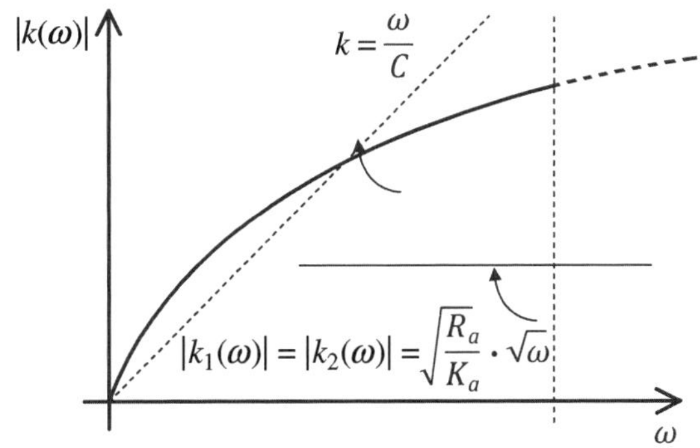

Figure 10.2: Dispersion relation summands with large attenuation and/or low frequencies.

The phase component of the solution has the same exponential factor as the amplitude component (Fig. 10.2):

$$k_1(\omega) = k_2(\omega) \simeq \sqrt{\omega} \cdot \sqrt{\frac{R_a}{K_a}} \tag{10.49}$$

Based on the reference data for a pipe with the radius r, the following friction coefficient formula is recommended:

$$R = \pi r \cdot \sqrt{2\rho \cdot \eta \cdot \omega} \tag{10.50}$$

which is consistent with the derived formula for low frequencies and low attenuations.

For the signals from two PPSs located at the distance L at fixed frequency ω_0, phase progression will be

$$\varphi(\omega_0) \simeq \sqrt{\omega_0} \cdot \sqrt{\frac{R_a}{K_a}} \cdot L \tag{10.51}$$

and transfer function modulus will be

$$|H(\omega_0)| = \sqrt{\frac{S_{\text{PPS1}}}{S_{\text{PPS2}}}} = e^{-\sqrt{\omega_0} \cdot \sqrt{\frac{R_a}{K_a}} \cdot L} \tag{10.52}$$

This leads us to the criterion for applicability of the method for any low frequency, such as $\omega_0 \cdot \tau_R \ll 1$:

$$\frac{\ln |H(\omega_0)|}{\varphi(\omega_0)} = 1 \tag{10.53}$$

The phase and amplitude characteristics (10.51), and (10.52) are independent of acoustic mass, since in the case with high frequencies, the experiment with two PPSs enables us to get a diagnostic indicator $\sqrt{\frac{R_a}{K_a}}$, which may be used to monitor the thermal and physical state of the coolant contained in the volume between two PPSs.

This method is applicable only for the monitoring of dissipation and stiffness properties of the medium. For example, when vapor phase appears, it will increase the friction coefficient R_a and decrease stiffness K_a. Both factors will increase the diagnostic sign $\sqrt{\frac{R_a}{K_a}}$.

If the considered range of pressure pulsation frequencies can be split into high-frequency and low-frequency ones, where the criteria derived in this section are valid, then we may try to estimate the parameters of the original model. For example, there may be up to ten different types of acoustic standing waves in the VVER coolant within the frequency range below 50 Hz. This is quite a representative set for the estimation of the three parameters of the model M_a, R_a, K_a assuming that M_a, R_a, K_a are constant for the considered frequency range. This idea proposed by the authors still needs to be proved during commissioning tests of a new VVER unit as VMS at the existing units have no PPS instrument channels.

10.4 Forced Oscillations of Dissipative Media

Let us supplement the idealized wave equation

$$\frac{\partial^2 U}{\partial t^2} - C^2 \cdot \frac{\partial^2 U}{\partial x^2} = 0$$

with a summand describing internal friction and proportionate to the friction coefficient R at the velocity $\frac{\partial U}{\partial t}$, and an external disturbing force in the harmonic form

$$f(x) \cdot e^{j\omega t} \tag{10.54}$$

This gives us another generalization of the wave equation, which considers both dissipation and an external effect:

$$\frac{\partial^2 U(x,t)}{\partial t^2} + R \cdot \frac{\partial U(x,t)}{\partial t} - C^2 \cdot \frac{\partial^2 U(x,t)}{\partial x^2} = f(x) \cdot e^{j\omega t} \tag{10.55}$$

The total energy of the system remains unchanged, being the sum of the potential and the kinetic energy which are periodically transformed one into the other. The summands $\frac{\partial^2 U}{\partial t^2}$, $C^2 \cdot \frac{\partial^2 U}{\partial x^2}$ represent the conservative part of the system, which does not consume any external energy, but only transforms potential energy into kinetic energy and back. The energy consumed (irreversibly absorbed) by the dissipative medium is replenished due to external energy of an external force.

$$f(x) \cdot e^{j\omega t} \longrightarrow R \cdot \frac{\partial U}{\partial t}$$

The solution to a homogeneous equation,

$$\frac{\partial^2 U}{\partial t^2} + R \cdot \frac{\partial U}{\partial t} - C^2 \cdot \frac{\partial^2 U}{\partial x^2} = 0$$

as it was demonstrated above, is a damped sinusoid:

$$U_1(x,t) = e^{-Rt} \cdot \cos(\omega t - \varphi) \tag{10.56}$$

After some time, the function $U_1(x,t)$ goes down to zero and may be neglected. Let us find the solution of Eq. (10.55) in a harmonic form

$$U_1(x,t) = U(x) \cdot e^{j\omega t}$$

which turns the partial differential equation to the ordinary differential equation

$$C^2 \cdot \frac{d^2 U(x)}{dx^2} + (\omega^2 - j2\sigma\omega) \cdot U(x) + f(x) = 0 \tag{10.57}$$

or

$$\frac{d^2 U}{dx^2} + (k^2 - j\alpha \cdot k) \cdot U = -\frac{1}{C^2} \cdot f(x) \tag{10.58}$$

where $\lambda = \frac{1}{k}$[m] is the spatial length of the wave related to frequency via the wave number k. Medium dissipation ($\sigma \neq 0$) leads to the complex function of spatial distribution of amplitude $U(x)$.

The transfer function obtained from a Fourier transformation of Eq. (10.58) is a complex function of two arguments — spatial and frequency ones:

$$H\left(\lambda, \frac{\omega}{C}\right) = \frac{1/C^2}{\lambda^2 - \left(\frac{\omega}{C}\right)^2 + j\alpha \cdot \frac{\omega}{C}} = \frac{U(\lambda)}{F(\lambda)} \tag{10.59}$$

where $U(\lambda), F(\lambda)$ are the corresponding Fourier transforms of spatial functions $U(x), f(x)$.

Exciting the medium with a spatially distributed harmonic force

$$f(x,t) = f(x) \cdot e^{j\omega t}$$

with spatial spectrum described by the function $F(\lambda)$ and time dependence in the form of a travelling wave $e^{j\omega t}$, the elastic dissipative medium performs filtering both in time and in space with the transfer function (10.59). The transfer function represented in the exponential form

$$H(\lambda, k) = \frac{1}{\sqrt{(\lambda^2 - k^2) + (\alpha \cdot k)^2}} \cdot e^{-j\frac{\alpha k}{\lambda^2 - k^2}}$$

is related to the dispersion relation

$$k(\omega) = jk_1(\omega) + k_2(\omega) = j \cdot \frac{1}{2} \cdot \ln[(\lambda^2 - k^2)^2 + (\alpha k)^2] + \frac{\alpha k}{\lambda^2 - k^2}$$

which establishes the following dissipation and phase progression:

$$|H(\lambda, k)| = \frac{1}{\sqrt{(\lambda^2 - k^2)^2 + (\alpha k)^2}}, \quad \varphi(\lambda, k) = \frac{\alpha k}{\lambda^2 - k^2}$$

When spatial frequency of an external disturbing force coincides with the spatial natural frequency of the medium $\lambda = k$, a resonance effect appears, with amplitude $U(x,t)$ limited by the dispersion properties of the medium.

$$U(x,t)_{\text{res}} = \frac{1}{\alpha k} = \frac{C}{\alpha \omega} \tag{10.60}$$

The simple expression, (10.60) is not an obvious result. Under a resonance excitation we usually understand the coincidence of the “temporary” frequency of an external disturbing force with the “temporary” natural frequency of the system. For an idealized dispersionless medium, the wave number $k = \frac{\omega}{C} = \frac{2\pi}{C} \cdot f$ which bounds the frequency parameters of the space $\lambda = \frac{1}{k}$ and time ω, makes them indistinguishable in meaning. For a dispersion medium, the spatial natural frequency is the first harmonic of the function $U(x)$ (spatial amplitude) in a space–time solution $U_1(x,t) = U(x) \cdot e^{j\omega t}$ In order to reach resonance excitation, the disturbing force $f(x) \cdot e^{j\omega t}$ should generate a wave of the same (wave) length, so that it coincides in form with the first harmonic of the function $U(x)$.

The larger the damping ratio α, the lesser the resonance amplitude we have (10.60), which is obvious. However, this is not obvious that for the media with high sound velocity "C" at low natural frequencies, the resonance amplitude will take high values.

Note that the law of variation of amplitude of spatial harmonics in case of forced oscillations $\frac{1}{\sqrt{(\lambda^2-k^2)^2+(\alpha k)^2}}$ — differs significantly from the behavior of natural oscillations amplitude envelope in the absence of an external force – $e^{-\alpha x}$:

$$U_1(x,t) = e^{-\alpha x} \cdot \cos(\omega t - \varphi)$$

If we set this task as a task with zero boundary conditions — a string with two fixed ends, or a pipeline with two dead ends:

$$U(0,t) = U(l,t) = 0$$

we shall get the sum of harmonics

$$U(x) = \sum_{m=1}^{\infty} \left[\frac{f_m}{\frac{m^2 \cdot \pi^2 \cdot C^2}{l} - \omega^2 + j2\sigma\omega} \cdot \sin\left(\frac{m\pi}{l} \cdot x\right) \right]$$

where

$$f_m = \frac{2}{l} \cdot \int_0^l f(x) \cdot \sin\left(\frac{m\pi}{l} \cdot x\right) dx \tag{10.61}$$

and

$$U(x,t) = \left[\sum_{m=1}^{\infty} \frac{f_m}{\frac{m^2 \cdot \pi^2 \cdot C^2}{l} - \omega^2 + j2\sigma\omega} \cdot \sin\left(\frac{m\pi}{l} \cdot x\right) \right] \cdot e^{j\omega t}$$

$$\omega_m = \frac{m\pi C}{l}$$

10.5 Examples of Constraint Forces

The spatially distributed amplitude of an external constraint force $f(x)$ in the expression (10.51) collapses into a single number which, on the one hand, defines the amplitude of the medium response and, on the other hand, is a Fourier coefficient.

$$f_1 = \frac{2}{l}\int_0^l f(\xi)\cdot\sin\left(\frac{\pi}{l}\cdot\xi\right)\cdot d\xi \tag{10.62}$$

An important example of an idealized spatially distributed external source is the δ-function in point x_0:

$$f(x) = f_0\cdot\delta(x - x_0) \tag{10.63}$$

Using the properties of δ-function

$$\int_a^b g(\xi)\cdot\delta(\xi - x)\cdot d\xi = \begin{cases} 0, & x < a, x > b \\ g(x), & a < x < b \end{cases}$$

we get the coefficient for the first harmonic of ASW:

$$f_1 = \frac{2}{l}\cdot\int_0^l f_0\cdot\delta(\xi - x_0)\cdot\sin\left(\frac{\pi}{l}\cdot\xi\right)\cdot d\xi = \frac{2\cdot f_0}{l}\cdot\sin\frac{\pi\cdot x_0}{l} \tag{10.64}$$

Note that the effect of an infinite amplitude of an external force gives a sinusoidal spatial response (sinusoid half-wave)

$$\begin{aligned} U(x,t) &= \frac{2}{l}\cdot f_0\cdot\frac{\sin\frac{\pi x_0}{l}}{\omega_1^2 - \omega^2 + j2\sigma\omega}\cdot\sin\frac{\pi x}{l}\cdot\cos\omega t \\ &= \mathrm{const}(\omega,t)\cdot\sin\frac{\pi}{l}x \end{aligned}$$

with max amplitude at the coincidence of frequencies $\omega = \omega_1$:

$$|U_{\max}(x,t)| = \frac{f_0}{l}\cdot\frac{1}{\sigma\cdot\omega_1} \tag{10.65}$$

The displacement of a single point x_0 was forced, but in response all points of the elastic medium have got displaced, except for the fixed boundaries.

The maximum amplitude of vibration response at any frequency of the external harmonic influence ω is reached if δ-function is applied exactly in the middle of the interval $[0, l]$:

$$\max U(x,t) = \frac{2 \cdot f_0}{l} \cdot \frac{1}{\omega_1^2 - \omega^2 + j2\sigma\omega} \tag{10.66}$$

Similar result has already been obtained for an idealized elastic medium without energy absorption, and it is shown here that dissipation has not changed the spatial point of amplitude maximum, but it has changed its amplitude. The closer the impact of δ-function to the fixed ends of the string, the lesser the amplitude response it causes. At the string fixing points, or at the ends of the pipeline, even a δ-type impact with an infinite amplitude causes a zero response.

For the first harmonic of ASW, its two nodes coincide with the ends of the interval $[0, l]$. For ASWs with ordinal numbers higher than one, nodes will also be found within the interval $[0, l]$. If δ-type impact is applied exactly to the coordinate of a node, the integral turns into zero. This is also a well-known qualitative result, and the dissipation does not change it.

δ-type impact is a mathematical abstraction, while an actual external force always has some spatial extension $-\Delta x$. So, let us consider an external impact in the form of time rectangle.

$$f(x) = \begin{cases} f_0, & x \in [x_1, x_1 + \Delta x] \\ 0, & x \bar{\in} [x_1, x_1 + \Delta x] \end{cases} \tag{10.67}$$

This is the case of a finite function which in the limit of $\Delta x \to 0$ approaches δ-function.

By integration of the formula (10.62), we get an approximate value of the response

$$f_1 \simeq \frac{4 \cdot f_0}{l} \cdot \Delta x \cdot \sin\left[\frac{\pi}{l}\left(x_1 + \frac{\Delta x}{2}\right)\right] \tag{10.68}$$

As in the previous case, amplitude may be equal to zero if the disturbance is not exactly in the ASW node, but is close to it in the coordinate $x_1 = l - \frac{\Delta x}{2}$. Note also that the amplitude increases together with the increase of the external source extension Δx.

The worst possible location of MCP set in terms of the amplitude of forced oscillations at the lowest-order standing waves is the

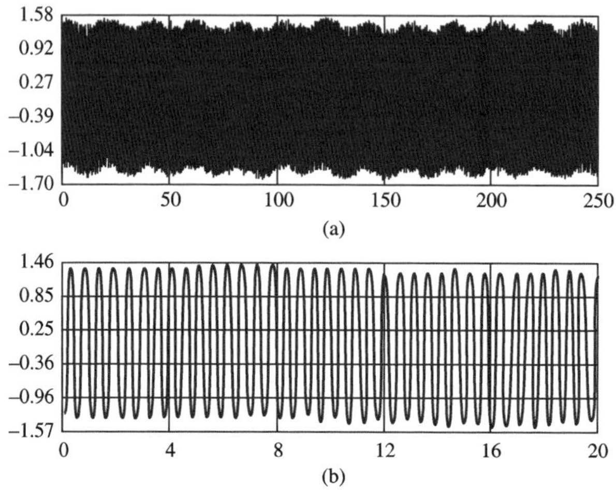

Figure 10.3: Realization of PPS amplitude-modified signal with the duration of 250 ms (a), and its 20-sec fragment (b). The carrier frequency coincides with the pressurizer resonance frequency (about 0.6 Hz).

middle of the heat exchange loop in point $\frac{l}{2}$. In VVER-1200, MCP set is located at the distance of exactly three-fourths of the length of the lowest-order half-wave ASW1p from the reactor vessel. This coincides with the coordinate of the node of ASW second harmonic, which makes the amplitude almost equal to zero, and this is proven experimentally (Fig. 10.3). However, there are such high-order types of standing waves, e.g. of the fourth harmonic, which will be excited with relatively high amplitude.

10.6 Resonance Excitation in Case of Coincidence of MCP Set Rotational Frequency with ASW Frequency

When coolant temperature is changing (due to the heating up or cooling down of the unit), ASW central frequency changes as well. The monotonic increase of coolant temperature causes the monotonic shifting of ASW central frequency towards zero Hz (Fig. 10.4). The tendency applies to the frequencies of all standing waves and, most important, each ASW central frequency moves along the frequency

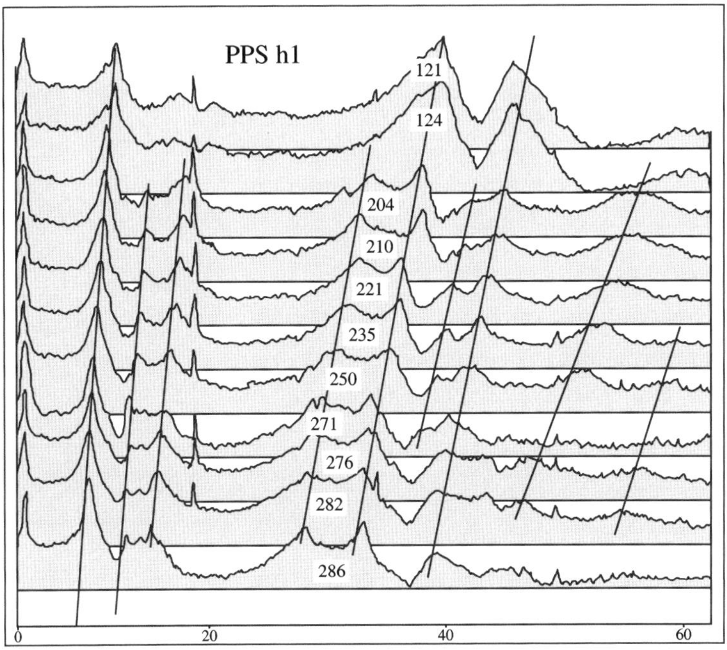

Figure 10.4: Pressure pulsations spectra in the hot leg of the primary loop vs. temperature (122–286°). At 276°, ASW coincides with the second harmonic of MCP rotational frequency. At low temperatures, coincidence is seen with the lowest- and high-order ASWs.

axis with its own speed [Hz/degree]. When the range of coolant temperature change exceeds 300°C, it is inevitable that some harmonics of MCP set rotational frequency coincide with the central frequency of some ASW (or its harmonic) at some intermediate temperature of the coolant.

In the nominal mode at 100% of reactor power, such coincidence shall be avoided, while during the dynamic modes it is fated. In VVER-1000 and VVER-1200 at 100% power, the lowest-order standing waves take on the values which are far less than the rotational frequency of MCP-set [380–382]. At the same time, it may happen that the frequencies of high-order harmonics of MCP-set rotational frequency coincide with the frequencies of high-order harmonics of ASW. Such phenomena are less dangerous than the coincidence of the lowest-order ASWs and the first harmonic of MCP-set rotational frequency. As follows from formula (10.66), the higher the standing wave frequency, the lower the amplitude: $\max[U(x,t)] \sim \frac{1}{\omega_{\mathrm{ASW}}}$.

Strictly speaking, each vibration resonance in a certain frequency range corresponds to its own equation of motion with its own triplet of parameters (M, C, K), and damping parameter $\ll C \gg$ will depend on frequency. In a quite narrow frequency range, which may coincide with the operational frequency range, it is acceptable to assume that the three parameters (M, C, K) remain unchanged. Unit heating up may be considered as the test mode in which the phenomena of resonance excitation are observed. The only component of the sum (10.61), which corresponds to the number of the resonant ASW, becomes the dominating one. Neglecting all other components and taking into account the coincidence $\omega = \omega_1$, we will have a simple ratio for relative damping $\sigma = \frac{C}{2M}$:

$$U(x,t) = \left(\sum_{m=1}^{\infty} \frac{f_m}{\omega_m^2 - \omega^2 + 2j\sigma\omega} \cdot \sin\frac{m\pi x}{l}\right) \cdot \cos\omega t \simeq \frac{f_m}{2j\sigma\omega_{\mathrm{ASW}}} \cdot \mathrm{const}(x,t)$$

Heating up is the test mode in which it is possible to get a simple estimation of the damping parameter — σ — during the resonant coincidence of the standing wave frequency with the harmonic of MCP set rotational frequency. APSD amplitude of the corresponding ASW resonance may be taken as a diagnostic sign — $S(f_{\mathrm{res}})$:

$$\sqrt{S(f_{\mathrm{res}})} \sim \frac{1}{C} \sim \frac{1}{\sigma} \sim \frac{1}{\alpha}; \quad \left(\alpha = \frac{\sigma}{\omega_0}\right)$$

As it would be difficult to get any absolute (metrologically supported) vibration amplitude measurements in the reactor, we may get an indicative damping estimation only, i.e. measure the relative damping in time units [Hz] — σ — or the dimensionless normalized damping — α. Apart from the resonance amplitude, there is another efficient diagnostic parameter, Q-factor. Q: resonance frequency detuning — $\Delta\omega$ — at resonance half-height in ASPD is equal to relative damping — σ

$$\frac{\Delta\omega}{\omega_0} = \frac{\sigma}{\omega_0} = \alpha = \frac{1}{2Q}$$

During commissioning measurements, it is necessary to find all coolant temperatures at which MCP-set harmonic frequencies coincide with ASW harmonics, and then, each time when similar temperature takes effect, one can perform spectral analysis of vibration measurements in order to estimate damping at different frequencies. It is also important to have such a test right after the scheduled preventive maintenance so that in case of an abnormality, the damage due to the unit shutdown would be reduced to a minimum.

Damping can be changed in the following cases:

- change of equipment mounting conditions (e.g. wear of reactor internals fasteners);
- wear or jamming of moveable base of MCP-set or SG;
- high steam content in the coolant.

These phenomena "spread" the vibration resonances and, as a rule, lead to the decrease in vibration level. In practice, however, alarms are generated when vibration level increases. The amplitude of vibration resonance may increase together with the increase of amplitude of the external disturbing force. As a rule, such abnormality appears due to an imbalance of MCP-set rotor. However, there might be such resonance phenomena when the wear of a structure leads to the change of resonance frequency of its component, and the new resonance frequency coincides with the frequency of the external disturbing force.

10.7 Wave Amplitude Attenuation Coefficient. Kirchhoff–Stokes Formula

Exponential spatial attenuation of a sound wave in a viscous elastic medium is only a postulate. It is based on a formal, to some extent experimentally confirmed assumption that the change of wave amplitude is proportional to the amplitude itself through an attenuation coefficient α:

$$dU(x) = -\alpha \cdot U(x) \cdot dx \tag{10.69}$$

or

$$\ln U(x) = -\alpha \cdot x + C$$

$$U(x) = U_0 \cdot e^{-\alpha x} \tag{10.70}$$

Such models are called phenomenological as they rely on experimental observations. An equation of motion, a wave equation, acoustics equations, etc., are fundamental physical laws, not phenomenological dependencies (10.69). Even the Navier–Stokes equation, which considers the viscosity of an elastic medium, is a fundamental law, as it is the generalization of a fundamental equation of motion (in acoustics, this is the Euler equation, in which a "viscous" component is added).

If in the above equations we take into account the fluid viscosity in one or another way, we get a frequency dependence of the attenuation parameter, the explicit form of which is not always confirmed experimentally for a wide range of frequencies. Probably, this is the reason why strict physical laws are combined with phenomenological models.

One of the widely spread ways to consider viscosity is the Kirchhoff–Stokes formula

$$\alpha = \frac{\omega^2}{2\rho C^3}\left[\frac{4}{3}\eta + \chi\left(\frac{1}{C_v} - \frac{1}{C_p}\right)\right] \tag{10.71}$$

which is derived from a simplified Navier–Stokes equation. Let us first look at the components of this formula corresponding to various physical phenomena, and in the next section we will derive it analytically. In formula (10.71):

- χ — thermal diffusivity,
- C_v — thermal capacity of an elastic medium at constant volume,
- C_p — thermal capacity of an elastic medium at constant pressure,
- η — viscosity coefficient $\left[\text{Pa} \cdot \text{s} = \frac{\text{N}}{\text{m}^2} \cdot \text{s} = \frac{\text{kg}}{\text{m} \cdot \text{s}}\right]$.

Through the thermal diffusivity coefficient — χ — a heat transfer is taken into account which was obtained as a result of work against viscous forces from the compression area to the expansion area.

Shear viscosity and volume viscosity are distinguished, and the latter may be neglected for nearly non-compressible fluid, which is why under the coefficient η we shall refer to shear viscosity only. Shear is present when different layers of fluid move with different speeds, causing an internal friction mechanism. This is why in a spatial representation of shear viscosity it is not enough to consider only the ≪X≫ component, along which the wave travels. A transverse component — ≪Z≫ or, in cylindrical coordinates, the radial component ≪R≫ — shall also be part of the description of shear viscosity mechanism. For VVER water coolant, volume viscosity is almost equal to zero, but it should not be neglected in reactor circulation loops with liquid–metal coolant.

The component in the expression (10.71) introduced by Kirchhoff

$$\alpha_{\chi} = \frac{\omega^2}{2\rho C^3} \cdot \chi \left(\frac{1}{C_v} - \frac{1}{C_p} \right) \tag{10.72}$$

is negligibly small for water coolant compared to the component introduced by Stokes:

$$\alpha_{\eta} = \frac{\omega^2}{2\rho C^3} \cdot \frac{4}{3} \cdot \eta \tag{10.73}$$

As it follows from formula (10.73), power-law dependences of attenuation coefficient

$$\alpha_{\eta} \sim \omega^2, \quad \frac{1}{C^3}$$

may change significantly the dissipative properties of a viscous elastic medium.

Quadratic dependence of attenuation α_{η} on frequency ($\alpha_{\eta} \sim \omega^2$) means that in a spectral representation an elastic viscous medium narrows the frequency range rapidly due to absorption of high-frequency components. In the time domain, this leads to signal "stretching" with simultaneous decrease of its amplitude as long as it is traveling in a viscous medium.

Example. Let us estimate approximately how many folds the attenuation coefficient α changes when the VVER-1000 coolant is heated up from 30°C to 300°C. The monograph [382] derives quite accurate linear dependence of acoustic standing wave central frequency on temperature empirically from multiple experimental heating-up of VVER-1000 coolant.

$$F_{\mathrm{ASW}_{1\mathrm{P}}} = 11.36 - \frac{1.60}{100} \cdot T^{\circ}$$
$$[\mathrm{Hz}]\ [^{\circ}\mathrm{C}]$$

While the coolant temperature increases from 30°C to 300°C, resonance frequency changes from 10.9 Hz to 6.6 Hz, approaching the low-frequency region. Thus, frequency ω^2 serves as the factor which reduces attenuation. Actually, viscosity is also a temperature-dependent parameter. Its change is approximately two-fold at the temperature change from 30°C to 300°C. Two factors remain: density ρ and sound velocity C, which increase attenuation as the temperature rises. The dependence of coolant density on temperature is well known, and within the given temperature range, the density reduces by 16%. The change of sound velocity is interconnected with the frequency change through the expression: $F+\Delta F = \frac{C+\Delta C}{\Lambda}$. Upon consideration of all factors, we get

$$\frac{\alpha_\eta(30^{\circ})}{\alpha_\eta(300^{\circ})} = 0.98$$

That is, only for this acoustic standing wave, all these differently directed factors compensate one another almost completely, leaving the attenuation unchanged. Unfortunately, temperature dependences for higher-frequency waves in VVER are unknown. There are some objective reasons for that. Beginning from certain intermediate temperatures, high-frequency acoustic resonances may vanish as their frequencies move towards 0 Hz (Fig. 10.4). Most probably, this phenomenon is caused by increasing of attenuation when the temperature increases.

For a two-phase medium, a separate study is needed. It is known that even minor volume fractions of steam in the coolant cause significant decrease in sound velocity. In the formula for a single-phase medium (10.73), there are very strong power-law dependences $\alpha_\eta \sim \omega^2$, $\frac{1}{C^3}$. It is understood that this formula is not applicable to two-phase coolant flows, and another dependence has to be found. The only thing we expect is that the steam phase will exacerbate the dependences (10.73). Resonance frequency will reduce faster, and for high-order standing waves this dependence will be even steeper.

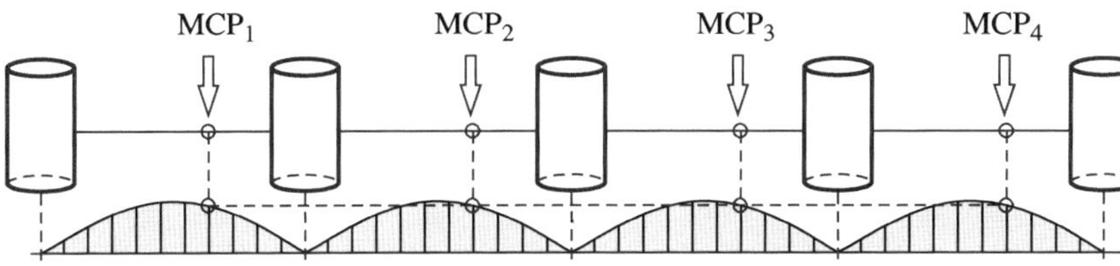

Figure 10.5: Idealized symmetry of amplitudes in distribution of a standing wave across the loops in case of synchronous operation of four pumps.

The high-frequency resonance will not only reduce in amplitude, it will also degrade in Q-factor, meaning that the resonance will get broader and broader until it vanishes. Both parameters, amplitude and Q-factor, are the diagnostic signs which may be used to monitor the coolant viscosity and, therefore, detect the steam phase in the coolant.

Four synchronously operating MCP-sets produce coherent pressure pulsations and thus make the picture of standing wave amplitude distribution in the circuit symmetrical by loops (Fig. 10.5). When MCP-sets are not identical, e.g. due to abnormal power supply, or due to the differences in their slip, each loop has its own amplitude distribution of the standing wave field. Thus, the asymmetrical field of ASW amplitudes is also a diagnostic indicator which integrally characterizes the non-identity of the four MCP-sets.

10.8 Derivation of Kirchhoff–Stokes Formula

Equation of motion, as a balance of forces affecting an elementary volume of the medium, takes into account the viscosity of the medium through friction force. It emerges when velocity is changing, and there will be an individual balance of forces for each of the Cartesian axes (X, Y, Z). That is why the Navier–Stokes equation is written in the form of a balance of projections of forces on each spatial axis, and here we speak about a system of Navier–Stokes equations. It can be written in a single equation using vector–matrix generalizations. For one-dimensional propagation of a flat wave in a viscous medium with the viscosity coefficient η, the Navier–Stokes equation is simplified in the following way.

In order to consider the fluid viscosity, the equation of motion (the Euler equation)

$$\rho \cdot \frac{\partial V(x,t)}{\partial t} + \frac{\partial P(x,t)}{\partial x} = 0 \tag{10.74}$$

is generalized with the gradient of viscous force (the right part of the equation):

$$\rho \cdot \frac{\partial V(x,t)}{\partial t} + \frac{\partial P(x,t)}{\partial x} = \frac{4}{3}\eta \cdot \frac{\partial^2 V}{\partial x^2} \tag{10.75}$$

To have the velocity as a single variable, let us make use of the continuity equation:

$$\beta \cdot \frac{\partial P(x,t)}{\partial t} + \frac{\partial V(x,t)}{\partial x} = 0, \tag{10.76}$$

which will enable us to eliminate the pressure. For this purpose, let us differentiate the equation of motion (10.74) with respect to time, and the continuity equation (10.75) — with respect to space; let us also replace the mixed second-order derivative $\frac{\partial^2 P(x,t)}{\partial t \partial x}$ with derivatives of velocity:

$$\begin{cases} \rho \cdot \dfrac{\partial^2 V}{\partial t^2} + \dfrac{\partial^2 P}{\partial x \partial t} = \dfrac{4}{3} \cdot \eta \cdot \dfrac{\partial}{\partial t}\left[\dfrac{\partial^2 V}{\partial x^2}\right] \\ \beta \cdot \dfrac{\partial^2 P}{\partial t \partial x} = -\dfrac{\partial^2 V}{\partial x^2} \end{cases} \tag{10.77}$$

As a result, we get the well-known equation

$$\frac{\partial^2 V}{\partial t^2} - C^2 \cdot \frac{\partial^2 V}{\partial x^2} = \frac{4}{3\rho} \cdot \eta \cdot \frac{\partial}{\partial t}\left[\frac{\partial^2 V}{\partial x^2}\right] \tag{10.78}$$

which can be interpreted as a generalized wave equation of an idealized elastic medium (without viscosity ($\eta = 0$)

$$\frac{\partial^2 V}{\partial t^2} - C^2 \cdot \frac{\partial^2 V}{\partial x^2} = 0 \tag{10.79}$$

In formula (10.78), viscosity is considered at the level of the mixed third-order derivative of velocity. Up to this moment, we considered the mechanisms of wave energy dissipation at the level of coefficients before the first-order derivative of velocity, or a complex coefficient before the second-order derivative of velocity. The corresponding coefficients before the derivatives have different physical meanings: acoustic resistance, complex velocity and viscosity coefficient. They can be interconnected, as they all take part in the corresponding dispersion relations.

After a complex substitution

$$V(x,t) = V(x) \cdot e^{j\omega t} \tag{10.80}$$

the mixed third-order derivative gives a complex coefficient $-j\omega \cdot \frac{4}{3\rho} \cdot \eta$, which defines amplitude attenuation and phase progress introduced by an elastic viscous medium. Equation (10.78) is reduced to the equation of a harmonic spatial oscillator

$$\frac{\partial^2 V(x)}{\partial x^2} + \frac{k^2}{1 + \frac{4}{3} jk \cdot \frac{\eta}{\rho \cdot C}} \cdot V(x) = 0 \tag{10.81}$$

where $k = \frac{\omega}{C}$, and the coefficient obtained from this equation

$$\alpha = \sqrt{\frac{k}{\sqrt{1 + \frac{4}{3} jk \cdot \frac{\eta}{\rho \cdot C}}}} \tag{10.82}$$

defines its particular solution

$$V(x) = e^{-j\alpha x}$$

Taking into account the smallness of the viscosity coefficient η, we have

$$\alpha \simeq k\left(1 - \frac{4}{6} jk \cdot \frac{\eta}{\rho \cdot C}\right)$$

which leads us to the solution:

$$V(x,t) = V_0 \cdot e^{j\omega t} \cdot e^{j\frac{\omega}{C}x} \cdot e^{-\frac{4}{6} \cdot \frac{\eta}{\rho \cdot C^3} \cdot \omega^2 \cdot x} \tag{10.83}$$

So, the velocity of a longitudinal wave, as the function of time and coordinates, is a travelling wave with the spatial phase progress being a linear function of frequency

$$\varphi(\omega, x) = \frac{\omega}{C} \cdot x \tag{10.84}$$

and the spatial attenuation of amplitude is defined by the exponential factor α (or the Kirchhoff-Stokes formula):

$$\alpha = \frac{4}{6} \cdot \frac{\eta}{\rho \cdot C^3} \cdot \omega^2 \tag{10.85}$$

Thus, elastic media with low viscosity and high sound velocity absorb high-frequency waves intensively and cause almost no resistance to low-frequency waves.

10.9 Damping and Viscosity

The two equations obtained in the previous sections describe one and the same phenomenon, longitudinal wave attenuation, in different terms, namely, with viscosity η and damping σ

$$\begin{gathered}\frac{\partial^2 V(x,t)}{\partial t^2} - C^2 \cdot \frac{\partial^2 V(x,t)}{\partial x^2} = \frac{4}{3} \cdot \mu \cdot \frac{\partial}{\partial t}\left[\frac{\partial^2 V(x,t)}{\partial x^2}\right] \\ \frac{\partial^2 V(x,t)}{\partial t^2} - C^2 \cdot \frac{\partial^2 V(x,t)}{\partial x^2} = -2\sigma \cdot \frac{\partial V}{\partial t}\end{gathered} \tag{10.86}$$

where

- $\mu = \frac{\eta}{\rho}$ is kinematic viscosity, $\left[\frac{\mathrm{m}^2}{\mathrm{s}}\right]$,
- η is viscosity coefficient, $\left[\mathrm{Pa} \cdot \mathrm{s} = \frac{\mathrm{N} \cdot \mathrm{s}}{\mathrm{m}^2} = \frac{\mathrm{kg}}{\mathrm{s} \cdot \mathrm{m}}\right]$,
- ρ is density, $\left[\frac{\mathrm{kg}}{\mathrm{m}^3}\right]$,
- $\tau = \frac{1}{2\sigma}$ is the time constant of damping, [s].

The notions of "viscosity" and "damping" are sometimes considered equal, which is wrong. Both coefficients describe internal friction in an elastic medium, but they have different physical meanings as they are coefficients before different functions. Damping coefficient σ is similar to friction coefficient in mechanics, since it is found before velocity $\frac{\partial V}{\partial t}$ (or rather, before the rate of velocity change, or before acceleration). Kinematic viscosity coefficient μ is more complex, as it is found before the function which takes into account the change of velocity both in time and in space. Odd derivatives with respect to velocity, namely, first-order derivative $\frac{\partial V}{\partial t}$ and third-order mixed derivative $\frac{\partial}{\partial t}\left[\frac{\partial^2 V}{\partial x^2}\right]$, after a complex substitution will give the corresponding complex coefficients. When those coefficients are found, it means that the components of equation related to energy absorption are present there.

This becomes obvious since the equation which interconnects the parameters μ and σ is obtained by equalizing the right sides of Eqs. (10.86):

$$\frac{4}{3} \cdot \mu \cdot \frac{\partial}{\partial t}\left[\frac{\partial^2 V}{\partial x^2}\right] = -2\sigma \cdot \frac{\partial V}{\partial t} \tag{10.87}$$

After integration with respect to time, it takes the form of the equation of a spatial harmonic oscillator

$$\frac{\partial^2 V(x)}{\partial x^2} - \left(j\sqrt{\frac{3}{2} \cdot \frac{\sigma}{\mu}}\right)^2 \cdot V(x) = 0$$

or

$$\frac{\partial^2 V}{\partial x^2} - \alpha^2 \cdot V = 0 \tag{10.88}$$

with particular solution

$$V(x) = V_0 \cdot e^{-j\alpha \cdot x} = V_0 \cdot e^{-\sqrt{\frac{3}{2} \cdot \frac{\sigma}{\mu}} \cdot x}$$

and

$$\alpha = \sqrt{\frac{3}{2} \cdot \frac{\sigma}{\mu}} \tag{10.89}$$

Thus, damping coefficient σ and kinematic viscosity μ are proportional to each other via the square of the exponential factor α:

$$\sigma = \frac{2}{3} \cdot \mu \cdot \alpha^2$$

When considering viscosity, errors may occur due to the following reasons:

- the wrong choice of viscosity mechanism (shear, volume, thermal conduction);
- neglected wall effects and bulk Stokes waves for small-diameter pipelines;
- the wrong choice of flow character (laminar, turbulent).

10.10 Description of Viscous Medium via Diffusion Equation

The two homogeneous wave equations obtained in the previous sections:

$$U_{tt} - C_{\alpha}^{2} \cdot U_{xx} = 0$$

$$U_{tt} + \frac{1}{\tau_R} \cdot U_t - C^2 \cdot U_{xx} = 0$$

describe one and the same phenomenon of exponential attenuation of the amplitude of a longitudinal wave in a viscous medium, where

$$C_{\alpha} = \frac{C}{1 - j\frac{\alpha C}{\omega}}$$

$$C = \sqrt{\frac{K_a}{M_a}} \tag{10.90}$$

Let us subtract one equation from the other in order to exclude the summand U_{tt}:

$$\frac{1}{\tau_R} \cdot U_t - C^2 \left[1 - \frac{1}{\left(1 - j\frac{\alpha C}{\omega}\right)^2}\right] \cdot U_{xx} = 0 \tag{10.91}$$

As a result, we get a diffusion equation (also known as heat conduction equation):

$$\frac{\partial U}{\partial t} = D \cdot \frac{\partial^2 U}{\partial x^2} \tag{10.92}$$

where diffusion coefficient D is the complex function of frequency

$$D = \tau_R \cdot C^2 \left[1 - \frac{1}{\left(1 - j\frac{\alpha C}{\omega}\right)^2}\right] \simeq -2j\tau_R \cdot \alpha \cdot \frac{C^3}{\omega} \tag{10.93}$$

For relatively high frequencies ω and low attenuations α, the diffusion coefficient D is proportional to the attenuation coefficient — α — and to the time constant of dissipation — τ_R. That is, the presence of viscosity in the elastic medium is equivalent to a diffusion process along the longitudinal coordinate $\ll X \gg$. Diffusion as the process of equalization of concentration of a substance is close to viscosity, which may be represented as the process of equalization of velocities of rubbing fluid layers in the whole fluid volume.

Viscosity as a volumetric property of an elastic medium, strictly speaking, cannot be described by a single number, that is why the Navier–Stokes equation has a vector–matrix form, and viscosity tensors play the role of coefficients in the equation. Giving preference to a single viscosity mechanism — shear viscosity solely — and neglecting the volume viscosity related to heat conductivity, it is still possible to stay within simple but non-one-dimensional spatial representations. Now, we have to consider also at least the spatial coordinate ≪Z≫ perpendicular to the direction of longitudinal wave propagation along the axis ≪X≫.

Let us demonstrate that the effect of shear viscosity also leads us to a diffusion analogy when measuring ≪Z≫. Let us assume that the transversal component of velocity has a harmonic form

$$V(z,t) = V_0 \cdot e^{j(\omega t - kz)} \tag{10.94}$$

After differentiation with respect to time

$$\frac{\partial V(z,t)}{\partial t} = j\omega \cdot V(z,t)$$

and two rounds of differentiation of Eq. (10.94) with respect to the spatial coordinate

$$\frac{\partial^2 V(z,t)}{\partial z^2} = k^2 \cdot V(z,t)$$

we will see that the harmonic transversal wave also satisfies the diffusion equation

$$\frac{\partial V(z,t)}{\partial t} = D \cdot \frac{\partial^2 V(z,t)}{\partial z^2}$$

if the diffusion coefficient is a complex number

$$D = -j\frac{\omega}{k^2}$$

Thus, the dispersion relation is written as

$$k(\omega) = \sqrt{e^{-j\frac{\pi}{2}} \cdot \frac{\omega}{D}} = e^{-j\frac{\pi}{4}} \cdot \sqrt{\frac{\omega}{D}} = \sqrt{\frac{\omega}{2D}} \cdot (1-j)$$

If internal friction between the liquid layers (or shear viscosity) is considered as diffusion transversal to the direction of travelling wave, we get a nonlinear dependence of transverse velocity on frequency, and a nonlinear dependence on the diffusion coefficient. It is natural to assume that the exponential attenuation in Z direction is related to viscosity:

$$V(z,t) = V_0 \cdot e^{j\omega t} \cdot e^{-jk(\omega)\cdot z} = V_0 \cdot e^{j\omega t} \cdot e^{-j\sqrt{\frac{\omega}{2D}}\cdot z} \cdot e^{-\sqrt{\frac{\omega}{2D}}\cdot z}$$
$$= V_0 \cdot e^{-\alpha \cdot z} \cdot e^{j\omega\left[t - \frac{z}{\sqrt{D\omega}}\right]} \qquad (10.95)$$

where

$$\alpha = \sqrt{\frac{\omega}{2D}} \qquad (10.96)$$

The complex harmonic wave also satisfies the diffusion equation when the transverse spatial coordinate $\ll Z \gg$ is replaced with the longitudinal coordinate $\ll X \gg$. Thus, viscosity as the diffusion mechanism leads to spatial attenuation of the wave amplitude in any direction of space.

10.11 Viscosity as the Reason of Pressure Drop in Transversal Direction

When boundary conditions are imposed on a viscous elastic medium, the velocity of a longitudinal wave cannot be represented only by a spatial coordinate $\ll X \gg$. For example, for a round pipe with the radius r and length L (Fig. 10.6), the presence of viscosity leads to pressure drop

$$\Delta P = P_1 - P_2 = \frac{\partial P}{\partial x} \cdot L \qquad (10.97)$$

which is taken into account in the right part of the simplified Navier–Stokes equation

$$\eta \cdot \frac{\partial^2 V_x(z)}{\partial z^2} = \frac{\partial P}{\partial x} \cdot L \qquad (10.98)$$

This means that pressure drop in X direction takes place due to viscosity of the medium and distribution of longitudinal velocity in Z direction.

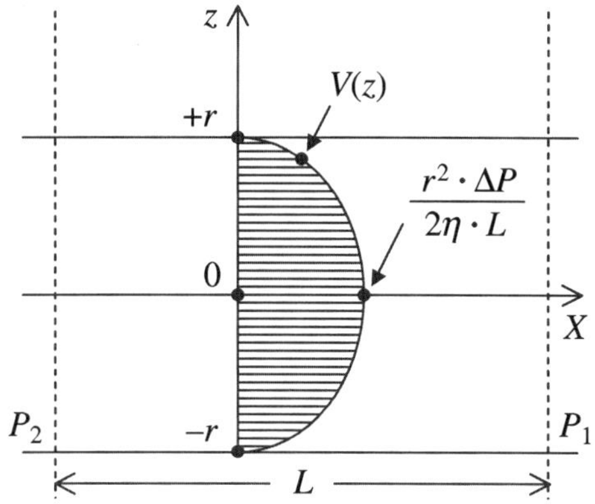

Figure 10.6: Transversal distribution of longitudinal wave velocity in a round pipeline.

The boundary condition is that the longitudinal wave velocity at the pipe wall is zero: $V_x(z = \pm r) = 0$. Then the Navier–Stokes equation (10.98) has an exact solution in the form of a parabolic function of the coordinate ≪Z≫:

$$V_x(z) = \frac{\Delta P}{2 \cdot \eta \cdot L}(r^2 - z^2), \quad |z| \leq r \tag{10.99}$$

This formula is true for laminar flows of an elastic viscous medium. When averaged by pipe section, velocity becomes the linear function of pressure drop

$$V_0 = \langle V \rangle_z = \text{const}(\eta, L) \cdot \Delta P \tag{10.100}$$

Formula (10.100) written as

$$\Delta P = \frac{8\eta \cdot L \cdot \pi}{S^2} \cdot V \tag{10.101}$$

is called Poiseuille law, and the coefficient $r_a = \frac{8\eta \cdot L \cdot \pi}{S^2}$ is the cylindrical pipe resistance with section S^2 and length L. This formula resembles the formula for electrical resistance of a conductor $R = \rho \cdot \frac{L}{S^2}$, with the role of specific electrical resistance ρ played by viscosity coefficient η.

Expressions (10.99) and (10.100) are true both for the phase velocity of a longitudinal wave and for the velocity of a laminar flow of liquid (i.e. the speed at which liquid is pumped through a pipeline).

The difference is that in one case ΔP is the macro pressure drop in pipe length L, and in the other case ΔP is the micro drop of pressure pulsations along the length dx. Such interpretation of viscosity is acceptable only for laminar flow as the linear dependence $\Delta P = r_a \cdot V$ is similar to Ohm's law: $\Delta U = R \cdot I$. Again, pay attention to the linearity of the dependence (10.100), however, for turbulent flows, the dependence of velocity on pressure drop is quadratic, and that is called the Darcy–Weisbach formula:

$$\Delta P = \text{const} \cdot \rho \cdot V_0^2 \tag{10.102}$$

Strictly speaking, it is incorrect to apply this to micro-pulsations of pressure.

The Darcy–Weisbach formula is most often represented via a dimensionless coefficient of local resistance — ξ:

$$\Delta P = \xi \cdot \frac{\rho \cdot V^2}{2}$$

Then, the quadratic factor $\frac{\rho \cdot V^2}{2}$ is interpreted as the quantity which is proportional to deceleration energy.

10.12 Coolant Pressure Perturbation in MCC

Spectral representation of pressure pulsations, even if nonlinear, in the form of complex dispersion relations, requires that the considered time realizations were stationary random processes. At the same time, the reactor, during nominal operating mode at constant power, with its automatic regulation systems generates minor perturbations of the reactor parameters. They are non-stationary acts short-lived by their nature, localized in space and have the information on the properties of the elastic viscous medium of the coolant. Based on the frequency representation of pressure pulsations solely, we can reconstruct these time parameters of the medium perturbation.

The nonlinear dispersion relation

$$k(\omega) = \frac{\omega}{C} + j \cdot a \cdot \omega^2 \tag{10.103}$$

in which the imaginary part

$$\gamma(\omega) = \frac{\omega^2}{2\rho C^3} \cdot \frac{4}{3} \cdot \eta \quad \text{and} \quad a = \frac{2 \cdot \eta}{3\rho \cdot C^3} \tag{10.104}$$

containing three parameters of elastic medium, density — ρ, viscosity — η and sound velocity — C enables us to reconstruct the initial differential equations.

For this purpose, the idealized equation for a travelling pressure wave

$$C \cdot \frac{\partial P}{\partial x} + \frac{\partial P}{\partial t} = 0 \tag{10.105}$$

with the solution $P(x,t) = P(x - C \cdot t)$ and, particularly, with the complex solution

$$P(x,t) = e^{j(kx-\omega t)}$$

shall be generalized by the non-zero right side of equation

$$C \cdot \frac{\partial P}{\partial x} + \frac{\partial P}{\partial t} = C \cdot a \cdot \frac{\partial^2 P}{\partial t^2} \tag{10.106}$$

for the solution is being generalized in case of a nonlinear dispersion relation (10.103). Indeed, by placing the dispersion relation (10.103) into the exponential complex solution, we have:

$$P_1(x,t) = e^{j(kx-\omega t)} = e^{j[\omega(\frac{x}{C}-t)-ja\omega^2 x]} = P(x,t) \cdot f(x)$$

where $f(x) = e^{-ja\omega^2 x}$. Now, the idealized equation (10.105) after insertion of function $P_1(x,t)$ takes the form

$$C \cdot \frac{\partial P}{\partial x} + \frac{\partial P}{\partial t} = -C \cdot \frac{f'(x)}{f(x)} \cdot P$$

or

$$C \cdot \frac{\partial P}{\partial x} + \frac{\partial P}{\partial t} = -C \cdot [\ln f(x)]' \cdot P$$

or

$$C \cdot \frac{\partial P}{\partial x} + \frac{\partial P}{\partial t} = jC \cdot a \cdot \omega^2 \cdot P$$

which is equivalent to Eq. (10.106).

Replacement

$$jC \cdot a \cdot \omega^2 \cdot P \longrightarrow C \cdot a \cdot \frac{\partial^2 P}{\partial t^2}$$

is acceptable for harmonic solutions only (complex or sinusoidal real one). That is, the factor for the complex solution is equivalent to the second-order derivative with respect to time, which leads us to the equation

$$C \cdot \frac{\partial P}{\partial x} + \frac{\partial P}{\partial t} = C \cdot a \cdot \frac{\partial^2 P}{\partial t^2}$$

Via substitution of variables $\tau = t - \frac{x}{C}$, it again goes to a diffusion equation (or a thermal conductivity equation)

$$\frac{\partial P}{\partial x} = a \cdot \frac{\partial^2 P}{\partial \tau^2}$$

with the well-known solution

$$P(x, \tau) = \frac{1}{2\sqrt{\pi a x}} \cdot \int P_0(\xi) \cdot e^{-\frac{(\xi-\tau)^2}{4ax}} \cdot d\xi \tag{10.107}$$

where $P_0(\xi) = P(0, \xi)$.

For long distances from the acoustic source, the solution (10.107) will approach in shape the Gaussian curve

$$P(x, t) = \frac{1}{2\sqrt{\pi a x}} \cdot e^{-\frac{\tau^2}{4ax}} \cdot \int P_0(\xi) \cdot d\xi \tag{10.108}$$

Irrespective of the initial coolant pressure perturbation in point $x = 0$, in the remote point there will be a nearly Gaussian shape pressure variation with time (10.108). The nonlinear frequency dependence of dispersion relation (10.104) demonstrates higher attenuation of high-frequency components of the signal compared to the low-frequency ones. In the time domain, this means that the initial perturbation of a viscous medium, as it travels in space, will reduce in amplitude $A = \frac{1}{2\sqrt{\pi a x}}$ and "spread" in width $\sigma = \sqrt{2ax}$.

In terms of Gaussian distribution, the role of dispersion is played by the parameter σ^2, depending on the parameter $a = \frac{2}{3} \cdot \frac{\eta}{\rho \cdot C^3}$, which is a global set of parameters of the medium between the perturbation source and the sensor (PPS, in this case). Thus, the global parameter "a" is the diagnostic feature, which may be estimated both from the amplitude ($A = \frac{1}{2\sqrt{\pi a x}}$), and from the pulse width ($\sigma = \sqrt{2ax}$) (Fig. 10.7).

When non-stationary part of PPS signal is detected, the corresponding part of its realization is approximated by a Gaussian

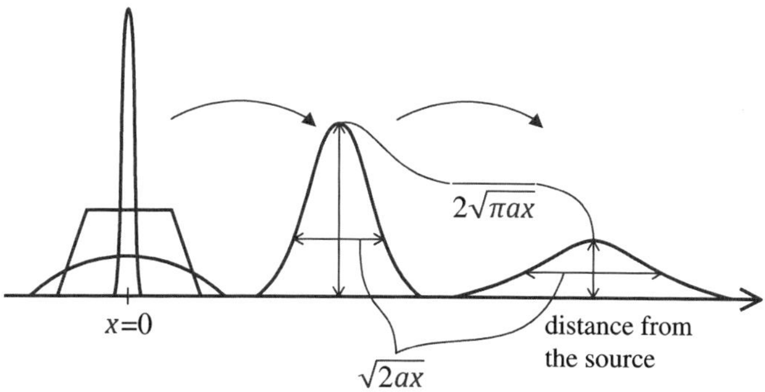

Figure 10.7: Spread of pressure perturbation in viscous elastic medium of the coolant.

curve from which the maximum amplitude and variance are derived. It is also possible to use an envelope of the pressure signal, also approximating its parts by a Gaussian curve. Two PPSs can also be used so that correlating their signals one could, first, detect the significant non-stationarities of their signals, and then, estimate the parameter "a" from both signals separately.

The product of amplitude by pulse width

$$A \cdot \sigma = \frac{1}{\sqrt{2\pi}} = \text{const}$$

is the invariant of the medium parameters, i.e. is independent of the parameter "a". The change of viscosity, medium, density or sound velocity in the medium has no impact on the invariant $A \cdot \sigma$, which will also be a diagnostic sign. When it changes in time, it indicates a new quality of the medium.

Example. Up to now, vibroacoustic measurements in VVER were related to the noise component of a pressure signal, or to pressure pulsations. Pressure perturbation as a result of the pressurizer system operation may cause major deviation from the average value, much higher than the background stationary broadband pressure pulsations in the MCC. The measurements of such pressure perturbations are not of the noise type anymore. They are related to the so-called dynamic measurements. The pressurizer system generates two types of pressure regulations: injection and start/stop of electric heaters. Pressure sensors installed in any point of the MCC

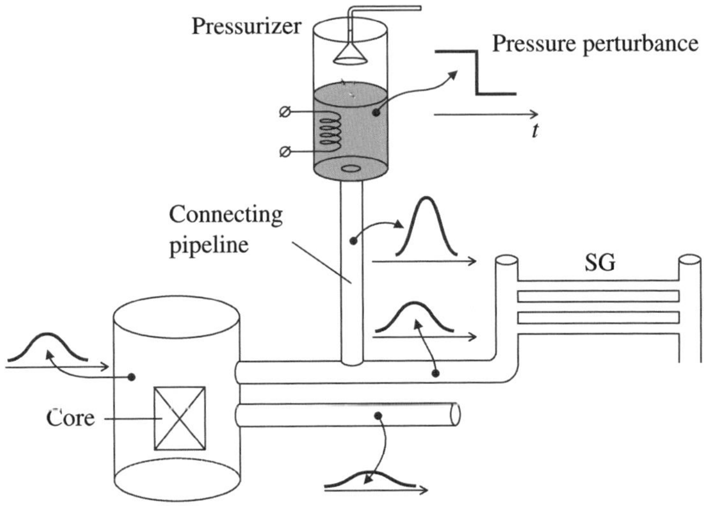

Figure 10.8: Injection in the pressurizer as an impulse perturbation of coolant pressure.

are sensitive to such level of pressure perturbation. The perturbation source is located within the pressurizer vessel. Injection is a kind of design-basis test input to the coolant from the pressurizer (Fig. 10.8).

The response of the coolant pressure to the injection is close in its shape to step-wise pressure change, and heaters start/stop, obviously, is a highly-inertial process, which in principle has totally different time characteristics of pressure change. The medium in the connecting pipeline between the pressurizer and the MCC is quite special; it is different from the MCC medium and the pressurizer vessel medium. Coolant pressure regulation causes the back-and-forth movement of the medium in the connecting pipeline ("slow" oscillations of the medium). During one half-period, hot water is pushed into the MCC, during the other, relatively cold water comes from the MCC to the connecting pipeline. Obviously, this is accompanied by the change of parameter a. That is, it characterizes the state of this medium in the nominal mode of VVER operation. The coolant is injected into the pressurizer many times during the day, that is why pressure response to such perturbation can be identified easily. That is, it is feasible to install PPSs not only in the VVER coolant circuit, but also in the connecting pipeline in order to characterize the pressurizer system itself.

Pressure perturbation in the MCC has quite a different character. The stationarity of the parameter a will characterize the stationarity (stability) of density, viscosity and sound velocity within the volume of the medium between two PPSs, or between the pressurizer vessel and PPS. In order to confirm the authors' hypothesis and, particularly, the Gaussian character of the response of a remote PPS, a series of dynamic experiments with the pressurizer system is needed. As a matter of fact, there is a standard commissioning procedure for adjustment of the rate of injection into the pressurizer, which is just the series of experiments we need. The only thing which is left to do is to register signals of a certain number of PPSs within certain periods of time, identifying the corresponding correlations and estimating the diagnostic parameters using the formula (10.108).

10.13 Stokes' Waves — Dispersion Properties of the Coolant in Steam Generator Heat Exchange Tubes

For large-diameter MCC pipelines and a liquid elastic medium with small attenuation coefficient, the viscosity coefficient is defined solely by shear viscosity, which is the only prevailing mechanism. For thin heat-exchange tubes of the steam generator, another mechanism of viscosity has to be taken into account. The key role here belongs to the wall effect. Its influence on an acoustic wave reduces with distance from the pipe wall to the pipe axis. This phenomenon was analytically described for the first time by George Gabriel Stokes, that is why such waves are called transverse Stokes' waves. When the pipe diameter is so small that the wall effect is not reduced to zero at the horizontal axis of the pipe (as in HET), this is a bulk wave covering the whole cross-section. If we denote the oscillation velocity of liquid close to the pipe wall as $V(z,t)$, the simplest form of Navier–Stokes equation will be

$$\eta \cdot \frac{\partial^2 V(z,t)}{\partial z^2} = \rho \cdot \frac{\partial V}{\partial t} \tag{10.109}$$

where z is the transverse coordinate.

Obviously, the Navier–Stokes equation has degenerated into a diffusion equation with the diffusion coefficient equal to kinematic viscosity — μ:

$$D = \frac{\eta}{\rho} = \mu \tag{10.110}$$

Let us present the solution in the complex form, just like we did before:

$$V(z,t) = V(z) \cdot e^{j\omega t} \tag{10.111}$$

Upon its insertion into Eq. (10.109), we come to the harmonic oscillator equation:

$$\frac{d^2V(z)}{dz^2} - j\frac{\omega\rho}{\eta} \cdot V(z) = 0 \tag{10.112}$$

with the solution $V(z) = V_0 \cdot e^{jkx}$ where

$$k = \sqrt{-j\frac{\omega\rho}{\eta}} = \beta(1-j), \quad \beta = \sqrt{\frac{\omega\rho}{2\eta}}$$

The velocity of a transverse Stokes wave is also called volume velocity, which is defined through the modulus

$$C_k = \frac{\omega}{|k|} = \sqrt{\frac{\eta}{\rho} \cdot \omega} = \sqrt{\mu \cdot \omega}$$

where μ is kinematic viscosity.

Thus, consideration of viscosity in the steam generator tubes leads to the frequency dependence, which differs significantly from the one for larger diameter pipelines.

10.14 Acoustic Impedance for Small-Diameter Pipelines

When a longitudinal wave travels in a small-diameter pipe, the effect of volume viscosity is equivalent to the effect of lumped acoustic resistance. If at the ends of a pipe with length L pressure drops $-\Delta P = \frac{\partial P}{\partial x} \cdot L$,the acoustic resistance (acoustic impedance) is by default:

$$z(L) = \frac{\partial P}{\partial x} \cdot \frac{L}{\langle V \rangle_z} \tag{10.113}$$

where $\langle V \rangle_z$ is the velocity averaged by pipe section, $\frac{\partial P}{\partial x}$ is the pressure gradient at the length L.

The Navier–Stokes equation in cylindrical coordinates for a round pipe

$$\frac{\partial V}{\partial t} + \frac{1}{\rho} \cdot \frac{\partial P}{\partial x} = \frac{\eta}{\rho} \left(\frac{\partial^2 V}{\partial t^2} + \frac{1}{2} \cdot \frac{\partial V}{\partial r} \right) \tag{10.114}$$

takes into account such pressure drop due to the second summand in the left side. That (pressure gradient $\frac{\partial P}{\partial x}$) may be replaced with the expression

$$\rho \cdot \frac{\partial V}{\partial t} = \eta \left(\frac{\partial^2 V}{\partial t^2} + \frac{1}{2} \cdot \frac{\partial V}{\partial r} \right) - \frac{1}{L} \cdot z(L) \cdot \langle V \rangle_z \tag{10.115}$$

The exact solution of this equation, written with Bessel functions, is quite cumbersome, and it can be simplified using a dispersion relation for a bulk wave

$$k(\omega) \sim \sqrt{\frac{\eta}{\rho}} \cdot \sqrt{\omega} \tag{10.116}$$

Omitting the intermediate steps, the approximate expression for impedance will be

$$z(L) \simeq \frac{8 \cdot \eta \cdot L}{r^2} + j\omega \cdot \frac{4}{3} \rho \cdot L \tag{10.117}$$

The active component of impedance depends only on viscosity η and pipe dimensions r, L. The reactive component $\omega \cdot \frac{4\rho \cdot L}{3}$ is equivalent to inductive resistance. It is independent of viscosity and is measured in mass units. While coolant mass regardless of viscosity in the pipe was defined by the product $\rho \cdot L$, the non-zero viscosity makes the mass sort of $\frac{4}{3}$ times "heavier".

The electrical equivalent of a SG heat exchange tube is a serial circuit (see Fig. 10.9).

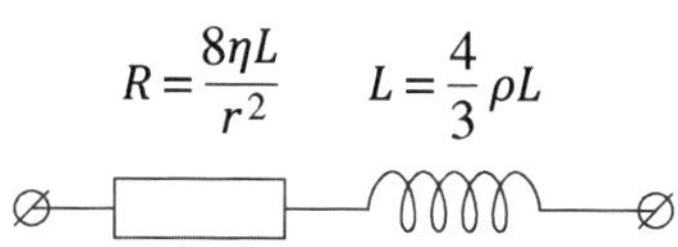

Figure 10.9: Serial circuit of inductance and active resistance as an analogy of SG HET.

The change of coolant density in SG HET depends on the processes in both primary and secondary circuits. The deposits on the inner walls of the tube from the primary circuit reduce the flow area, thus reducing the mass. The deposits on the outer walls of SG HET impair heat transfer to the secondary circuit, thus preventing the coolant density from reducing to the design-basis level. In the first case, the equivalent mass of the coolant reduces, in the second one — it increases. These two oppositely directed trends increase or reduce the reactive component of the impedance, which may be used to monitor the two processes of deposition of impurities on the inner and outer surfaces of SG HET.

The phase of complex impedance (10.117)

$$\varphi(\omega) = \text{arctg}\frac{I_{m_z}}{R_{l_z}} = \text{arctg}\frac{\omega \cdot \rho \cdot r^2}{6\eta} \sim eq\frac{\omega \cdot \rho \cdot r^2}{6\eta}$$

for small pipeline diameters ($r \sim 0$) and low frequencies ($\omega \sim 0$) will be a linear function of frequency with the slope due to the three parameters ρ, η, r. The factors of coolant density — ρ — change have already been considered for RP operation, coolant viscosity — η — is temperature-dependent, and it also changes in case of coolant boiling, pipeline radius — r — changes due to the deposition of impurities from the primary coolant. The power-law dependence $\varphi \sim r^2$ defines the process of HET radius change as the main operational factor of HET acoustic impedance change. The authors' hypothesis states that the long-term measurement of the phase of complex impedance (10.117), during at least one VVER RP campaign, will allow selection of the dominant resource factors of HET "aging".

Table 10.1 gives the summary of expressions for dispersion relations considering the viscosity of elastic medium.

10.15 Pressure Pulsation Sensor Signal Modulation Parameters

Pressure pulsation sensors applied in reactor measurements have practically no limits in terms of low-frequency measurement. It can be arbitrarily close to zero. In particular, low resonance frequency of a pressurizer in VVER (about 0.6 Hz) is monitored primarily with

Table 10.1: Dispersion relations for wave equations considering viscous and damping properties of an elastic medium.

No.	Equation	Solution and dispersion relation
1.	Idealized case, zero viscosity $U_{tt} - C^2 \cdot U_{xx} = 0$	$U(t,x) = U_0 \cdot e^{(\omega t + kx)}$ $k = \frac{\omega}{C}$
2.	Complex sound velocity $-C_\alpha$ $U_{tt} - C_\alpha^2 \cdot U_{xx} = 0$ $C_\alpha = \frac{\omega}{k - j\alpha}$	$U(t,x) = U_0 \cdot e^{j(\omega t + k_\alpha x)}$ $k_\alpha = k - j\alpha = \frac{\omega}{C} - j\alpha$
3.	With Navier–Stokes damping coefficient $\sigma \ll \omega$ $U_{tt} - C^2 \cdot U_{xx} = -2\sigma \cdot U_t$	$U(t,x) = U_0 \cdot e^{j(\omega t + k(\omega) \cdot x)}$ $k(\omega) = k \cdot \sqrt{1 - j\frac{2\sigma}{\omega}} \simeq \frac{\omega}{C} - j\frac{\sigma}{C},$ $\sigma \ll \omega$
4.	With Navier–Stokes damping coefficient $\sigma \gg \omega$ $U_{tt} - C^2 \cdot U_{xx} = -2\sigma \cdot U_t$	$U(t,x) = U_0 \cdot e^{j(\omega t + k(\omega) \cdot x)}$ $k(\omega) \simeq$ $-\sqrt{2\sigma} \cdot \frac{1}{C} \cdot \sqrt{\omega} \cdot (1 - j),\ \sigma \gg \omega$
5.	$U_{tt} - C^2 \cdot U_{xx} = \frac{4}{3}\mu \cdot \frac{\partial}{\partial t}\left[\frac{\partial^2 U}{\partial x^2}\right]$	$U(t,x) = U_0 \cdot e^{j[\omega t - k(\omega) \cdot x]}$ $k(\omega) \simeq \frac{\omega}{C} - j \cdot \frac{4}{6} \cdot \frac{\mu\omega^2}{C^3},\ \sigma \gg \omega$
6.	In transverse direction $\frac{d^2 V_x(z)}{dz^2} - j\frac{\omega}{\mu} \cdot V_x(z) = 0$ $\eta \cdot \frac{\partial^2 V_x(z,t)}{\partial z^2} = \rho \cdot \frac{\partial V_x(z,t)}{\partial t}$	$V_x(z) = V_0 \cdot e^{jk(\omega) \cdot x}$ $V_x(z,t) = V_x(z) \cdot e^{j\omega t}$ $k(\omega) = \sqrt{\frac{\omega}{2 \cdot \mu}} \cdot (1 - j)$

PPSs. Due to their ability to register ultra-low frequencies, PPS signals are often used to discover the so-called beats (Figs. 10.3 and 10.10).

Most often, beats are represented by nearly periodical amplitude modulations of the signal. Modulation frequency and depth are the two parameters which characterize periodical modulation. PPS signal modulations in VVER RPs may differ from loop to loop, leading to the change of modulation parameters.

The simplest known model of an amplitude-modulated signal is the sum of two harmonic processes with closely-spaced frequencies

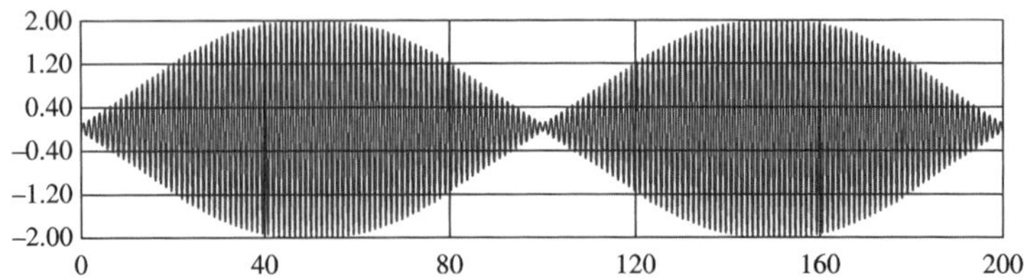

Figure 10.10: Beats as an amplitude-modulated signal. An example of periodical modulation with maximum modulation depth — θ.

ω_0 and $\omega_0 + \Delta\omega$:

$$x(t) = \sin \omega_0 t + \sin (\omega_0 + \Delta\omega)t$$
$$= 2 \cdot \cos\left(\frac{\Delta\omega}{2} \cdot t\right) \cdot \sin\left(\omega_0 + \frac{\Delta\omega}{2}\right) t \qquad (10.118)$$

The first component $\cos\left(\frac{\Delta\omega}{2} \cdot t\right)$ defines the signal envelope, while the second one — $\sin\left(\omega_0 + \frac{\Delta\omega}{2}\right) t$ — defines its content at the carrier frequency $\omega_0 + \frac{\Delta\omega}{2}$. If the envelope is periodical, it is characterized with modulation frequency (in this case $\frac{\Delta\omega}{2}$) and depth — θ — the difference between the maximum $x(t)$ and the minimum of the envelope modulus.

Harmonic components of PPS signal with closely-spaced frequencies may appear due to various physical reasons. For example, due to the so-called shaft precession effect of a vertical pump (or a spinning top effect), which is accompanied by oscillations at the frequency f_1, close to the shaft rotational frequency f_0. Half-difference of these frequencies $\left(\frac{|f_1 - f_0|}{2}\right)$ corresponds to the "slow" precession frequency.

The so-called sidebands close to the rotational frequency are also a vibration cause of PPS signal modulation (and also vibration sensor signal modulation). Apart from vibration causes, the signal of a pressure pulsation sensor may be modulated due to the characteristics of the measurement circuit. For example, if a PPS is mounted on a long impulse line with its natural frequency close to the frequency of some resonance phenomenon in the monitored object, it also leads to beats.

In applications, each resonance effect has its finite Q-factor. That is, close to the central frequency, there are also spectral components with significant amplitudes, though somewhat smaller compared to the amplitude at the central frequency. They comprise a continuum set of frequencies around the central frequency, that is why the addition of such spectral components, actually, does not necessarily lead to periodical modulation. The envelope may be some stationary function of time and, particularly, some random process. Its peak-to-peak amplitude, or RMS value, or modulation depth will be related to the Q-factor of this resonance.

Let us take another exaggerated case as an example. Suppose that, unlike the previous example, this time we have two sinusoids with close frequencies but different amplitudes. Let the amplitude at the central frequency still be equal to one, and the amplitude A at a close frequency — $\omega_0 + \Delta\omega$ — below one. By representing the initial process in the identical form

$$\begin{aligned} x(t) &= \sin\omega_0 t + A \cdot \sin(\omega_0 + \Delta\omega)t \\ &= (1 - A) \cdot \sin\omega_0 t + A \cdot \sin\omega_0 t + \cdot \sin(\omega_0 + \Delta\omega)t \\ &= (1 - A) \cdot \sin\omega_0 t + 2A \cdot \cos\left(\frac{\Delta\omega}{2} t\right) \cdot \sin\left(\omega_0 + \frac{\Delta\omega}{2}\right) t \end{aligned} \tag{10.119}$$

we get that the signal maximum $(1 + A)$, and the minimum of the envelope modulus $(1 - A)$, i.e. modulation depth θ, will be equal to $2A$.

As long as the parameter A decreases, the modulation depth $\theta = 2A$ decreases linearly, and modulation frequency remains the same (Fig. 10.11).

If a modulated signal consisting of two sinusoids with closely spaced frequencies goes through a viscous dispersion medium, its modulation parameters change. Applying the Kirchhoff–Stokes formula,

$$\alpha(\omega) = \frac{2\eta}{3 \cdot \rho \cdot C^3} \cdot \omega^2 = \text{const}(\rho, C, \eta) \cdot \omega^2$$

we get that the amplitude of the second component of the signal (10.119) with higher frequency reduces by a higher value. If the modulated signal at the inlet to the viscous medium had two spectral

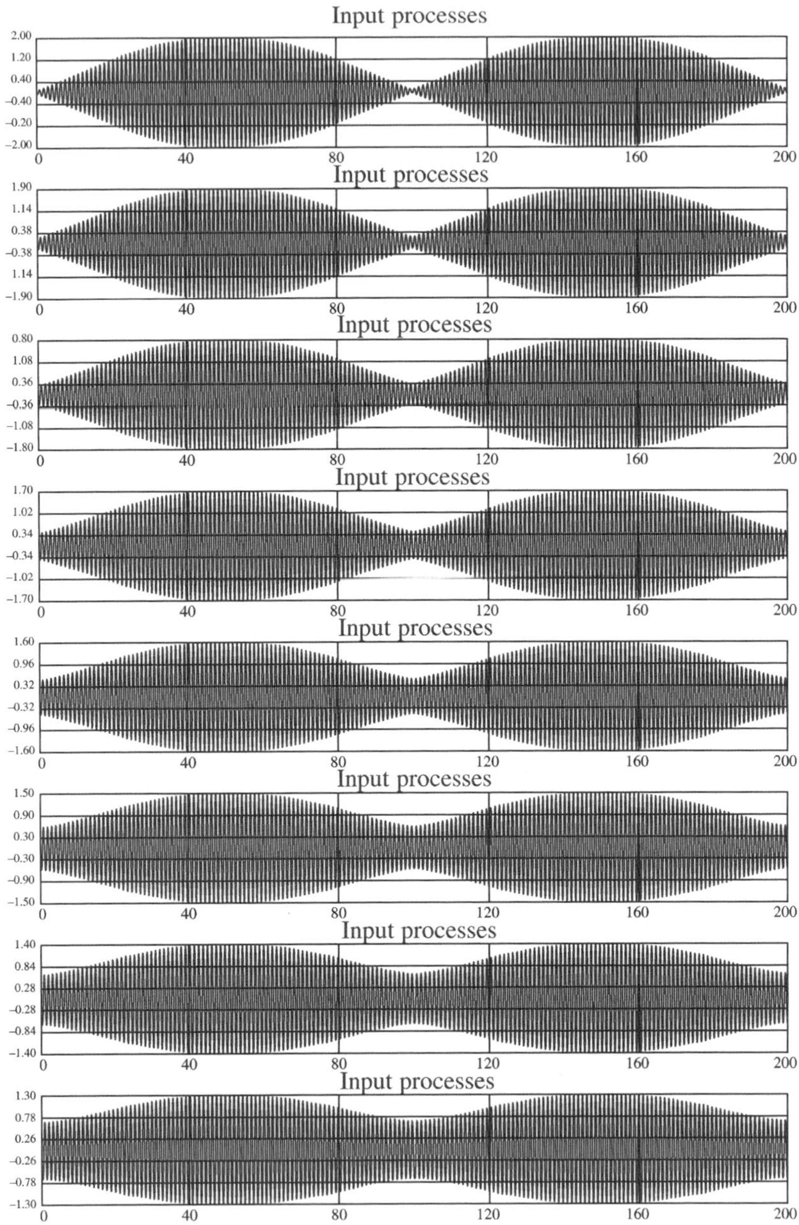

Figure 10.11: Numerical experiment: dependence of modulation depth with the change of parameter $A = 1$; 0.9; 0.8; 0.7; 0.6; 0.5; 0.4; 0.3 (top-down) in the signal $x(t) = \sin \omega_0 t + A \cdot \sin(\omega_0 + \Delta\omega) \cdot t$.

components

$$x(t) = f(t, \omega_0) + a \cdot f(t, \omega_0 + \Delta\omega)$$

i.e. it had the modulation depth $2A$, after it goes along its way L, it reduces in amplitude and takes on the following form:

$$\begin{aligned} x_L(t) &= e^{-\alpha(\omega_0) \cdot L} \cdot f(t, \omega_0) + A \cdot e^{-\alpha(\omega_0 + \Delta\omega) \cdot L} \cdot f(t, \omega_0 + \Delta\omega) \\ &= e^{-\alpha(\omega_0) \cdot L} \cdot [f(t, \omega_0) + A \cdot e^{[-\alpha(\omega_0 + \Delta\omega) - \alpha(\omega_0)] \cdot L} \cdot f(t, \omega_0 + \Delta\omega)] \end{aligned}$$

This means that the modulation depth reduces and becomes equal to

$$\theta = 2 \cdot A \cdot e^{[-\alpha(\omega_0 + \Delta\omega) + \alpha(\omega_0)] \cdot L} \tag{10.120}$$

Let us consider the exponential factor

$$\begin{aligned} &[-\alpha(\omega_0 + \Delta\omega) + \alpha(\omega_0)] \cdot L \\ &\quad = -\mathrm{const}(\rho, C, \eta)[(\omega_0 + \Delta\omega)^2 - \omega_0^2] \cdot L \\ &\quad = -\mathrm{const}(\rho, C, \eta) \cdot \omega_0^2 \cdot [(1 + \frac{\Delta\omega}{\omega_0})^2 - 1] \cdot L \\ &\quad \approx 2 \cdot \mathrm{const}(\rho, C, \eta) \cdot \Delta\omega \cdot \omega_0 \cdot L \end{aligned}$$

We get an exponential dependence of modulation depth on the parameters of the medium, which, among others, include viscosity η:

$$\begin{aligned} \ln\theta &= \mathrm{const}_1 - 2 \cdot \Delta\omega \cdot \omega_0 \cdot L \cdot \mathrm{const}(\rho, C, \eta) \\ &= \mathrm{const}_1 - 2 \cdot \Delta\omega \cdot \omega_0 \cdot \frac{2\eta}{3 \cdot \rho \cdot C^3} \cdot L \end{aligned} \tag{10.121}$$

This technique developed by the authors may be applied only if there is an a priori known point source of modulated pressure pulsations. One of the possible sources of modulated pressure pulsations in VVER is the pressurizer. It acts as a Helmholtz resonator with the central frequency ω_0 and resonance Q-factor $\sim \frac{1}{\Delta\omega}$, which are defined by the properties of the two-phase medium in the pressurizer. The pressurizer pipeline is connected to the MCC in a known point of the hot leg of the 4th loop. This is the reference point for length L estimation in the formula (10.121). A single PPS may be used to monitor the medium parameters ρ, C, η by measuring the modulation depth of its signal. PPS signal modulation depth — θ, which is a diagnostic sign, has a number of special properties:

- the applied modulation model, with two sinusoids, is too rough to use the formula (10.121) for the medium parameters measurement;
- the modulation depth of PPS signals shall be used to build indication methods for monitoring the parameters of the medium, and not their metrologically supported noise measurements;
- high sensitivity to the change of the medium parameters (due to the exponential dependence (10.120));
- dependence on many parameters of the medium ρ, C, η;
- dependence on both MCC coolant parameters and the two-phase pressurizer medium parameters $\omega_0, \Delta\omega$.

In connection with the last property of the above list, it is reasonable to monitor the thermophysical properties of the MCC coolant with several PPS instrument channels (or several accelerometer channels), with one sensor mounted right next to the nozzle of the pressurizer connecting pipeline. This way we shall have the PPS input perturbation signal, and all other PPS signals will be considered as output responses.

It is also reasonable to have a pair of PPSs mounted on the so-called passive part of the MCC pipeline. Between such PPSs there are no acoustic sources, such as MCPs, SG or any inhomogeneities of the MCC path, e.g. tees, bends, pipe nozzles. Such passive part of the pipeline shall be long enough, on the one hand, to reveal the viscosity mechanism in the form of significant attenuation of pressure pulsations amplitude, and short enough, on the other hand, to stay in line with the initial point assumptions of the acoustic model.

Described sector can be used for calibration purposes.

10.16 Typical Errors in Application of Analytical Models of Elastic Media

Let us set a task to formalize an acoustic medium in a segment of the length [*0*–*l*]. It may be a fuel assembly, as a heat generating channel with a single-phase or two-phase coolant, or a steam line with a single- or two-phase medium, or a pipeline, or sector of MCC with some inhomogeneity, or even the reactor core, or the steam generator. The properties of an acoustic medium are often studied just to find its natural resonance frequencies and then match them with natural vibration frequencies of some element. As a very rough approximation, the lowest natural frequency of a single-phase acoustic medium can be found using the formula for the frequency of the lowest-order ASW: $F_{0_a} = \frac{C}{2l}$. For example, for a fuel assembly medium with the sound velocity at about 800 m/s, $F_{0_a} = 160\,\mathrm{Hz}$. If we limit ourselves by the fuel assembly length only, then frequencies above 160 Hz may be studied. This high-frequency range will surely include natural frequencies of fuel assembly elements (part of fuel element between two spacer grids, spacer grids themselves), but will not include natural frequencies of FA itself, or that of RC, PTU, reactor vessel and the frequencies of loop ASWs. In spite of the fact that the acoustical model is true on the sector $[0\text{–}l]$, it is still not correct to interpret all frequencies observed in an experiment, even if they were obtained by measuring pressure pulsations on that very same segment $[0\text{–}l]$. The origin of a low-frequency resonance of pressure pulsations may have a global character, and for its interpretation one has to take into account the whole of the MCC, while the segment $[0\text{–}l]$ may be just a very short and insignificant fragment of it.

If we apply a known point model presented in Chapter 5

$$m_a \cdot \ddot{\xi}(t) + r_a \cdot \dot{\xi}(t) + \frac{1}{C_a} \cdot \xi(t) = P(t)$$

in lumped parameters with the numeric values m_a, r_a, C_a, where $\xi(t)$ is the volume displaced under pressure $P(t)$, we have to separate the inertial part spatially from the elastic part. The segment $[0\text{–}l]$ is usually presented as an aggregate of two adjacent segments: $[0\text{–}l_1]$ and $[l_1\text{–}l_2]$: the first one — inertial, represented by acoustic mass — m_a, while the second one — elastic, represented by acoustic capacitance — C_a. Damping properties represented by the parameter r_a, may be localized in any of the two segments. The only thing that matters is that the volume has to travel along that segment with non-zero velocity and some friction. This model has no spatial coordinates, the whole modeled segment $[0\text{–}l]$ is compressed into a point. This segment is characterized by a resonance frequency

$$\omega_{0_a} = \frac{1}{\sqrt{C_a \cdot m_a}}.$$

The following rule shall be applied as a criterion: the point model is applicable as far as the modeled acoustic medium may be represented by a Helmholtz resonator.

A typical mistake is made when a point model is applied to non-point acoustic media. For example, it is stated that at the segment [0–l], inertial and elastic properties are distributed evenly, but the whole acoustic medium is modeled by a point serial L–C–R circuit with the same resonance frequency ω_{0_a} Moreover, in case of a round pipeline with the diameter "d", the following reference formulas are applied: $m_a = \frac{4\rho l}{\pi d^2}$– for acoustic mass, while for acoustic capacitance there is the formula $C_a = \frac{\pi d^2 l}{4\rho C^2}$. Note that both parameters are defined at the same length l. Thus we can get the natural frequency of acoustic medium oscillation $\omega_{0_a} = \frac{1}{\sqrt{C_a \cdot m_a}} = \frac{C}{l}$, which resembles the formula for acoustic standing wave frequency $F_{0_a} = \frac{C}{2l}$, but differs from it.

If it is assumed that the heat generating channel has the steam phase in its upper part, the point model is eligible. If it is stated that the steam phase is evenly mixed with the water phase of the coolant along the whole length of the heat generating channel, we need to use distributed models. The same is true for steam pipelines with the distributed steam medium.

If distributed parameters are introduced into the acoustic model, its electrical equivalent is represented by long lines described by a set of equations (telegraph equations):

$$\begin{cases} ₵ \cdot \dfrac{\partial U(x,t)}{\partial t} + \dfrac{\partial I(x,t)}{\partial x} = 0 \\ L \cdot \dfrac{\partial I(x,t)}{\partial t} + \dfrac{\partial U(x,t)}{\partial x} = 0 \end{cases}$$

with two line parameters — ₵ and L. The equivalent equations of fluid and gas dynamics

$$\begin{cases} \beta \cdot \dfrac{\partial P(x,t)}{\partial t} + \dfrac{\partial V(x,t)}{\partial x} = 0 \\ \rho \cdot \dfrac{\partial V(x,t)}{\partial t} + \dfrac{\partial P(x,t)}{\partial x} = 0 \end{cases}$$

have two point parameters — ρ, β, and the replacement

$$\rho \leftrightarrow m_a$$
$$\frac{1}{\beta} \leftrightarrow C_a$$

is false. It is also a mistake to represent a long line in the form of L–C–R circuit in lumped parameters. The product and the quotient of the parameters ($\not{C}$–L) and (ρ, β). Play a role in the solution of the above equations. They have the meaning of wave resistances, complex resistances, propagation constants, natural frequencies, but they are not acoustic mass and acoustic capacitance.

Now, the acoustic medium is described in other terms, namely, in terms of standing and travelling waves, and the resonance phenomenon is that the length of ASW half-wave coincides with the segment-length $[0$–$l]$. It is also important to define what is the load of the long line (the acoustic medium), i.e. to represent two complex resistances in the limiting points of the segment-length $[0$–$l]$. It will have an impact on the reflection coefficient of travelling waves and their ratios with standing waves.

The continuity equation

$$\frac{\partial \rho}{\partial t} + \operatorname{div}(\rho \cdot v) = 0$$

and the motion equation (Euler equation)

$$\rho \frac{\partial v}{\partial t} = -\operatorname{grad} P$$

through which the system of linearized equations of fluid and gas dynamics was obtained, are also applied in hydraulics in order to describe the process of pumping a medium with variable density $\rho(x,t)$ through a pipeline of variable cross-section $S(x,t)$ with variable speed $v(x,t)$ at variable pressure in every point of the liquid $P(x,t)$.

For a non-compressible medium $\rho = \text{const}$ at constant cross-section, the system reduces into a simple condition of invariability of flow W:

$$\frac{\partial W(x,t)}{\partial x} = 0 \quad \text{and} \quad W = v \cdot S = \text{const.}$$

However, sometimes the continuity and motion equations are erroneously represented in the following form:

$$\begin{cases} C_a \cdot \dfrac{\partial \Delta P(x,t)}{\partial t} + \dfrac{\partial W(x,t)}{\partial x} = 0 \\ m_a \cdot \dfrac{\partial W(x,t)}{\partial t} + \dfrac{\partial \Delta P(x,t)}{\partial x} = 0 \end{cases} \tag{10.122}$$

where the point acoustic parameters m_a and C_a of an elastic medium are the coefficients before the derivatives of functions describing fluid pumping with the flow rate $W(x,t)$ and pressure drop $\Delta P(x,t)$ Formal replacement of fluid velocity with fluid flow rate and replacement of pressure with pressure drop ΔP are wrong, for one thing, because in the first equation of the system (10.122), the dimension of the parameter C_a does not correspond to the dimension of acoustic capacitance.

Propagation of a longitudinal wave in an elastic acoustic medium, characterized by sound velocity — C, and pumping this medium with low velocity — v — are essentially different physical phenomena. Coolant velocity in VVER-1200 RP can be as high as 10 m/s, while a longitudinal wave travels in the elastic medium of the coolant with sound velocity, which is about 800 m/s. Pressure drop — $\Delta P(x,t)$, which is one of the main global characteristics of the circuit in hydraulics, is almost never applied in distributed acoustic media. For them, instead of pressure drop (difference of pressures) in two points of space or at two points of time, pressure differential is applied (micro-changes of pressure drop at the length dx), or the corresponding partial derivative.

In hydraulics, it is acceptable to draw electrical analogies between flow rate and current ($W \Longleftrightarrow I$) and between pressure drop and voltage ($\Delta P \Longleftrightarrow U$). Two Kirchhoff laws from electrical engineering, when generalized, are also applicable to a hydraulic circuit:

- the sum of flow rates in the node is zero (the total rate of the inlet coolant flows is equal to the total rate of the outlet coolant flows);
- the sum of pressure drops in a closed circuit is zero (taking into account the signs of the pressure drops).

However, Ohm's law, which has a linear form in electrical engineering, $U = \text{const} \cdot I = R \cdot I$ is similar to a quadratic dependence between the flow rate and the pressure differential at a turbulent flow of a fluid:

$$\Delta P = \text{const} \cdot W^2 \tag{10.123}$$

This electrical analogy has nothing to do with the long line analogy expressed by differential equations. Hydraulics are similar to DC electric circuits, while hydro- and gas dynamics are similar to AC circuits, including super-high-frequency circuits (long lines). Sometimes this equation is modified by using a sinusoidal function of time with frequency F instead of flow rate, which enables one, due to the nonlinear (quadratic) relation, to find harmonics with frequency $n \cdot F$ and, in particular, double frequency $2 \cdot F$, in the pressure drop. The corresponding harmonic may coincide with some natural vibration frequency of a structural element, which leads to the false conclusion on the resonance excitation of oscillations. Actually, the formula (10.123) is applicable only for numeric values of flow rate and pressure drop at a steady flow of a fluid. In electrical analogy it is true only for direct current and voltage. While coolant flow rate is the function of time, pressure drop as the function of time shall be obtained from differential equations, not from the algebraic equation (10.123).

Consideration of acoustic losses using point models is extremely limited. Complex point resistance of a serial resonance L–C–R circuit:

$$z_a = j\omega m_a + r_a + \frac{1}{j\omega C_a}$$

does not reveal the physical mechanism of energy loss, described by active resistance r_a, it is only declared. That is why the resonance Q-factor

$$Q_a = \frac{1}{r_a}\sqrt{\frac{m_a}{C_a}}$$

is not represented in the form of a numeric parameter even with the known numeric values m_a and C_a.

Both in a point model and in a space–time model with linear (distributed) parameters M_a, R_a, K_a

$$M_a \cdot \frac{\partial^2 V}{\partial t^2} + R_a \cdot \frac{\partial V}{\partial t} - K_a \cdot \frac{\partial^2 V}{\partial x^2} = 0$$

with electrical analogy,

$$L \cdot \frac{\partial^2 i}{\partial t^2} + R \cdot \frac{\partial i}{\partial t} - \frac{1}{\not{C}} \cdot \frac{\partial^2 i}{\partial x^2} = 0$$

and in a long line model with losses,

$$\begin{cases} \dfrac{\partial^2 U}{\partial x^2} = L \cdot \not{C} \cdot \dfrac{\partial^2 U}{\partial t^2} + \left(R \cdot \not{C} + G \cdot L\right) \dfrac{\partial U}{\partial t} + G \cdot R \cdot U \\ \dfrac{\partial^2 I}{\partial x^2} = L \cdot \not{C} \cdot \dfrac{\partial^2 I}{\partial t^2} + \left(R \cdot \not{C} + G \cdot L\right) \dfrac{\partial I}{\partial t} + G \cdot R \cdot I \end{cases}$$

it is necessary to represent the corresponding energy loss parameters via additional physical equations. Active parameters of the models r_a, R_a, R, G shall be related via analytical dependences to the viscosity parameters of an acoustic medium and/or to some other friction mechanisms, as demonstrated in this chapter.

Chapter 11

ASW Globality–Locality

11.1 ASWs in Loop and its Harmonics

This chapter elaborates ASW phenomenology in VVER MCC started in [361] based on experimental data of VVER-440 and continued in the monographs [382, 384] based on those of VVER-1000. Now, according to the extensive experimental material of VVER-1200, it is possible to establish ASW properties as stable regularities for the reactor plant MCC of any VVER project.

Four main features allow distinguishing an acoustic standing wave from other physical effects. First, in two PPS signals, oscillations at the ASW frequency will be, as a rule, either in phase (zero phase) or out of phase (180° phase). In rare cases, this ASW property (in phase or out of phase) in different MCC points can be broken. This occurs in branched circulation loops because of the interference of waves propagating along different paths and having different phase progressions at the points where they, summing up, again form a single ASW.

Second, in some vicinity of the ASW central frequency the CPSD phases of two PPS signals stay constant in frequency dependence. That is, the ASW resonances are always "blurred" or have low Q-factor compared to the resonances of a vibration origin. Other types of resonances caused, for example, by forced oscillations on the harmonics of MCPs rotational frequency have permanently a high Q-factor. The Q-factor of oscillation resonances at natural frequencies depends on the damping properties of the structure and, as a rule, it is not worse than the Q-factor of the ASW resonances.

Frequency sub-bands of phase uniformity (0° or 180°) can be up to 20% of the value of the ASW central frequency.

Third, with a monotonous coolant temperature change, the ASW central frequency also changes monotonically. This effect is associated with a change in the speed of sound from temperature in the coolant. As it was shown above, in case of signals of two accelerometers mounted directly on the loop pipeline, these three features also fully exhibit themselves.

The fourth ASW property is that the low ASW type is accompanied by an almost harmonic series of ASW. The use of this feature for ASW identification in practice is very difficult. Some ASW harmonics may be absent and their amplitude does not necessarily decline with the frequency.

Restoration of the field of different ASW types arising even on a small number of acoustic inhomogeneities is a complex combinatory problem. A standing wave can be localized as a single half-wave between any two inhomogeneities of the acoustic path.

Let us introduce the following classification. We shall call an ASW consisting of only one half-wave, simple. For example, if a pipeline segment without branching contains n inhomogeneities, then it is possible to have $n(n-1)/2$ simple ASW types with a natural frequency proportional to the inverse value of the distance between two inhomogeneities which are not necessarily adjacent (Fig. 11.1).

This number is obtained as the number of items in the upper half of the square matrix with a size of $(n \times n)$, that is, a total of $n(n-1)/2$. However, at the same time, only two inhomogeneities are enough to form infinite number of ASW types. Indeed, if between the two inhomogeneities a simple ASW as a half-wave with a frequency $f = \frac{C}{2L}$ is formed, then two half-waves having double frequency $2f$ will fit into the same distance (i.e. one spatial period), and also three half-waves with a frequency $3f$, etc. (Fig. 11.2). Such a harmonic series has only one simple wave — the first one. It is obvious that each simple ASW generates its infinite harmonic ASW series and we will call it ASW harmonics.

Infinite multiplication of ASW harmonics leads to the fact that there will certainly be a coincidence of the frequencies of some ASW harmonics with some frequencies of external driving forces, for example, with the harmonics of the MCPs rotational frequency or with some natural frequencies of vibrations of the RP structural elements.

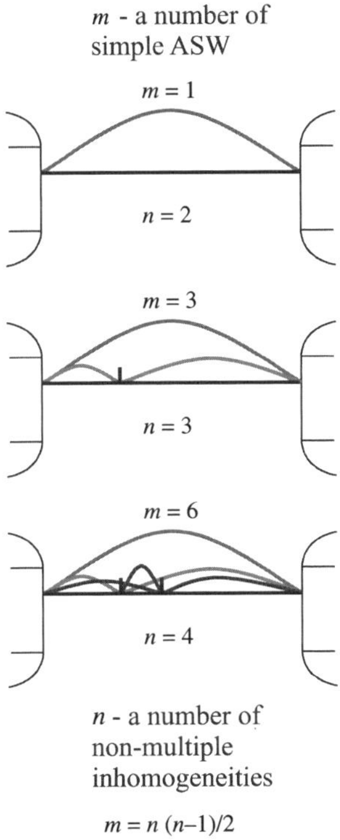

Figure 11.1: ≪m≫ simple ASW on ≪n≫ non-multiple inhomogeneities.

The natural oscillation frequencies in a nonlinear medium are also subjected to harmonic propagation, which again leads to all kinds of coincidences of frequencies of different physical effects. Sometimes they are declared catastrophic phenomena that lead to accidents, although the amplitude of the oscillations in the resonant excitation is limited by the damping properties of the structure. Structural adequacy of damping shall be justified at the design level, and, what is more important, during commissioning tests. Unfortunately, the calculations of the acoustic wave dissipation effects have not reached the level when the full-scale experiment can be replaced by a numerical experiment. Therefore, any computational models shall be verified by full-scale measurements.

In what follows, we will adhere to the following terms. An ASW registered in more than one loop will be called global, unlike local ASW which is registered only within one loop. As for applicants for

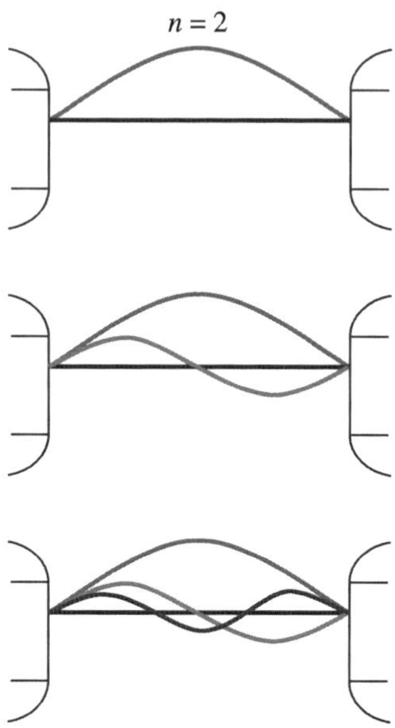

Figure 11.2: Simple loop ASW and its loop harmonics (harmonics 2 and 3).

globalism, we select the simple loop $\mathrm{ASW}_{1\mathrm{loop}}$ half-wave which covers that loop, and all its harmonics are also localized in that loop. This ASW series will also be called as loop series. It can go from the loop status into the global one if there are correlations at the relevant frequencies with at least one PPS signal from the other loop.

A simple ASW also acts as a local ASW. The longest simple local $\mathrm{ASW}_{1\mathrm{loop}}$ is affected by all of the loop inhomogeneities and on each of them it loses some of its amplitude due to reflection and absorption. Its spatial half-wave length is the sum of all distances between $(n-1)$ adjacent inhomogeneities. However, the round passages of the incident and reflected waves with unchanged phase parameters result in an increase of the resulting amplitude of the low type of ASW due to an infinite number of additions of the numerical series. Let us illustrate the effect of amplification of the ASW amplitude on the following idealized example.

Let the resulting spatial distribution of the amplitudes of a simple ASW be obtained as the sum of an infinite number of incident and reflected traveling waves of the coolant pressure. After each act of reflection, the direction of the traveling wave motion changes to the opposite one and its amplitude decreases due to the semi-permeability of the reflecting surface and its absorption of part of the wave energy. When both reflecting surfaces have the same reflection coefficient in the range "$0 < K < 1$", the decreasing amplitude of the traveling wave depending on the number of reflections occurs by the law of geometric progression. If at some fixed moment the incident wave has a spatial distribution $A_0(x) = A \cdot \sin \frac{2\pi x}{L}$, then after the first reflection the amplitude will decrease by "K" times: $A_1(x) = A \cdot K \cdot \sin \frac{2\pi x}{L}$, after the second one — by K^2 times: $A_2(x) = A \cdot K^2 \cdot \sin \frac{2\pi x}{L}$, etc. As a result, the spatial distribution of the ASW amplitude takes a form of geometrical progression:

$$A(x) = A \cdot \left(\sin \frac{2\pi x}{L}\right) \cdot (1 + K + K^2 + K^3 + \cdots)$$
$$= \frac{A}{1-K} \cdot \left(\sin \frac{2\pi x}{L}\right) \tag{11.1}$$

Thus, there is an increase in the amplitude of the incident wave by $\frac{1}{1-K}$ times. The extreme cases when $K = 0$ imply being total absorption and $K = 1$ imply total reflection are idealized and hardly possible.

For short waves localized between two adjacent inhomogeneities (Fig. 11.1), the so-called multiplicity conditions are not met. According to that condition, the integer number of half-waves of a given frequency should fit in that spatial interval and reflections from boundary end inhomogeneities should not distort the phase half-wave distribution of its amplitude. That is, such short waves cannot exist outside the simple interval under consideration, since other inhomogeneities are located at non-multiple distances. In contrast, a simple wave stops being simple, when, overcoming its semi-permeable boundaries, it finds other inhomogeneities located at distances which are multiples of its half-wave length.

If it turns out that some inhomogeneities are located at multiple intervals, it is possible that the frequency of a simple ASW matches the harmonic of some other ASW. Because of summing their amplitudes, there can be both strengthening of a simple ASW or its weakening or even disappearance. This is a pure linear interference effect. The last case occurs when the amplitudes of the simple wave and its harmonics coincide in frequency, and their moduli are out of phase.

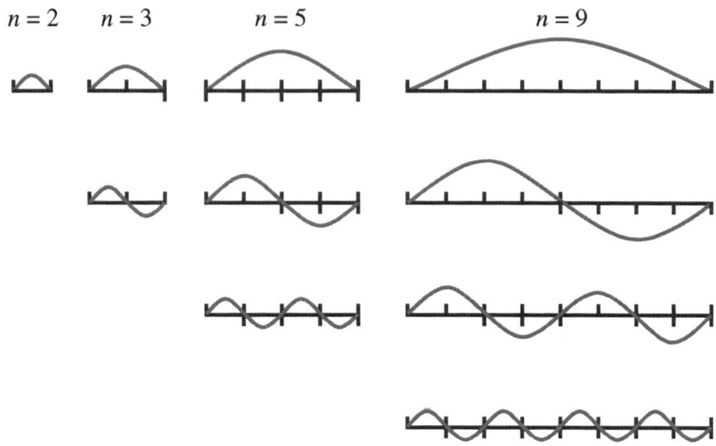

Figure 11.3: ASW types when the number of equal intervals covering the length of the pipeline is the power of two.

Let us consider a hypothetical case, the ideal multiplicity, when all "n" acoustic inhomogeneities with the same spatial step $L/(n-1)$ exactly divide the length of the pipeline, so that the two edge inhomogeneities lie on the left and right boundaries of the pipeline. If the number "$n-1$" is a power of two ($n = 2, 3, 5, 9, 17, \ldots$), then all the ASW types at a given length will be multiples of even harmonics (Fig. 11.3). There will be a number of ASW harmonics in powers of two: first harmonic, second harmonic, fourth, eighth, etc.

As a result, we will get a kind of multi-section resonator, which destroys the third, fifth, sixth, seventh, etc. harmonics, but, on the contrary, enhances the harmonics with numbers that coincide with the power of two. Multiple arrangements of inhomogeneities can be created in design of the main coolant loop, placing the main equipment of the loops (SG and MCPs) in certain places.

For example, if the main source of RP vibration is MCPs at a frequency of 16.6 Hz, then the vibration can be suppressed by placing the MCPs so that the ASW would be formed with the same frequency, but out of phase. That is, we obtain a kind of natural phase inverter. However, there is a danger of amplifying the MCPs harmonics $(16.6 \cdot n)$ Hz. It is also possible to use a method when artificial inhomogeneities in the form of diaphragms are introduced into the acoustic path.

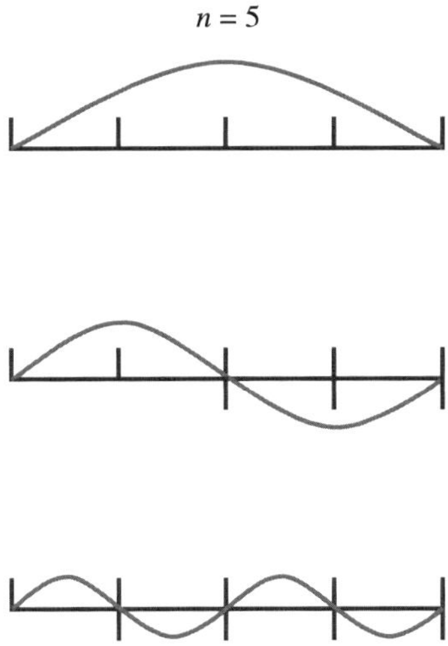

Figure 11.4: The first consecutive spatial harmonics, the second and fourth, with four multiple loop inhomogeneities.

Let us again imagine our hypothetical case of an even spatial step, but with five inhomogeneities of the loop (Fig. 11.4), that is, having four equal intervals.

Unlike the previous case, here there are only two consecutive harmonics $\text{ASW}_{2\text{loop}}$ and $\text{ASW}_{4\text{loop}}$. Not only is the existence of these harmonics acceptable, but also their amplification in such a sectional spatial resonator. There will be other simple ASW, for example, between the first and third inhomogeneities or between the second and fifth ones, etc., but they will not be amplified by the loop design. At such four intervals, two items of the main equipment of the VVER main coolant loop, namely, the steam generator and MCPs representing acoustic inhomogeneities can divide the length of the loop into parts like a sectional resonator, with a danger of amplifying the amplitude of the $\text{ASW}_{1\text{loop}}$ harmonic.

11.2 *A Priori* Qualitative Comparison of Vibration–Acoustic Properties of VVER-1200 and VVER-1000

The new VVER-1200 reactor plant (RP) differing from the VVER-1000 RP by weight and dimension properties, obviously, shall have different parameters of the natural oscillations of its structure and coolant. The risk of coincidence of frequencies of natural and forced oscillations or the risk of the so-called resonance excitation pose a threat of an increase of oscillation amplitudes, which reduces the RP vibration reserve. The amplitude of the resonance oscillations is limited by the damping capacities of the structure, which can be estimated experimentally. They also, strictly speaking, differ from the damping capacities of the VVER-1000 RP structural items.

The average coolant temperature in the VVER-1200 core is higher than that of VVER-1000 and the so-called "subcooled" boiling of the coolant may shift the ASW frequencies considerably more. Even a small increase in the coolant steam fraction can significantly reduce the ASW frequency. Under this factor, reduction of ASW frequencies of VVER-1200 in comparison with VVER-1000 is expected. It is almost impossible to estimate the resultant effect of changes in the ASW properties of VVER-1200 compared to VVER-1000 due to the above factors by calculation.

If the frequency series of natural oscillations of a certain structure is represented as a set of spectral lines or a kind of frequency comb and another frequency comb represents a series of spectral lines of forced oscillations, the problem of resonance excitation is posed as a search for coincidences of teeth in these combs. The comb of harmonics of the MCP rotational frequency has equidistant (located at the same distance from each other) teeth. They are the same as in the VVER-1000, and the MCP rotational frequency remained unchanged: it is equal to 16.6 Hz. Thus, one comb of harmonics (MCP) remained unchanged, and the other two — the comb of natural frequencies and the ASW comb — began to move, changing the number of teeth, their width and the distance between adjacent teeth. To take into account all of the above vibration properties of VVER1200 RP by calculation are almost impossible.

In this regard, the role of vibration measurements in the process of VVER-1200 commissioning activities gets more important. Practically valuable is the approach according to which the vibration state of a new VVER-1200 reactor plant is similar to that of VVER-1000, in the sense that the above frequencies will change slightly. This assumption is based on the fact that the weight and dimensions of the structure and the coolant heat and hydraulic characteristics changed slightly in VVER-1200 as compared to VVER-1000, following the principle of structural conservatism. Nevertheless, the effect of triple coincidence of frequencies of forced and natural oscillations and MCP harmonics cannot be excluded, despite the fact that this effect is not present in the VVER-1000 reactor plant in nominal operating modes.

Changing the parameters of a spectral property peak depends on its nature of origin. If there is a resonance of natural oscillations, it is necessary to detect small changes in the absolute value of the natural frequency. With natural oscillations, the amplitude, frequency and quality factor of the resonance are associated with the mass, stiffness and damping of a certain structural element. Natural frequency changing after a small change in stiffness, can comply with significant changes in the properties of the supports of the structure. Already a small change in the natural frequency can indicate the beginning of the defect and, thus, it is an important diagnostic feature.

If the vibration resonance is caused by an external driving force, the change in its parameters can correspond both to a change in the properties of the structure and the properties of the external force. Here, the limits of permissible changes in the amplitude, frequency and quality factor of the resonance are much greater than in the case of natural oscillations.

The broadband effect on the design of hydrodynamic forces (pressure pulsations) of the coolant is estimated by integral characteristics (most often the RMS value). Pressure pulsations can change slightly due to small changes in local hydraulic resistances of individual flow sections of the VVER-1200 path. MCPs' modernization implemented in the VVER-1200 project and affecting the design of its flow part can also change the spectrum of driving vibration forces in the broadband turbulent component of pressure pulsations. Therefore, the integral criteria for the normal vibration state of VVER-1200, such as RMS or the amplitude of oscillations in a broad frequency band at well-known points of the structure may differ slightly from similar characteristics of VVER-1000.

Strength standards [388], establishing the design life of the structure, take into account two factors: the number of loading cycles (that is, the frequency of vibration loads) and their amplitude. Significant changes in the natural frequencies of the VVER-1200 MCC compared to the VVER-1000 MCC are not expected; therefore, in the experimental evaluation of the vibration strength, more attention should be paid to the amplitude factor, as a consequence of the possible effect of resonance excitation.

Extensional sources of oscillations in the MCC equipment and pipelines as well as internal and fuel assemblies are acoustic standing waves. The excitation power of oscillations of the equipment from the ASW is the largest in comparison with other sources of oscillations. For example, the oscillation energy of any loop equipment (including the MCP) at the MCP rotational frequency is much less than the oscillation energy at the frequencies of acoustic standing waves.

To explain the cause of a particular ASW, it is sufficient to demonstrate two reflecting surfaces, thereby determining the resonator generating ASW, and to link ASW frequency with the acoustic distance between these two reflecting surfaces. Thus, the actual resonator is replaced by a cylindrical (tube) resonator or an equivalent pipeline with two dead ends, each with its own acoustic reflecting and absorbing properties. The distance between these ends — L — is explicitly related to the spatial ASW wavelength $\lambda = 2L$ and inversely proportional to the ASW frequency F while the speed of sound — C — is a coefficient linking the wavelength and its frequency: $F = \frac{C}{2L}$.

Depending on the acoustic properties of the ends on the pipeline, either an ASW half-wave is laid with perfectly reflecting surfaces or a quarter of the wave with one perfectly reflecting end and the other one open. These are two extreme cases of the idealized representation of reflections of incident traveling waves of the coolant density. The metal surfaces at the pipeline ends are almost ideal reflective surfaces for acoustic waves of infrasonic frequency range, on which a node (zero amplitude) of the standing wave is formed. Accordingly, in the middle of the pipeline there is maximum (anti-node) ASW amplitude.

The pipeline with one end absent or half-open pipeline is called $\lambda/4$-transformer. It is able to generate the lowest ASW type where the node is located on the metal surface of one end, and the maximum is on the open end (or tee), resulting in a quarter of the ASW period on the length of the pipeline. $\lambda/4$-transformers can act as acoustic vibration isolators at one frequency, or, on the contrary, as external oscillators at certain frequency. Such $\lambda/4$-transformers represent ECCS pipelines, and the injection pipeline and the connecting pipeline of the pressurizer system.

As it has been shown in the previous sections, the acoustic inhomogeneities in MCC include the following:

- pipe bending and branching;
- structural elements-adapters connecting different types of MCC equipment (pipe nozzles connecting the pipelines to the reactor vessel, SG and MCP, SG headers connecting the pipelines to the heat exchange tubes);
- coaxial structures (so-called pipe shells), arranging the downward and upward coolant flows in the RP vessel;
- structural items transverse to the direction of the coolant movement (the reactor vessel bottom and head, the core barrel, in particular, its cylindrical part, the bottom of the reactor core barrel, the PTU inner surface in the area of the coolant flow turn into the hot MCC pipelines, support plates in the reactor vessel and the lids of SG headers);
- small-scale items (support columns, anti-debris filters, spacer grids, heads of fuel assembly, etc.).

The inhomogeneity of the hydraulic path, introducing little change into its local hydraulic resistance, causes a reflected wave of small amplitude in comparison with the amplitude of the incident wave. Accordingly, the standing wave, resulting from the superposition of the incident and reflected traveling waves, will then have small amplitude.

As initial *a priori* information for vibration and acoustic parameters of VVER-1200, the values of resonance frequencies of vibration signals of VVER-1000 of a standard V-320 plant at 100% power level are as follows [382, 384]:

~0.6 Hz — the pressurizer frequency;
~6.8 Hz — the frequency of the first loop ASW (ASW_{1loop});
~9.2 Hz — the frequency of the first vessel ASW ($ASW_{1vessel}$);
~13.2 Hz — the frequency of the second loop ASW (ASW_{2loop});
~16.6n Hz — the harmonics of the MCP rotational frequency;
~13.2 Hz — the frequency of the fourth loop ASW, (ASW_{4loop});
~5.5 Hz — the natural frequency of the low frequency oscillation type of the reactor core barrel (its type belongs to pendulum of a bar with the fixed upper end);

~10.0 Hz — the natural frequency of the reactor core barrel oscillations (which is of the type oscillation of a bar with the two fixed ends);

It is necessary to find the same resonance features of VVER-1200 near the above frequencies, and then, with the help of auto-, cross- and multidimensional spectral features, to prove their identity.

11.3 Basic ASW Types of VVER-1000 and VVER-1200

The following ASW types have been identified according to the four distinctive features specified in Section 11.1. ASW_{1loop}, the first loop ASW, is the simple local ASW with the lowest frequency covering the entire loop. ASW_{1loop} has one node on the wall of the reactor core barrel in the "hot" nozzle of the reactor vessel and another one on the wall of the reactor core barrel in the "cold" nozzle of the reactor vessel, and the anti-node is approximately in the middle of the loop on the U–shaped pipeline.

One end of the equivalent pipeline is the reactor core barrel visible from the side of the cold pipeline. After the reflection from the reactor core barrel, the wave runs in the opposite direction to the other equivalent end, which is the same core barrel, only visible from the "hot" pipeline. Between the two areas of the reflecting core barrel surfaces, there is only the flow separator. All the loop sensors register the ASW_{1loop} oscillation frequency. Figure 11.5 schematically shows the location of two nodes and a single ASW_{1loop} antinode relative to the main equipment, as well as zero phase characteristics between any pairs of signals from PPS installed in the loop.

Zero phase between the signals of any loop sensors indicate the invariance of the sign of amplitude of ASW_{1loop}, that is, the presence of a single half-wave of ASW_{1loop} in the loop.

Measuring the length of the ASW_{1loop} half-wave on the mean acoustic trajectory (MAT), we obtain a value of 57.6 m. Using that and the known central frequency of the ASW one can calculate the velocity of the sound. In the isothermal mode of heating (cooling) of the reactor, for each coolant temperature we can find its sound velocity and ASW frequency.

The power operation mode is a more ambiguous situation. One ASW_{1loop} quarter will be localized in the "cold" pipeline, the other

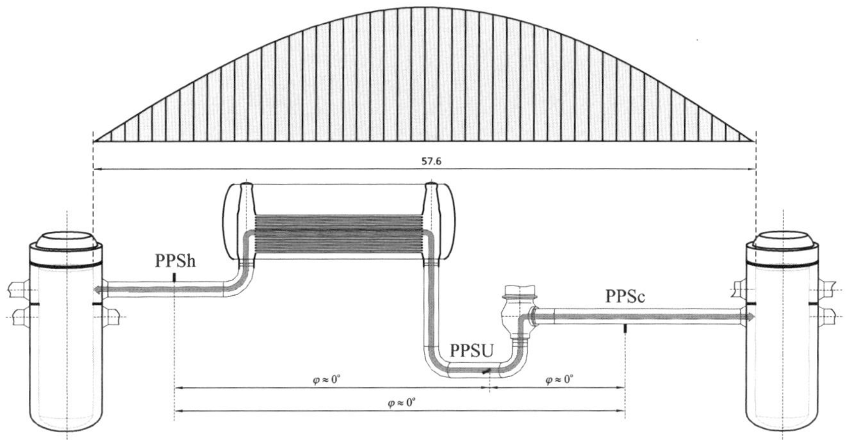

Figure 11.5: Distribution of ASW_{1loop} amplitudes on the scan of the VVER-1200 MCC heat exchange loop with the phase values of the pairs of signals from PPS_h, PPS_U, PPS_c.

quarter — in the "hot" pipeline. The coolant having different temperatures is anisotropic in density medium with different sound velocities. The spatial ASW wavelength as well as its frequency is a point parameter relating to the entire period of ASW oscillations, both in space and in time. The question arises: can the ASW frequency be considered a value depending on the spatial coordinate? Or in other words: can we expect that the ASW_{1loop} will have different frequencies in the "cold" and "hot" pipelines with the transition from one frequency (sound velocity) to another in the SG and in the core? Or, on the contrary, is a standing wave some integrator in the period or in the half-period, reducing the distributed properties of the medium to some average values, for example, to the average velocity of sound, and hence to one ASW frequency?

Let us answer these questions experimentally when PPS are mounted in both hot and cold pipelines, although for this purpose it is necessary to obtain reliable high-resolution estimates of the ASW frequencies differing only little from each other. Unfortunately, this experiment was not carried out at the VVER-1200, since the PPS were not designed for power operation modes and all-mode accelerometers were only installed on "cold" pipelines. The effect of frequency anisotropy at 100% power level is presented below by the example of signals of PPS_h, PPS_U and PPS_c in VVER-1000 [382].

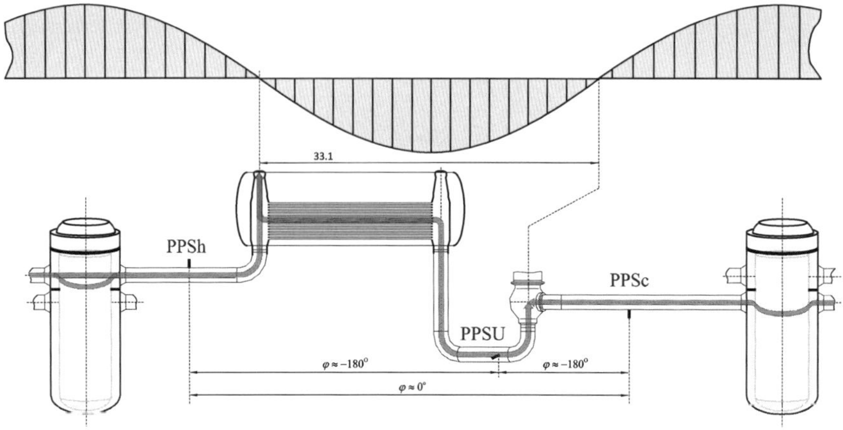

Figure 11.6: Distribution of $ASW_{1vessel}$ amplitudes on the scan of the VVER-1200 MCC heat exchange loop with the phase values of the pairs of signals from PPS_h, PPS_U, PPS_c. $ASW_{1vessel}$ is not a simple ASW; it has other half-waves.

The next standing wave in increasing frequency range after ASW_{1loop} has zero phases in any pair of signals of four PPSs between four "hot" pipelines (Fig. 11.6). That is, this wave is global and covers the reactor vessel with its half-wave. Let's denote it as $ASW_{1vessel}$ and call it the first vessel ASW. Judging by the fact that its frequency is almost by 1.5 times longer than the ASW_{1loop} frequency, the spatial $ASW_{1vessel}$ half-wave will be nearly 1.5 times shorter than the loop length. The length of the VVER-1200 $ASW_{1vessel}$ half-wave based on its frequency, the ASW_{1loop} frequency and wavelength will be equal to 33.1 m. If the MAT for $ASW_{1vessel}$ is calculated from the lid of one "hot" SG header to the lid of the other opposite "hot" SG header, the distance will also be the same value of 33.1 m. As a result, the half-wave of this ASW fits into the distance from the "hot" header of one SG to the "hot" header of the opposite SG, passing through the reactor vessel. That is, its half-waves extend from SG1 to SG3 and from SG2 to SG4. It can also be represented as four quarter-waves of the "hot" pipelines of each loop, which are combined at the points of their anti-nodes on the vertical axis of symmetry of the reactor into a single global wave. As in all the "hot" pipelines and the core volume, there is one and the same sign of the $ASW_{1vessel}$ amplitude, we can observe a powerful barometric effect on the neutron flux.

The resonance in the $ASW_{1vessel}$ frequency is dominant in both excore and in-core neutron noise.

This is confirmed by the PPS_h and PPS_U anti-phases in any loop at high coherence values of $ASW_{1vessel}$ frequency. MCPs are located exactly at a distance of 33.1 m along the mean acoustic trajectory from the "hot" SG header. At this length, there is a negative $ASW_{1vessel}$ half-wave. Moving further along the loop, we come across weak anti-phase correlations in the $ASW_{1vessel}$ frequency between PPS_U and PPS_c signals. It is possible that the reason is the $ASW_{1vessel}$ node located on the MCPs, and, therefore, we have small values of the $ASW_{1vessel}$ amplitudes on the left and right of the node at PPS_U and PPS_c sensor coordinates. The next $ASW_{1vessel}$ half-wave is located starting from MCPs and further on the "cold" pipeline in the direction of the reactor vessel. In contrast to the positive $ASW_{1vessel}$ half-wave covering the reactor together with the "hot" pipelines, this positive half-wave covers the "cold" pipelines, walking round the reactor core barrel and penetrating into the adjacent loops and the opposite loop. The distance between opposite MCPs (MCP1 and MCP3, MCP2 and MCP4) is also equal to about 33.1 m. This positive half-wave is also volumetric, it also penetrates into the reactor vessel, as well as other positive half-waves, but it does not penetrate into the core.

Thus, we managed to close $ASW_{1vessel}$ into a single entity of a global acoustic field with four nodes in the four "hot" headers of four steam generators, with four nodes in the MCPs and with anti-nodes in the "cold" SG headers and in the reactor vessel. All the major MCC items are involved in generation of $ASW_{1vessel}$: SG, MCPs, the reactor vessel. Their spatial arrangement turned out to be such that the initial simple ASW in the form of one half-wave covering the reactor has been multiplied by multiple distances into a closed, exactly periodic in space acoustic field of $ASW_{1vessel}$ covering the entire MCC.

The $ASW_{1vessel}$ node in the "hot" SG header as well as a node in MCPs are the unfavorable factors in terms of vibration loads. In the node of the standing wave, there is the maximum spatial gradient of the acoustic field, which defines the amplitude of forced oscillations of the structure at this frequency. Perhaps this explains the so-called torsion oscillations of the SG around the "hot" header detected in

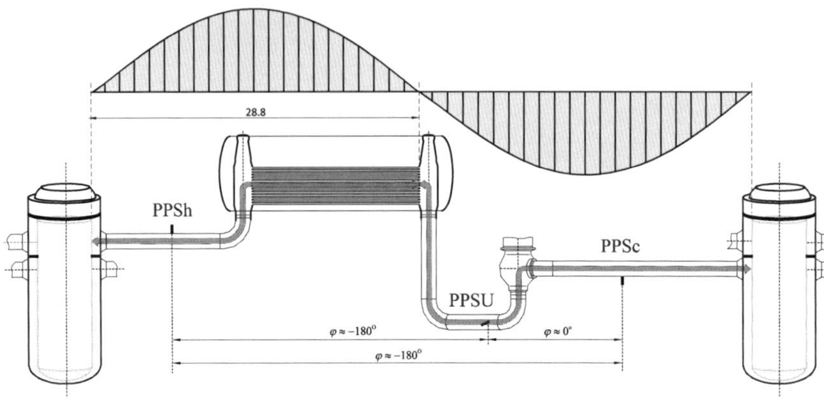

Figure 11.7: Distribution of ASW_{2loop} amplitudes on the scan of the VVER-1200 MCC heat exchange loop with the phase values of the pairs of signals from PPS_h, PPS_U, PPS_c.

the vibration measurements [384], which cause increased operational loads on the welded joints of the SG header.

According to the known distribution of ASW_{1loop} amplitudes, it is easy to solve the distribution of amplitudes of its harmonics (Fig. 11.7). We shall name the second harmonic of ASW_{1loop} through ASW_{2loop} or, in other words, the second, loop-ASW.

Its node is exactly on the "cold" SG header, thereby increasing its vibration loading. The second harmonic is small in amplitude compared to the amplitude of the first, loop-ASW, so we can hardly expect any vibration anomalies at this frequency.

The third harmonic of ASW_{1loop} is not observed in the experiment, and this was predicted by a simple model of alternating inhomogeneities, when their number is equal to the power of two.

It is also easy to build the fourth harmonic of ASW_{1loop}. We will express it through ASW_{4loop} and call it the fourth, loop-ASW (Fig. 11.8). Its frequency in case of the Minimum Controllable Power Level is in the range of (30–33) Hz.

Its node is located near the "hot" SG header and the amplitude of the fourth harmonic is commensurate with the ASW_{1loop} amplitude. Another important ASW_{4loop} feature requiring monitoring is its closeness to the second MCP set rotational frequency (33.2 Hz). Changing the speed of sound when the power unit is heating up / cooling down leads to the coincidence of these frequencies. Actually,

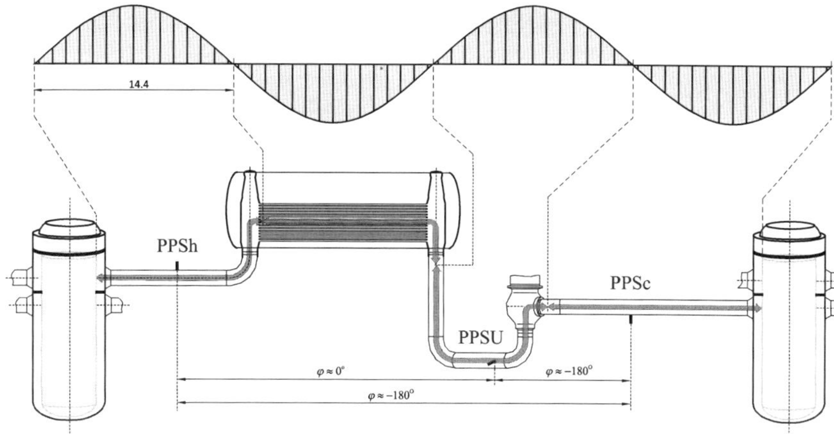

Figure 11.8: Distribution of ASW_{4loop} amplitudes on the scan of the VVER-1200 MCC main coolant loop with the phase values of the pairs of signals from PPS_h, PPS_U, PPS_c.

this event always occurs at some intermediate coolant temperature during the power unit heating up and cooling down. This coincidence is fraught with amplification of oscillations on the second harmonic of the MCPs rotational frequency and, as a consequence, with a possible water hammer. However, a favorable effect of reducing the forced oscillations at the MCPs frequency and its harmonics is also possible. Exactly on these predicted coincidences of ASW harmonic frequencies with the harmonics of the MCPs rotational frequency, we have to carry out the evaluation of the dissipative properties of the reactor plant design. At this coolant temperature, the calculation code for ASW fields can be verified.

Simple local ASWs covering with their half-waves the hot and cold MCC pipelines (Fig. 11.9) are also obvious. ASW_{1h} is a simple "hot" ASW. ASW_{1h} has one node on the wall of the reactor core barrel in the "hot" nozzle of the reactor vessel and another one on the lid of "hot" SG header. The MAT is 15.8 m (slightly longer than the ASW_{4loop} half-wave). Its frequency in case of the Minimum Controllable Power Level is in the range of 29–32 Hz. Thus, a coincidence of ASW1hot and ASW4loop with the strengthening/weakening of the resultant ASW amplitude in the hot pipeline are theoretically possible.

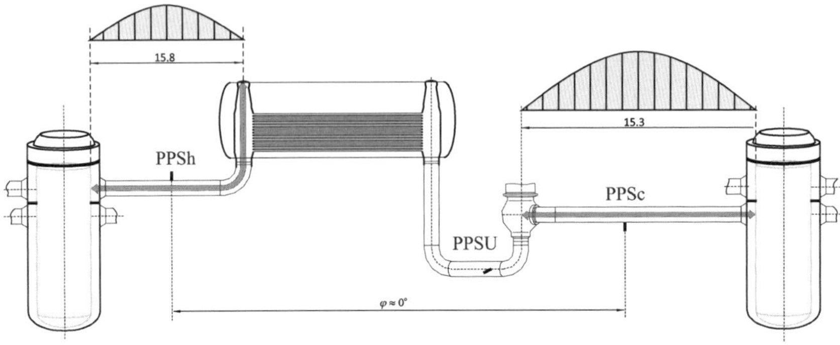

Figure 11.9: Distribution of simple ASW_h and ASW_c amplitudes.

ASW_{1cold} is a simple ≪cold≫ ASW (Fig. 11.9), similar to ASW_{1h}. ASW_{1c} has one node in the MCPs center and another one on the wall of the reactor core barrel on the cold nozzle side. Its MAT is 15.3 m long and its frequency is in the range of 30–33 Hz. That is, once again, there is a probability of coincidence of ASW_{1cold} and ASW_{4loop} frequencies.

Given that the acoustic inhomogeneities of the loop are three-dimensional, it is possible that simple ASW_{1hot} and ASW_{1cold} are integrated due to interference in ASW_{4loop}. As a result, the respective half-waves can both increase the amplitude or self-compensate to zero amplitude. With a different number of MCPs in operation, this effect may increase or decrease.

Figure 11.10 shows all the ASW presented in this section on the length of the "stretched" loop.

11.4 Experimental Loop Components of Coolant Pressure Pulsations in the VVER Main Coolant Loop

In the standard scheme of the detection equipment of the commissioning measurement system and vibration noise diagnostic system, three pressure pulsation sensors, mounted on "hot" pipeline — PPS_h, a U-shaped pipeline — PPS_U and "cold" pipeline — PPS_c, play the role of input and output sensors for SG and MCPs (Fig. 11.11).

Upon a PPS_h signal as input perturbation for an SG, as well as upon a PPS_U signal as the resulting SG response, cross-spectral

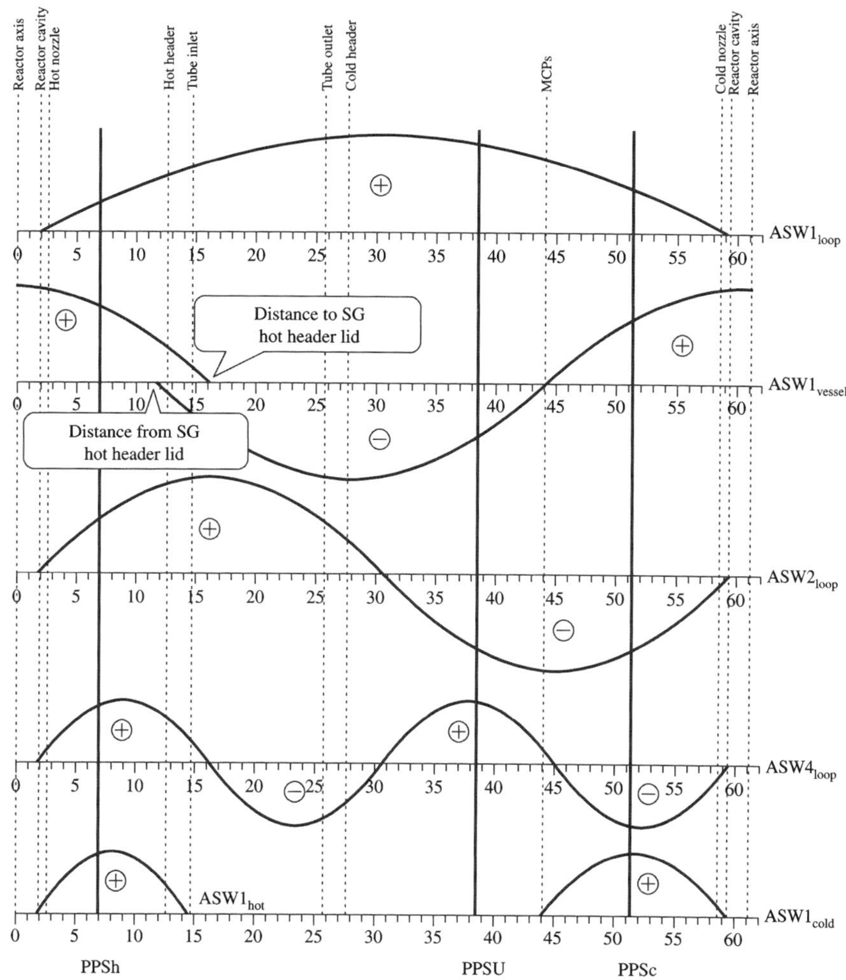

Figure 11.10: Amplitude distributions of global ASW in the VVER-1200 heat exchange loop.

functions that characterize the SG as acoustic inhomogeneity can be estimated. These frequency characteristics are the coherence and phase of the cross-spectral density (CPSD) or a complex transfer function.

If a trivial equivalent circuit replacing the SG with induction in accordance with the electrical analogy were valid (Fig. 11.12), that is the SG complex resistance would be equivalent to the expression

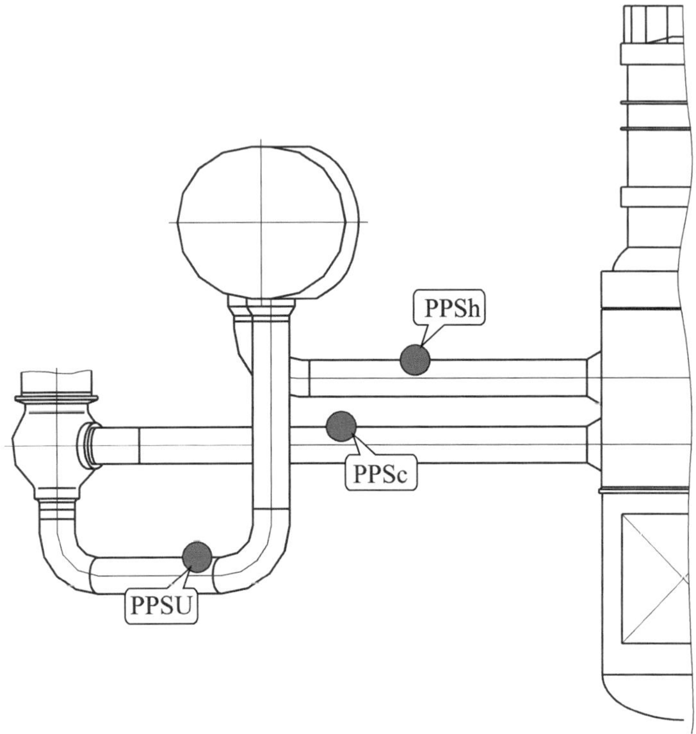

Figure 11.11: Schematic diagram of the heat-exchange loop equipment (SG and MCPs) and the location of the three PPS.

$Z_{\mathrm{SG}} = j \cdot \omega L = e^{j\frac{\pi}{2}} \cdot \omega L$, the phase of the transfer function $\varphi_{\mathrm{SG}}(\omega)$ would not depend on the frequency and everywhere would be equal to 90°, and the coherence would be equal identically to unity.

If an equivalent circuit replacing the SG with a straight pipe with a length of ≪l≫ and section "S", loaded on a complex resistance Z_0, is valid, then the phase of the transfer function would vary in space in accordance with the formula

$$Z_{\mathrm{SG}}(x) = \frac{\rho \cdot C}{S} \cdot \frac{Z_0 \cdot ch\,\gamma \cdot (l-x) + sh\,\gamma \cdot (l-x)}{Z_0 \cdot sh\,\gamma \cdot (l-x) + ch\,\gamma \cdot (l-x)}$$

And with $x = l$ would coincide with the phase characteristic of the load

$$Z_{\mathrm{SG}}(x) = \frac{\rho \cdot C}{S} \cdot Z_0; \quad \varphi_{\mathrm{SG}} = \varphi_{Z_0}(\omega)$$

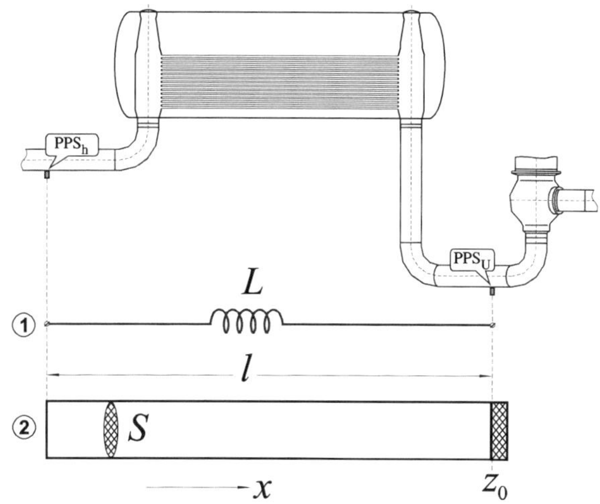

$$X_{SG} = j\omega L \quad \varphi_{SG}^{(\omega)} \equiv 90^0$$

$$Z_{SG}(x) = \frac{\rho \cdot C}{S} \cdot \frac{Z_0 \cdot ch\,\gamma \cdot (l-x) + sh\,\gamma \cdot (l-x)}{Z_0 \cdot sh\,\gamma \cdot (l-x) + ch\,\gamma \cdot (l-x)}$$

$$Z_{SG}(x) = \frac{\rho \cdot C}{S} \cdot Z_0 \qquad \varphi_{SG} = \varphi_{Z_0}(\omega)$$

③ $$\begin{bmatrix} \delta P_{input} \\ \delta V_{input} \end{bmatrix} = \begin{bmatrix} A & B \\ C & D \end{bmatrix} \begin{bmatrix} \delta P_{output} \\ \delta P_{output} \end{bmatrix} \qquad \varphi_{SG} = \varphi(A, B, C, D, \omega)$$

Figure 11.12: Conventional representation of SG with the input and output PPS and its equivalent circuit: 1 — based on a trivial electrical analogy, 2 — replaced by an equivalent pipeline with a load, 3 — replaced by an equivalent quadrupole.

In such a model, the SG acoustic features should be incorporated into the equivalent complex load resistance Z_0, as a certain point inhomogeneity. It represents an abrupt change in the acoustic path properties, while the equivalent pipeline exhibits a continuous, non-abrupt change in the acoustic resistance, most often by sinusoidal law. The equivalent pipeline in combination with the equivalent load can set, generally speaking, any degree of complexity for a phase dependence on frequency.

If based on the electrical analogy with a long line the SG is represented as a quadrupole with a spatial transfer matrix $\left\| \begin{matrix} AB \\ CD \end{matrix} \right\|$, the phase characteristic will depend on complex spatial functions "A, B, C, D", which, as the coefficients for frequency, are, in turn, arguments

0.67906	-03922	-4.1319	-1.735
7.7284	-6.6895	-4.5461	-42504
83104	-42502	-3.4909	-4.4074
11.479	-133.65	-167.87	17526
12.158	164.08	160.89	156.91
15.812	–172.61	-177.57	178.49
16.556	179.44	-174.08	–167.86
30.881	17021	175.62	175.9
32.757	173.74	169.75	15512

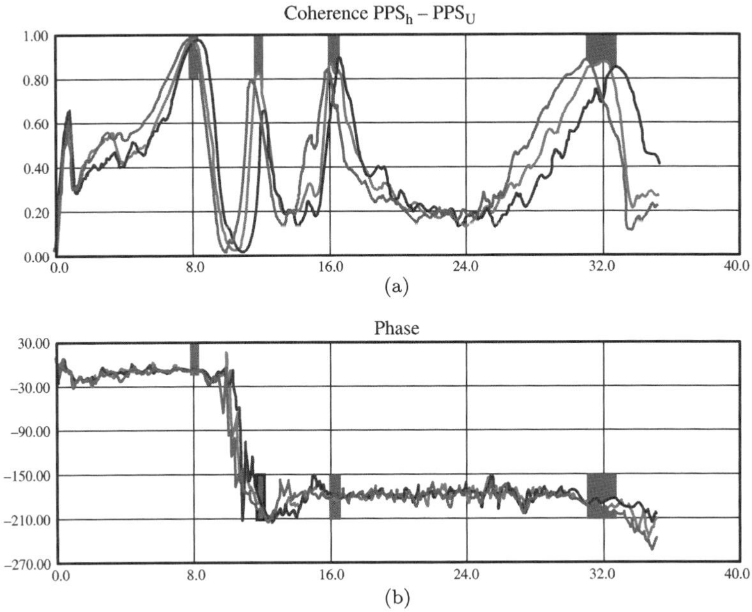

Figure 11.13: Coherence Functions of PPS Signals at the SG Inlet and Outlet: PPS_h-PPS_U at three different coolant temperatures with a phase table at characteristic frequencies (a) and a family of corresponding phase characteristics (b).

of hyperbolic functions. In this model, the phase characteristic can have any degree of complexity for the frequency function.

In practice [380], the experimental phase properties $\varphi_{SG}(\omega)$ only demonstrate characteristic phase values, namely, 0° or 180° on the resonance features (Fig. 11.13). The coherence functions of PPS_h-PPS_U signals registered at four coolant temperatures (204°C, 215°C, 235°C, 250°C) have five highly coherent resonances in the frequency subranges:

- (0.6–0.7) Hz — the pressurizer resonance with a zero phase;
- (7.7–8.3) Hz — the ASW_{1loop} resonance with a zero phase;
- (11.5–12.2) Hz — the $ASW_{1vessel}$ resonance with an antiphase;

- (15.8–16.6) Hz — the ASW_{2loop} resonance with an antiphase;
- (30.9–32.8) Hz — the ASW_{4loop} resonance with an antiphase;

When the coolant temperature changes causing the frequency drift of these resonances, the above-mentioned phase characteristics remain unchanged.

ASW_{2loop} central frequency at a coolant temperature of 204°C coincided with the current MCPs frequency of 16.6 Hz. Such coincidence of frequencies of two physically different sources of driving forces is one of the necessary conditions of resonance excitation of oscillations. It is possible that for this reason the amplitude of pressure pulsations at the rotational frequency has increased (Fig. 11.14) at a coolant temperature of 204°C. However, there is nothing abnormal in this, since it does not exceed the amplitude of the pressurizer resonance and the ASW_{1loop} resonance amplitude. In this case, the phase characteristic does not manifest significant deviations from the anti-phase which indicates the insignificance of the contribution of the MCPs forced oscillations on the rotational frequency into the coolant pressure pulsations compared to those at the same central frequency from ASW_{2loop}.

On the MCPs rotational frequency, impact on the SG is minor even though there has been a coincidence of ASW_{2loop} frequency and the MCPs rotational frequency. This means that the designed damping mechanisms for SG oscillations do not allow developing resonantly excited oscillations to any significant amplitude. We observe the same five resonant features in APSD of PPS_h and PPS_U (Fig. 11.14).

They are similar to each other, having no individual resonance features. That is, the PPS signals exhibit prevailing effects with a length commensurate with the length of the loop, and the other local ones inherent only to the PPS signal in the given point of measurement are much smaller in power. Thus, the SG input and output PPS register mainly acoustic effects, the length of which is much larger than the SG overall dimensions and the amplitude of the oscillations is dominated by the pressurizer resonance at the SG inlet, but at the SG outlet the biggest is the ASW_{1loop} resonance. This ratio of the amplitudes of the pressurizer and ASW_{1loop} resonances is in full agreement with their spatial distribution along the MCC for the case of an ideal pipeline, when no inhomogeneities in their path occur.

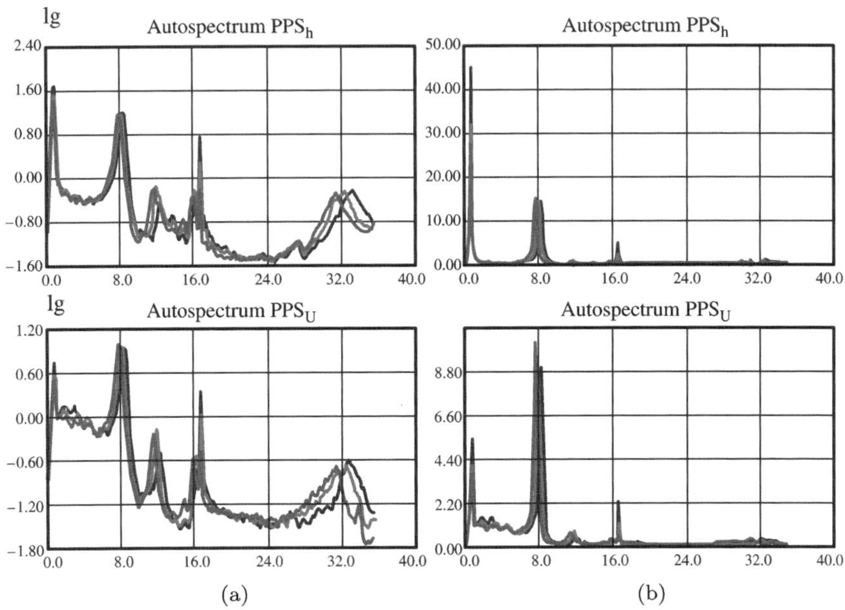

Figure 11.14: APSD of two PPS signals in logarithmic scale (a) and in linear scale (b).

It turns out that the SG is only an active acoustic resistance for ASW that changes the amplitudes of the input spectral components of pressure pulsations differently, without affecting the phase characteristics of the "SG input–output" type.

The conclusion about the loop length of the coolant pressure pulsations on the SG does not mean that the SG as an acoustic inhomogeneity is not a generator of local pressure pulsations. As a passive acoustic element, first of all the SG can amplify or weaken the above-mentioned ASW acting as a resonator. Secondly, the SG generates small-scale turbulent coolant pressure pulsations called traveling waves with a spatial length less than the distance between the two PPS: PPS_h and PPS_U. Overcoming this distance, small-scale traveling waves are damped, without making any contribution to the coherence function of the pair of PPS_h–PPS_U signals. To study the local effects of pressure pulsations caused by the SG, it is necessary to increase the PPS number in the heat exchange loop. Such measurements should be carried out at the stage of commissioning measurements of a new VVER power unit.

In a simplified form, the SG can be represented as a combination of three acoustic elements: two headers and the SG tubes (HET) or in the form of two abrupt acoustic inhomogeneities:

- the "hot" SG header — the SG tubes,
- the SG tubes — "cold" SG header.

These inhomogeneities act as reflecting and simultaneously permeable obstacles for longitudinal acoustic waves, capable of generating their simple ASW, which, by definition, cannot be global. Such a simple ASW is the half-wave which is located between:

- the "hot" nozzle of the reactor vessel (core barrel) and the inhomogeneity of the "hot" SG header — the SG tubes;
- the SG headers occupying the SG tube volume;
- the "cold" nozzle of the reactor vessel (core barrel) and the inhomogeneity of the "cold" SG header — the SG tubes.

The first of the above simple ASW is localized in the "hot" pipeline and is named as ASW_h, the second ASW with a half-wave coinciding with the HET average length of 11.1 m will be called ASW_{hot}, the third ASW is localized in the "cold" pipeline and is named as ASW_c. Simple ASW_h with its half-wave on the "hot" pipeline, practically coincides in frequency with the ASW_{4loop} central frequency, which leads to a change in the amplitude of its spectral component in the frequency range (30–33 Hz) in the PPS_h signal.

For MCPs, the PPS_U signal can be considered as an input and the PPS_c signal — as an output. Their coherence is largely similar to the above-mentioned coherence between SG signals: PPS_h–PPS_U (Fig. 11.15).

We observe the same five resonance features. The difference is that $ASW_{1vessel}$ (11.5–12.2 Hz) has a very small amplitude both in the coherence function between PPS_U–PPS_c and in APSD of PPS_h signal. This is a consequence of the $ASW_{1vessel}$ amplitude distribution which has a node close to MCPs.

Another characteristic feature of the PPS_U–PPS_c coherence function (Fig. 11.16) is the presence of significant coherences in the low-frequency range between the pressurizer resonance (0.6–0.7 Hz) and the ASW_{1loop} resonance (7.7–8.3 Hz). Pressure pulsations between the MCPs inlet and outlet as well as for the ASW take characteristic antiphase values. In this frequency sub-range with a central frequency

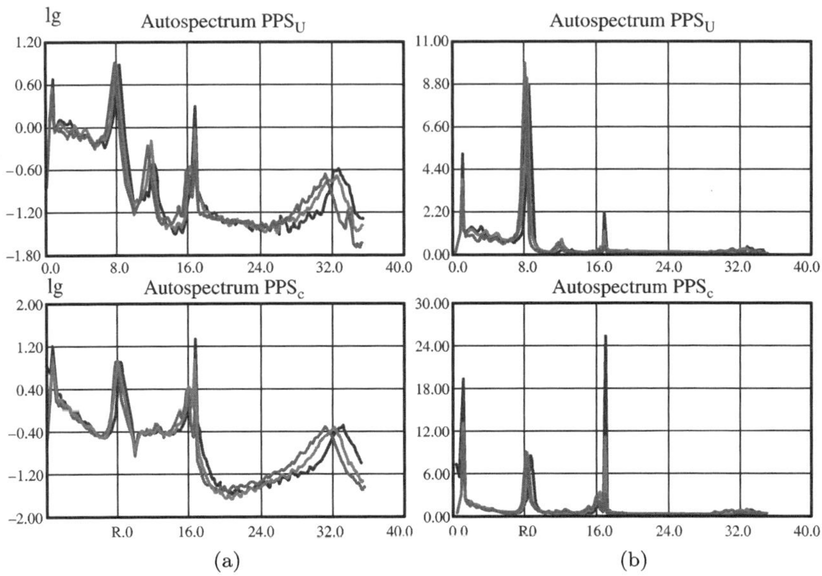

Figure 11.15: APSD of two PPS signals in logarithmic scale (a) and linear scale (b).

of about 4.0 Hz and a width of about 7 Hz, there can be no ASW with a corresponding half-wave length exceeding the MCC dimensions.

Most likely, such pressure pulsations correspond to long-wave (with large *wavelength*) turbulent traveling waves. APSD of PPS_U and PPS_c signals have no resonant features in this frequency subrange, which also indicates a broadband, and hence, turbulent origin of these pressure pulsations. Their antiphase nature between the MCPs inlet and outlet corresponds to depression of an elastic medium in the MCPs suction pipe and simultaneous compression of the elastic medium on the MCPs head. The considered low-frequency effect is associated with the MCPs operation and corresponds to its intended purpose: generation pressure for the coolant pressure drop. It is usually interpreted at the level of the constant components of the coolant pressure before and after the MCPs, that is, at the level of the constant pressure drop at the MCPs. Turbulent pressure pulsations of low frequency of around 0 Hz are a kind of continuation of the constant pressure drop on the MCPs and thus characterize its performance.

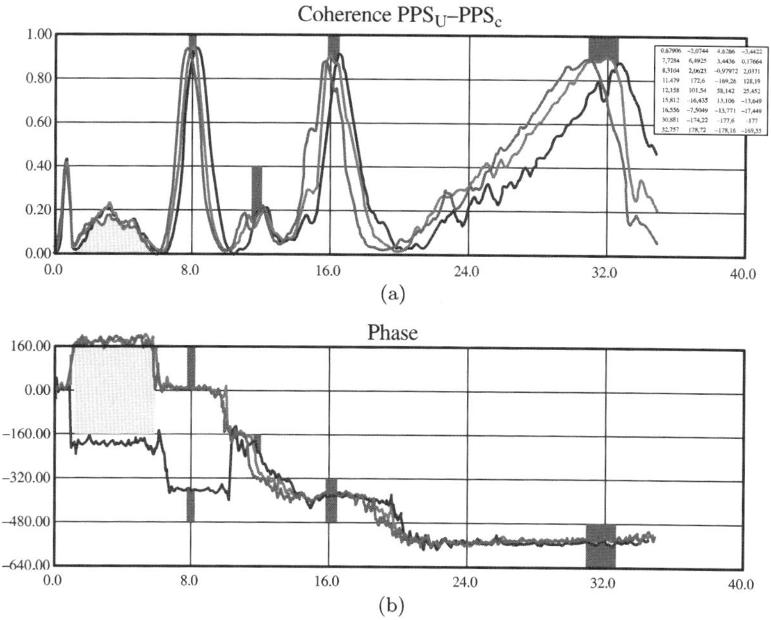

Figure 11.16: Coherence functions of PPS signals at the SG inlet and outlet: PPS$_U$–PPS$_c$ at three different coolant temperatures with a phase table at characteristic frequencies (a) and a family of corresponding phase characteristics (b).

11.5 Globality of Low ASW Types

Correlation of PPS signals of different loops allowed to estimate the distribution of the ASW$_{1\text{vessel}}$ amplitudes and define it as a global ASW. It has proved to the global, like the other two, at least, low ASW types, namely, ASW$_{1\text{loop}}$ and ASW$_{2\text{loop}}$. Let us remind you that the necessary condition for globality is the presence of ASW resonance in more than one loop. At the frequency of this resonance, the coherence of the PPS signals of different loops should significantly differ from zero, and the phase should be either 0° or 180° (deviation from these characteristic phase values is a rare exception). So, Figs. 11.17 and 11.18 present the families of the coherence function of different loop PPS signals having the same name: for adjacent loops — PPS1h and PPS2h, and the opposite loops — PPS1h and PPS3h — with the phase characteristics.

These properties of mutual characteristics of PPS signals at ASW frequencies mean that the pressure pulsations in different loops are

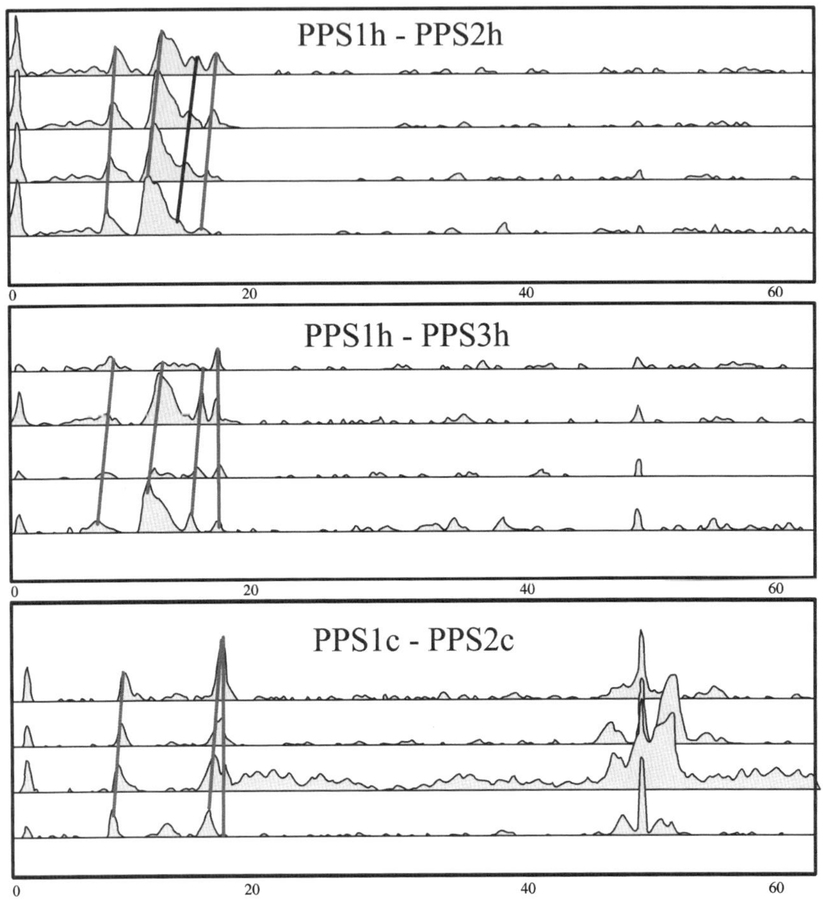

Figure 11.17: Families of coherence functions of PPS signals of different loops but same name at four coolant temperature values.

synchronized with each other and represent a single global acoustic field of the MCC. It can force synchronous oscillations of the main equipment of different loops.

Different MCC loops are interconnected through the volume of the reactor vessel. There are two main ways of transitioning of pressure pulsations from one loop to another (Fig. 11.19):

- flowing around the reactor core barrel on the annular region "reactor vessel–reactor core barrel" on the left and right;
- penetrating through the reactor core barrel holes and PTU.

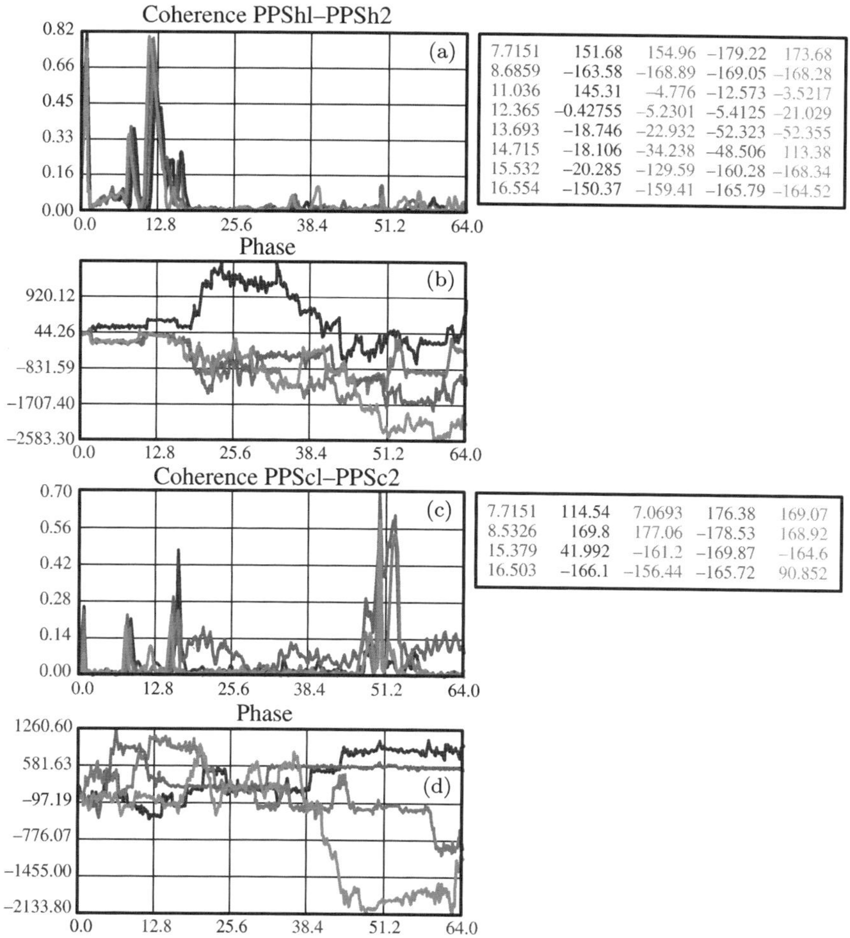

Figure 11.18: Coherence functions and phases for PPS1h–PPS2h (a), and (b), and for PPS1c–PPS2c (c), and (d).

Figure 11.19 in a simplified way shows the paths of pressure pulsations' propagation from one loop into another. The same pattern exists for the other pair of opposite loops. Adjacent loops can have similar paths. The above paths can have transversal obstacles in the form of internal and external cylindrical surfaces of reactor core barrel and PTU, and PTU tubes. The given set of paths contains the shortest, straight path from one loop to the opposite one.

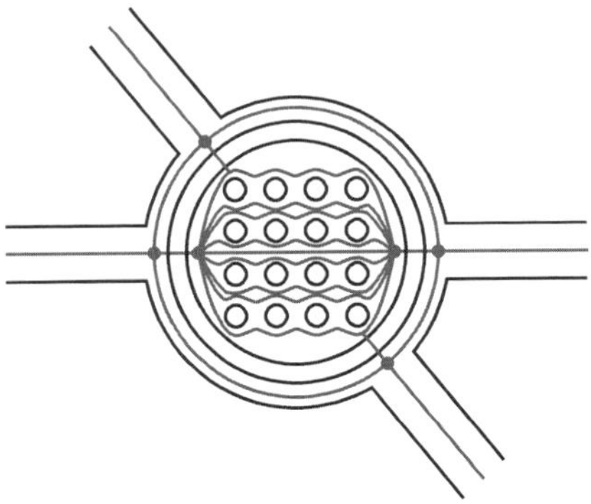

Figure 11.19: Schematic diagram of pressure pulsations paths from one loop into another.

The presence of a continuum set of paths predetermines the non-zero Q-factor of the $\text{ASW}_{1\text{vessel}}$ resonance.

In addition, it is possible to have several adjacent resonances of the same ASW type, each of which corresponds to a dominant acoustic path. For example, the first path is flowing around the reactor cavity through the annular region and the second dominant acoustic path is straight propagation from one loop to the opposite loop along the diameter of the reactor vessel, and both should generate two adjacent $\text{ASW}_{1\text{vessel}}$ resonances. Their difference in frequency is determined by the difference between the length of the reactor core barrel semicircle and its diameter. This is the so-called effect of splitting the frequency of the resonance of one type of ASW. Along the straight path, the coolant has a temperature higher than along the path around the cavity. On the first path, there may be the coolant steam phase, which shifts the $\text{ASW}_{1\text{vessel}}$ resonance frequency toward 0 Hz.

Obviously, a very high spectral resolution is required to resolve two such resonances in frequency. It can also be observed for other ASW. For example, $\text{ASW}_{1\text{loop}}$ can have frequency resonance splitting due to reflections from the external and internal surfaces of the reactor core barrel (Fig. 11.20).

The existence of several acoustic paths of different lengths leads to wave interference or interaction of waves of the same frequency (the interaction of coherent waves). Let us illustrate the interference effect by a simple example in the operation of one MCP, which obviously fulfills the condition for coherence.

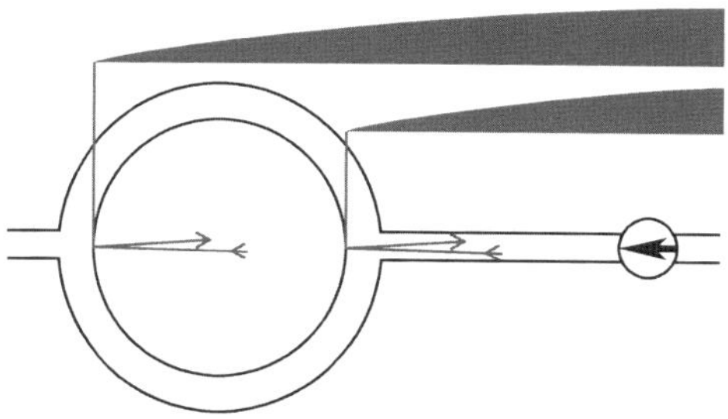

Figure 11.20: Reflections from the external and internal surfaces of the reactor core.

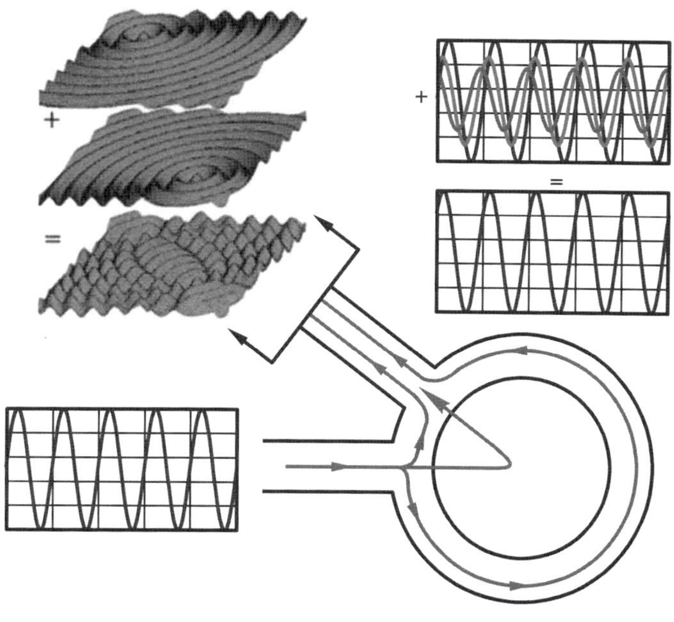

Figure 11.21: Conventional representations:

- single incident traveling wave (lower sine wave);
- different paths of its transition to the adjacent loop (lower right corner);
- interference pattern of alternating pressure extremes in the cross-section of the pipeline (upper left corner);
- the sum of sinusoids of the same frequency but different amplitudes and phases (top right corner).

The traveling wave of the coolant pressure coming from the active loop has several paths along which it enters the adjacent passive loop (Fig. 11.21).

The result is a classical pattern of interference obtained when a passive loop includes traveling waves of the same frequency, but with different phases and amplitudes. Their superposition in any cross-section of the pipeline gives an interference pattern of alternation of local maxima and minima of ASW amplitude (Fig. 11.22). The difference in the lengths of acoustic paths within one ASW type compared to the spatial wavelength is negligible, so the difference in the amplitude of local extremes from each other in the pipeline diameter will be small. The difference of phases and amplitudes of several waves of the same frequency in the longitudinal direction will eventually lead again to sinusoidal oscillations of the same frequency, but with some phase shift in the spatial distribution of the ASW amplitude along the length of the pipeline.

In mathematical terms, the sum of an arbitrary number of sinusoids of one frequency, but with arbitrary amplitudes and phases, will again give a sinusoid of the same frequency with some amplitude and some phase:

$$\sum_i A_i \cdot \sin(\omega t + \varphi_i) = A \cdot \sin(\omega t + \varphi)$$

This trivial expression is illustrated by a numerical example in Fig. 11.22.

As previously stated, the phase relationship of two PPS signals mounted in various MCC locations can only have co-phased (0°) or anti-phase (180°) values at the ASW frequency. This is true if the ASW is composed of a single sinusoid. In the case of interaction of several waves of the same frequency, arbitrary ASW phases are possible. Significant differences of the phase value from the characteristic values (0° and 180°) indicate that the given ASW was formed under the influence of several acoustic waves of the same frequency.

For the low $\text{ASW}_{\text{1loop}}$ type, the external surface of the reactor core barrel is a wave-generating obstacle. Reflection of a traveling wave of pressure from the reactor core barrel close to "hot" nozzle of the reactor vessel and reflection from core barrel close to the "cold" nozzle generate, in fact, the standing $\text{ASW}_{\text{1loop}}$. The volume of the "hot" nozzle is modeled by a flat horizontal tee. All "hot" MCC pipelines are connected to each other by four of such tees. The pressure wave falling on the surface of the reactor core barrel from the "hot" pipeline undergoes a reflection in the tee that forms $\text{ASW}_{\text{1loop}}$

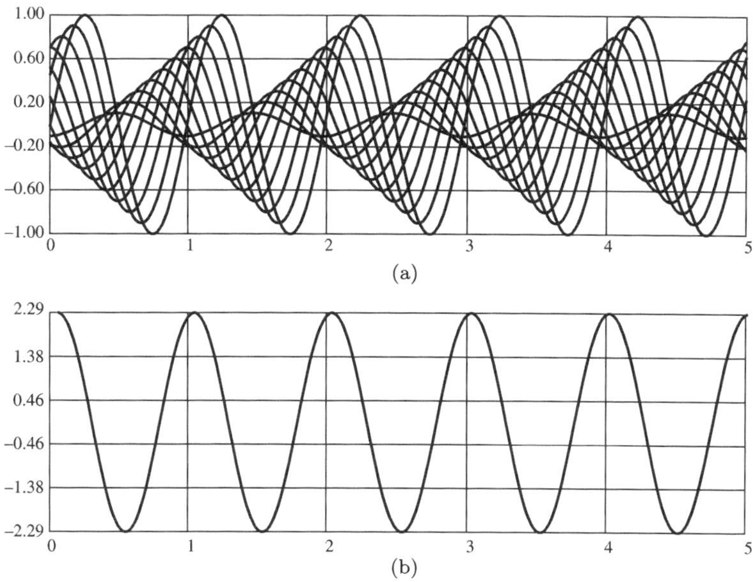

Figure 11.22: Model experiment: the resulting sinusoid (b) is the sum of 10 sinusoids of the same frequency (a), but monotonically decreasing amplitudes and different phases. $x(t) = \sum_{k=0}^{9} (1 - 0,1k) \cdot \sin[2\pi ft + \frac{\pi}{3}k] = A \cdot \sin(2\pi ft + \varphi)$.

and branches in two opposite horizontal directions. This branching arranges communication of ASW_{1loop} with the loops. Note that in elastic medium the pressure wave propagate both in the direction of the flow and against the flow into the direction of the current and the opposite direction as well. In the hot nozzles of the reactor vessel, propagation of pressure pulsations into adjacent loops occurs against the coolant flow.

The "cold" nozzle can no longer be represented by a simple horizontal tee. It turns the coolant flow downward to annular region of "reactor core barrel – reactor vessel". Incident pressure wave from the "cold" pipeline will branch into three directions: in two opposite horizontal directions (from one loop to another) and vertically down to the third direction. If in each direction propagation of acoustic waves is affected by some equivalent acoustic resistance, it turns out that the interaction between the "cold" pipelines is shunted by a small value of the third resistance. As a result, the ASW interaction via "cold" pipelines will be much weaker than similar ASW interactions via "hot" pipelines. This explains the high coherence between

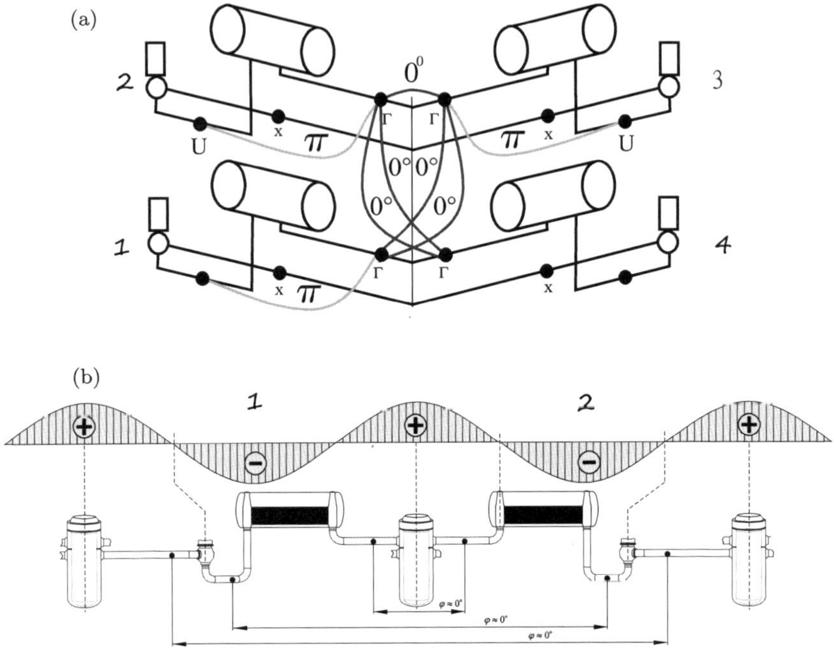

Figure 11.23: $ASW_{1vessel}$ — experimental phase ratios (a) and the distribution of the amplitudes by the loops (b).

the PPS_h signals of different loops and the relatively low coherence between PPS_c signals of different loops.

For the vessel type $ASW_{1vessel}$, the reactor core barrel is a kind of volumetric antinode "blurring" $ASW_{1vessel}$ resonance. Incident traveling pressure waves from the "hot" pipelines against the coolant flow decrease in amplitude, overcoming the holes (perforation) of the reactor core barrel and PTU, but the final $ASW_{1vessel}$ antinode on the vertical symmetry axis of the reactor plant increases as a result of summing of four quarter-waves from four loops.

Near the "cold" nozzles there should also be observed $ASW_{1vessel}$ antinode, as its nearest nodes are on the opposite MCPs (MCP1–MCP3 and MCP2–MCP4), but it will be a lot smaller in amplitude than the $ASW_{1vessel}$ antinode on "hot" piping due to the shunting effects of the descending annular region "reactor core barrel – reactor vessel". Figure 11.23 schematically shows the distribution of the amplitude of pressure pulsations for $ASW_{1vessel}$.

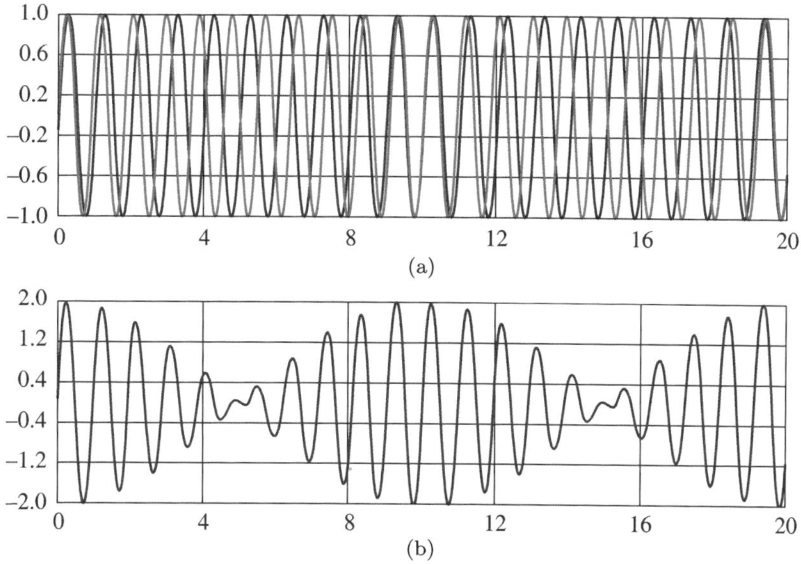

Figure 11.24: Beating (b) arising as a sum of two sinusoids with close frequencies $x(t) = \sin[2\pi ft] + \sin[2\pi(1.1) \cdot ft]$.

Thus, there is a strong interaction of ASW of the same type (ASW_{1loop}, $ASW_{1vessel}$, ASW_{2loop}) between "hot" pipelines and weak interactions between "cold" pipelines, despite the fact that in the latter case, pressure pulsations propagate down the coolant flow.

Interference of waves of the same frequency in loops coexist with beating effect caused by interaction of waves with different but close frequencies. The above-described effect of splitting the resonance of one type of ASW means the coexistence of ASW with very close central frequencies. Due to their individual characteristics, each of the heat exchange loops generates half-wave of ASW_{1loop} in frequency slightly different from ASW_{1loop} of another loop. Their superposition leads to the so-called beating effect. It is illustrated by a model example in Fig. 11.24.

The sum of the sinusoids is represented as a product

$$\begin{aligned} x(t) &= \sin[2\pi ft] + \sin[2\pi(1.1) \cdot ft] \\ &= 2[\cos(2\pi \cdot 0.05 \cdot ft)] \cdot \sin[2\pi(1.05) \cdot ft], \end{aligned}$$

one of the factors of which is a slowly changing function of time. It is often erroneously referred to as a modulating factor, bearing in mind that there is a low frequency component of the signal. The sum of two sinusoids results in the final spectral composition represented by two spectral lines at frequencies of one and the other sinusoid, and no more. There will be no third resonance in the spectrum at a low beating difference frequency. Lower graph in Fig. 11.24 creates the illusion of beat frequency. It can only occur when there is a nonlinear item in the loop, for example, a quadratic element where new frequencies arise.

The PPS signals installed in different loops demonstrate beat effect (Fig. 11.25).

Their envelopes can very much differ from one another. As shown in the previous sections, beats may have several physical causes. One of the beats generating mechanisms is interaction of ASW of different loops. It is easy to extract the beats frequency equal to the half-difference of the frequency of the same type ASW. After obtaining the ASW_{1loop} frequencies for each loop, we can propose a hypothesis about the origin of a specific, observed beat effect.

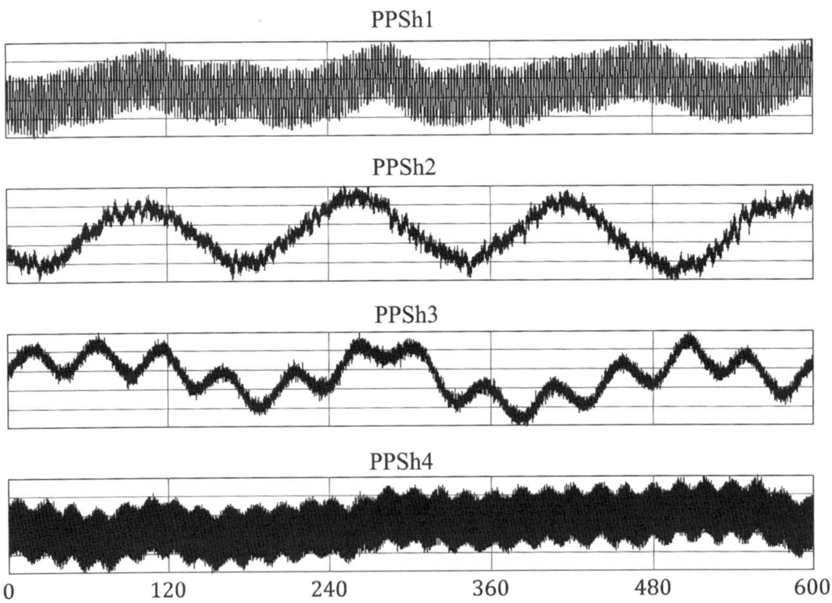

Figure 11.25: Same PPS signals PPShi ($i = 1, 2, 3, 4$) in four loops.

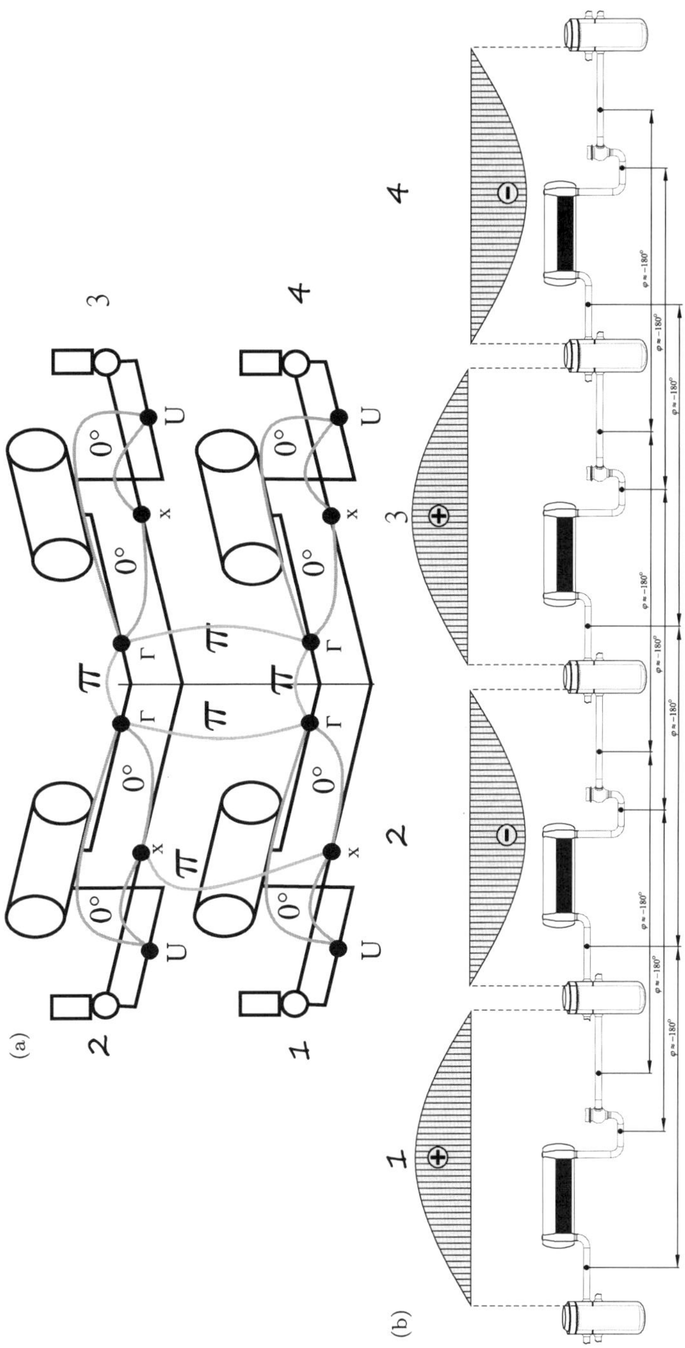

Figure 11.26: ASW_{1loop} — experimental phase ratios (a) and the distribution of the amplitudes by the loops (b).

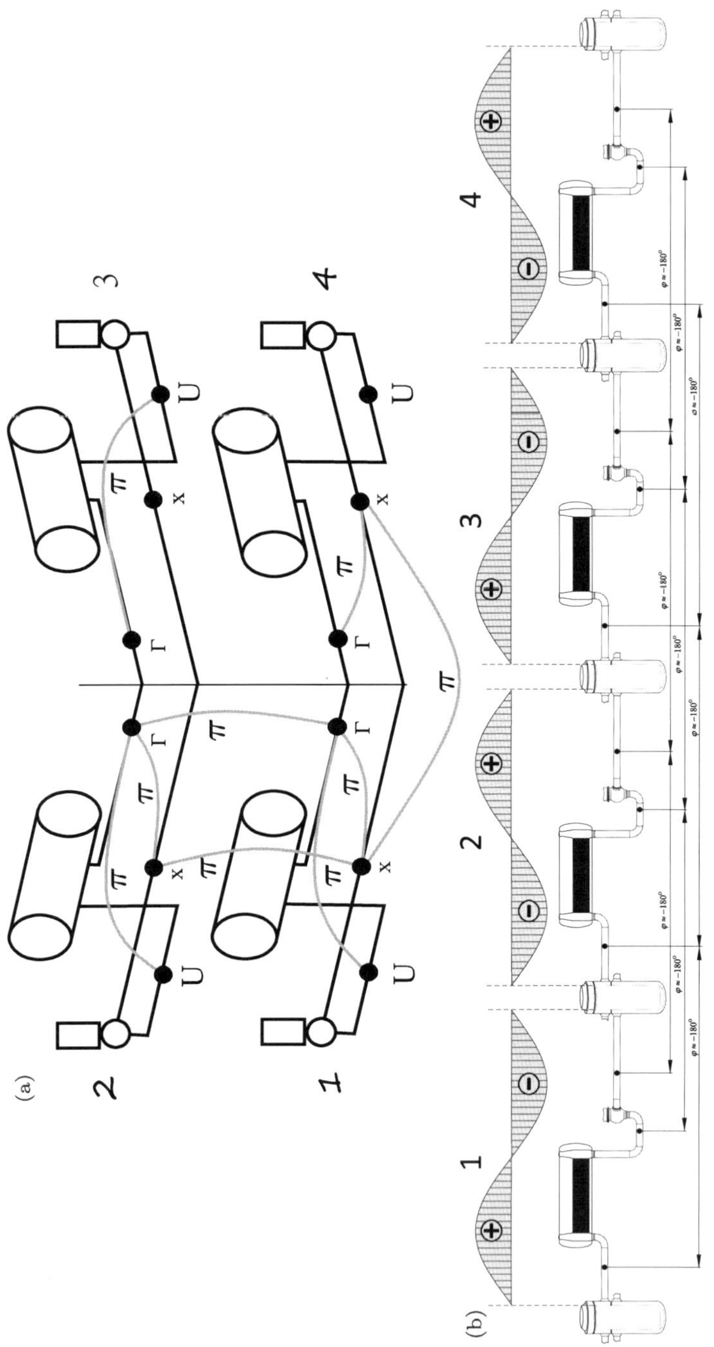

Figure 11.27: ASW_{2loop} — experimental phase ratios (a) and the distribution of the amplitudes by the loops (b).

ASW_{1loop} is a global phenomenon (common for all the loops), there are four half-waves, interconnected through a common spatial (volumetric) node, which is the reactor core barrel cylindrical surface. That is, the MCC is a four-section resonator (by the number of loops). At the same time, the half-wave length of ASW_{1loop} in the different MCC loops can be slightly different. Thus, the following experimental result is steadily observed in VVER-1200: the ASW_{1loop} central frequencies in different loops are slightly different from each other. Figure 11.26 shows in a simplified form the ratio of phase characteristics for different PPS pairs for ASW_{1loop}, and Fig. 11.27 — for ASW_{2loop} and approximate distribution of its amplitude along MCC.

It is statistically significant for signals of PPS installed in "hot" pipelines, the central ASW frequencies are ranked in loops by the following inequalities: $F4 < F2 < F1 < F3$. ASW_{1loop} frequency ranges for the loops overlap each other and all have one common part, which makes ASW_{1loop} a common loop phenomenon. The smallest central ASW_{1loop} frequency, and hence the largest spatial wavelength, is in the fourth loop. It is well known that, there is a nozzle of the pressurizer connecting pipeline. This tee is equivalent to "lengthening" of the acoustic path of the traveling wave. It is also possible that the true loop dimensions slightly vary from each other in length within the mounting tolerances. It is also possible that additional circulation loops having individual characteristics for each loop are equal to different increments of the acoustic path.

Chapter 12

ASW Phenomenology

12.1 Identification of Resonance Spectral Features of Vibration Sensors and PPSs During Unit Heating Up and Archiving Full Power Operation

In vibration diagnostics systems of the first generation, vibration monitoring and thermal displacement monitoring were implemented as two subsequent tasks divided in time. The vibration condition of the power unit in such dynamic states as heating up or cooling down, and also during power increase/decrease, changes very much, that is why during these periods only thermal displacement of the equipment was monitored using RDS signals. As soon as the power unit reaches 100% power, thermal displacements become negligibly small, and vibration images become stationary. This is when the same RDS instrument channels may be used for vibration condition monitoring (however, this time the amplification factors of the channels shall be increased). The practice of application of diagnostic systems in normal operational conditions has shown that micro-displacements in stationary modes shall also be monitored, and vibration shall be monitored in the dynamic regimes of RP operation [112, 119, 134, 171, 220–225, 259, 266, 270, 271, 278–280, 284, 363, 366, 368, 377–384].

In what follows, we intend to develop the non-stationary vibration analysis, which has been missing in VNDS of the first generation. Nowadays, such an analysis is the mandatory part of commissioning testing for the subsystems subject to vibrations. However, commissioning testing has nothing to do with research — this is just a series of measurements according to the approved program, which serve to prove that the vibrations are within the design-basis range.

Nevertheless, vibration diagnostic indications in dynamic modes are quite valuable. If abnormal vibration is detected at an early stage right after the power unit begins to heat up after the scheduled outage, it may prevent its further development. As you will see in the following, the scope of diagnostics data in the RP dynamic regimes is much bigger than in the stationary modes. In particular, temperature dependences of vibration resonance spectra parameters as diagnostic feature are available in dynamic modes only.

The number of MCP-sets in operation — n, coolant temperature — Tc, and, starting from a certain Tc, RP power — N — are the three critical parameters, which define the vibration state of the RP in different nominal regimes of the power unit operation. In a three-dimensional space of the parameters (n, Tc, N), we may identify the curves corresponding to the unit heating up, and a point (or, strictly speaking, a volume) of these coordinates-parameters corresponding to the nominal stationary state of the RP (including the nominal design-basis 100% power state). During the unit heating up/cooling down, it is easier to study the effect of parameters (n, Tc) on the vibration frequencies and amplitudes separately. In that way, we can keep the coolant temperature constant, as far as possible, and change the number of MCP sets, and then, fix the number of MCP-sets in operation and change the coolant temperature as monotonically as possible. After that, with four MCP sets in operation, we shall increase the RP power monotonically, starting from the minimum controllable level. Such a classical approach to the experiment planning is limited by the procedural constraints and the dependence of parameters (n, Tc, N) on each other.

It is prohibited to put all four MCP sets in operation at relatively low coolant temperatures (below $\mathrm{Tc} = 200°\mathrm{C}$). Otherwise, reactor internals and fuel assemblies may ascent due to the high density of the "cold" coolant. During the unit heating up, after all four MCP sets have been started, there is relatively narrow space for coolant temperature rise (200–290°C) up to the level corresponding to the minimum controllable level (MiCL) of the RP. When the RP power increases from MiCL to 100%, the change of the coolant temperature is also relatively small: in the "hot" pipeline (at the core outlet), the coolant temperature changes within the range (290–330°C), in the "cold" pipeline (at the core inlet), the coolant temperature changes within the range (290–300°C).

If we preliminarily interpret the spectral features of the signals after we change the number of MCP-sets in operation with the constant coolant temperature, further on, when we change the coolant temperature monotonically, this interpretation may be supported by extra evidence. For this purpose, we have to monitor the central frequencies of resonances. It is important to select a continuous monotonic dependence of the change of frequency of a particular resonance on Tc. If the resonance is interpreted as ASW resonance, its frequency will decrease monotonically as long as the coolant temperature increases monotonically.

It is also important to record the coolant temperatures at which the abrupt changes of spectral characteristics are observed. Frequency drop may be related to the appearance of a steam phase in the coolant. The resonances of vibration origin do not necessarily change their frequencies monotonically with the monotonic change of temperature. Thermal extension of a structural node may reduce the natural frequency of its oscillations; however, the same frequency may increase due to the thermal increase of stiffness of this node.

While the monotonic shift of resonances along the frequency axis is an expected phenomenon, the disappearance of the old resonances and appearance of the new ones have to be related to reaching that particular coolant temperature. The structural changes in RI fixtures due to thermal extension of structural elements: e.g. reaching the design-basis condition in the couple "vibration damper bracket on the reactor vessel — vibration damper slot on the core barrel" at a certain coolant temperature may be the reason for a jump in the spectral image of the measured signal. This phenomenon matches the change of the model from natural oscillations of the core barrel from oscillations of a beam with one fixed end to oscillations of a beam with two fixed ends. The appearance of natural frequencies in the coolant thermal-hydraulic instability may also lead to an abrupt change in the spectral image.

If the resonance remains in nearly the same frequency position, it may be the case that its origin is related to the natural frequency of oscillations of the structure or that of the rotational frequency of MCP-set. In the latter case, it may be equal to the precise harmonic or sub-harmonic of MCP-set rotational frequency. Unlike the resonance of the natural frequency of oscillations, resonances at harmonics and sub-harmonics of MCP-set rotational frequency have the same Q-factor as the resonance of the first harmonic at rotational frequency Fmcp.

Strictly speaking, the dependence of sound velocity on the medium temperature is nonlinear, but close to linear. The monograph [384] provides the following approximate experimental linear dependences for a commercial power unit with VVER-1000 reactor plant (V-320 design) for $\mathrm{ASW}_{1\mathrm{loop}}$ and $\mathrm{ASW}_{1\mathrm{vessel}}$ frequencies and temperature range 240–320°C (frequency in Hz, temperature in °C):

$$\begin{aligned} F_{\mathrm{ASW}_{1\mathrm{loop}}} &= 12.20\text{–}0.0175 \cdot T \\ F_{\mathrm{ASW}_{1\mathrm{vessel}}} &= 16.60\text{–}0.0215 \cdot T \end{aligned} \tag{12.1}$$

These dependences were obtained for a wide range of coolant temperature changes, including the temperatures of unit heating up and the temperatures or RP operation at non-zero power. Let us demonstrate that different slopes of straight-line dependences (12.1) are actually a natural phenomenon. For ASW frequency as the function of coolant temperature

$$F(T) = \frac{C(T)}{\lambda(T)}$$

in the assumption of linearity of dependences of sound velocity and ASW spatial length on the coolant temperature:

$$\begin{aligned} C(T) &= C(T_0) \cdot (1 - \alpha_c \cdot \Delta T) \\ \lambda(T) &= \lambda(T_0) \cdot (1 + \alpha_\lambda \cdot \Delta T) \end{aligned} \tag{12.2}$$

where α_c, α_λ — coefficients of linear change of sound velocity and ASW length correspondingly; $\Delta T = T - T_0$ — coolant temperature change; $C(T_0)$, $\lambda(T_0)$ — initial sound velocity and initial ASW length at temperature T_0, we have:

$$\begin{aligned} F(T) = \frac{C(T)}{\lambda(T)} &= F(T_0) \cdot \frac{1 - \alpha_c \cdot \Delta T}{1 + \alpha_\lambda \cdot \Delta T} \\ &\approx F(T_0) \cdot [1 - (\alpha_c + \alpha_\lambda)\Delta T] \\ &= F(T_0) \cdot [1 - \alpha_F \Delta T], \end{aligned}$$

where $F(T_0) = \frac{C(T_0)}{\lambda(T_0)}$ – initial ASW frequency, $\alpha_F = \alpha_c + \alpha_\lambda$ – resulting slope line coefficient.

In the formation of ASW frequency, the main role belongs to the temperature dependence of sound velocity, not to MCC thermal expansion, which causes an insignificant increase in the spatial length of ASW. Indeed, the linear expansion coefficient — α_λ — for steel is within the range $(1.2\text{–}1.8) \cdot 10^{-5}[\text{deg}]^{-1}$, while, as follows from experimental dependences (12.1), the coefficient of temperature-related frequency change α_F is within the range $(1.75\text{–}2.15) \cdot 10^{-2}[\text{deg}]^{-1}$, which is almost three orders of magnitude more. That is why we may consider that

$$\alpha_F = \frac{dF(T)}{dT} \approx \frac{1}{\lambda} \cdot \frac{dC(T)}{dT} = \alpha_c,$$

and the temperature dependence of ASW frequency takes the linear form

$$F(T) = \frac{C(T)}{\lambda} \approx F(T_0) \cdot [1 - \alpha_c(T - T_0)]$$

Thus, in the experiment, we may expect a linear dependence of ASW frequency on the coolant temperature in the following form:

$$F(T) = b - a \cdot T = F(T_0) \cdot (1 + \alpha_c T_0) - F(T_0) \cdot \alpha_c \cdot T \qquad (12.3)$$

with the parameters

$$a = F(T_0) \cdot \alpha_c \qquad (12.4)$$

$$b = F(T_0) \cdot (1 + \alpha_c T_0) \qquad (12.5)$$

For different ASWs, there will be a different slope of linear dependences. The higher the ASW central frequency, the bigger the increment in frequency that will be seen at one and the same temperature change.

The slope of temperature dependences like (12.4) carry additional information on the ASW. If any pair of inhomogeneities in the acoustic path may result in a half-wave of some ASW, one may find the half-wave length of an unknown ASW, knowing the true half-wave length of one ASW and the temperature dependences of two ASWs.

As follows from the expression (12.4), the relation of line slopes for two ASWs is a constant value equal to the relation of initial frequencies, which, in its turn, is equal to an inverse relation of wave lengths of these ASWs.

$$\frac{a_1}{a_2} = \frac{F_1(T_0)}{F_2(T_0)} = \frac{\lambda_2}{\lambda_1} \tag{12.6}$$

The same is true for b parameters in the equations for lines (12.5)

$$\frac{b_1}{b_2} = \frac{F_1(T_0)}{F_2(T_0)} = \frac{\lambda_2}{\lambda_1} \tag{12.7}$$

For ASW temperature dependences in VVER-1000 (12.1), the conditions (12.7) are fulfilled approximately: $\frac{a_1}{a_2} = 0.81$; $\frac{b_1}{b_2} = 0.73$. The difference between them, apart from the measurement error and nonlinearity, which has not been taken into account, is due to the fact that the dependences (12.1) have been obtained not only for isothermal modes, but also for power operation modes, in which the coolant temperatures at the inlet and at outlet of the reactor core are different.

12.2 Linear Approximations of Temperature Dependences of ASW Parameters in VVER-1200

In vibroacoustic applications, it is important to get some exact figures in any calculation and experimental analysis. Such a requirement exists, for example, in justification of vibration resources of a reactor plant [84–86, 94, 114, 123, 124, 127, 132, 133, 144, 149, 154, 156–158, 171–177, 375, 378, 382, 388]. The materials needed for such calculations include the series of frequencies and amplitudes of natural and forced oscillations of the structure and the coolant. If we go back to all kinds of analytical models of elastic media, none of them exist without sound velocity. Sound velocity also defines the numeric value of ASW frequency. However, different papers give different estimations of sound velocity, with some of the figures several times higher than the others [6, 7, 18, 23, 30, 32–34, 37, 38, 56, 145, 170, 206]. In Refs. [348–353], which use numeric values of ASW frequencies in the VVER coolant, numeric values of sound velocity can also be found in an implicit form, but they also disagree with the values given in the above papers.

On the one hand, this can be explained by the multi-factor nature of sound velocity in the coolant. Indeed, it is defined not only by the parameters of a single-phase water medium of the coolant (density, temperature, pressure, compressibility), but also by the dissolved gases, salts (e.g. borates) and other impurities. The presence of gas–vapor fractions in the coolant, even an insignificant amount, may change the sound velocity dramatically. Boiling (including subcooled boiling), radiolysis, coolant cavitations, which are in one way or another always present in the VVER primary circuit, may contribute significantly to the real value of sound velocity. The parameters of coolant flow turbulence, its viscosity parameters, as well as the dispersion effect (dependence of the speed of perturbation propagation in an elastic medium on frequency) also have their influence on the estimated sound velocity.

On the other hand, the experimental evidences gained by the authors at various VVER power units [382, 382] show high repeatability of numeric values of ASW resonance frequencies, both in the time statistics demonstrated in many fuel campaigns of the same VVER unit, and in a range of similar VVER units. For example, multiple factors listed above remain the same due to the procedural and operational limitations.

The task of experimental estimation of sound velocity has to be put as a commissioning test task. During commissioning tests, the required dependences of sound velocity shall be obtained, as well as the tabulated values, which will be used further to build the diagnostic features during commercial operation of the reactor plant. As an example, let us look at the family of APSD estimations of PPSh1 obtained during commissioning tests at Novovoronezh NPP Unit 6 with VVER-1200 RP, when the number of MCP-sets in operation was changed with the simultaneous increase of the coolant temperature. We select only four estimated ASPD having equal noise sources (see Fig. 12.1). They were obtained with all four MCP-sets in operation and nearly equal monotone change of the coolant temperature in values: 204°C, 221°C, 235°C, 250°C.

These four APSDs have an advantage over the rest of the APSD family as they enable us to highlight seven nearly linear temperature dependences. They definitely correspond to seven ASW types.

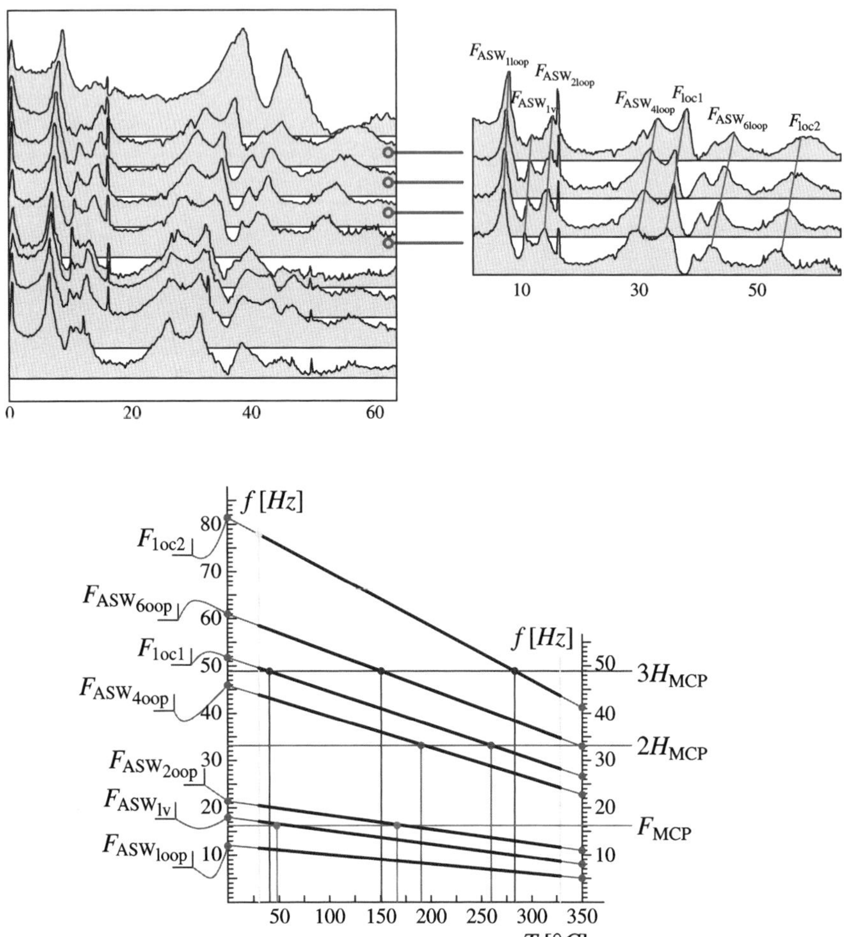

Figure 12.1: APSD family of PPSh1 signals. Experiment with monotonic increase of coolant temperature from 120° to 280° with simultaneous change of the number of MCP-sets in operation. Selection of 4 APSDs with 4 MCP-sets in operation with the change of coolant temperature: 204°, 221°, 235°, 250°.

The first three lowest-order ASWs have already been interpreted as ASW_{1loop}, $ASW_{1vessel}$, $ASW_{2vessel}$, while the remaining ones, the highest-order ASWs, are still to be identified. ASW half-wave has to rest on (begin and end at) some two inhomogeneities of the acoustic path, and if this is a simple ASW, it has a single antinode.

If the highest-order ASW is a higher harmonic, which already includes several half-waves, its internal nodes do not necessarily coincide with inhomogeneities, but its first and last nodes should be at inhomogeneities only.

Figure 12.1 also shows linear approximations of seven temperature dependences of ASW frequencies. It also demonstrates the points where some ASW frequency coincides with the first three harmonics of MCP rotational frequency. There are quite a lot of such points – seven of them – and this is only for the lowest-type ASWs and MCP harmonics. That is, the frequencies of disturbing forces (harmonics of MCP rotational frequencies) and natural frequencies of coolant oscillations (ASW frequencies) coincide many times during the power unit heating up/cooling down within the coolant operating temperature range without any visible impact on the vibration characteristics. Thus, the nominal dynamic modes are sort of natural tests for sufficiency of RP structure damping. The straight-line analytical dependences (Fig. 12.1 and Table 12.1) have been obtained using the least square method for resonances of the selected APSDs [67].

Parameters "a" and "b" of the straight-lines change monotonically with the change of ASW number (or ASW initial frequency) in accordance with the dependences (Table 12.1): the moduli of coefficients of linear dependences increase with the increase of ASW initial frequency. For the first lowest-order types in this table, $\mathrm{ASW}_{1\mathrm{loop}}$ and $\mathrm{ASW}_{1\mathrm{vessel}}$, the values nearly coincide with the dependences (12.1) for VVER-1000. Although they were obtained for different temperature ranges, the coincidence of the relevant coefficient remains within the measurement error tolerance. This is an expected result, as the dimensions of heat exchange loops in VVER-1000 and VVER-1200 are almost the same. The SG volume size in VVER-1200 was changed in comparison with VVER-1000, but the parameters of SG tube bundle in VVER-1200 remained the same, meaning that the acoustic path for the loop ASWs also remained the same.

Extrapolated up and down ASW frequencies in Table 12.1 show good coincidence with the experimental values of the corresponding frequencies obtained at the temperatures of 40°C and 290°C, which again proves that the quality of the linear temperature approximations is satisfactory. In the previous sections, we justified the experimental fact of aliquant frequencies of ASW harmonics due to

Table 12.1: Experimental temperature dependences of the parameters of seven ASWs.

	ASW central frequency $F(T) = F(T_0) \cdot (1 + \alpha_c T_0) - F(T_0) \cdot \alpha_c \cdot T$	Temperature coefficient $\alpha_c[°C^{-1}] \cdot 10^{-3}$	Extrapolated ASW central frequency [Hz] for the temperature of 290°C	Extrapolated ASW central frequency [Hz] for the temperature of 40°C
1	$F_{1\text{loop}} = 12.31 - 1.80 \cdot 10^{-2} \cdot T$	1.42	7.09	11.59
2	$F_{1\text{vessel}} = 16.80 - 2.36 \cdot 10^{-2} \cdot T$	1.40	9.96	15.86
3	$F_{2\text{loop}} = 22.73 - 3.17 \cdot 10^{-2} \cdot T$	1.40	13.53	21.46
4	$F_{4\text{loop}} = 43.30 - 5.93 \cdot 10^{-2} \cdot T$	1.40	26.10	40.93
5	$F_{\text{loc1}} = 51.82 - 6.98 \cdot 10^{-2} \cdot T$	1.38	—	—
6	$F_{6\text{loop}} = 61.20 - 7.96 \cdot 10^{-2} \cdot T$	1.36	38.17	58.02
7	$F_{\text{loc2}} = 81.30 - 10.57 \cdot 10^{-2} \cdot T$	1.36	—	—

non-zero pipeline diameter. The same result also clearly follows from Table 12.1 with the experimental "least square methods" approximations. For example, the relation $\frac{F_{2\text{loop}}}{F_{1\text{loop}}}$ is not equal to two, but, as follows from the table, $\frac{F_{2\text{loop}}}{F_{1\text{loop}}} = 1.85$ and $\frac{F_{4\text{loop}}}{F_{2\text{loop}}} = 1.90$. With the increase of the number of ASW harmonics, or the increase of the number of ASW nodes and antinodes, the continuum of the possible acoustic paths grows, and ASW resonances get broader and broader with the possible shift of ASW central frequency. In this case, the aliquant nature of ASW harmonics is the consequence of a systematic shift of central frequency of ASW harmonics. That is why the lowest error in the estimation of temperature dependences of sound velocity by the value of ASW central frequency can be expected for the lowest-type ASWs. Knowing the spatial length of a certain ASW for any of the dependences given in Table 12.1, it is possible to obtain sound velocity and, moreover, the temperature linear dependence of sound velocity. The dependence for the first ASW of MCCs will have the lowest error.

$$F_{1\text{loop}} = 12.31 - 1.80 \cdot 10^{-2} \cdot T$$

Using the half-wave length equal to 57.6 m, we get the temperature dependence of sound velocity in VVER-1200 acoustic path.

$$C(T) = 1420 - 2.07 \cdot T \tag{12.8}$$

Many papers state that sound velocity increases with the increase of the coolant temperature, pressure and concentration of the dissolved salts. It is also stated that water is a unique liquid that has positive coefficients of temperature dependences, unlike all other liquids. However, the experimental estimations of sound velocity are mostly related to sea water within relative narrow ranges of temperature and pressure change, and the conclusions on the special properties of water are also related to sea water. It is dangerous to use such references for the VVER coolant. At the constant coolant temperature — "T" — and constant MCC pressure — "P" — sound velocity — "C" — has also a constant value. The dependence $C(P, T)$ for water is nonlinear for both arguments. It is also necessary to take into account the volume of gases dissolved in water and the weight of the dissolved salts and impurities. These factors are usually not taken into account, but they may have a significant impact on the value of sound velocity in this medium.

The authors tried to find the dependences of sound velocity on temperature and pressure for infrasonic frequencies of the lowest-type ASWs in the above-mentioned papers, but failed. Having the ASW frequency with high accuracy and knowing the acoustic distance between the reflecting surfaces (half-wave length), it is reasonable to have an experiment in real MCC and get the value of sound velocity. That is, MCC is a kind of experimental test bench with an adequate metrological support for the measurement of temperature, pressure and coolant chemical parameters. Due to very high values of wave lengths of the lowest-type ASWs, it is impossible to conduct such measurements in laboratories with small-scale prototypes. The most popular method of sound velocity measurement in the laboratory conditions is the method of standing waves excitation in a resonator.

It would be reasonable to include the identification of dependency of sound velocity on the above factors in the commissioning test program. The specific features of a certain RP design — particularly, the individual water chemistry properties — will lead to the situation when sound velocity is not seen as a reference constant anymore.

Furthermore, when specific vibroacoustic diagnostic features are built for Novovoronezh NPP-2 Unit 1, the particular linear dependence will be used (12.8). That is, at the temperature of 20°C (and the pressure of 16 MPa), the extrapolated sound velocity will be equal to 1379 m/s, close to the reference value — 1450 m/a at atmospheric pressure. In MCC (temperature Tc = 290°C) the sound velocity will be equal to 820 m/s, and at the max design-basis temperature Tc = 340°C, it will be equal to 716 m/s. The deviation of sound velocity in operational conditions from the dependence (12.8) will indicate the change of thermal and physical properties of the medium, e.g. the presence of an abnormal amount of vapor–gas fraction in the coolant.

Different local ASW types exist in different parts of the MCC (including the reactor vessel). The ASWs located in loops only may be additionally called "one-dimensional", or "pipeline", or "loop", and those existing in the reactor vessel may be called "volume ASWs". Pipeline ASWs have only one spatial coordinate coinciding with MAT, while volume ASWs, as seen from the definition, have three dimensions. In power operation modes, ASWs of different types include different volumes of "hot" and "cold" water. Correspondingly, for each ASW, actually, there will be a separate resulting equivalent sound velocity.

12.3 Common Signal Sources of Pressure Pulsation Sensors and Accelerometers

The spectral characteristics of signals from PPSs and accelerometers mounted on MCC pipelines and main equipment, presented in the previous chapters, show that these signals are closely interrelated. Accelerations of the pipeline wall are mainly defined by the effects of vibration nature. Their main sources are vibrations of MCP sets at the harmonics of rotational frequency, vibrations of MCC elements at natural frequencies, and forced oscillations of MCC due to standing waves in the coolant. At the same time, vibration perturbation may spread both in the coolant and in the metal of the pipeline wall.

The coolant, as an elastic acoustic medium in a limited volume, like in the acoustic resonator, generates natural oscillations – acoustic standing waves. The coolant is also a medium which conducts vibration perturbations occurring due to MCP operation and vibration of MCC elements at their natural frequencies. The listed sources are common both for accelerometer signal generation, and for PPS signal generation.

PPSs have their own unique class of sources. PPSs are sensitive to the thermal and physical state of the coolant, characterized not only by temperature, pressure and coolant density, but also by turbulence, flow viscosity, sound velocity and the presence of two phases. The coolant may become two-phased due to the energy generation processes in the reactor causing different stages of coolant boiling and radiolysis, and due to cavitation effects occurring in local volumes of the coolant close to rotating elements of rotary equipment.

Table 12.2 provides APSDs of PPS signals installed in two loops: the left column — for loop one (PPSh1, PPSU1, PPSc1), the right column — for loop two (PPSh2, PPSU2, PPSc2).

It is obvious that the corresponding APSD pairs are identical. They include the same ASW types as the dominant sources, but other sources of pressure pulsations are different. For example, the signals of PPSU1 and PPSc1 located to the left and to the right of MCP-set No1, include the additive white noise of significant amplitude, which is missing in the relevant signals of the second loop: PPSU2 and PPSc2. It also defines the resonance values in the "PPS–PPS" coherence functions (Table 12.3). The correlation can be seen not only in ASW frequencies, but also in the broadband noise contamination of pressure pulsations in the first loop close to the MCP-set. Cavitation effect is the most probable common source of pressure pulsations for these two PPSs.

The accelerometer mounted on the thick-wall MCC structure may not be sensitive to the change of thermal–hydraulic characteristics of the coolant. PPSs and accelerometers located close to each other demonstrate the domination of the same lowest-type ASWs (compare Table 12.2 and Fig. 12.2) in APSD of their signals.

In APSD of accelerometer signals, they are sharper than in APSD of PPS signals. For instance, $ASW_{1vessel}$, located close to 10 Hz, in PPS_{c1} signal is almost immersed in the background, but it stands out of the background noticeably in Acc c1 signal. In PPSU1 signal,

Table 12.2: APSDs of PPS signals in hot, U-shaped and cold pipelines of loops 1 and 2.

	Loop 1	Loop 2
Hot	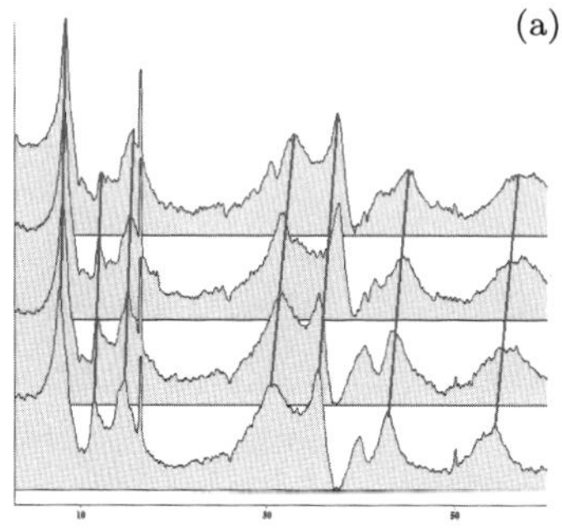	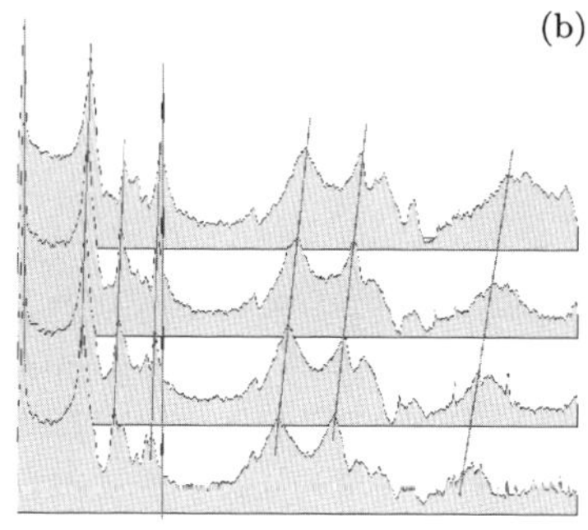
U-shaped	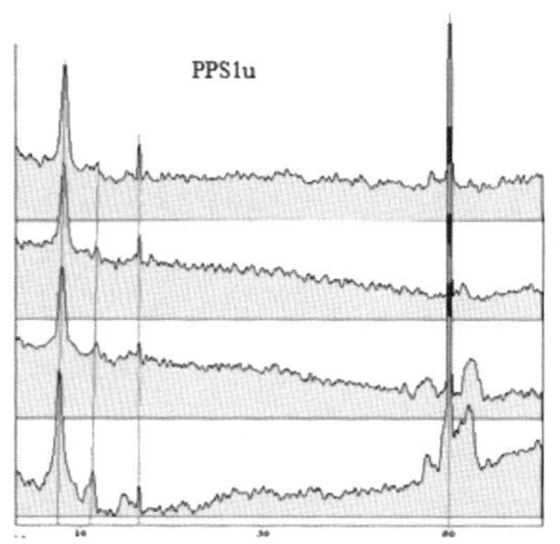	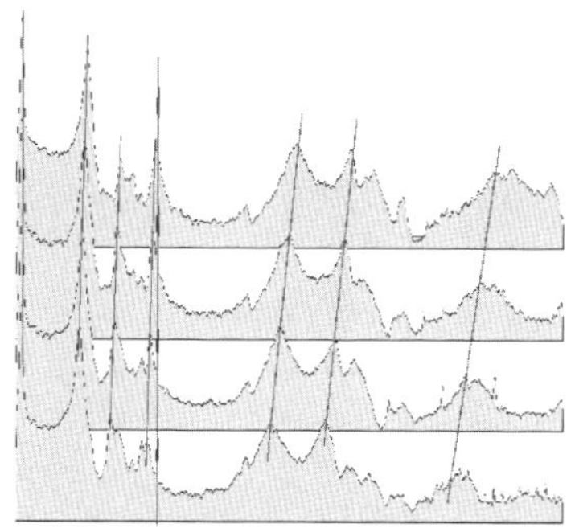
Cold	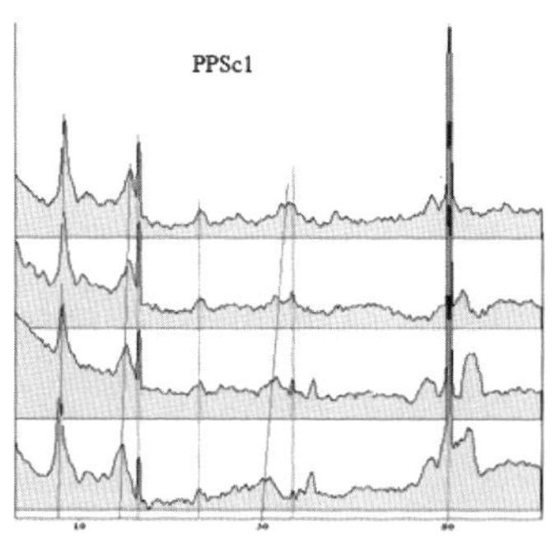	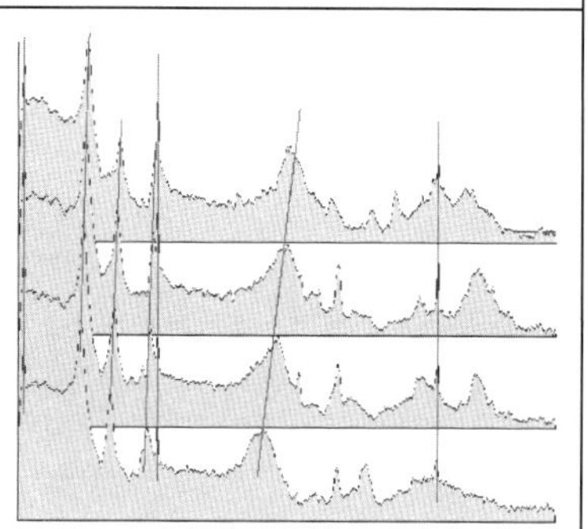

for frequencies above 20 Hz, broadband thermal-hydraulic noise contamination is dominant, while in the signal of the neighbouring accelerometer Acc U1, resonance features of highest-order ASWs are reflected.

Table 12.3: Coherence functions between PPS signals of the first loop (a) and between PPS signals of the second loop (b).

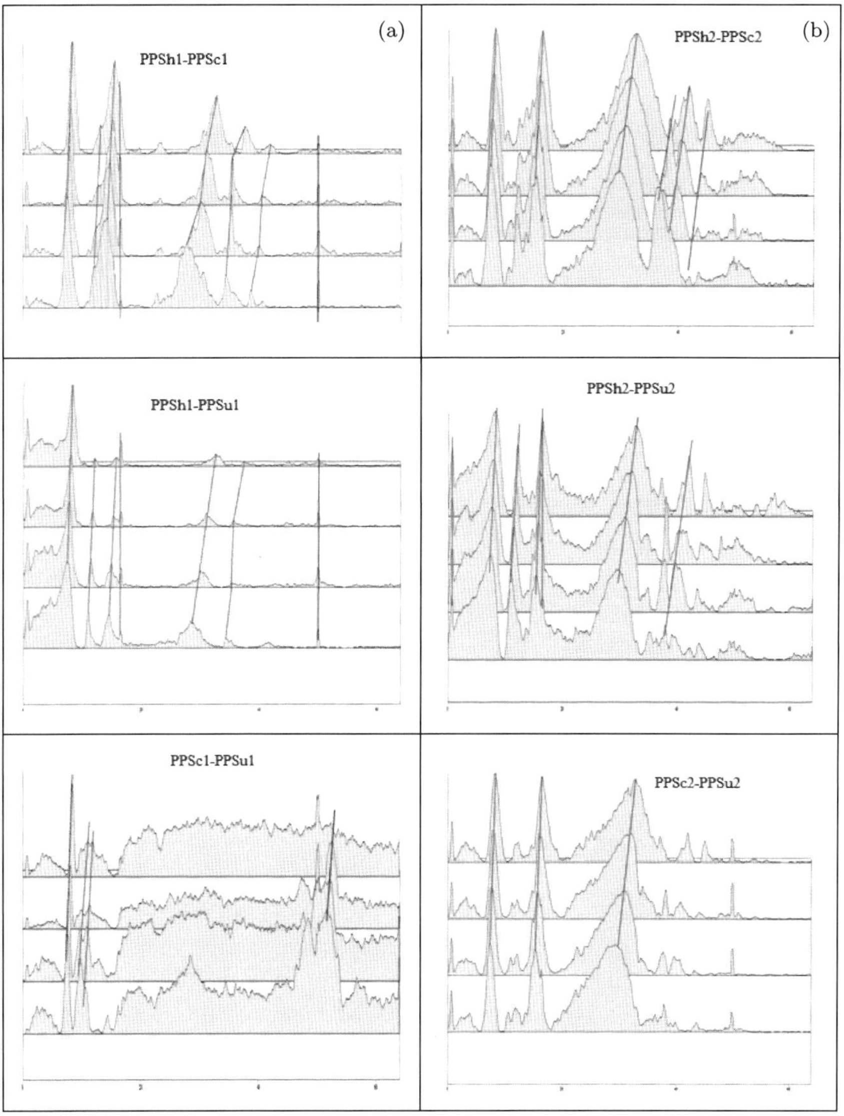

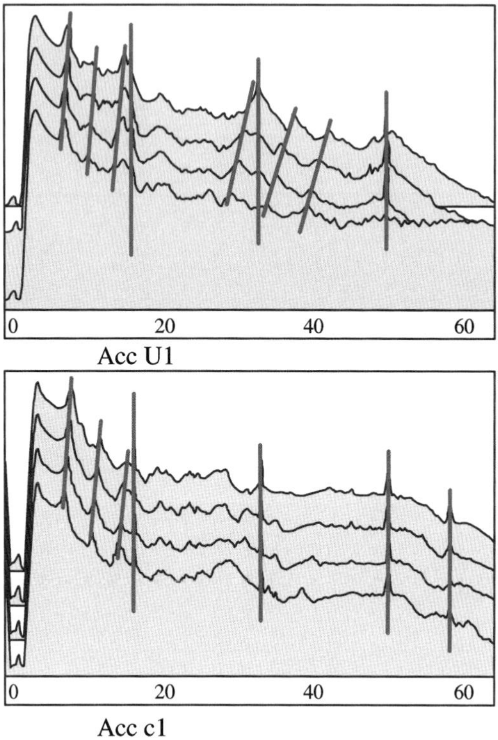

Figure 12.2: APSD family of accelerometer signals in U-shaped and cold pipelines.

The functions of coherency of signal pairs "PPS-Accelerometer" single out ASWs and harmonics of MCP rotational frequency, but they are not sensitive to thermal-hydraulic sources of pressure pulsations, which are not sensed only by accelerometers (Fig. 12.3). Accelerometers are hi-tech, relatively cheap passive sensors widely used in various reactor monitoring, control and diagnostic systems. Metrological support of accelerometer instrument channels is usually implemented in the form of calibration vibration machines, which are also widely spread.

PPSs are special sensors in reactor applications. Most often, they are active sensors requiring external high-frequency power supply. The sensing element of PPS is cut into the wall of reactor primary loop and thus it is related to high safety class. PPSs require the so-called purge of impulse lines, which makes their maintenance

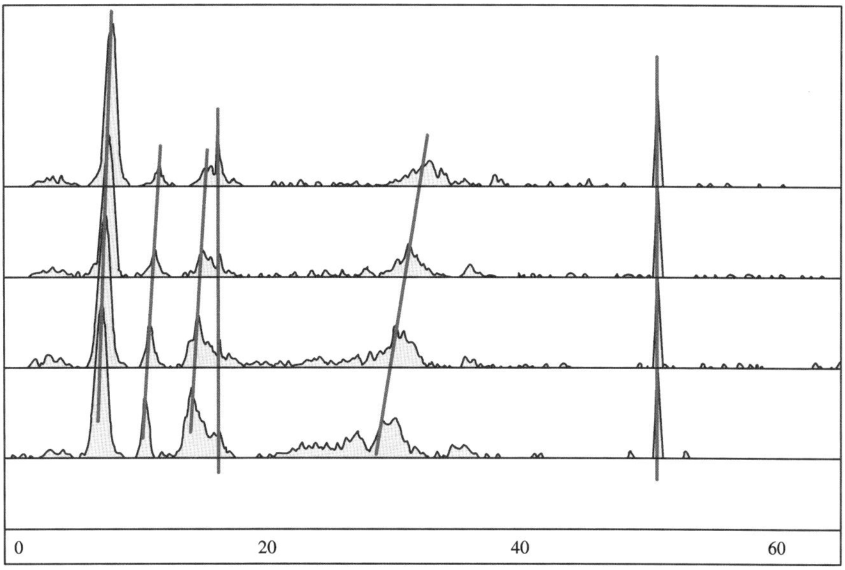

PSS U1 - Acc U1

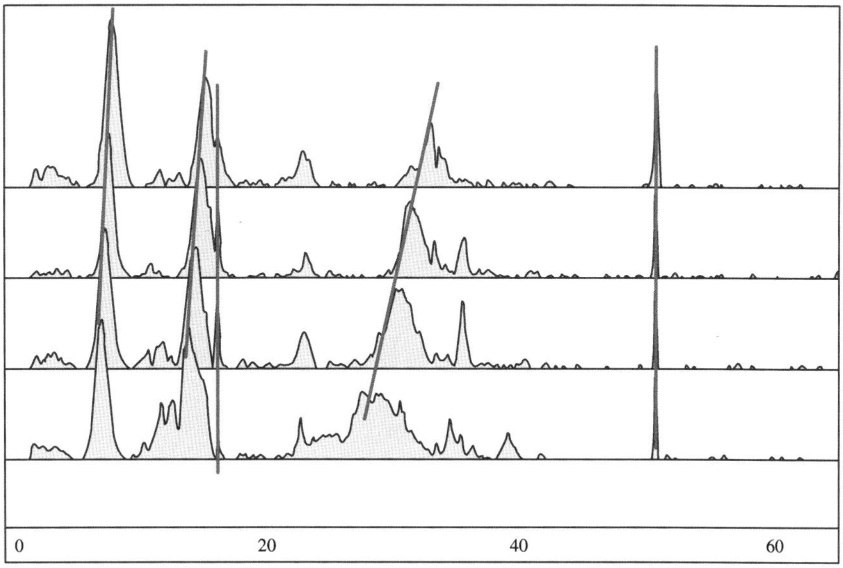

PSS c1 - Acc c1

Figure 12.3: Families of signal coherence functions of neighboring PPSs and accelerometers.

more complicated. Once installed, it is difficult to dismantle a PPS for periodic metrological testing. PPS has a major advantage over an accelerometer: it has almost unlimited lower operating range of frequencies. Accelerometers have always had trouble in measurement of low frequencies. For example, the accelerometer instrument channels, considered in this section, have the lower limit of the operating range of about 3 Hz.

Measurements as part of commissioning tests are made both with PPSs and accelerometers. In accordance with the commissioning test program, PPSs are used at primary stages at zero reactor power, while accelerometers are used not only at every stage of commissioning measurement and testing, but also during further commercial operation of the power unit as part of stationary vibration noise diagnostics system. After the accelerometer channels have been calibrated by PPS channels in zero reactor power modes, further on, during power operation, the absence of PPSs may be compensated to a certain extent by them. Obviously, this is true for major vibration effects, including MCC forced oscillations at ASW frequencies.

12.4 Forced Oscillations of Reactor Vessel and MCP at ASW Frequencies

The lowest-order ASWs — ASW_{1loop}, $ASW_{1vessel}$, ASW_{2loop}, ASW_{4loop} — which dominate in the loops, are the main sources of vibration of the reactor vessel. As follows from Fig. 12.4, at the frequencies of these ASWs we can see the maximum coherence between PPS signal and the signal of the accelerometer installed in the upper unit: "PPS–ACCvessel".

Vessel oscillations are also induced both from the "hot" pipeline (see coherence of signals "PPSh–ACCvessel" in Fig. 12.4) and from the "cold" pipeline (signals "PPSc–ACCvessel" in Fig. 12.4). In these two coherences, the monotonic decrease of ASW frequencies with the increase of the coolant temperature takes place simultaneously for ASW_{1loop} and ASW_{2loop}. The slope of the straight-line dependences of ASW1loop frequency on temperature is the same for both functions of coherence in Fig. 12.4.

We can also see the same slope for ASW_{2loop} wave in both families of coherence functions. Paradoxically, the vessel wave $ASW_{1vessel}$

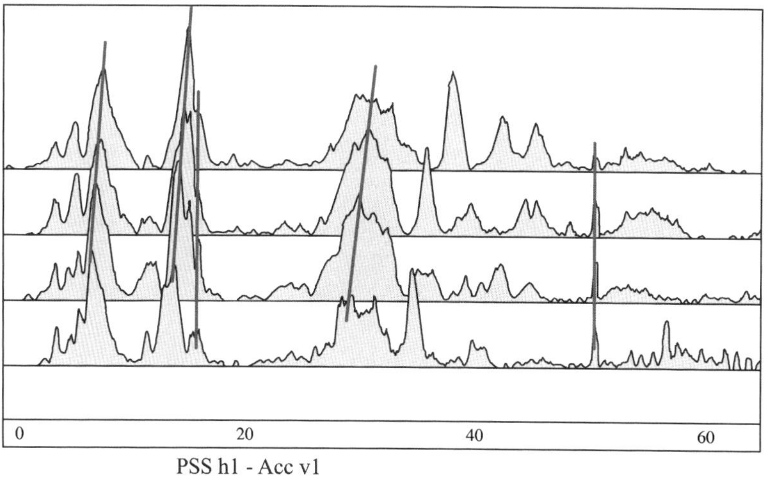

(a)

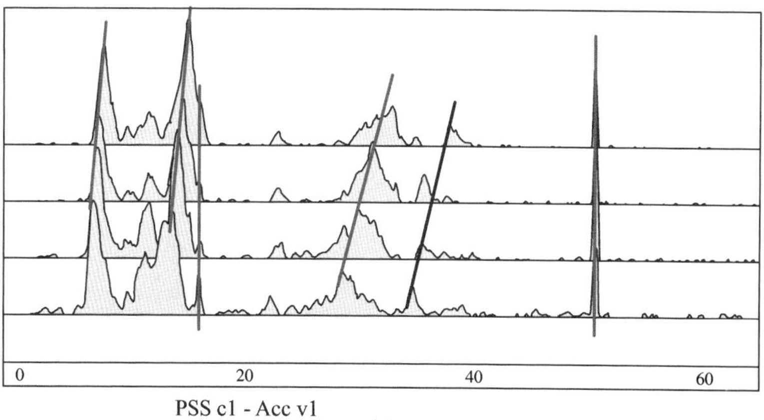

(b)

Figure 12.4: Functions of coherence of signals "PPSh1–accelerometer in reactor vessel" (a) and "PPSc1 — accelerometer in reactor vessel" (b) at monotonic increase of coolant temperature.

has lower impact on the reactor vessel oscillations compared to the loop waves ASW_{1loop} and ASW_{2loop}. Loop ASWs have nodes of their amplitude distribution close to the reactor vessel and thus have maximum spatial gradient here, unlike $ASW_{1vessel}$, which has the maximum amplitude at the vertical symmetry axis of the vessel with zero spatial gradient.

The slope of frequency lines in Fig. 12.4 within the range of 30–33 Hz is significantly different for the "hot" and "cold" pipelines of the loop. It means that different types of the highest-order ASWs have an impact on the reactor vessel from different pipelines. This range includes ASW_{4loop} half-waves and simple local half-waves ASWh, ASWc.

Linear phases observed in the signal pairs "PPSh–ACCvessel" in the subranges of frequencies of ASW_{1loop}, ASW_{2loop}, ASW_{4loop} (Fig. 12.5) carry information on the time delay of vibration displacement of the vessel due to the impact of a periodic force on the vessel (PPS signal in the loop). The delay time (the slope of the linear phase) depends on the state of the reactor vessel and internal fixtures and, thus, constitutes a diagnostic information. Each ASW, actually, has its own delay time.

Similar effects of forced oscillations from ASWs are also observed for MCP-sets (Fig. 12.6). The same frequencies, ASW_{1loop}, ASW_{2loop}, ASW_{4loop}, with the same temperature dependence are identified by the auto characteristic of accelerometer signals installed on MCP motor (Fig. 12.7), and by the mutual characteristics of "PPS–Acc MCP" signals (Fig. 12.8). The monotonic change of ASW oscillation frequency with the monotonic change of the coolant temperature does not change the linear phase dependence on frequency between "PPS–Acc MCP" signals. They also play the role of diagnostic indicators, which characterize the dynamic stiffness of MCP structure.

The linear phase characteristic in a certain subrange of frequencies (here, in the subrange of ASW frequencies), between the external force and the response, is a typical approach applied by the authors in building new diagnostic indicators. In order to use this method in practice, we need to exclude the ambiguity in phase estimation. In the proprietary spectral analysis software developed by the authors, phase is additionally determined by continuity, thus excluding instant jumps $\pm 2\pi$ at neighbouring frequencies, which enables us to conduct its linear approximation (see phase at Fig. 12.5).

Apart from ASWs and harmonics of MCP rotational frequency, in Fig. 12.7 we can see MCP low-frequency oscillations with the central frequency of 4 Hz, which are almost independent of the coolant temperature, i.e. do not belong to the class of ASWs. Earlier it was believed that they are related to long-wave turbulent traveling waves

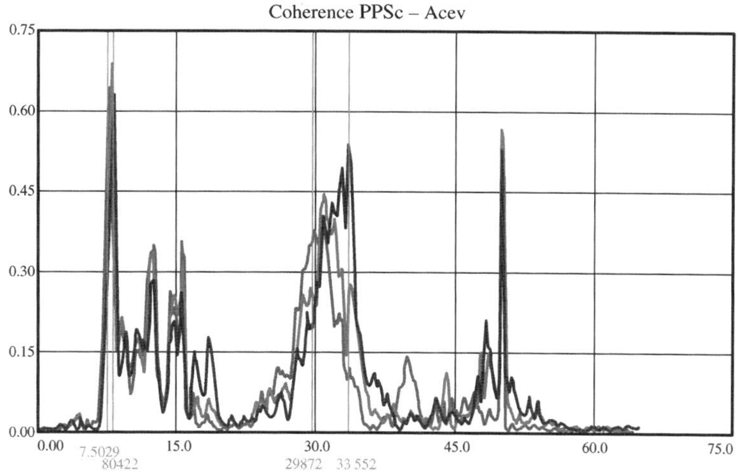

(a)

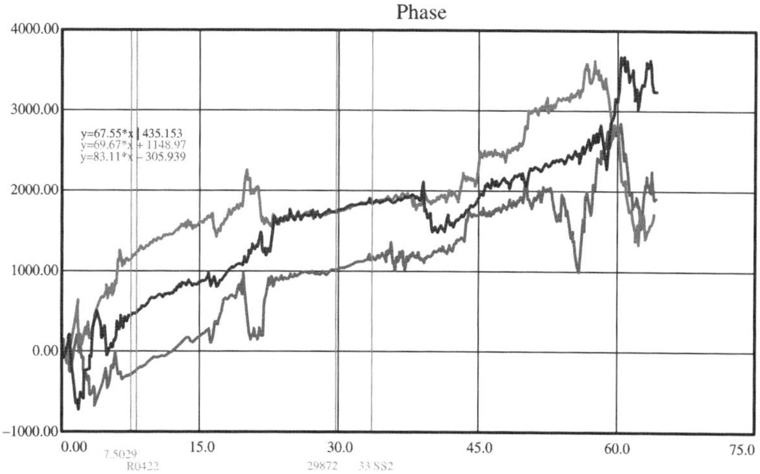

(b)

Figure 12.5: Functions of coherence between "PPSc" and "ACCvessel" signals at various coolant temperatures (a), and the family of the corresponding phase characteristics (b) with linear fragments near ASW frequencies.

in the coolant, which are formed when MCP creates a constant flow. Now, we can state that MCP itself also has the same low-frequency, low Q-factor oscillations, coherent with the coolant pressure pulsations, and MCP itself is the source of pressure pulsations of this type.

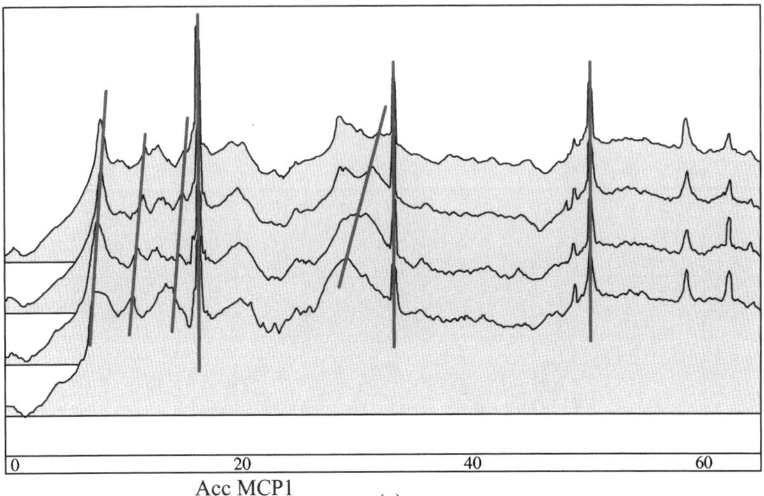

(a)

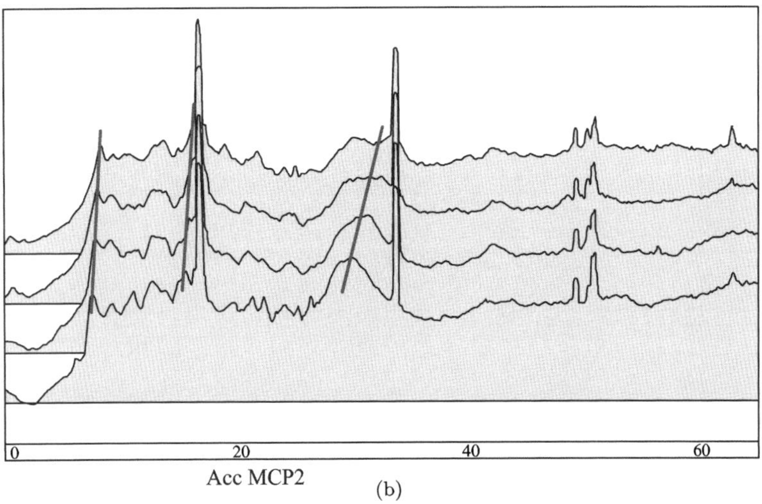

(b)

Figure 12.6: APSD families of signals of accelerometers mounted on the motor of MCP1 (a) and MCP2 (b).

Oscillations of MCP and reactor vessel are induced by the same source, which is a kind of intermediate between the oscillations of these units of MCC equipments: by standing waves in the coolant (Fig. 12.8). However, the existence of interconnection between the oscillating pairs "coolant–MCP" and "coolant–reactor vessel" does not mean that the reactor vessel and the MCP have mutual

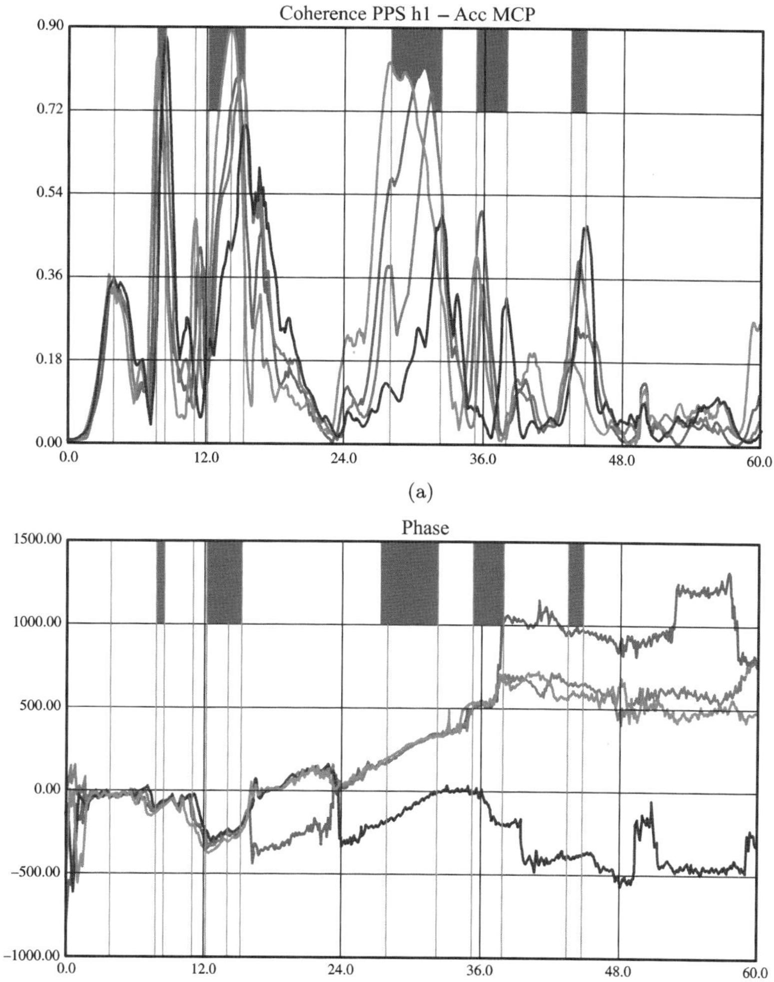

(b)

Figure 12.7: Functions of coherence of "PPSh" and "Acc MCP" signals at various coolant temperatures (a), and the family of the corresponding phase characteristics (b) with linear fragments in the vicinity of ASW frequencies.

oscillations. The proof of synchronous mutual oscillations of the MCP and the reactor vessel is the significant values of coherence function between the signals of two accelerometers "ACC MCP–ACC vessel" (Figs. 12.8 and 12.9). These curves demonstrate the known linear temperature trends at the frequencies of the lowest-order ASWs — ASW_{1loop}, $ASW_{1vessel}$, ASW_{2loop} — which means that the

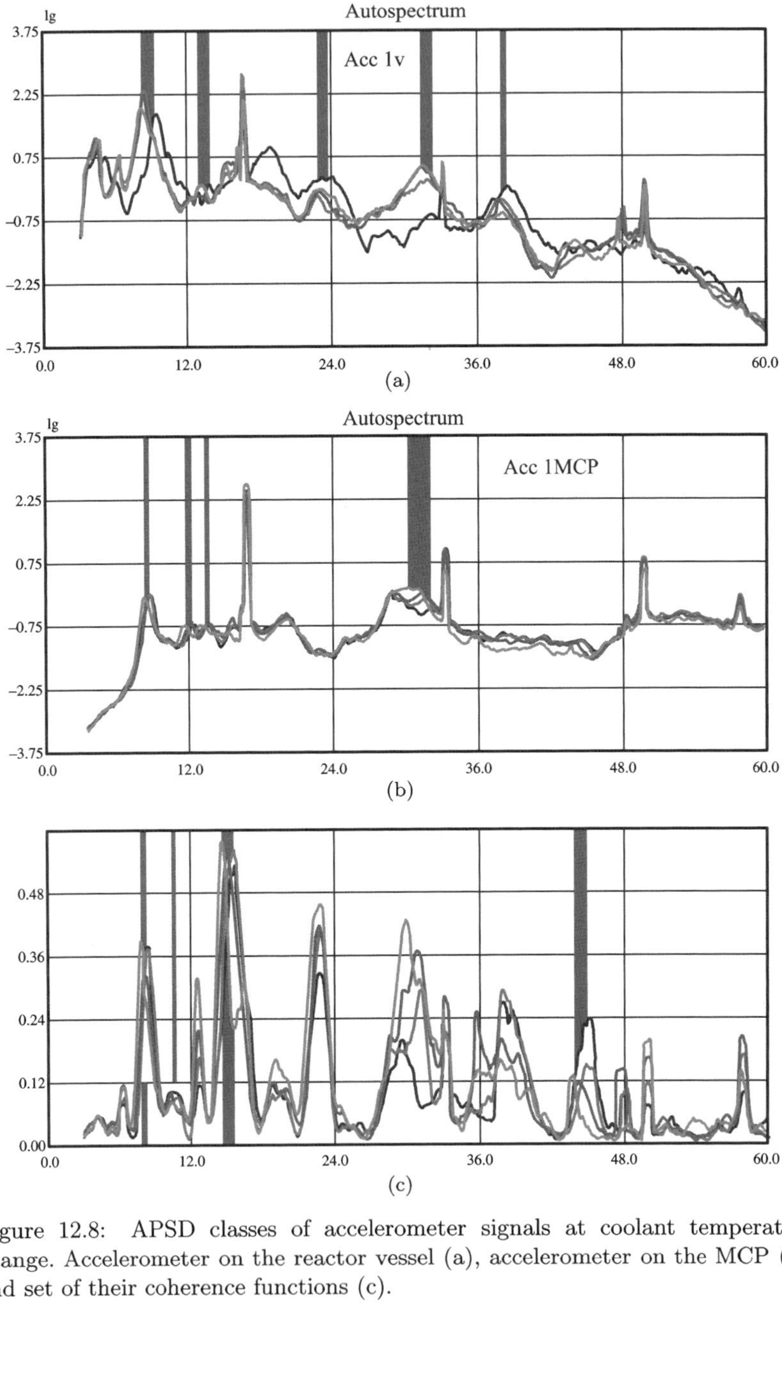

Figure 12.8: APSD classes of accelerometer signals at coolant temperature change. Accelerometer on the reactor vessel (a), accelerometer on the MCP (b), and set of their coherence functions (c).

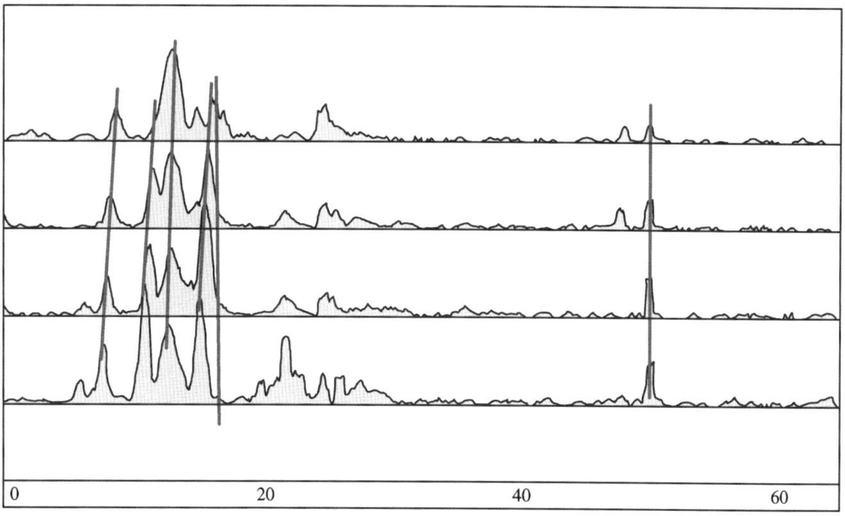

Figure 12.9: Functions of coherence of signals of two accelerometers mounted on different MCP-sets at different coolant temperatures.

lowest-order ASWs synchronize the oscillations of the MCP and the reactor vessel. MCP is more subject to forced oscillations at vessel $ASW_{1vessel}$ rather than the reactor vessel itself. Both of these units of MCC equipment are located in the special points of $ASW_{1vessel}$ distribution: MCP — in the node of vessel $ASW_{1vessel}$ distribution with the maximum gradient; reactor vessel — in the antinode of the same distribution with the minimum gradient.

As far as the highest-order ASWs are concerned, mutual oscillations of the reactor vessel and the MCP are observed, which are low in amplitude and thus are difficult to identify. However, we may state for sure that at the fourth and the sixth harmonics, ASW_{4loop} and ASW_{6loop}, synchronous mutual oscillations of the MCP and the reactor vessel also exist. That means that the location of the MCP in the node of ASW_{4loop} became the root cause for supplementary mutual oscillations of the MCP and the reactor vessel.

Ending with the topic of the global properties of the lowest-order ASWs in the MCC, we need to note an important feature identified by the authors for the first time during commissioning tests at Novovoronezh NPP Unit 6. These ASWs create a global coherent field of the lowest-order ASWs in the coolant in all MCC loops, which synchronizes the oscillations of equipment in different loops. Figure 12.8 shows the function of coherence of signals of accelerometers installed on MCP1 and MCP2 with significant resonances at ASW frequencies.

It turns out that these two MCP-sets, large and heavy components belonging to different loops and separated by significant distance, vibrate synchronously at the frequencies of four lowest-order ASWs. Synchronous oscillations of these MCP sets are also observed at the first and the third harmonics of rotational frequency.

The existence of interrelations at MCP rotational frequency is undoubted, as the synchronizing agent here is the frequency of the power supply network, which is common for all MCP-sets. For MCP vibrations at ASW frequencies, the coolant is the common agent. As a result, there is the mutual influence of MCP oscillations on each other. In the end, the source of the energy inherent in any ASW is MCP operation. If some MCP contributes too much energy to ASWs, e.g. due to the imbalance of its rotor, it will have an impact on the oscillations of three other MCP sets.

12.5 Change of the Global Field of the First Loop ASW with the Change of the Number of MCP in Operation

Let us consider again the ten modes with different numbers of MCPs in operation registered during commissioning tests at Novovoronezh NPP Unit 6 (Table 12.4). Now, we will analyze only PPS signals in different loops, but only at one frequency of the first loop ASW — $\text{ASW}_{\text{1loop}}$. Remember that it is carries a global character, i.e. $\text{ASW}_{\text{1loop}}$ exists in all loops, and it has the highest energy (power) compared to other ASWs. Every time another MCP is started/stopped, it will change the amount of energy accumulated in $\text{ASW}_{\text{1loop}}$. Depending on the stationary distribution of the field where $\text{ASW}_{\text{1loop}}$ belongs, and depending on the sequence of MCP starts/stops, the amplitude of $\text{ASW}_{\text{1loop}}$ will change in the loops differently. It is also possible that MCP starts/stops will cause significant rearrangement of $\text{ASW}_{\text{1loop}}$ distribution. At the same time, a rapid release of energy accumulated in one or another $\text{ASW}_{\text{1loop}}$ half-wave is possible, which may lead to hydraulic shock. Let us evaluate the phases between signals of all possible PPS pairs in different loops

Table 12.4: Phase relations and distribution of ASW_{1loop} amplitude across the loops at different number of MCP sets in operation.

Mode number	MCP sets in operation (filled circles indicate the first start of each MCP)	All possible "PPS-PPS" phases of different loops	ASW_{1loop} amplitude signs in different phases
1 10	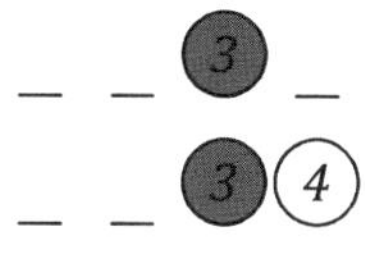	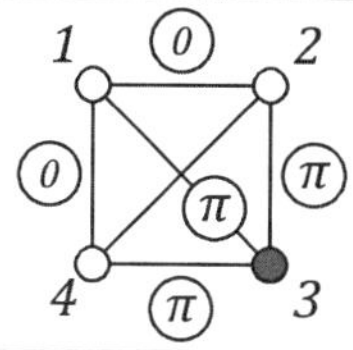	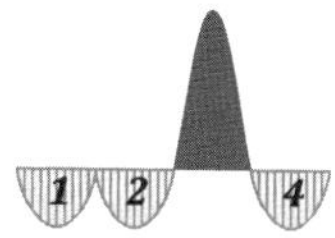
2 6	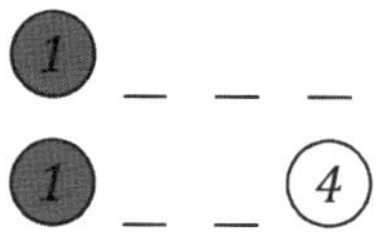	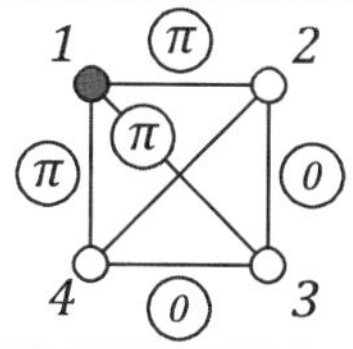	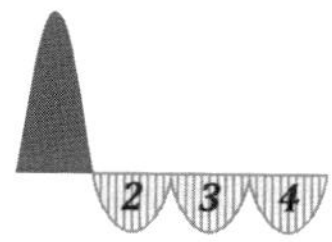
4 3	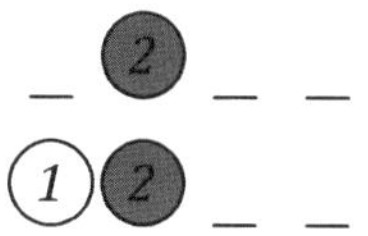	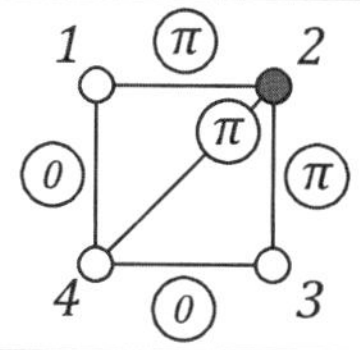	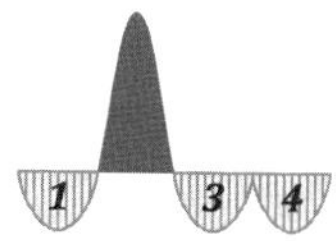
9		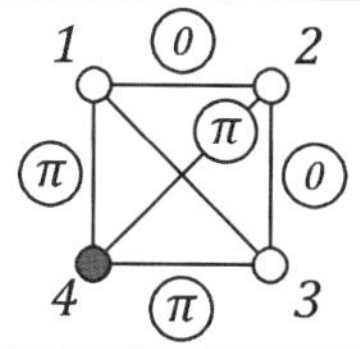	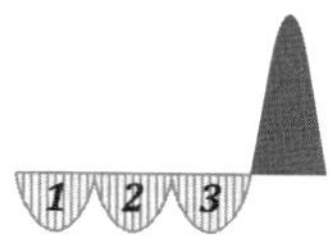
5 7 8	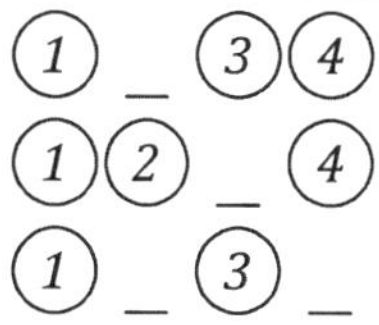	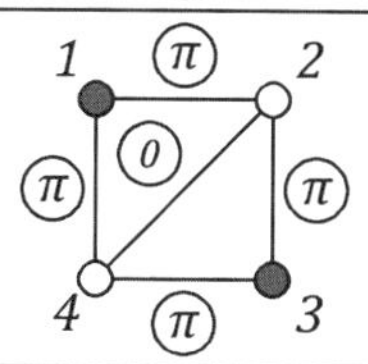	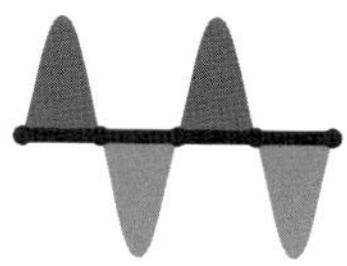

and, based on that, build amplitude distributions for this ASW_{1loop}. Here we can see the following trends.

When there is only one MCP in operation (no matter which exactly) it induces the field of ASW_{1loop}, non-symmetrical by loops. Assuming that, in the active loop, the amplitude has a positive sign, in all other loops amplitudes will have a negative sign (see Table 12.4). There is a simple explanation for that. When only one MCP is switched on, and the traveling pressure wave comes from the active loop to the three passive loops via the volume of the reactor vessel, its power is divided into three equal parts, like in a parallel connection. Three passive loops, assumed equal to each other, will give equal response in the form of three waves reflected in the same phase. As a result, in the passive loops, there will be ASW_{1loop} half-waves of the same sign, and in the active loop — half-waves of the opposite sign.

In the figures in Table 12.4, filled areas indicate active loops with positive half-wave amplitudes, and hatched areas stand for half-waves with negative amplitude in passive loops. In experiments, estimations of all possible phases of PPS signals in different loops give just the same distribution of amplitudes of ASW_{1loop} half-waves. In all four cases when only one MCP-set was in operation, the phase image included the same qualitative entity of in-phase and out-of-phase values (see column "All possible phases" in Table 12.4): in relation to the active loop, all other passive loops were out of phase.

That is, one positive half-wave in the active loop is always accompanied by three negative half-waves in three passive loops. Taking into account the fact that ASW power within one spatial period is divided in half between one active and three passive half-waves, and assuming that, in the active loop, the power of half-wave is equal to an arbitrary unit, the distribution of ASW power across the loops will be the following $1 : \frac{1}{3} : \frac{1}{3} : \frac{1}{3}$. That is, it turns out that one MCP set generates only two arbitrary units of ASW_{1loop} power ($1+\frac{1}{3}+\frac{1}{3}+\frac{1}{3} = 2$). The distribution of ASW amplitudes across the loops will be described by the relation

$$(+1) : \left(-\frac{1}{\sqrt{3}}\right) : \left(-\frac{1}{\sqrt{3}}\right) : \left(-\frac{1}{\sqrt{3}}\right)$$

bearing in mind that ASW power is proportionate to the square of its amplitude. The arbitrary amplitude unit has got the (+) sign based on experimental phase estimations. Other amplitudes equal to $(-\frac{1}{\sqrt{3}})$ have got the $(-)$ sign because the phase of passive loops is opposite to the phase of the active loop, which has also been proven experimentally.

If the next (the second) MCP-set is started, and this MCP is in the adjacent loop (no matter, right or left), the phase image, as the relation of amplitude signs, never changes (see line one, two and three in Table 12.4). Based on the superposition principle, the second MCP set also "pumps" power into $\mathrm{ASW}_{1\mathrm{loop}}$ of different loops at the ratio of

$$\frac{1}{3} : 1 : \frac{1}{3} : \frac{1}{3}$$

In the end, the power from two MCP-sets is added in the loops, and now the distribution of power across the loops will be the following (see Table 12.5):

$$\frac{4}{3} : \frac{4}{3} : \frac{2}{3} : \frac{2}{3}$$

As the phase image remained the same, the amplitude signs in different loops also remain the same, and the amplitudes will be in the following relation:

$$\left(\frac{2}{\sqrt{3}}\right) : \left(-\frac{2}{\sqrt{3}}\right) : \left(-\sqrt{\frac{2}{3}}\right) : \left(-\sqrt{\frac{2}{3}}\right)$$

The start of a neighbouring MCP causes no significant change in the distribution of $\mathrm{ASW}_{1\mathrm{loop}}$ field: there is just a minor increase in absolute values of ASW amplitudes in all loops.

However, if an MCP of the opposite loop is started second (1–3 or 2–4), ASW distribution becomes alternate in sign for different loops: +,−,+,− (see line five in Table 12.4). This result has also been repeatedly proven in experiments. In order to switch from a non-symmetrical distribution +,−,−,− to a distribution alternate in sign +,−,+,−, in the third new active loop, we need to change the sign of $\mathrm{ASW}_{1\mathrm{loop}}$ amplitude to the opposite one. That is, after the second MCP in succession is started in this active loop, we need to switch from the negative amplitude of $\mathrm{ASW}_{1\mathrm{loop}}$ in relative units $(-\frac{1}{\sqrt{3}})$ to the positive amplitude $(\frac{2}{\sqrt{3}})$. As a result, the peak-to-peak

Table 12.5: ASW power in loops and ASW_{1loop} amplitude distribution across the loops.

No.	MCP sets in operation	Distribution of ASW_{1loop} power across the loops	Distribution of ASW_{1loop} amplitude across the loops
1	MCP1	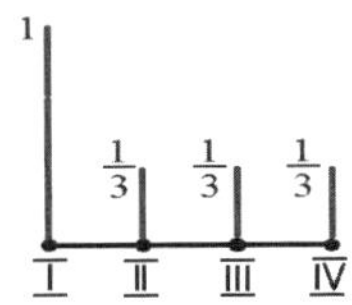	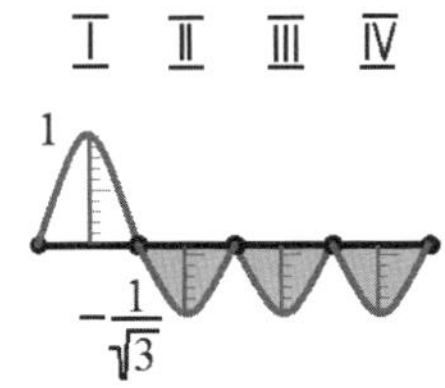
2	MCP1+ +MCP2	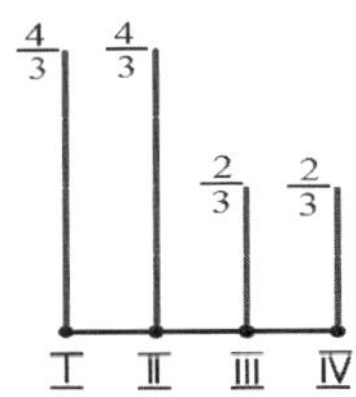	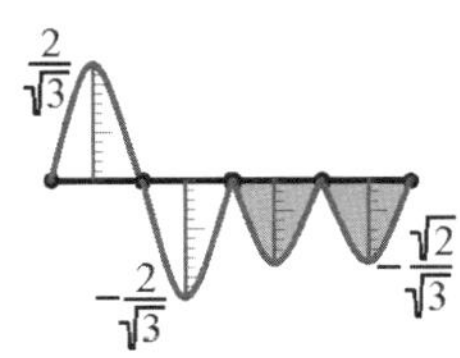
3	MCP1+ +MCP3	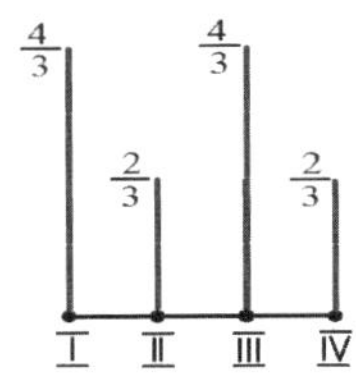	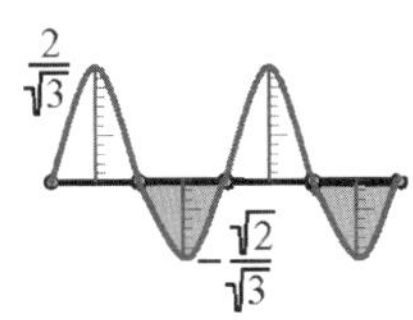
4	MCP1+ +MCP2+ +MCP3	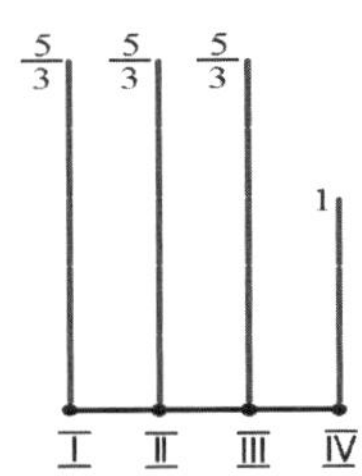	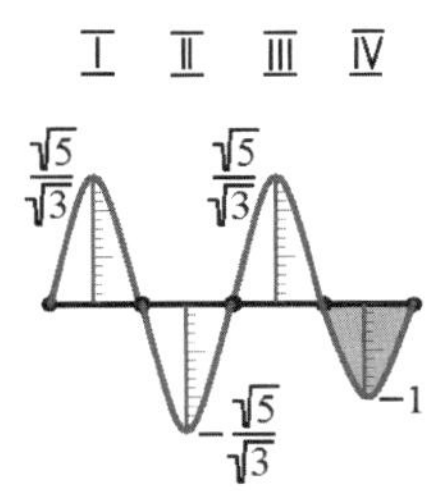
5	MCP1+ +MCP2+ +MCP3+ +MCP4	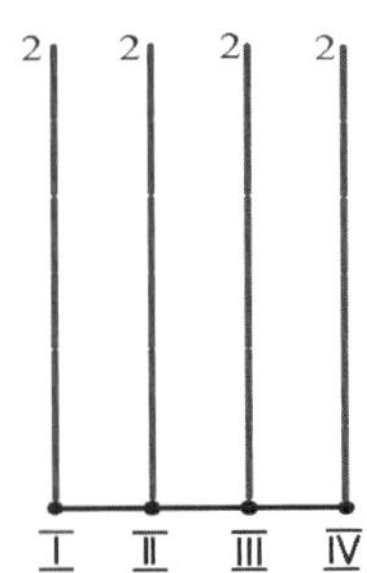	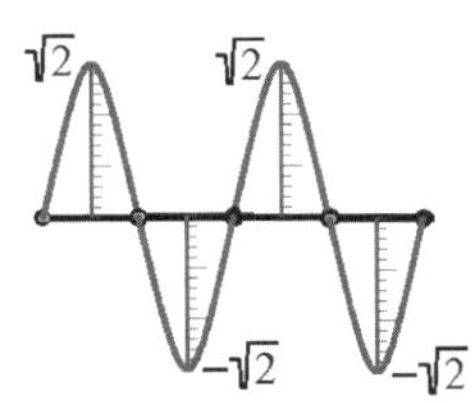

amplitude will be equal to $\frac{5}{3}$ units of power, whereas two MCP-sets together have generated four arbitrary units of ASW_{1loop} power. If this transfer from ASW negative amplitude to ASW positive amplitude happens instantly, this is equal to a hydraulic shock of significant amplitude.

Starting the MCP No. 3 always results in the change of the sign of ASW_{1loop} field, alternating in sign from one loop to another (see line four in Table 12.4). The amplitude modulus will be the same in the three active loops – $\sqrt{\frac{5}{3}}$, while in the passive loop it will be somewhat less: equal to just one arbitrary amplitude unit. If the transfer was from one alternate field (line 3) to another alternate field (line 4), nothing changes significantly, e.g. ASW does not change its sign. Otherwise, there is the transfer from $+,-,-,-$ state to $+,-,+,-$ alternate state, or from amplitude $(-\sqrt{\frac{2}{3}})$ to amplitude $(+\sqrt{\frac{5}{3}})$. For this purpose, we will need to change the power of ASW_{1loop} in the new active loop by $\frac{7}{3}$ units of power.

Thus, here is the evidence and explanation of the following phenomenology: if there are only two MCP sets in operation in the neighbouring loops, starting the third MCP in succession causes the most adverse change of ASW_{1loop} field. The remaining non-symmetrical ASW field after the second MCP is started in the neighbouring (left or right) loop, together with the increase in the energy potential of this field, may cause hydraulic shock of maximum power when the third MCP is started. On the contrary, if the second MCP started in succession belongs to the opposite loop, it results in much lower perturbance of ASW_{1loop}. The operation of only two MCP-sets belonging to the adjacent loops is unfavorable not only because it has an unbalanced hydrodynamic effect on the RP and reactor internals, but also due to caused hydraulic shock which have maximum power.

The transfer from three to four operating MCP-sets does not cause significant change on ASW_{1loop}. This is always a transfer from an alternate ASW_{1loop} field to the equally alternate field. All possible major rearrangements of ASW_{1loop} amplitude distribution have already happened when the second or the third MCP set had been started.

The result presented in this section is a simple phenomenological model primarily based on the experimentally obtained trend.

The trend includes the stable loop-by-loop phase images of ASW_{1loop} depending on the number of MCP-sets in operation. The ratios of ASW_{1loop} amplitudes given in Table 12.5 have also been proven in experiments. The experimental stability of ASW field is supplemented with a hypothesis on addition of power transferred to ASW_{1loop} when another MCP-set is started.

The energy approach adopted in this methodology ignores the phase relations between two independent ASWs induced by two different MCP sets. Meanwhile, different phases of two ASW_{1loop} may either double the resulting ASW amplitude, or compensate each other and make the amplitude equal to zero. That is, the powers generated by different MCP-sets would not be added as a direct sum. If we consider that there may be a random phase shift of the born ASW in relation to the existing ASW distribution, the resulting ASW field would be more changeable. However, every time we reproduced the same set of MCPs in operation, we got the same ASW_{1loop} distribution both in phase and in amplitude. There is a certain mechanism for synchronization of ASWs of the same type (the same natural frequency) in time, when a newly-born ASW is "adjusted" (synchronized) with the existing field, unless there is the need to change the amplitude sign of some ASW half-wave.

At multiple heating up of the RP, if the four MCP-sets are started in the same sequence, the potential hydraulic shock due to the significant change of ASW_{1loop} field will always happen in the same loop, thus accumulating the damage in its most vulnerable parts. In this sense, the optimal routine strategy is to change the number of MCP-set which is started first every time the unit is heated up. It is also better to proceed with the start of MCP belonging to the opposite loop — in that case, there will be no further significant change of ASW_{1loop} field.

Chapter 13

MCP Vibrations

13.1 Classes of Vibration Diagnostic Features of MCP

The MCP consists of components generating vibrations in various frequency ranges, so that MCP, being a single object of vibration diagnostics, is quite a broadband object [95, 98, 99, 105, 178–187]. The standards for monitoring of rotor equipment [393–396, 402–409] are used to perform vibration diagnostics of MCP. These standards specify the operating range [10-1000] Hz, and vibration velocity is a diagnostic parameter. First, the threshold limitations of GOST shall be met in the MCP test bench, breaking away from the real MCC. At this stage it is possible to identify the natural frequencies of MCP in the "rotor-shaft-movable bearings" system. The first real vibration loading of MCP occurs during commissioning measurements.

The MCC acoustics, as a vibration factor or global fluctuations of the entire MCC, have also the low-frequency components at natural frequencies. These components require registration of vibration frequencies below 10 Hz. The high frequency range of up to 1000 Hz is effective for control of the rotor equipment electric drives, while the asynchronous motor of MCP generates a vibration spectrum being much more effective than any other types of electric drives.

This section is not aimed at specifying a complete list of diagnostic indicators of various MCP defects. Only additional features of joint application of instrument channels of various diagnostic systems (VNDS, RCP process diagnostics system) and commissioning measurement subsystems in interpretation of vibration diagnostic parameters are demonstrated in what follows.

The MCP AEM (which is an abbreviation for asynchronous electrical motor), being the most broadband element, has vibration frequencies originated from two frequencies f_c, and f_{MCP}, where f_c is the frequency of three-phase AC power and f_{MCP} is rotational frequency of AEM rotor. In diagnostics of the rotor equipment, it is difficult to conceive of such a defect, which would not have changed the amplitude of vibrations on the rotational frequency f_{MCP}. A whole class of defects is detected by the amplitude of the second harmonic of the rotational frequency. For example, the attenuation of the magnetic iron ties of the stator or the ellipse of the gap between the stator and the rotor increase the amplitude of vibrations at a frequency $2 \cdot f_{\text{MCP}}$.

Electrical defects of AEM stator windings cause a harmonic series of frequencies $n \cdot f_c$, where n is an integer. And again, the second harmonic $2 \cdot f_{\text{p}}$ will dominate, and the number of observed harmonics of the power frequency — n — can be more than ten. Electrical defects of the rotor cause another harmonic series $m \cdot f_{\text{MCP}}$, where the integer m can also be over ten.

The interaction of electromagnetic fields of AEM rotor and stator is nonlinear. It can cause both f_{p} and f_{MCP}–($n \cdot f_{\text{p}}$ and $m \cdot f_{\text{MCP}}$) harmonics and their combination frequencies $n \cdot f_{\text{p}} \pm \text{m} \cdot f_{\text{MCP}}$ even in the defect-free MCP. In addition, each of the harmonics can be surrounded by the so-called side lines and bands to the right and to the left. Their physical nature of origin can be quite diverse and determined by the following:

- design features of the rotor and stator;
- fluctuations at natural frequencies of MCP and MCC;
- excessive friction of the rotating MCP parts, etc.

Such effects can often be demonstrated visually in the form of low frequency amplitude modulations of the measured process and cause the whip and precession effects (the so-called whirling effect, which refers to a slow rotation along the cone generatrix of axis of AEM rotor and/or MCP shaft as a modulating rotation effect on the rotational frequency).

Two main frequencies f_{p} and f_{MCP} forming the spectrum are connected by quite small number called slips – s (for MCP it is about 2% and less):

$$s = 1 - P\frac{f_{\mathrm{MCP}}}{f_c} \tag{13.1}$$

where P is a number of pairs of AEM poles ($P = 3$ for VVER-1000/1200 MCP). The frequency of MCP rotation of 1000 rpm (or 16, (6) Hz), which is the most important passport characteristic of MCP, is obtained as a starting point at $f_c = 50\,\mathrm{Hz}$ and at zero slip.

Slip is also a diagnostic parameter (feature) defining the state of the air gap between the AEM rotor and stator, quality of supply voltage of the three-phase circuit, state of the stator and rotor windings, etc. Slip causes both ultra-low AEM vibration frequencies ($f_s = s \cdot f_c$ — the so-called sliding frequency) and high-order harmonics close to harmonics of the mains frequency. For example, non-harmonic components in mains voltage supplied to AEM increases vibrations on harmonics of the mains frequency, by multiples of six: $6 \cdot k \cdot f_c$. Breaking the bar in the so-called squirrel cage of the rotor produces the lateral lines $m \cdot f_{\mathrm{MCP}} \pm n \cdot f_s$. AEM vibration effects are classified into electromagnetic, electrodynamic and magneto-striction effects.

The AEM rotor in conjunction with shaft and impeller of MCP together with the bearings are the sources of mechanical vibrations. For example, the static and dynamic imbalances of the rotor and various defects of bearings and coupling also increase the amplitude of MCP vibrations at harmonics of the rotational frequency $m \cdot f_{\mathrm{MCP}}$. The mutual impact of interfacing AEM rotor components and MCP shaft, bearings and movable supports causes vibrations at frequencies of sub-harmonics of the rotational frequency $\frac{p}{q} \cdot f_{\mathrm{MCP}}$ (where p, q are integers). Any AEM defect most often appears as a certain ratio of amplitude of successive harmonics of $m \cdot f_{\mathrm{MCP}}$ series.

The coolant medium pumped by the MCP transmits hydrodynamic sources of vibrations to the mechanical part of the MCP. In addition to the above ASW, the interaction of the impeller and coolant generates the non-integral harmonics of the so-called "oil whirl": $(0.42\text{–}0.48) \cdot k \cdot f_{\mathrm{MCP}}$. The impeller of the MCP shaft with six blades is a source of vibrations at the so-called blade frequency $f_b = 6 \cdot f_{\mathrm{MCP}}$ and its harmonics $6 \cdot k \cdot f_{\mathrm{MCP}}$. Due to the nonlinearity of this interaction a half sub-harmonic of the blade frequency $3 \cdot k \cdot f_{\mathrm{MCP}}$ is often observed.

Two related harmonic series $n \cdot f_p$ and $m \cdot f_{\text{MCP}}$ have quite close frequencies with a multiplicity of three, i.e. the successive frequencies of one series $n \cdot f_c, n = 1, 2, 3, \ldots$ are quite close to frequencies $3 \cdot n \cdot f_{\text{MCP}}$ of the other series due to a small value of the slip parameter. The preferred harmonics of the rotational frequency with "3" coefficient are ultimately determined by the three-phase AEM power supply and three pairs of AEM poles. This multiplicity is further enhanced due to the multiple number of impeller blades — six pieces.

The difference between the adjacent harmonics of the two listed series will grow in absolute value with increasing numbers of harmonics. This is a base for the method of assessing the slip according to pairs of the corresponding high-frequency harmonics.

In addition to the harmonic series of MCP vibrations at the circuit frequency $n \cdot f_c, n = 1, 2, 3, \ldots$, the same harmonic series of the circuit parasite components, not quite accurately referred to as interference, is summed up. They always accompany the low-voltage measurements against the high power currents of electric drive power consumption. For example, in this case the measurement of low-voltage currents of accelerometers against the kiloampere currents of AEM power consumption. The power supply circuit interference can occur in the measurement circuits in different ways:

- through the "earthing" buses and instrument channel power supply circuits;
- through the induction links of the adjacent instrument and power circuits;
- through electromagnetic interference (antenna effect).

The parasitic mains components not only increase the amplitudes of the signal on their harmonics, but also shift the phase relations between the signals of two instrument channels. Only in case when all phase relations, are zero, one can draw a conclusion on the dominance of the mains interference over all vibrations at the same frequencies regardless of vibration sensor location. In all other cases of phase relations, it is difficult to interpret the ratio of interferences and vibrations on the circuit harmonics.

The design of the spherical part of MCP casing has another source of high-frequency oscillations, the so-called guide vane. It is a stationary ring structural component, attached to the MCP spherical casing and generating the coolant flow by its fixed 14 blades. They generate the 14th harmonic of the MCP rotation frequency.

In addition to the high-quality harmonic components, broadband phenomena are observed in MCP vibration spectra. They are roughly classified in terms of "cavitation" and "turbulence". The cavitation is generated by the MCP impeller in the coolant medium, and the "turbulence" sources are localized not only in the MCC coolant, but also in auxiliary MCP subsystems, operating in the liquid operating environment (auxiliary subsystems of the seal water and systems of cooling loops). The broadband phenomena of cavitation and turbulence depend on the reactor plant mode parameters. In particular, during heat ups of the reactor plant at low temperatures the so-called gas cavitation is observed, referring to the process of origination and collapse of small bubbles from gases dissolved in the coolant. The effects of gas cavitation decrease with increasing coolant temperature, but they are replaced by a steam cavitation, referring to the local boiling of the coolant due to local fluctuations of coolant pressure. The vibration effects of cavitation are negligible in the nominal operating mode of the unit at 100% power.

The above-mentioned auxiliary systems, as part of MCP, are controlled by vibration channels of the standard MCP Technical Diagnostic System and by "slow" channels of standard measurements of temperature, level, flow, etc., which are not direct objects of vibration diagnostics, but serve as vibration sources as well.

Electromagnetic, mechanical and hydrodynamic sources of MCP vibrations can be combined in their permissible ranges, forming some integral spectral images of the defect-free MCP. The actual spectral portrait of each defect-free MCP has its own individual features and it shall not exactly coincide with the same spectral portrait of another MCP. The presence of simultaneously operating four MCPs of the same design can improve quality of their diagnostics. In [333–340], the formal approaches to the design of diagnostic features based on measurements of the standard MCP non-vibration signals were developed. These diagnostic features of defect-free MCPs slightly deviate from each other and can indicate the presence of anomalies in case of deviation of one of them from the remaining three. A low-quality power supply of MCP (for example, non-sinusoidal voltage) causes the same change of the relevant diagnostic feature in all MCPs by affecting all MCPs at the same time. This method of informal detection of the group change of the same diagnostic feature under the "E" scheme in all MCPs belongs to the so-called complex diagnostics method.

13.2 Vibration Amplitude and Power

The spectral diagnostic features listed in the previous section are defined by various parameters. We will consider both the power and amplitude of the vibration source to compare vibration sources. The vibration power of a source is calculated through the integral of APSD–$S(f)$ in the sub-band of frequencies $[f_1, f_2]$, where this source dominates (Fig. 13.1):

$$\sigma^2(f_1, f_2) = \int_{f_1}^{f_2} S(f)df$$

The high-quality resonance in the discrete spectrum can be represented sometimes by a few number of points (up to a single point); in this case an estimate of the area under the resonance is complicated, and, on the contrary, the amplitude estimation of a vibration level will be natural (Fig. 13.1). If the vibration displacements were measured, for example, in microns [μm], then after the integration of APSD we obtain the functionality $\sigma^2(f_1, f_2)$ in [μm]^2 units. It is also called a dispersion of the process in the selected sub-band of frequencies. Dispersion is an energy feature of the measured process. The square root of this value provides the root-mean-square value (RMS) of the process in a given sub-band of frequencies $\sigma(f_1, f_2)$ with the same dimensions as the measured value [μm].

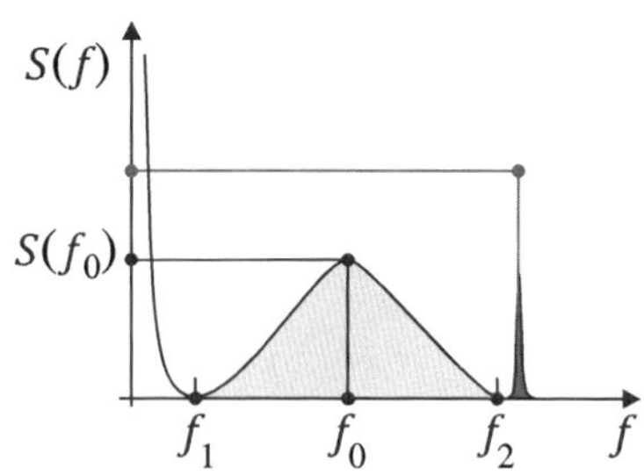

Figure 13.1: Broadband (low-quality) and high-quality resonance effects.

If we consider the spectral APSD amplitude – $S(f_0)$ (with dimension $[\mu\text{m}]^2/\text{Hz}$), it is commensurable neither with the RMS nor with variance. It is impossible to provide the ordinate "Y" of APSD graph with dimension $[\mu\text{m}]^2$ or $[\mu\text{m}]$. Such errors are quite common in application vibration studies of the reactor equipment.

The amplitude of one resonance is commensurate with the amplitude of another resonance, as a measure of vibration level, within the same spectral estimation and within APSD of signals of the same physical nature. Moreover, the parameters of the spectral estimation of various APSDs should coincide, as the amplitude of APSD estimation depends on the spectral resolution. The operations of APSD smoothing with regard to frequency and/or estimation of APSD with low resolution reduce the APSD resonance amplitude and lower its quality (widen the resonance). In this case, the area under the resonance (power of vibration effect) remains unchanged. In this sense, the variance (or RMS) in the frequency band of this resonance is preferable to its amplitude.

One resonant vibration effect is not necessarily higher than the other when there is higher amplitude. For example, there is a typical misconception about the dominance of MCP vibrations at the rotational frequency over the forced oscillations of MCP at ASW frequencies. Indeed, the resonance amplitude of the vibration sensor signal at the rotational frequency is usually the global maximum of the spectrum. However, the area under this resonance is less than that under the ASW resonance due to the high Q-factor, so the ASW and vibration source power exceeds the power of vibration sources at the rotational frequency.

Therefore, it is possible to compare both high-quality and low-quality resonance effects by transferring them into power. The high-quality resonances of harmonic series of the MCP rotational frequency shall be integrated in a narrow band of frequencies in the vicinity of the resonance frequency, while the low-quality ASW resonances shall be integrated in the wide frequency band. The integral criteria of vibration state of the rotor equipment, established by GOST, are based on the RMS of the wide frequency range (10–1000) Hz. This diagnostic feature involves numerous vibration effects without highlighting the main contributions to this parameter (RMS).

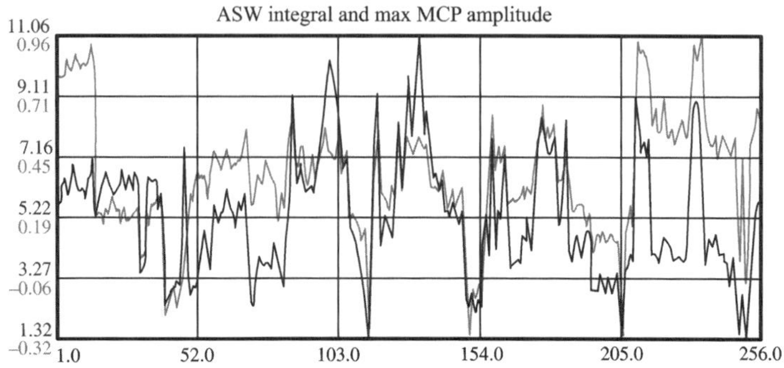

Figure 13.2: The time variation of the resonance amplitude at the rotational frequency (16.6 Hz) — and the variance in the vicinity of the ASW_{1loop} frequency sub-band in arbitrary units of semi-logarithmic scale.

Having several spectral estimations of the same sensor signal, successive in time, some of its numerical diagnostic features (for example, the amplitude or variance in a frequency sub-band) turn to be a function of time, providing additional diagnostic information. Figure 13.2 shows the time variations of the resonance amplitude at MCP rotational frequency and variance in a wide sub-band of frequencies in the vicinity of the central frequency ASW_{1loop}.

We have an obvious strong correlation of these processes. It is obtained because of more than 250 consecutive measurements of PPS signals performed during more than a month. Hence, the MCP vibration at a rotational frequency "pumps up" the ASW_{1loop} standing wave with energy, which has a second-order dependence on the amplitude of oscillations. The energy contained in the standing wave comes from the MCP operation. Its operation can be described as a continuous generation of pressure pulsations, which fall on the inhomogeneities, of the acoustic path in a form of the traveling wave. After being reflected from these inhomogeneities, they form a standing wave. The coolant pressure pulsations are generated as a result of MCP shaft rotation and do not depend on whether the MCP is subject to vibrations or not.

That is, the MCP vibrations create an additional source of coolant pressure pulsations. The vibrations are some kind of resonance amplitude modulator at MCP rotational frequency. The contribution of MCP vibrations to ASW power is insignificant, but it can be used to control the physical mechanism of converting the MCP rotation energy into coolant pulsations (into the ASW energy). This contribution serves as a diagnostic feature and its immutability in time specifies normal operation of MCP. The mutual characteristic, for example, a regression coefficient between considered amplitude and variance (or RMS) in different frequency sub-bands of the same APSD, obtained during long vibration measurements of the same PPS signal, will be a diagnostic feature.

13.3 Pressure Pulsations of Coolant in MCC and MCP Vibrations in a Wide Frequency Range

MCP vibration tests, performed in specialized stands prior to their delivery to NPP, are not able to exactly reproduce the hydrodynamic disturbing forces on the side of the coolant either in amplitude or in frequency composition. It is even more difficult to simulate the vibration loading of MCP on the side of the MCC pipelines. At the same time, each MCP itself is a source of MCC vibrations and coolant pressure pulsations. The joint operation of four MCPs creates both a common vibration field and common field of coolant pressure pulsations. These fields interact with each other through direct and inverse connections, and it is almost impossible to reproduce them either in test conditions, or by calculations.

Therefore, while applying the real vibration loads to all equipment during commissioning measurements, it is very important to consider the MCC, equipped with PPS, and MCP, equipped with vibration sensors as a single vibration object, in which the initial vibration sources shall be established.

Let us consider three types of sensors to identify the main vibration effects of MCP by joint analysis of signal relations:

- pressure pulse sensors;
- accelerometers, installed in the upper cross-piece of AEM of MCP-set (Fig. 13.3);
- proximity transducers, installed to monitor vibration displacements of MCP shaft.

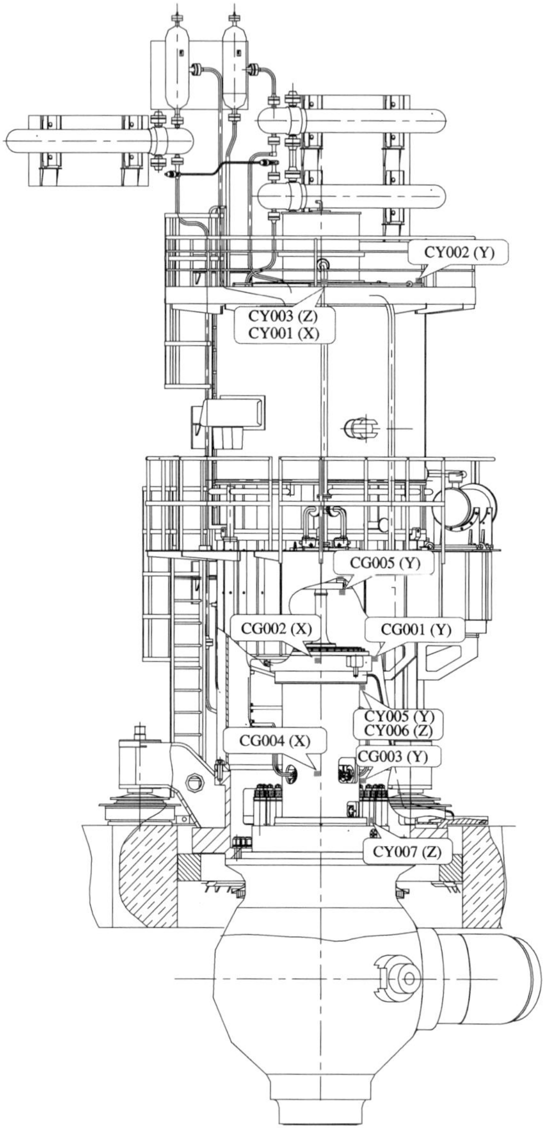

Figure 13.3: Measurement Points of MCP Technical Diagnostic System. CY001, CY002, CY003 are accelerometers on the upper spreader part of AEM, CG001, CG002 are proximity transducers in mutually perpendicular directions.

We consider only one mode, namely, the zero reactor plant power, in which the coolant temperature is 280°C and all 4 MCPs are operating. The listed sensors are classified as high-frequency sensors and their signals will be considered in the frequency range up to 250 Hz. Analysis of the high-frequency (50–250 Hz) range allows to identify not only the highest types of MCP oscillations, but also to detect the harmonic series of disturbing forces. If the series $f_{\mathrm{MCP}}(n) \approx n \cdot 16.6\,\mathrm{Hz}$ is expected as a sequence of the high-quality resonances due to the fact that their primary source, being the resonance at the frequency of the first harmonic $f_{\mathrm{MCP}} \approx 16.6\,\mathrm{Hz}$, is a high-quality resonance, then the ASW harmonic series are low-quality by definition.

While the series $f_{\mathrm{MCP}}(n) \approx n \cdot 16.6\,\mathrm{Hz}$ is a so-called series of exact harmonics, the frequency of which can be exactly calculated based on the number of the harmonic, the number of ASW harmonic series is only approximately linked to the ASW frequency. For example, as shown in Fig. 13.4, at a value of the first ASW frequency of the loop $\mathrm{ASW}_{1\mathrm{loop}} = 7.0\,\mathrm{Hz}$, the maximum of the second harmonic is $\mathrm{ASW}_{2\mathrm{loop}} = 12.9\,\mathrm{Hz}$ and the maximum of the fourth harmonic is $\mathrm{ASW}_{4\mathrm{loop}} = 26.9\,\mathrm{Hz}$.

If the main source of the series $f_{\mathrm{MCP}}(n) \approx n \cdot 16.6\,\mathrm{Hz}$ is the imbalance of the MCP rotating parts and it affects the coolant pressure pulsations, then the harmonic ASW series will force the MCP vibrations. The resonance of the coolant pressure pulsations with a significant amplitude will not necessarily cause the response MCP vibration at the same frequency. The opposite effect is also possible: small amplitudes of external forces excite the structure oscillations with significant amplitude at natural frequencies. The mechanism of exciting the oscillations by the hydroelastic coolant medium is quite complex; its computational modeling requires a three-dimensional field of disturbing forces.

The response vibration will be mainly determined by the damping properties of the structure itself, which are difficult to obtain by calculations.

As follows from APSD and coherence function of PPSh1 and PPSc1 (Fig. 13.4), whose transducers are spaced at a considerable distance from each other (about 40 m), the upper ASW with a frequency of higher than 50 Hz does not show the property of globalism. This means, that the amplitude of the upper ASW is comparable or

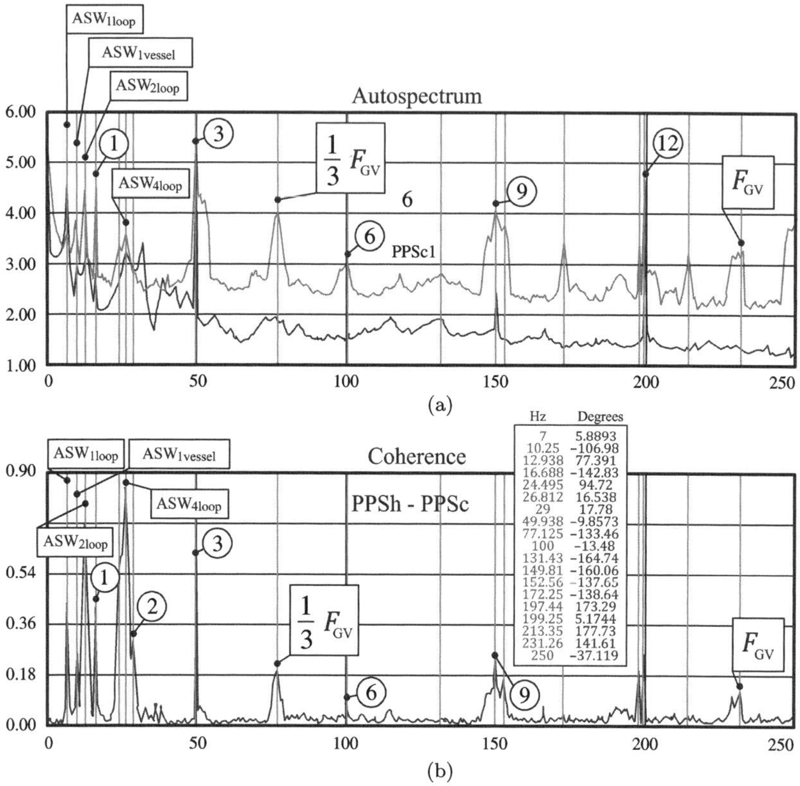

Figure 13.4: APSD of signals of pressure pulsation sensors — PPSc1 and PPSh1 (a) and their coherence function (b). The standing waves and the guide vane frequencies are marked with rectangles, and harmonics of the reverse MCP frequency are marked with circles. The table on the graph includes marks of the resonance frequencies and the corresponding phases.

less than the RMS of the white noise turbulent components of the pressure pulsations (the so-called APSD pedestal).

The harmonics $f_{\mathrm{MCP}}(n) \approx n \cdot 16.6\,\mathrm{Hz}$ shown in Fig. 13.4 are marked with a number, enclosed in a circle, and it is obvious that in pressure pulsations

- there are no high amplitudes in the upper harmonics at frequencies higher than 50 Hz;
- there are only harmonics aliquot by three.

Quite weak interactions are observed at the frequency of the MCP guide vane — F_{GV}. The F_{GV} frequency completely coincides with the value $14 \times 16.6\,\text{Hz} = 232.4\,\text{Hz}$. Also a 1/3 sub-harmonic F_{GV} exactly equal to 77.47 Hz is observed and this is an evidence of nonlinear effects produced by the guide vane when forming the coolant flow in the "cold" pipeline.

The coolant, as a hydroelastic medium which forces MCP oscillations, has the following properties in a wide frequency range:

- The main power of pressure pulsations (PPS signal) consists of the lower-type ASW power below 50 Hz and harmonics of the MCP rotational frequency divisible by three;
- The broadband white noise turbulent component of the coolant in the "cold" pipeline has a significantly higher power than that of the similar component in the "hot" pipeline.

The APSD of accelerometer signals Ax, Ay (Fig. 13.5) unlike APSD of PPS signals (Fig. 13.4) are quite broadband and has no tendency for decreasing with frequency. This is not only due to the

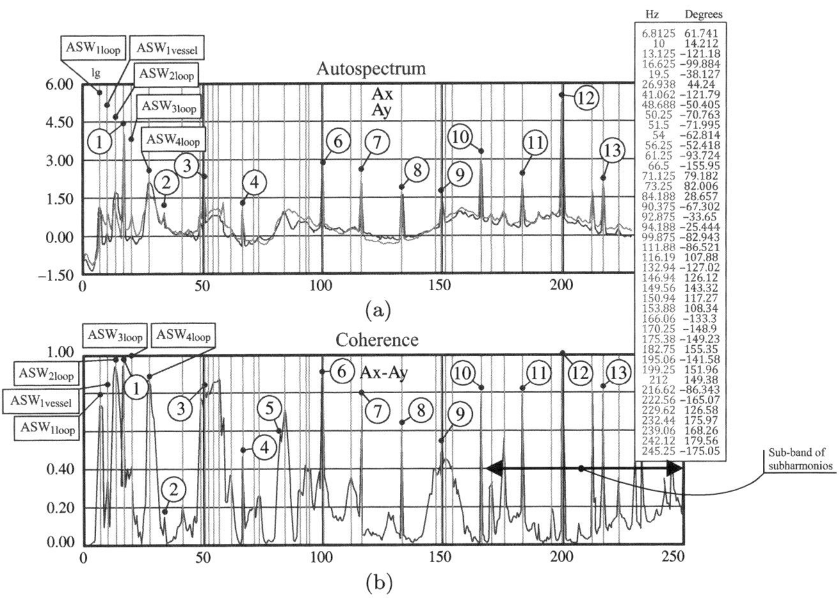

Hz	Degrees
6.8125	61.741
10	14.212
13.125	−121.18
16.625	−99.884
19.5	−38.127
26.938	44.24
41.062	−121.79
48.688	−50.405
50.25	−70.763
51.5	−71.995
54	−62.814
56.25	−52.418
61.25	−93.724
66.5	−155.95
71.125	79.182
73.25	82.006
84.188	28.657
90.375	−67.302
92.875	−33.65
94.188	−25.444
99.875	−82.943
111.88	−86.521
116.19	107.88
132.94	−127.02
146.94	126.12
149.56	143.32
150.94	117.27
153.88	108.34
166.06	−133.3
170.25	−148.9
175.38	−149.23
182.75	155.35
195.06	−141.58
199.25	151.96
212	149.38
216.62	−86.343
222.56	−165.07
229.62	126.58
232.44	175.97
239.06	168.26
242.12	179.56
245.25	−175.05

Figure 13.5: APSD of signals of accelerometers, installed in MCP AEM — Ax1 and Ay1 (a) and their coherence function (b). The standing waves are marked with rectangles, and harmonics of the reverse MCP frequency are marked with circles. The table on the graph includes marks of the resonance frequencies and the corresponding phases.

high-frequency nature of AEM vibrations, but also due to the fact that the accelerometer instrument channels differ from the vibration displacement channels by parabolic frequency weight, which amplify the high-frequency spectral components. Unlike the APSD of PPS signals, the APSD of Ax and Ay signals have all successive harmonics of the rotational frequency and starting from the tenth harmonic (166 Hz) the sub-harmonics of the rotational frequency are observed.

Ac and Ay accelerometers register the spectral components of the lower-type ASW: ASW_{1loop}, $ASW_{1vessel}$, ASW_{2loop}, which in terms of vibration will specify a dominance of ASW vibration sources in high-frequency range (e.g. by double integration of accelerometer signal).

The spectral picture of AEM vibrations up to 30 Hz (Fig. 13.4) is very similar to that of the pressure pulsations (Fig. 13.5). Once again, the ASW resonances are dominating: ASW_{1loop}, $ASW_{1vessel}$, and ASW_{4loop}. It is possible to state that the pressure pulsations stop their impact on AEM starting from 100 Hz frequency and above.

The coherence function of Ax and Ay (Fig. 13.5), which has low sensitivity to parabolic spectral weight, exhibits high values at both high-quality (harmonics and sub-harmonics) and low-quality resonance frequencies. The low-quality resonances in the coherence function of "Ax–Ay" signals are observed in the vicinity of the "round" values of frequencies: 50, 100 and 150 Hz, where harmonics of the mains frequency almost coincide with harmonics of MCP rotational frequency in multiples of three. It is obvious that there is a superposition of vibrations on the mains harmonics with vibrations on the rotational frequency harmonics. It is very likely that the low-quality of the mains harmonics is caused by electromagnetic vibration sources of the stator (magnetic iron core) of MCP AEM.

MCP electric motor, rigidly fixed by cylindrical spacers to the spherical casing of MCP, which is rigidly connected to U-shaped and "cold" MCC pipelines, can be represented as an integrated unit with MCC. The coolant forces vibrations of MCC pipelines at ASW frequencies in low-frequency range and it also affects the MCP AEM vibrations. In addition, the MCP, having a structure with a vertical layout, due to its vertical shoulder, has more vibration displacements at the level of AEM upper spreader part compared to that of the pipelines themselves.

There are two ways to excite AEM vibrations by the coolant:

- Coolant → MCC pipelines → spherical MCP casing → spacers → AEM drive;
- Coolant → impeller → MCP shaft → torsional coupling → AEM rotor → AEM drive;

in the second case there is a relatively "soft" component which is spatially non-homogeneous in its stiffness properties — a coupling which blocks some disturbing forces on the side of the coolant. The maximum stiffness of connection between MCP shaft and AEM rotor through the torsion coupling is observed in the vertical direction, but AEM vibrations occur mainly in the horizontal plane. Figure 13.6 shows comparison of APSD signals of accelerometers installed in three directions. In the vertical direction (z-direction), the vibration power is two orders less than the power in horizontal directions. There are only harmonics of the rotational frequency, and the standing waves of the coolant impact AEM vibrations in z-direction rather weakly.

Also, there are no low-quality vibration sources in the vertical direction of the mains harmonics (presumably, because vibrations of the magnetic iron of AEM stator exist in the horizontal plane only), but there are harmonics with a multiplicity of number three (the third, the sixth, the ninth and the twelfth). So, the source of

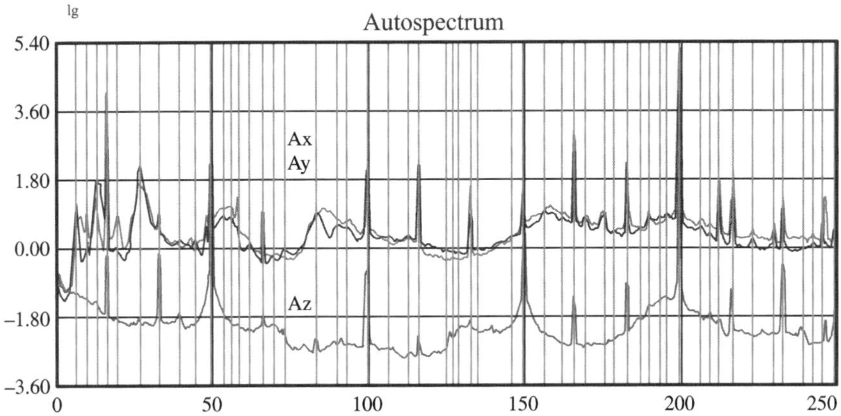

Figure 13.6: APSD of signals of accelerometers installed in MCP AED in horizontal plan — Ax1, Ay1 and in vertical direction – Az1.

vibrations at blade frequencies of the harmonics and sub-harmonics of the impeller dominate in z-direction. Due to the complex shape of the impeller blades, they create a non-zero vertical component of the force reaction, which is transmitted from the MCP shaft to AEM rotor via the coupling.

To find the relationship of the above vibration sources, let us consider the coherence functions of a signal pair of different types with sensors significantly spaced from each other: "PPS_h–accelerometer on MCP AEM" (Fig. 13.7). They confirm that the frequency range is conditionally divided into two parts:

- Up to 100 Hz — dominance of hydrodynamic sources of AEM oscillations in a form of ASW;
- Above 100 Hz — mutual oscillations of the impeller of the MCP shaft and MCP AEM on harmonics of the MCP rotational frequency, divisible by three.

The presence of the high AEM oscillations on the lower-type ASW frequencies and their absence in the MCP shaft indicate that the lower ASWs force vibrations of MCC pipelines, and consequently they force the MCP oscillations. That means, the mediation of pressure pulsations does not occur through the shaft of MCP to AEM, but through the MCC pipelines and spacers to AEM. The power of AEM vibrations originated from the sources of acoustic standing

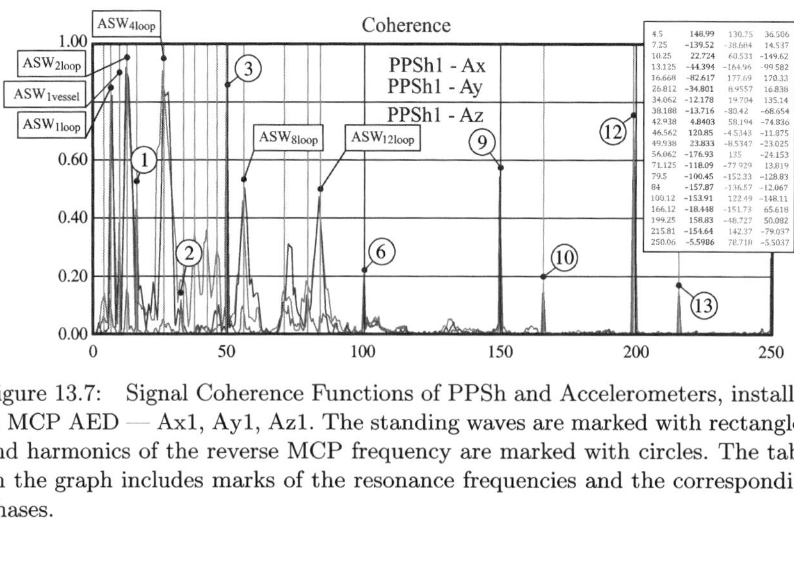

4.5	148.99	130.75	36.506
7.25	-139.52	-38.684	14.537
10.25	22.724	60.531	-149.62
13.125	-44.394	-164.96	-99.582
16.668	-82.617	177.69	170.33
26.812	-34.801	8.9557	16.838
34.062	-12.178	19.704	135.14
38.188	-13.716	-80.42	-68.654
42.938	4.8403	58.194	-74.836
46.562	120.85	-4.5343	-11.875
49.938	23.833	-8.5347	-23.025
56.062	-176.93	135	-24.153
71.125	-118.09	-77.929	13.819
79.5	-100.45	-152.33	-128.83
84	-157.87	-136.57	-12.067
100.12	-153.91	122.49	-148.11
166.12	-18.448	-151.73	65.618
199.25	158.83	-48.727	50.082
215.81	-154.64	142.37	-79.037
250.06	-5.5986	78.718	-5.5037

Figure 13.7: Signal Coherence Functions of PPSh and Accelerometers, installed in MCP AED — Ax1, Ay1, Az1. The standing waves are marked with rectangles, and harmonics of the reverse MCP frequency are marked with circles. The table on the graph includes marks of the resonance frequencies and the corresponding phases.

waves of the coolant is much higher than that coming from other sources. The ASW forces the AEM oscillations via the first path, through the spacers bypassing the torsion coupling. From the coherence functions, one can make an unexpected conclusion: not only the lower ASW types (the first types of the loop and casing ASW), but also $\text{ASW}_{4\text{loop}}$, $\text{ASW}_{8\text{loop}}$ and $\text{ASW}_{12\text{loop}}$ contribute to AEM vibrations. These three higher types of the loop ASW have a spatial joint in the MCP installation place along the MCC pipelines.

The proximity signal can be attributed to the class of polyharmonic processes, i.e. to the deterministic processes represented by the sum of sinusoidal processes with sinusoid frequencies being in harmonic relations (Fig. 13.8). In this case, all frequencies of this series refer to harmonics of the MCP rotational frequency. There are all successive harmonics of the MCP rotational frequency in both APSD

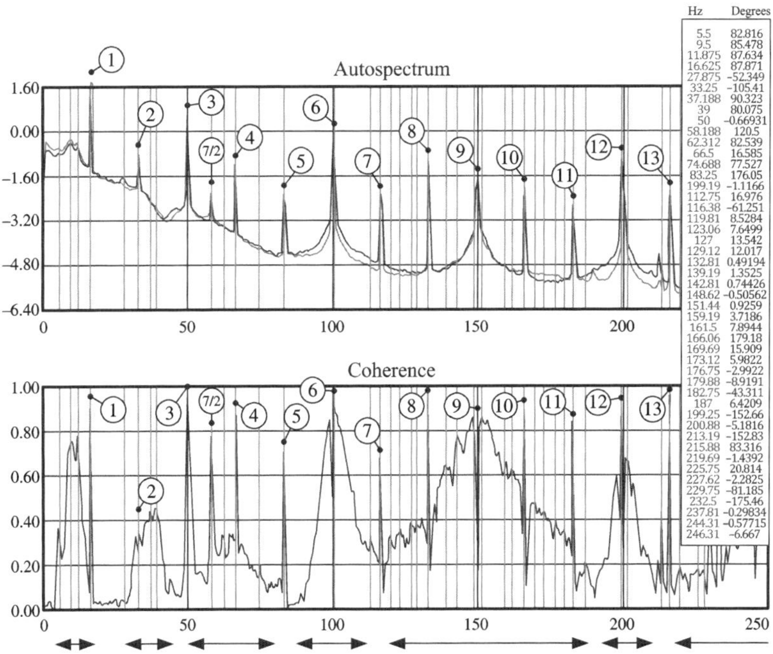

Figure 13.8: APSD of Signals of Accelerometers, installed to monitor MCP Vibration Displacement — Px1 and Py1 (Upper Graph) and their Coherence Function (Lower Graph). Harmonics of the reverse MCP frequency are marked with circles and the broadband effects are marked with double arrows. The table on the graph includes marks of the resonance frequencies and the corresponding phases.

of proximity signals (Fig. 13.8) and APSD of accelerometer signals (Fig. 13.6). The proximity sensor as a detector of the MCP shaft position, more precisely, as a shaft vibration displacement sensor relative to the MCP casing, is the most sensitive to the phenomena of the shaft imbalance.

This mechanical source causes vibrations on all successive harmonics of the rotational frequency.

Unlike MCP AEM, the MCP shaft is not sensitive to high magnitude of lower ASW: ASW_{1loop}, $ASW_{1vessel}$, ASW_{2loop}. At low frequencies, near 5 Hz and above, the APSD signals of proximity sensors contain highly coherent resonances in frequencies other than the central ASW frequencies. The lower type ASW, having spatial wavelengths much larger than the dimensions of the impeller, are almost unchanged in amplitude within the spherical MCP casing and they do not affect its shaft. The lower ASW types excite vibrations of the entire MCC loop by their large spatial wavelengths. These vibrations force AEM vibrations through the MCC pipelines bypassing the MCP shaft. Thus, the AEM stator which is subject to vibrations at ASW frequencies in horizontal plane affects AEM rotor vibrations via electromagnetic bonds. The AEM rotor is "softly" connected with the MCP shaft by the coupling in horizontal direction. As a result, a low-frequency modulation of oscillations at the rotational frequency of both AEM rotor and MCP shaft, causing their precessions, is possible. The adverse MCC pipeline vibrations at ASW frequencies can cause abnormal MCP vibrations even though the natural vibration properties of MCP comply with all design requirements.

An exclusive feature of the proximity signal spectrum is a resonance at the frequency of the loop ASW fourth harmonics about 28 Hz. This resonance may occur due to the fact that this ASW_{4loop} has a node in the space falling on the MCP shaft, i.e. the maximum gradient of its amplitude. The idea that the MCP shaft is under a strong influence from this acoustic wave of the highest type — ASW_{4loop} – can be concluded from the coherence function between "PPS–proximity" signals (Fig. 13.9). One can observe the similar effect to the ASW_{4loop}, having a node in the space of highest even ASWs on MCP: ASW_{6loop}, ASW_{8loop} and ASW_{12loop}.

Apparently, a sub-harmonic at frequency $\frac{7}{2} \cdot f_{MCP}$ was formed because of the nonlinear interaction of the coolant flow with 14 blades of the guide vane.

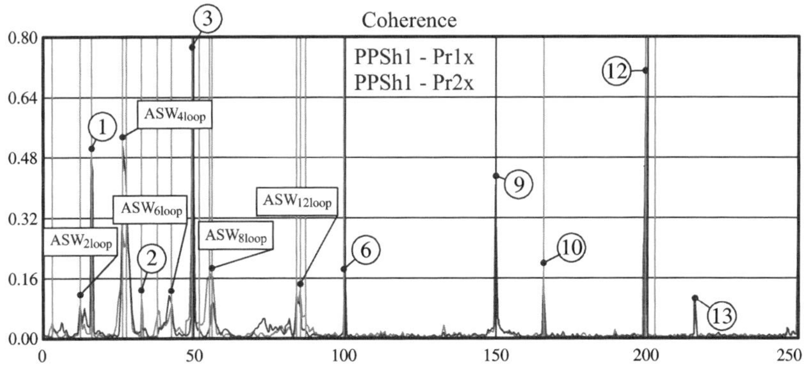

Figure 13.9: Coherence Function of Pressure Pulsation Sensors Signal – PPSh and Proximity sensors. The standing waves are marked with rectangles, and harmonics of the reverse MCP frequency are marked with circles. The table on the graph includes marks of the resonance frequencies and the corresponding phases.

The coherence function of proximity signals included broadband resonances, marked by double arrows in Fig. 13.8. They are in the vicinity of the frequencies of the "round" mains harmonics: 100 Hz, 150 Hz, 200 Hz. The overlay of low-quality and high-quality resonances of two effects can be observed again. However, this superposition has a fundamentally different character. If the high-quality and low-quality resonances were superimposed on one another and summed for a pair of "accelerometer–accelerometer" for the same type of sensors fixed on AEM (Figs. 13.4 and 13.5), then in case of a pair of "proximity–proximity" signals the high-quality resonance on the rotational frequency harmonic is subtracted from the broadband mains resonance. Now one source has a pump origin (but not MCP), the other shall be attributed to the "turbulence" class and their general localization shall be attributed to the shaft sealing system, where both sources are generated from. These sources are in the opposite phase, as their amplitudes are subtracted from each other, so the graph (Fig. 13.8) demonstrates valleys (local minimums) at the listed mains frequencies. An increase of the amplitude of effect causes an amplitude decrease of the other, or an increase of water sealing by the auxiliary loop pump reduces the coolant leaks through the sealings and vice versa.

The first four resonances of non-acoustic origin in the coherence function of proximity signals at frequencies 5.50; 9.50; 11.87; 16.62 Hz have the phases quite close to 90° (see table in Fig. 13.8). The sensors themselves (two proximity detectors) are also geometrically located at an angle of 90°. When the phase divergence of two sinusoidal signals of the one frequency and one amplitude is 90°, the corresponding Lissajous figure (motion trajectory) is exactly a circle. Since the total magnitude of vibration sources at the listed frequencies exceeds the total magnitude of all other sources, and the signal amplitudes in two horizontal directions are approximately equal, this results in the shaft trajectory being quite close to the circle one. All other spectral components with frequencies above 16.6 Hz (including the blade frequency close to 100 Hz), having a lower amplitude, contribute only to "thickness" of the line close to the circular trajectory.

As we emphasized, above the ASW are not observed in the proximity signals directly on the natural frequencies of the lower-type ASW. However, the ASW indirectly affect oscillations of the shaft through the combination of frequencies. Two high-magnitude resonances in APSD of proximity signals at frequencies of 5.50 Hz and 9.50 Hz are located at combination frequencies $f_{\mathrm{MCP}} - f_{\mathrm{ASW}_{1\mathrm{vessel}}}$ and $f_{\mathrm{MCP}} - f_{\mathrm{ASW}_{1\mathrm{loop}}}$. Thus, the low-frequency range of proximity signals before the MCP rotational frequency and including the MCP rotational frequency, on the one hand defines a motion trajectory of the MCP shaft, and on the other hand, depends on ASW which can cause shaft precession.

These conclusions are confirmed by the coherence functions of accelerometers and proximity sensors as well (Fig. 13.10). These two types of sensors in the horizontal plane are installed unidirectionally and in mutually perpendicular directions (Fig. 13.3). If their phase relations are zero, the AEM stator and MCP shaft move synchronously along their cylindrical generatrixes. If the signals are anti-phased, their trajectories correspond to the cone generatrix, which is typical for precession. The anti-phases in the table of Figure 13.10 are observed in two lower combination frequencies of modulation due to $\mathrm{ASW}_{1\mathrm{loop}}$ and $\mathrm{ASW}_{1\mathrm{vessel}}$ effects: $n \cdot f_c - f_{\mathrm{ASW}_{1\mathrm{n}}}$ and $n \cdot f_c - f_{\mathrm{ASW}_{1\mathrm{vessel}}}$. These coherence functions again confirm the previously identified conditional division of the frequency range into two parts: up to 100 Hz — impact of hydrodynamic sources (especially ASW) and above 100 Hz — the highest harmonics of the MCP rotational frequency. The AEM mains supply frequency is not exactly

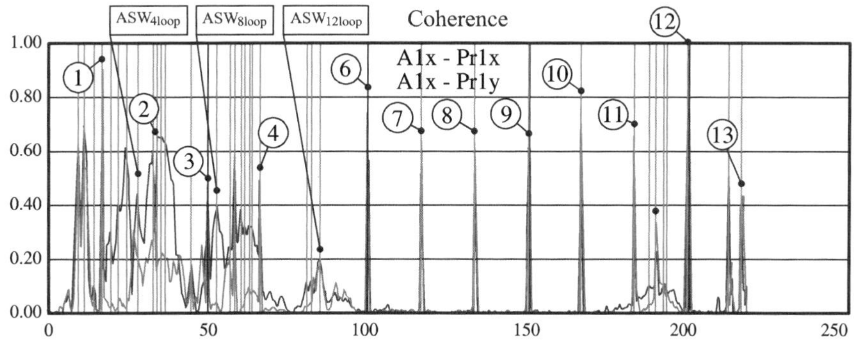

Figure 13.10: Coherence function of signals of accelerometers Ax, Ay and Proximity sensors Px, Py. The standing waves are marked with rectangles, and harmonics of the reverse MCP frequency are marked with circles. The table on the graph includes marks of the resonance frequencies and the corresponding phases.

50 Hz and it is *a priori* unknown. With an unknown slip parameter — s — the RCP rotational frequency is also *a priori* unknown. A non-zero frequency step or a finite spectral resolution of the APSD digital evaluation can lead to a situation where the respective adjacent resonances of two series $n \cdot f_c$ and $m \cdot f_{\mathrm{MCP}}$ will "merge" into a single resonance, i.e. they will be spectrally unresolved. Consequently, these series, having common frequencies and genetically related to each other, can correlate.

If one of the harmonic series has undergone some transformation resulting in the broadening of resonances, i.e. their reliabilities became worse, then the non-zero coherence will be observed at common frequencies of series $n \cdot f_c$ and $m \cdot f_{\mathrm{MCP}}$. Thus, the pump of auxiliary circuit of MCP sealing water subsystem has a resonance at its rotational frequency, but also received from the mains frequency. This resonance is diluted due to the turbulent throttling effects of the primary water "sealing" by excessive pressure from the auxiliary loop. As a result, the "wide" resonances can be observed in APSD of proximity signals at mains frequencies, correlated with resonances at the MCP rotational frequency (and not at the rotational frequency of the auxiliary pump). Figure 13.11 shows in a simplified form the main sources of MCP components' oscillations.

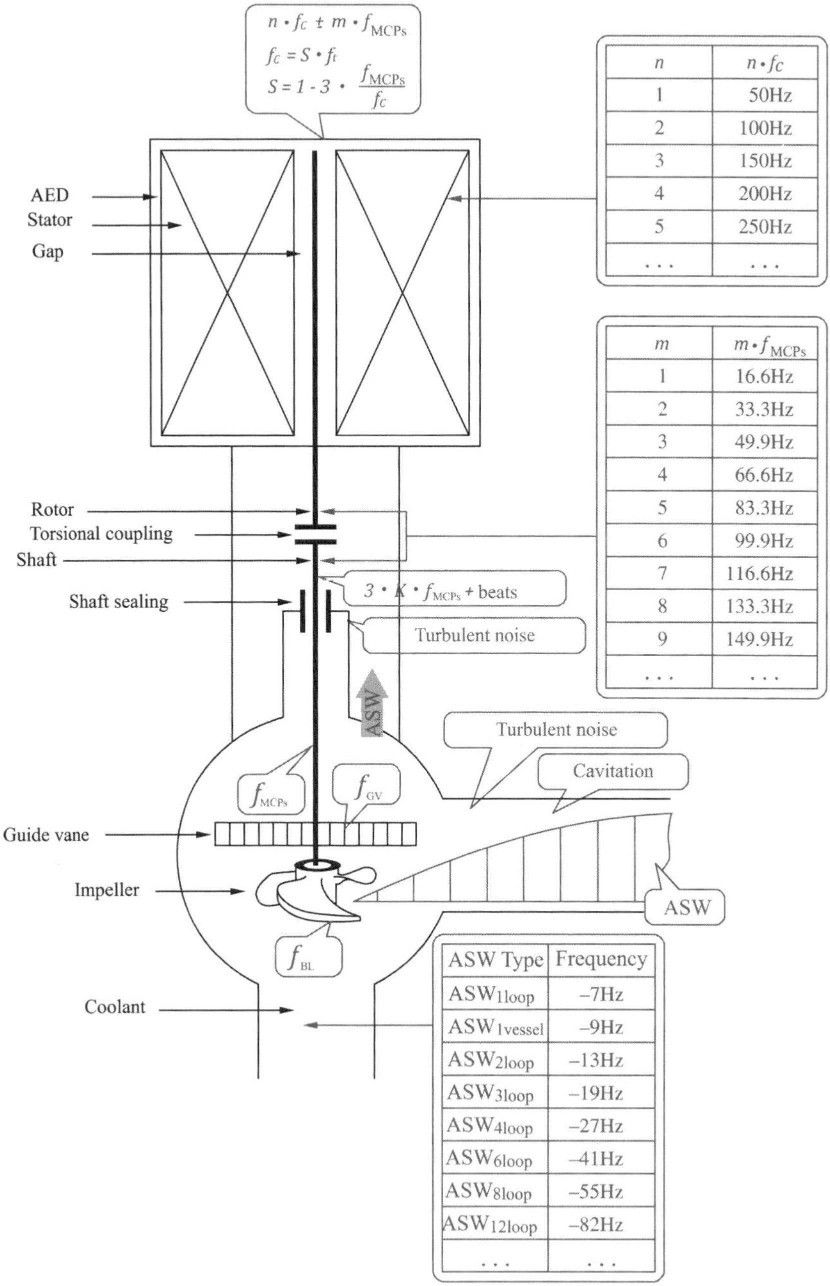

Figure 13.11: Localization of the main vibration sources of MCP.

13.4 Motion Trajectory of AEM and MCP Shaft

Mutually perpendicular arrangement of two accelerometers — Ax and Ay — in a horizontal plane and two proximity sensors — Px and Py — in the other horizontal plane (Fig. 13.3) allow to build displacement trajectories of AEM and MCP shaft in these two horizontal cross-sections. The trajectories for each MCP are reproduced twice for two difference time intervals (Table 13.1). These almost identical graphs according to the same corresponding cells of Table 13.1 illustrate the time stability of trajectories as diagnostic non-spectral attributes. All four MCPs comply with GOST requirements for the relevant vibration velocity RMS. At the same time, the differences of the oscillation magnitude of four MCP shafts are within 40%; the shapes of trajectories are fundamentally different from each other. If the shaft of the first MCP can be at any point within the ellipse (the first row, the first two columns of Table 13.1), then the shaft of the second MCP can be within the circle (the second row, the first two columns of Table 13.1). There are banned areas for shafts of the third and fourth MCP, so the trajectories form almost circular areas. This type of trajectory is the evidence of the low-frequency precession of the shaft, and high-frequency components of vibration process define the thickness of the ring.

Table 13.1 reflects not only the numerical values of the respective vibration swings and general nature of the trajectories, but also minor features. If the external boundary of the trajectories is a smooth function in the trajectory of proximity signals, the external boundary of trajectories according to accelerometer signals has numerous fractures of the small amplitude. They are evenly distributed along the outer boundary and correspond to the high-frequency components of accelerometer signals of electromagnetic origin. On the contrary, the angular anisotropy is observed in the trajectory of MCP shafts. For example, the shaft of the first MCP has a naturally large random component when passing the lower left quadrant. The shaft of the third MCP has a naturally small thickness in the upper left quadrant. The shape of the inner generatrix of the ring has been reproduced exactly in time. Regular breaks of trajectories are observed in the inner envelope of the ring at perfectly defined angular and amplitude coordinates, while the outer envelope is a quite smooth circular function. While the shaft approaches its

Table 13.1: Shaft trajectories and AEM of four MCPs in two experiments.

	Shaft trajectory according to proximity signals		Trajectory of AEM accelerations according to accelerometer signals	
	Experiment No. 1	Experiment No. 2	Experiment No. 1	Experiment No. 2
MCP1	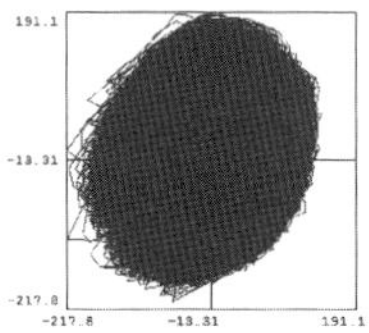	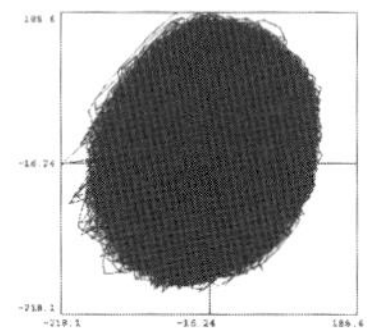	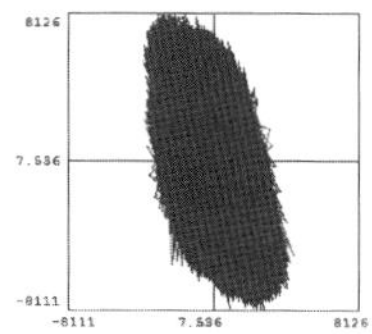	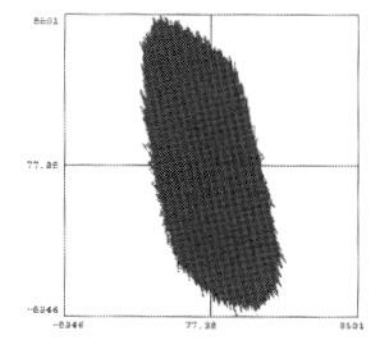
MCP2	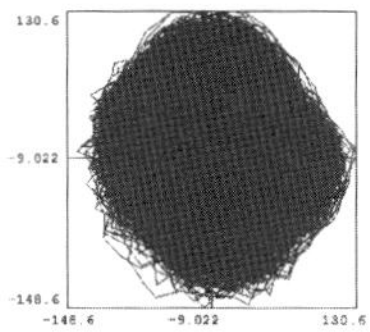	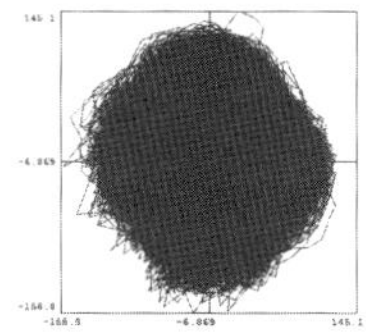	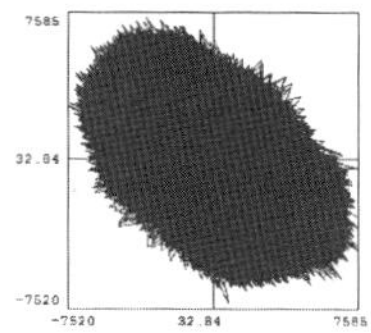	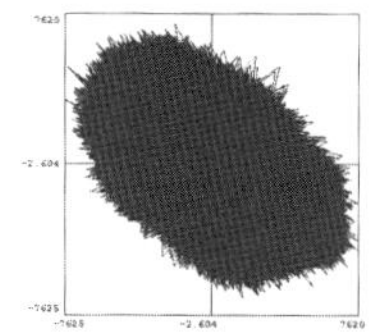
MCP3	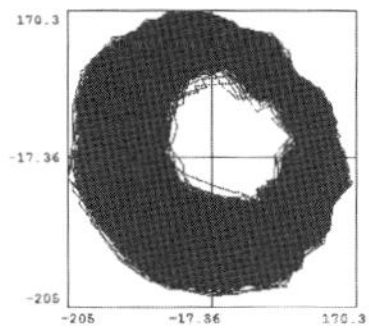	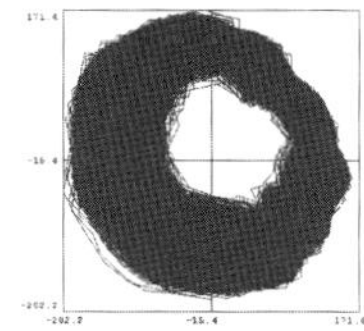	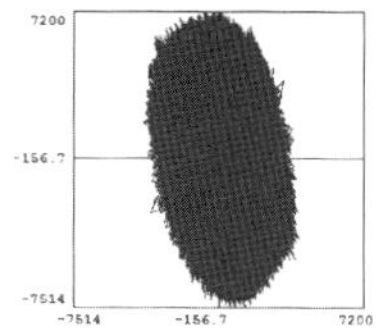	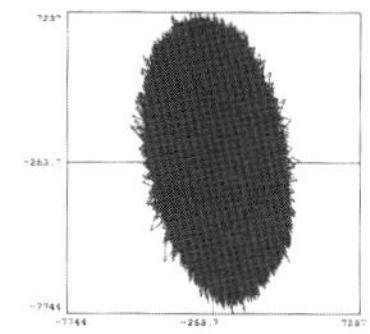
MCP4	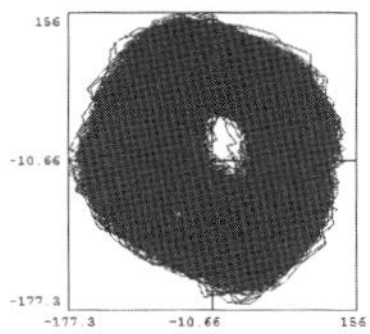	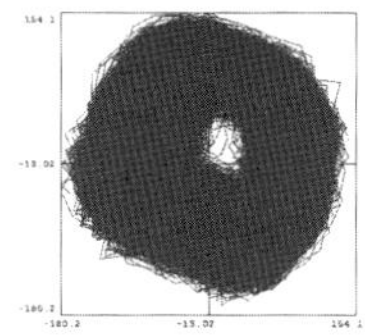	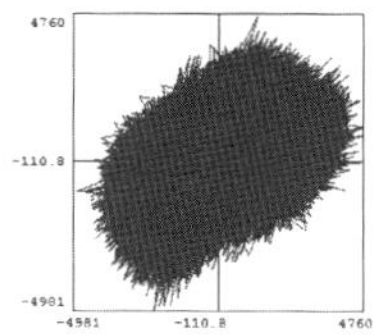	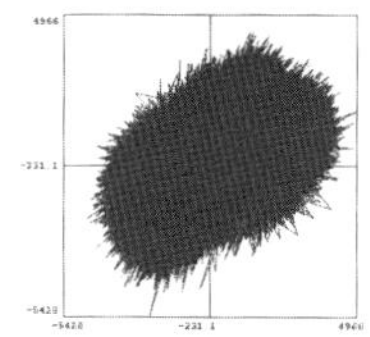

vertical equilibrium position, the obstacles occur, and the shaft colliding on these obstacles suddenly changes its trajectory. This is most typical for the shaft of the third MCP.

Distinctive features of the third MCP can also be detected by means of high-resolution spectral analysis. Figure 13.12 includes APSD of four unidirectional signals of proximity sensors — Px, installed in four MCPs. Only the APSD corresponding to the third MCP has low-frequency resonances of the small amplitude at combination frequencies. For all other spectral features and for harmonic series of the MCP rotational frequency, the APSDs of the same name almost coincide with each other.

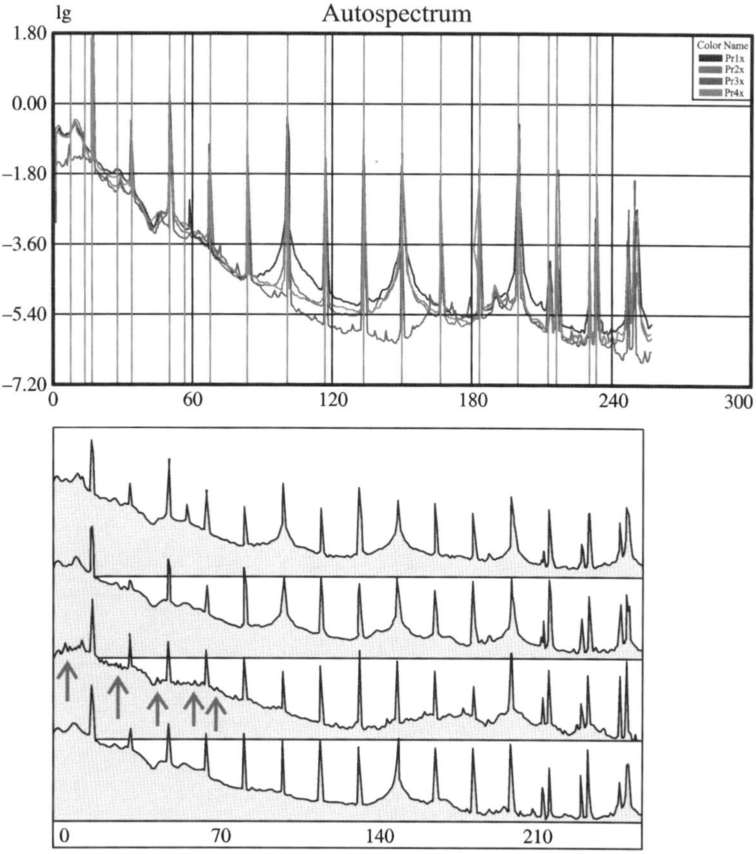

Figure 13.12: APSD of signals of four proximity sensors on four MCPs. The lower graph shows the waterfall of APSD spectra similar to that of the upper graph.

Thus, the autospectral characteristics of the proximity signals are less sensitive to the singularities of MCP shaft motion than the above trajectories of mutual amplitude characteristics of the same signals.

The same conclusion can be made for the autospectral characteristics of signals of accelerometers, installed in AEM of four MCPs (Fig. 13.13). Even though AEM trajectories of four MCPs are fundamentally different from each other, their APSD are identical to each other in terms of the spectral features. Notice that the considered accelerometers are high-frequency accelerometers, which are not designed for measurements below 10 Hz.

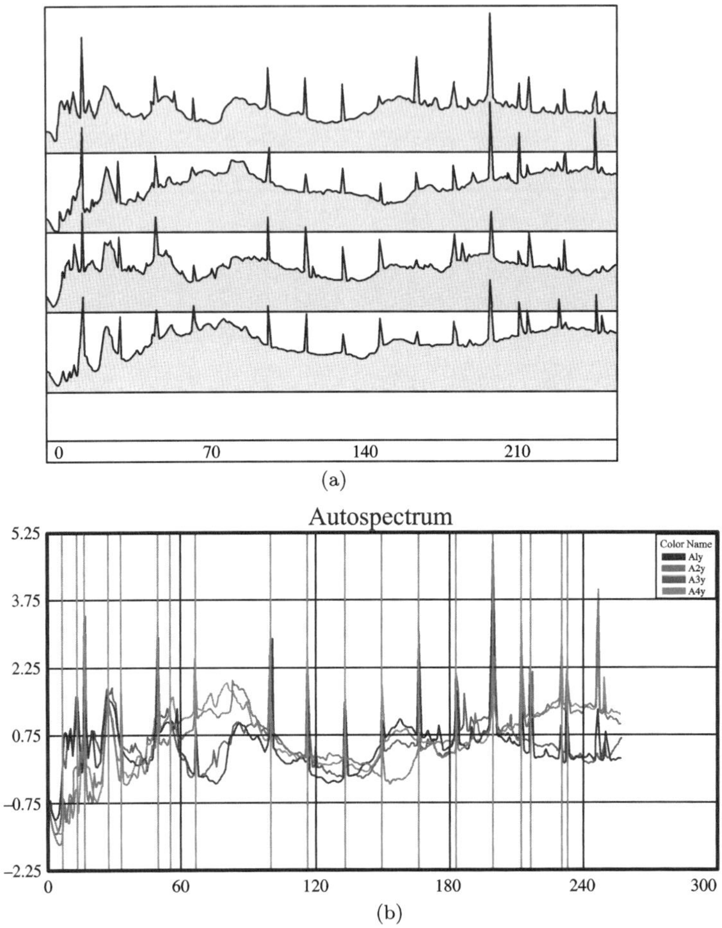

Figure 13.13: APSD of signals of four proximity sensors on four MCPs. The (a) shows the waterfall of APSD spectra similar to that of (b).

13.5 Vibrations of Operating MCP and Vibrations of Turned off MCP affected by Coolant Counterflow — The Single Operating MCP as a Source of Coherent Traveling Pressure Waves

The operating MCP being the main vibration source of the pipelines, the main MCC equipment, reactor plant vessel, reactor internals and fuel assemblies, it is itself an object that is susceptible to the forced vibrations of the coolant. MCC vibrations and coolant pressure pulsations in MCC interact, i.e. they are described by a direct and reverse connection.

During commissioning, different variations of operating MCP equipment of heat exchange MCC (SG, MCP itself, and pipelines) can exist in status of passive and active loops. The excitation of oscillations in the passive loop occurs due to the coolant moving in the so-called counterflow. The amount of counterflow in the given loop with any number of operating MCPs in other loops is always higher than that of the direct coolant flow through that loop with four MCPs operating in a standard mode. The coolant turbulence in the active loop is always higher than that of the passive loop, when the primary sources of the turbulence (MCPs of active loops) are at a significant distance from each other. On the one hand, the passive loop has much weaker vibration load compared to vibration load of the active loop equipment. On the other hand, the passive loop is subjected to different vibration loads at different numbers of the operating MCPs.

There is a well-known problem of the modal analysis — the excitation of all types of natural oscillations of the structure. Its response vibration depends on the amplitude, location and spectral composition of oscillations excited. It is obvious that the main source of oscillations or vibrations in the active loop is the pump itself. First, it is a source of polyharmonic perturbation on harmonics and sub-harmonics of the MCP rotational frequency. In this case, the polyharmonic processes are deterministic superposition of sinusoidal processes at multiple frequencies (harmonics) and fractional frequencies (sub-harmonics). The presence of sub-harmonics is a sign of nonlinear effects, especially shock interactions of moving and stationary components of MCP.

The MCP also generates random non-resonant almost "white" hydrodynamic pulsations of the coolant flow in a broadband of frequencies. Thus, the vibration energy generated by MCP is concentrated not only in deterministic harmonic processes, but also in purely random broadband processes. Turning off such a generator and switching to remote sources of exciting the oscillations significantly changes not only the amplitude properties, but the spatial and spectral picture of the disturbing forces.

The passive MCP is typically represented by a concentrated mass (unlike the pipelines, where a distributed and linear mass are handled) with its stiffness characteristics. The stiffness and mass are two parameters defining the natural frequency of MCP (the natural but not rotational frequency of MCP — f_{MCP}). However, such a simplified representation is not quite adequate even for simple diagnostic tasks. A broad range of oscillations at natural frequencies is observed in MCP. In this case, the concept of stiffness is generalized by the stiffness matrix, each component of which is responsible for its own type of oscillations. This or that type of oscillations has some individual features of its excitation. An operating MCP can mask some types of natural oscillations. Their amplitude can be negligible compared to the level of broadband noise of turbulence or cavitation origin, which are also quite broadband.

On the contrary, hydrodynamic excitation of the passive MCP can more precisely identify the resonances of the MCP natural oscillations at a low background masking.

Thus, switching the pumps on and off allows, on the one hand, determining the origin of a resonance, and on the other hand to identify mechanisms of oscillation's excitation in terms of their spatial distribution (globalism–locality).

13.6 Amplitude Modulation of Oscillations of the Switched-off MCP

During operation of other MCPs, the disabled MCP is a passive vibrating component of MCC under impact of the reversed current of the coolant. The oscillation amplitude of the passive MCP is much lower than the oscillation of the enabled MCP, but their amplitude modulation differs by form (see Table 13.2). It is a low-frequency modulation, which is most likely caused by the beating. It is a persistent effect, which is always observed regardless of the numbers of the operating MCPs. Except for the case when only a single MCP is operating, then the oscillations of three passive MCPs are free of this modulation effect. The strictly coherent perturbation of a single MCP does not cause this modulation, it is necessary to have two or three active sources (operating MCPs) with very close excitation frequencies, and due to their interaction the passive MCP would have the beating effect.

Table 13.2: Signals of accelerometers installed in MCPs.

	Implementation of signals of four accelerometers installed in MCP motor	On (+) and off (•) MCPs	Period of envelope [s]
MCP1 MCP2 MCP3 MCP4	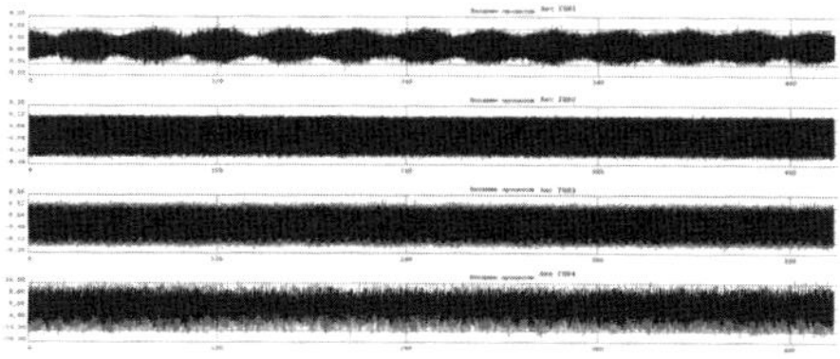	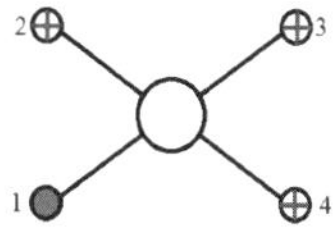	40 s
MCP1 MCP2 MCP3 MCP4	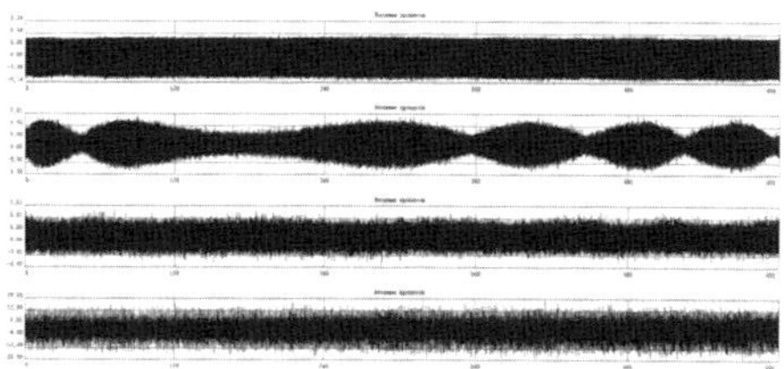	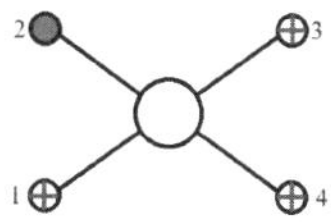	—
MCP1 MCP2 MCP3 MCP4	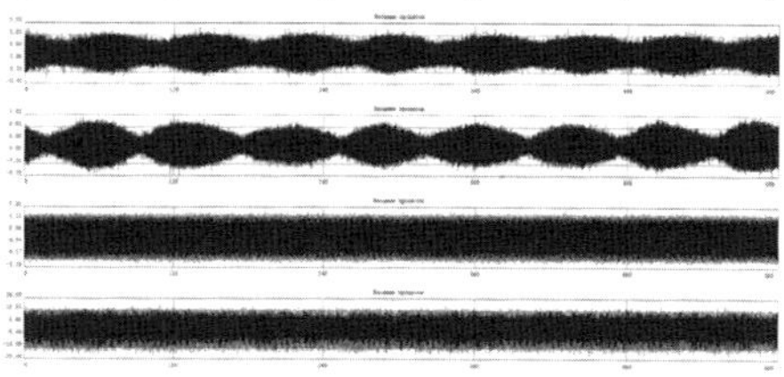	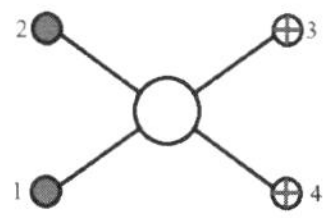	75
MCP1 MCP2 MCP3 MCP4	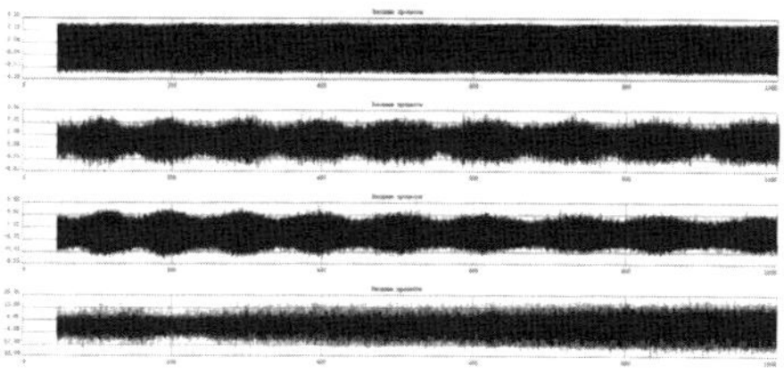	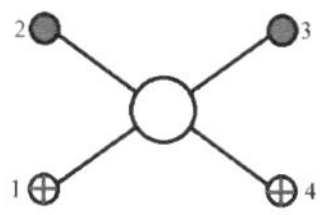	100
MCP1 MCP2 MCP3 MCP4	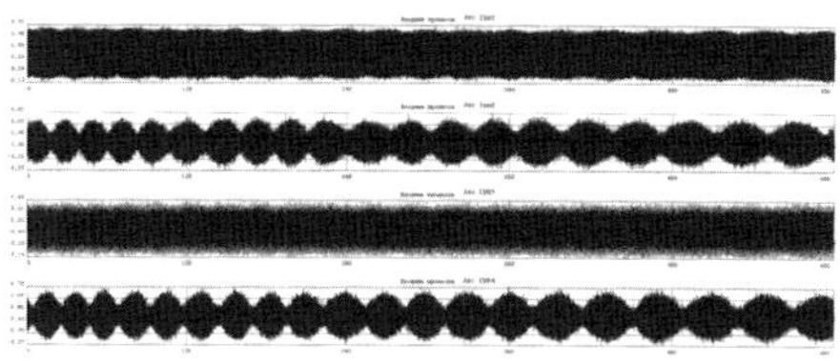	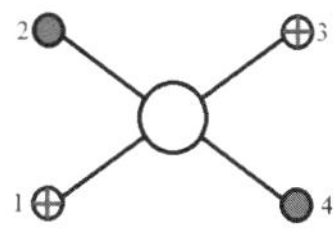	30

It is possible that the observed effect is akin to the effect of the MCP axis precession (whirling effect), when the rotation axis deviates from the symmetry axis by its constant force, and the force itself is a result of interaction between the coolant pressure pulsations from several sources.

The observed modulation is periodic, although it is non-stationary, and its beat frequency can change in time (see the third and last rows of Table 13.2). The phase characteristics of modulations of each passive MCP may differ, but they are the same in frequency, even when the frequency changes in time.

Figure 13.14 shows signals of accelerometers of two passive MCPs with two other MCPs in operation. This is the case presented in the last row of Table 13.2. The frequency of modulations changes synchronously: at the end of implementation, the modulation frequency synchronously decreases on two passive MCPs. At the same time, according to the envelope form of these signals (see Fig. 13.14(b)), the beating of two MCPs are always in opposite phases.

When two MCPs are operating and the other two MCPs are off, we have a classic example of beat frequency being the difference in rotor frequencies of two operating MCP $|f_{\text{MCP}\,i} - f_{\text{MCP}\,j}|$. The difference of these frequencies occurs only due to the difference of the slip parameter of these two MCPs — s:

$$s = 1 - P\frac{f_{\text{MCP}}}{f_c}$$

where f_c is a frequency of power supply circuit ($f_c = 50\,Hz$ in an ideal case), P is a number of pairs of AEM poles ($P = 3$ for VVER-1000 and VVER-1200).

That is, the beat frequency is defined by the difference between the slips of the two active MCPs:

$$|s_j - s_i| = 3 \cdot \frac{|f_{\text{MCP}\,i} - f_{\text{MCP}\,j}|}{f_c}$$

The s-parameter is an important diagnostic indicator of MCP operability and it has quite a small value, which is difficult to estimate directly by means of the frequency analysis.

The actual deviation of the RCP rotor frequency from the ideal value 16.(6) Hz = 50:3 with an acceptable error is difficult to define

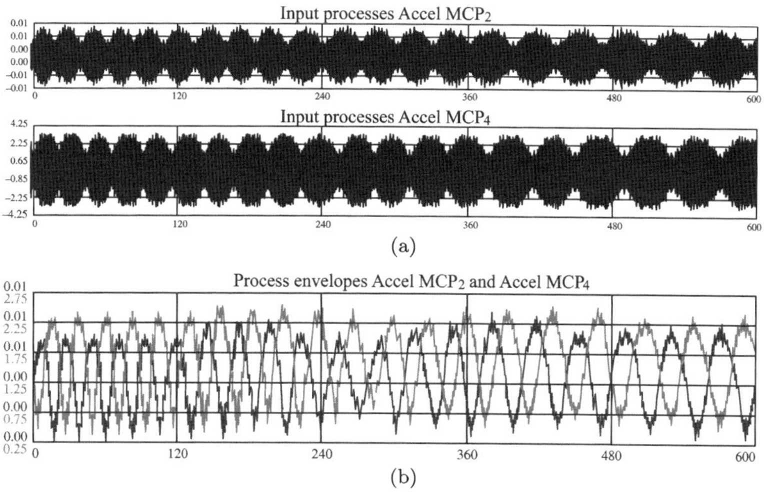

Figure 13.14: Beating of passive MCPs (MCP_2 and MCP_4) in operation of MCP_1 and MCP_3 (a) and envelopes of these signals (b).

by experiment. This requires long measuring times and a high-resolution spectral analysis.

While performing the commissioning measurements, various combinations of number of operating MCPs are tested. Only six of them are the cases when two of four available MCPs operate simultaneously. As a result, we obtain six linear equations with four unknowns — $s_i, i = 1, 2, 3, 4$:

$$\left\{\begin{array}{l} |s_j - s_i| = \ \ \Delta_{ij}; \\ i < j \\ i = 1,\ 2,\ 3 \\ j = 2,\ 3,\ 4 \end{array}\right\} \tag{13.2}$$

The equations in the system are linearly dependent, and the rank of the system matrix is three, i.e. it is impossible to obtain a solution in the form of estimation of four slides $s_i, i = 1, 2, 3, 4$. However, it is possible to rank the slips of four RCPs. The more the difference of rotational frequencies of two operating MCPs $f_{\mathrm{MCP}\,i}$ and $f_{\mathrm{MCP}\,j}$, i.e. the more the Δ_{ij} parameters, the lesser the beating period of two passive MCPs. Very short beating periods correspond to very different rotational frequencies of MCP. The extension of the beats period up to the absence of beats corresponds to the rotational frequencies, close to each other. If we rank six numbers Δ_{ij} in descending order, the first elements of this series correspond to a large difference of the slip parameters of two MCPs, and the last elements correspond to MCP pairs with sliding parameters being close to each other. If the same MCP number appears among the first elements of the series, then it most likely has the worst slip. If the same MCP number appears among the last elements of the series, then it most likely has a minimum slip rate.

For example, three of six possible combinations of operating MCP pairs are given in the last rows of Table 13.2. The shortest periods of beats are observed in pairs with operating MCP3 and the longest periods are observed in pairs with operating MCP4. Therefore, a preliminary conclusion can be made that the slip in MCP4 is less than that of MCP3. This proprietary method has an estimating and indicating character, ranking four MCPs in ascending (descending) slip. However, it is enough to numerically determine the slip of only one MCP, e.g. the high-resolution spectral analysis to find the numeric values of slip of other MCPs from the system (13.2). Thus, the measurements of the beat periods of the signals of accelerometers, installed in MCPs, generate diagnostic features of MCP operability with two operating and two passive MCPs.

13.7 Detection of Cavitation Effects according to PPS Signals

To some extent the MCP operation is always accompanied with the cavitation phenomenon, involving generation of gas micro-volumes (micro-bubbles) from gases, dissolved in the coolant with subsequent destruction of these micro-bubbles as a result of the coolant pressure micro-pulsations in close vicinity of the rotating surfaces of the pump. The water vapors can serve as gas and then the cavitation is called a vaporous cavitation, which occurs as a result of coolant boiling in pressure drop.

The cavitation can be caused not only by the turbulent pressure pulsations, but also by the negative ASW half-waves, corresponding to periodic decrease of the coolant pressure. The assumption on the identity of heat exchange loops is valid only for the lowest ASW_{1loop} type. Therefore, there are adverse loops in considering excitation of the cavitation effect by means of the higher ASW types.

The negative effects of cavitation are well known, and the main ones are the wear of the pump impeller surfaces and generation of pressure shock waves, which may cause hydraulic shocks in the circulation circuit. The cavitation level is defined by both the microscopic coolant parameters — pressure, temperature, density, velocity, concentration of dissolved salts and gases, and by the effects affecting the coolant pressure pulsations (ASW, turbulence parameters). It is obvious that the above parameters have high variability in various modes of reactor plant operation, but for any design modes of reactor plant operation, a low and negligible cavitation level is established by calculations. For example, the coolant temperature in the loops and pressure in MCC in any standard modes are selected so that the margin to the coolant boiling temperature never depletes in any pressure fluctuations due to ASW or turbulent effects.

However, it is advisable to experimentally confirm a negligible coolant cavitation level during commissioning measurements by interpreting the characteristic features of the pressure pulsations. From that two main features shall be allocated: locality and broadband behavior. The cavitation is a local phenomenon, meaning not only the birth of gas bubbles, but also their collapse is caused by the local changes of coolant pressure. The cavitation is always higher where the source of turbulence of the coolant flow is located, e.g. in the local vicinity of MCP flow part volume. If the average life of the gas bubbles is long enough, then before collapsing it can be transported by the coolant flow to a certain average path length and thereby increase the volume of cavitation.

In addition, the origination and death (collapse) of the gas bubble causes a shock wave in pressure, which propagates through the elastic medium, along the flow and against the coolant flow to some distances, which also increases the volumes of cavitation. So, in order to register cavitation, it is unnecessary to have PPS directly within the volume, where the gas micro-bubbles are generated, for example, in the flow in the pump body of MCP. Of course, the sensitivity of PPS channels to cavitation will be higher, the closer they are to the place of bubbles' origin. In this sense, the standard location of three PPSs in the loop during commissioning measurements is not quite efficient for cavitation control. The PPS installed on the U-shaped pipeline under the MCP is closest to MCP. This PPS can register the shock waves, propagating against the coolant flow. The PPS installed in the "cold" pipeline after the MCP can register the cavitation effects transported by the coolant flow and/or propagation of the shock wave caused by origination/collapse of the bubbles. The PPS installed in the "hot" pipeline can register only shock waves.

The broadband effects of cavitation are caused by the sudden origination and collapse of the bubble. A stepwise micro-change of the coolant pressure causes an infinite spectrum of pressure pulsations. The life time of the bubble is connected with the duration of the rectangular pressure pulse of the coolant. When these events massively occur in random (due to turbulence) moments of time with random duration in some local volume, the resulting effect becomes not only broadband, but also gains the so-called white noise property. The pressure pulsations spectrum becomes close to the "white" spectrum, i.e. to some constant with regard to the amplitude in a broad frequency band without visible resonances. Let us again turn to the modes with the first MCP operating at coolant temperature of about 120°C, considered in the above sections. Figure 13.15 shows all APSD of three PPS signals of the first loop, and each class of APSD is presented twice to compare them with each other in pairs. The signal of PPS_U sensor installed in the U-shaped pipeline has a maximum amplitude. Its APSD is characterized by a large white-noise component, which "absorbs" the small resonance features so that only high-amplitude resonances of Pressurizer ~0.8 Hz, ASW_{1loop} ~9.4 Hz and the highest type ASW ~51.0 Hz can be seen.

The white-noise component is also observed in PPS_{c1} APSD (Fig. 13.16). It has much lower amplitude than the PPS_{3U1} APSD, but it is correlated with it. This is an important result, two white-noise sources do not correlate with each other, i.e. they have an auto-correlation function — δ-function, and the mutual correlation function and coherence function are identical – zeros. The correlation of the two white noises means that in each moment of time there is a common source of their origin. Such is the phenomenon of cavitation. The origination and subsequent collapse of a small gas (or steam) bubble causes the coolant pressure pulsations, registered by both PPSs, while the cavitation mechanism is a white noise itself.

The coherence function of PPS_c–PPS_U signals reaches high values in a wide frequency range from 10 Hz to 50 Hz without visible resonances (Fig. 13.16).

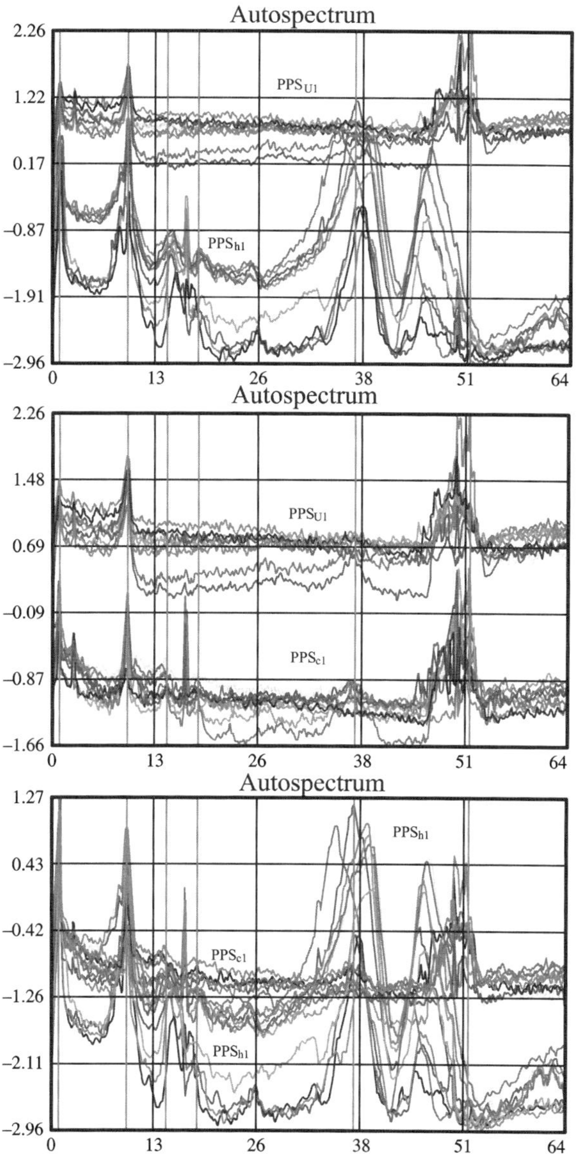

Figure 13.15: APSD of PPS_{h1}, PPS_{U1}, PPS_{c1} during MCP_1 operation. $T_{coolant} = 120°C$.

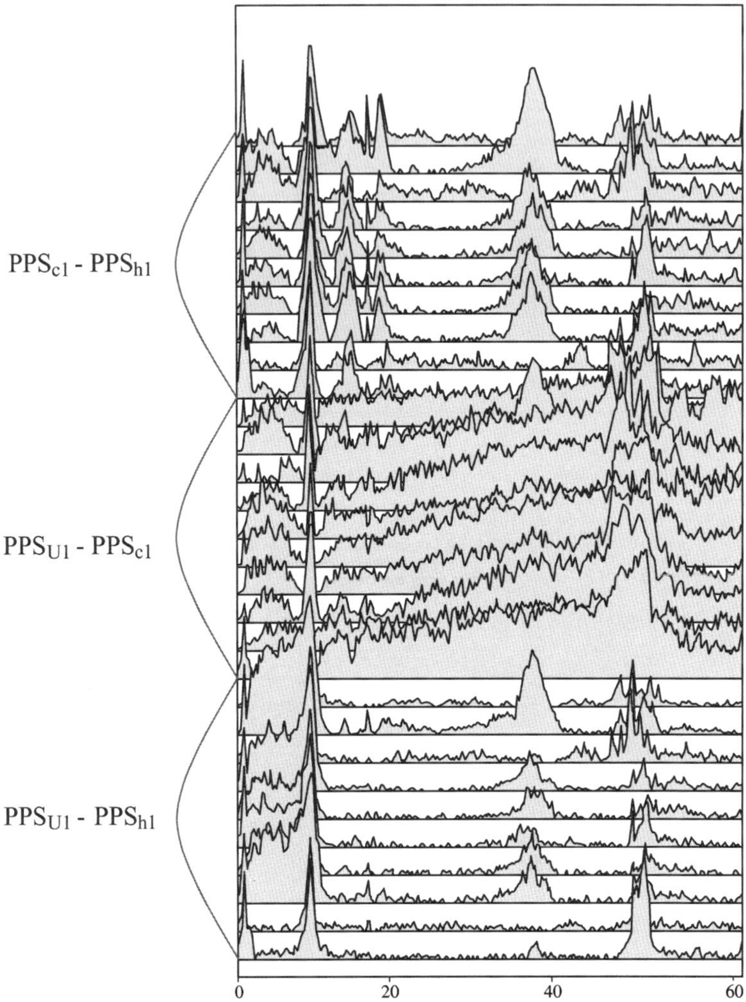

Figure 13.16: Coherence functions of PPS signals: The pair PPS_{c1}–PPS_{h1} from 10 measurement scan be seen in the upper segment. The coherence pairs of PPS_{U1}–PPS_{c1} from the same measurements are shown in the middle segment, and the coherence pairs of PPS_{U1}–PPS_{h1} are shown in the bottom. $T_{\text{coolant}} = 120°$C.

The source of this broadband noise is obviously the MCP located between two PPSs: PPS_U and PPS_c. The noise is localized near the MCP, as its biggest value is observed at the closest PPS_U, and it is almost not observed in the faraway PPS_c. Some bubbles "collapse" in the coolant flow after overcoming a certain distance, therefore the PPS_c, located downstream of the coolant, is more sensitive to cavitation than the PPS_h. Coherence function of PPS_U–PPS_h signals (Fig. 13.16) are almost zero in the wide sub-range of frequencies, where there is a white noise of cavitation. Its absence here, being the additive component in only one PPS_h signal, contrasts with the coherence function of PPS_h–PPS_c signals of the most distant sensors.

Caviation effect has two main properties, the wide bandwidth and localization, to those were joined effect of correlation of signals of two faraway PPSs, between which the MCP is located.

Chapter 14

Experimental Results on VVER-1200 Vibration Acoustics

14.1 Vibration Acoustics Tasks during Commissioning Measurements

The commissioning measurements are characterized by a huge amount of diagnostic information of a temporary nature [188–200]. First of all, they are based on an extremely large amount of instrument channels, installed under the temporary scheme and dismantled upon completion of the commissioning measurements. In particular, the following sensors shall be installed for commissioning measurements:

- vibration sensors on the main equipment, pipelines and RI (reactor internals, core barrel, PTU), FA simulators [194];
- pressure pulsation sensors — in various MCC points and in FA imitator [199];
- displacement and deformation sensors outside the pipelines and main MCC equipment [189].

The composition of the instrument systems for commissioning measurements and standard layout of the sensors is described in [191, 198, 200]. When processing the registered signals during the commissioning measurements, spectral characteristics, RMS and amplitude of oscillations are evaluated. These characteristics with hundreds of instrument channels of commissioning measurement system and variety of reactor plant operating modes generate a flood of reporting data. They are compared with the corresponding characteristics obtained on the reference unit. If the measured characteristics are within the thresholds specified by the Chief Designer, a conclusion on the normal vibration state of the new unit is made. Some mutual characteristics may be given as illustrations but not as criteria functions. It is clear that the number of pairs of mutual characteristics (coherence phases) is so huge that it is almost impossible and unnecessary to sort them. It is necessary to perform a targeted selection of diagnostic-significant pairs of signals, supported by appropriate software (SW) for group and automatic spectral estimation [381, 383].

Characteristics achieved during commissioning measurements are not sufficient to address several vibration acoustic tasks of the new unit. All significant spectral features of the measured signals and resonances of auto and mutual spectral characteristics in particular shall be interpreted. The physical nature of their origin must be proved. After completing the commissioning measurements, the operating personnel of the new unit diagnostic division shall provide specific information about all significant resonances of the measured signals. It is necessary to specify not only the frequency and amplitude, but also the diagnostics thresholds, its origin and localization of the physical phenomenon that caused this resonance, its change in various standard operating modes and dynamics during the long-term operation of the unit.

At the design level, it is impossible to justify the vibration resource of the equipment of the new unit only by calculation, since it is *a priori* unknown which vibration amplitudes and frequencies of various reactor plant components will be achieved. These missing data will become obvious after commissioning measurements. After that you can calculate vibration resource using the real experimental data. If there is no understanding of the origin of a resonance, it is impossible to predict its dynamics during normal operation of the unit.

The regulatory documents, regulating equipment vibrations, require the frequencies of natural oscillations to be mistunes of perturbation frequencies. If the acoustic standing waves related to perturbation are identified in MCC, and natural frequencies of equipment vibrations are distinguished from the multitude of resonance frequencies, then it is not only interpretation of commissioning measurements, but also compliance with requirements of the regulatory documents. Wear of movable supports of the main MCC equipment or change of stiffness of RI or reactor plant vessel fixtures, or change of coolant steam concentration during reactor plant operation, etc. cause shift in frequencies of the relevant resonances. While justifying vibration resource, it is necessary to predict that this trend will not ruin the mistunes of frequencies. Thus, the commissioning measurements result in recording of initial vibration state of reactor plant, setting vibration thresholds according to the main vibration characteristics, interpretation of spectral characteristics and subsequent justification of vibration resource.

However, it is never possible to completely adjust frequencies. While heating up/cooling down the power unit, the frequencies of some ASW will always coincide with harmonics of the RCP rotational frequency and natural frequencies of reactor plant equipment oscillations in some intermediate temperatures of the coolant. It is necessary to demonstrate experimentally that the damping properties of the structure allow the oscillation amplitudes. Such frequency matches in commissioning measurements shall be recorded. It is also necessary to find values of the relevant temperatures of the coolant and vibration levels. During commissioning measurements, it is necessary to prove that no coinciding frequencies are possible at 100% power level of reactor plant.

It is also necessary to define locations (volumes and sections) of structures with the maximum vibration rate in the same regulatory documents. It is a computational and experimental task. The spatial extension of vibration phenomenon is proved not only by installation of sufficiently large number of vibration sensors in commissioning measurements, but also by interpretation of its global/local properties and their computational modeling.

It is insufficient to give references to results of large-scale bench experiments [142, 151, 158, 159, 171, 173, 194, 198] to interpret the full-scale commissioning measurements. It is extremely difficult to simulate relationship between acoustic effects of MCC and oscillations at natural frequencies of some prototype component on a large-scale stand. For a long time, the acoustic effects in MCC were not taken into account, and it was believed that the natural frequency oscillations of equipment components dominate. A huge amount of bench experiments were devoted to them. After long-term field measurements, it became clear that the acoustic effects in MCC mainly determine vibration condition of its components. Firstly, natural oscillations of a single element is an incorrect concept in conditions of mutual oscillations of components of the actual structures. Secondly, the natural oscillations of the structure component, obtained on the stand with other components isolated, cannot be used to justify the frequency adjustment.

Due to huge background of the most powerful acoustic effects of MCC, it is almost impossible to distinguish the natural oscillations of some of the elements of MCC, since they have negligible amplitude. Most often this demonstrates the high damping qualities of the MCC structure.

There is a risk that phenomena, recorded in the bench conditions, will not appear in natural conditions. Consequently, further interpretation of natural measurements will take a wrong way. For example, the shell types of oscillations during bench experiments on reactor internals core barrel (RIC) model were literally visualized by holographic methods. In natural conditions, there are beam types of oscillations, defined by the state of RIC fixture, but not by its oscillations as a thin cylindrical shell. If we verify experimental natural frequencies of RIC oscillations based on the model frequencies of the shell types of RIC oscillations, the resulting and seemingly verified calculation model will be simply harmful for applications.

Today, the commissioning measurement systems establish experimental fact of compliance of the actual vibration state of a new unit with design vibration state and vibration state of the reference NPP unit which is already operating. Experimental studies of vibration acoustic properties of the new unit are a much more extensive task with regard to interpretation of the recorded multi-channel processes.

14.2 Diagnostic Information, Obtained During Joint Measurements of Commissioning System Signals and Standard Diagnostic System Signals

Correlating signals of different physical nature is a key to understanding their physical origin. For example, the studies of the mutual characteristics of signals of pressure pulsation sensors and various vibration sensors allow to solve *in situ* problems of natural and forced oscillations of reactor plant components, and in particular, to determine the amplitude and frequency distribution of various ASWs.

Mutual characteristics of signals of different types of vibration sensors, installed in different equipment, allow the interpretation of their mutual oscillations. Pairs of components "RI–FA", "RI–reactor vessel", "reactor vessel – MCC equipment" are always in state of mutual oscillations. There is a transmitter in these pairs in a form of coolant pressure pulsations. It is necessary to determine a contribution of the transmitter to mutual oscillations of the listed pairs.

A typical example is the interpretation of the resonance of the spectral characteristic existing at the same frequency in the signals of vibration sensors mounted on reactor vessel and on MCP. This resonance is present in both auto- and cross-characteristics. It could be interpreted as natural oscillations of reactor vessel, as natural oscillations of MCP and as their mutual oscillations. In diagnostic applications, it is important to know the "origin" of this resonance. If the resonance is determined by vibration properties of MCP and only its remote effect (vibration response) is observed on reactor vessel, then the monitoring of this resonance (by continuously monitoring its amplitude, frequency and Q-factor) achieves the diagnostics objective of MCP, but not reactor plant vessel. For example, an increase of the stiffness of MCP movable supports due to their wear would increase frequency of this resonance.

On the contrary, when the resonance is defined by oscillations of reactor vessel and oscillations of this massive structure cause forced oscillations of MCP, the monitored resonance performs diagnostics of the reactor plant vessel. Its vibration condition is determined by the stiffness characteristics of vessel attachment points and attachment points of "vessel–RI" type. It is also possible that all weight and stiffness characteristics, of reactor plant vessels and MCPs, affect the frequency of this resonance.

When considering the simplified task of mutual oscillations in case of a single operating MCP and reactor plant vessel, it becomes possible to identify the source of these oscillations. It is difficult to solve a task of localizing the source of oscillatory disturbances in normal operating mode with four operating MCPs, when there are four powerful sources of oscillations, each of which has its own vibration identity.

Compared to the standard commercial operation of reactor plant, the reactor coolant pumps are switched on and off during commissioning measurements by a special program in various combinations.

It is obvious that hydrodynamic state of the coolant and vibration condition of the main equipment and pipelines will significantly differ in the passive and active loops. The operating MCP in the active loop is a powerful and almost point source of oscillations for both coolant and main equipment of the loops. In the passive loop, these disturbances are introduced from the outside by the coolant flow and/or vibrations of components of MCC structures. Two different excitation methods for interconnected MCC components facilitate the task of interpreting their sources.

The field of acoustic standing waves in MCC is generated by an infinite number of reflections, absorptions, diffractions and interference of pressure traveling waves. A single operating MCP is a coherent, almost point source of coolant pressure traveling waves, which further undergoes the above physical transformations. In this case, all traveling waves of a given frequency in various MCC points differ only in amplitude and phase. Their interaction provides a coherent picture of ASW field, when the corresponding ASW resonances have a maximum Q-factor (minimum "width" of the resonance).

Four operating MCPs can be considered as coherent wave sources only with some approximation. If rotational frequencies of four MCPs can be considered similar to each other, then the phase characteristics of traveling waves generated by four MCPs are arbitrary. Therefore, we can expect that at a qualitative level, the central ASW frequencies coincide at any number of operating MCPs, but the other resonance parameters, i.e. amplitude and Q-factor, will be different for each fixed group of operating MCPs. One, single operating MCP will provide the most contrasting picture of stationary (steady) ASW field due to its coherent acoustic broadcasting. This allows to easily identify different types of ASWs. Four sources of traveling waves, even if they are coherent in a form of four MCPs, will generate different resulting ASW amplitudes in any point of MCC, depending on amplitudes and phases of various waves, overcoming various acoustic paths.

During commissioning activities, the coolant changes its temperature in a wide range. The resonances, originated by natural frequencies of some MCC component oscillations, mostly are not affected by the monotonic temperature dependence. Some resonances of natural oscillations can appear or disappear when the coolant reaches a certain temperature. For example, the temperature oscillations change abruptly when the reactor core barrel is compressed by the flow separator starting from a certain temperature of the coolant and above. If before compression the type of core barrel oscillations fluctuated with one top fixed end (the top of core barrel attachment point on a flange of reactor plant vessel), the length of equivalent beam after compression will decrease, and the natural frequencies of cavity oscillations will increase, respectively. Essentially, the coolant temperature has changed the core barrel mounting conditions. The coolant temperature at which the natural frequency of core barrel oscillations abruptly changes shall also be recorded during commissioning measurements.

Core barrel mounting conditions are also changed in the lower horizon of vibration dampers (so-called dowels, bumpers or guide lag). Increasing coolant temperature, the core barrel length monotonically increases. Final mounting of the core barrel into the lower fixture (attachment to guide lag) shall be performed at coolant temperature corresponding to Minimum Controllable Level in accordance with design. Thus, it can be expected that the type of oscillations corresponding to vibration of the beam with two fixed ends will be finally formed at Minimum Controllable Level. Its frequency will be approximately twice as high as the natural frequency of cavity oscillations with one fixed end. RI fixtures are not rigid, so in practice there are several types of RI oscillations.

The upper core barrel fixture consisting of the so-called spring elements of expanded graphite also changes its stiffness and temperature. Therefore, the natural frequencies of core barrel oscillations are changed as well. When the coolant temperature changes, many other splined RI joint assemblies can also change the type of their oscillations and hereby complicate interpretation of vibrations, measured during commissioning measurements.

The same range of coolant temperature will be covered many times during heating up (cooling down) of the unit in commercial operation in accordance with established procedures. In this case, compared to commissioning measurements the power on/off of MCP will be strictly regulated depending on the coolant temperature and unit operating mode.

After sealing the MCC, the reactor vessel lid presses down the core barrel, the protective tube unit and the ensemble of fuel assemblies. Each of these structural components has its own spring units.

When the reactor is compressed, the reactor lid also deforms all the above spring units. Having different stiffnesses they have different pre-compression static forces for RI and FA. The resulting force must be applied strictly in a vertical direction. Otherwise, the loading (including vibration) of RI and FA will be distributed unevenly along the azimuthal coordinate. In addition, the static pre-compression force from the side of reactor vessel lid counteracts the lifting force from the coolant flow, thereby preventing RI and FA "lifting". Non-design static pre-compression forces slightly change the natural frequencies of their vibration, and vertical deviation of the static force results in varying stiffnesses of azimuthally symmetric splined connections, which also changes the natural frequencies of RI oscillations.

The problems of identifying the natural and forced oscillations shall be solved during commissioning measurements at the experimental level, for currently this issue is almost not addressed. The standard diagnostic systems require such information, since their interpretation of the diagnostic events is based on an *a priori* knowledge of the natural frequencies of components subject to diagnostics. Monitoring the changes of oscillation frequency of various reactor plant components during commissioning measurements allows correctly interpreting their nature of origin. This is also facilitated by a small number of excitation points of the structure oscillations in the form of operating MCPs.

Sensors of standard diagnostic system are involved in different combinations: vibrations transducers (accelerometers) mounted on pipelines of MCC and on the so-called Upper Unit on the reactor vessel and, during power operation, also neutron detectors.

Using the correlation of signals of temporary CMS sensors, mounted directly on the controlled equipment with the signals of the standard diagnostic system sensors, which are typically considerably far away, opens the way for some kind of calibration of the latter sensors. Synchronous multi-channel recording of signals of two systems (CMS and VNDS), one of which is temporary (CMS, designed for installation of the new NPP power unit to design parameters), and the other (VNDS) designed for long-term diagnostics of the power unit during commercial operation, shall ensure transmission of unique diagnostic information to VNDS. This information is *a priori* and setup for VNDS, and it has not been provided to VNDS earlier. The current practice of VNDS commissioning in operation provides adjustment and startup of systems only when the power unit reaches the 100% power level. This excludes their use at the stage of commissioning measurement and power ascension.

Mutual measurements of signals of commissioning measurement systems, conducted for the first time during hot and cold trial of Novovoronezh NPP 2 Unit 1, and mutual measurements of VNDS signals allow to obtain reliable experimental diagnostic data, which have not been earlier obtained during startup of the other NPP units in a form of specific spectral characteristics for the standard diagnostic systems.

14.3 Example of Identifying the Primary Source of Joint Oscillations

Multi-dimensional autoregressive analysis (MAR analysis) is an efficient tool for identifying the root cause of common oscillations. Its application in vibration diagnostics is described in the monograph [381]. The easiest way to present the essence of the MAR analysis is to describe it by two-dimensional case.

Let's suppose there are two interconnected signals — $x(t)$ and $y(t)$ (Fig. 14.1).

They interact via direct and inverse connections through transfer functions:

$W_{xy}(f)$ is a transfer function from signal $y(t)$ to signal $x(t)$;
$W_{yx}(f)$ is a transfer function from signal $x(t)$ to signal $y(t)$.

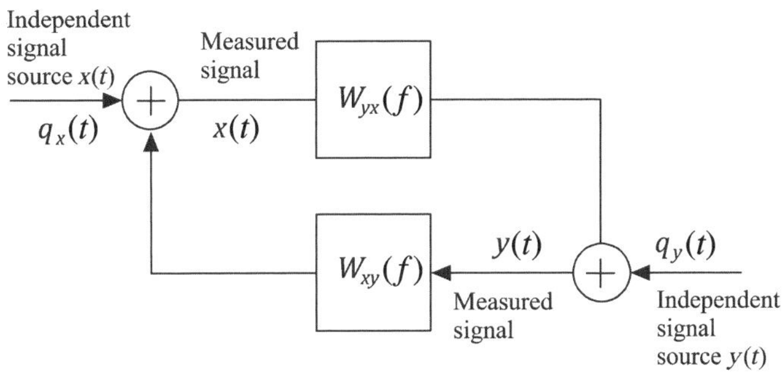

Figure 14.1: Second-order MAR model.

Each signal $x(t)$ and $y(t)$ has its own source $q_x(t)$ and $q_y(t)$. If these sources are independent from each other, i.e. their cross-power spectral density is equal to zero – $S_{q_xq_y}(f) \equiv 0$, the contribution of one signal source to another signal is defined as follows: $A_{xy}(f)$ and $A_{yx}(f)$.

Let us take it for granted that all spectral functions of MAR model (algorithms of their spectral estimation can be found in [381]) are determined by two measured signals, $x(t)$ and $y(t)$:

$$W_{xy}, W_{yx}, A_{xy}, A_{yx}, S_x, S_y, S_{q_x}, S_{q_y}$$

where functions A_{xy}, A_{yx} are defined by the term "normalized global contribution" in addition to the known spectral functions:

A_{xy} is a spectral function defining the parts where the independent source of a signal $y(t)$ affects the measured signal $x(t)$,
A_{yx} is a spectral function defining the parts where the independent source of a signal $x(t)$ affects the measured signal $y(t)$ (inverse contribution).

Based on comparison of the absolute values of these contributions it is possible to make a conclusion on the origin of this or that resonance. The corresponding measured signal ($x(t)$ or $y(t)$) will be the carrier of this resonance. It is possible to interpret only those frequencies where, firstly, the coherence function has a resonance, and secondly, its amplitude is significant. If both conditions are met and contribution in one direction is approximately equal to contribution in the reverse direction, then it is likely that there is a third unaccounted source and MAR model is not adequate. In this case, it is necessary to add other measured signals, i.e. to increase the dimension of the model by more than two.

Let us consider the simple two-dimensional MAR model, composed of signals from two accelerometers, one of which is installed in MCP — Acc_{MCP}, and the other — on reactor vessel — Acc_{V}. These signals are strongly related, as it can be seen from their coherence functions (Fig. 14.2). A simple physical conclusion is obvious here: MCP and reactor vessel are in a state of mutual oscillation. However, oscillations have different qualities as mutual oscillations at each frequency, where coherence function is significantly nonzero, have their own original sources. Forced mutual oscillations at ASW frequencies

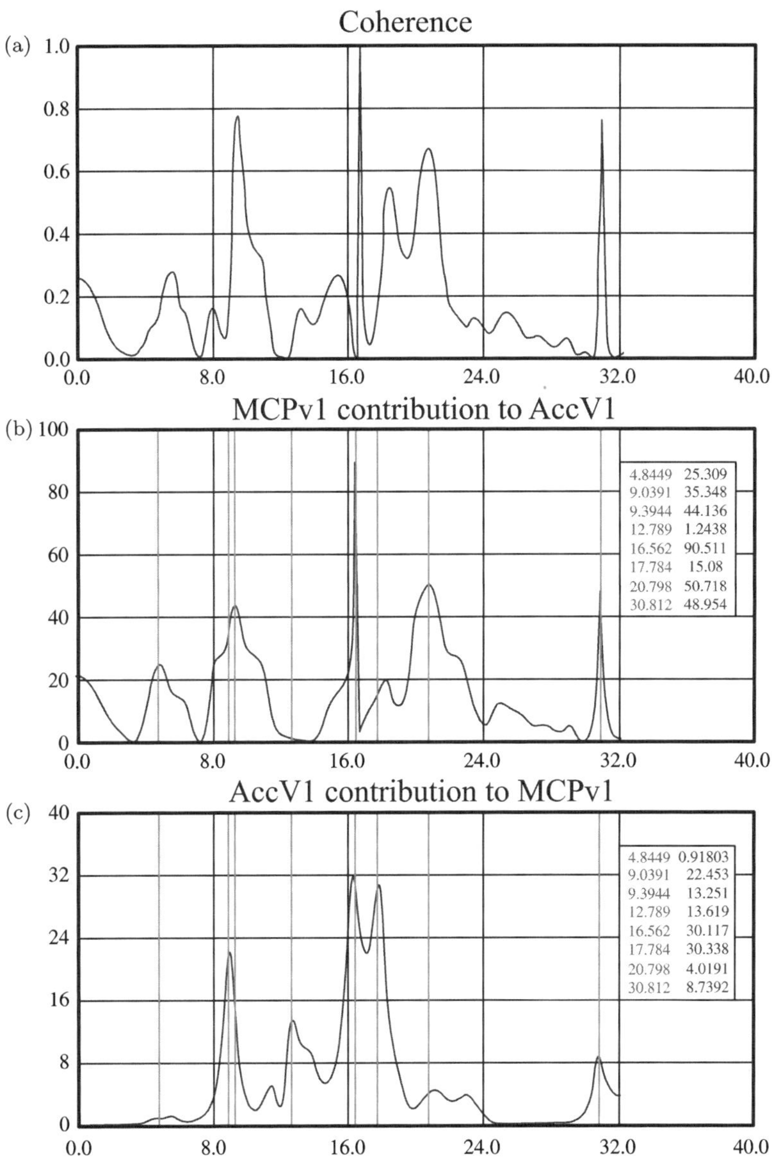

Figure 14.2: Coherence of signal of accelerometer installed on MCP — Acc_{MCP} — and signals of accelerometer installed on reactor vessel – Acc_v (a), contribution of MCP oscillations to vessel oscillations (b) and reverse contribution of vessel oscillations to MCP oscillations (c). Novovoronezh NPP-2 Unit 1, commissioning measurements.

are caused by extended spatial sources. They cannot be attributed either to MCP or to reactor vessel. They have their own origin independent from these elements, namely, coolant pressure pulsations.

Nevertheless, the mutual natural oscillations of MCP and reactor vessel as two connected masses with their own stiffness components can be interpreted in terms of their primary sources.

If stiffness of supports of one component (e.g. MCP) and its mass allow it to make vibration displacements of much higher amplitude than that of displacements of the other component (e.g. reactor vessel), it is considered that MCP is a primary source of mutual oscillations. They are often incorrectly referred to as natural oscillations of MCP only, but not mutual natural oscillations of "MCP–reactor vessel" pair.

An inverse problem is solved in diagnostic applications, which involves identification of the root cause of changes in the measured resonance parameters. For example, its frequency will depend on the mass of MCP and reactor vessel, on stiffness of MCP moving supports and fixtures of reactor vessel.

The mentioned above supports of MCP and reactor plant vessel shall be subject to diagnostics. The frequency of observed resonance of their mutual oscillations, as a comprehensive diagnostic feature, describing condition of supports, MCP and reactor vessel, will however be more sensitive to changes of mass and stiffness parameters of MCP.

There is an independent source of MCP signal at rotational MCP frequency. There is a sort of test of MAR model validity here: contribution from MCP to reactor plant vessel shall also dominate the reverse contribution, which is demonstrated on the second graph in Fig. 14.2. A conclusion shall be perceived as a control check: disturbance at 16.6 Hz frequency is caused by MCP, and not vice versa, although both MCP and reactor vessel are in a state of mutual forced oscillations.

It is useful to compare the contributions of one source of signal to another source and vice versa with coherence. If the shape of contribution curve is close to that of coherence curve, then this signal is the root cause on the other signal at all frequencies. The situation with contribution of MCP signal source to the signal of reactor vessel is almost the same. Table 14.1 summarizes mutual global contributions at frequencies of significant resonances of their coherence function.

Table 14.1: Mutual global contributions to MAR models (Acc_{MCP}, Acc_v).

	The contribution of pressure pulsations to vessel oscillations $Acc_{MCP} \rightarrow Acc_v$	The contribution of vessel oscillations to pressure pulsations $Acc_v) \rightarrow Acc_{MCP}$
4.8 Hz	25%	1%
9.0 Hz	35%	22%
9.4 Hz	44%	13%
16.6 Hz	90%	30%
20.8 Hz	51%	4%

Both contributions are comparable with each other only at ASW1loop frequency of 9.0 Hz, as the third source of pressure pulsations is not taken into account. In all other cases, MCP initiates the mutual oscillations. It is important that at frequency of 4.8 Hz the contribution of MCP vibrations to vibrations of reactor vessel (25%) is decisive compared to the reverse contribution (1%). A natural frequency of vessel oscillations is close to five Hz for VVER-1000 according to the oscillation type of the beam with a fixed upper end. This type is observed in case of a significant wear of core barrel joints.

14.4 Vibration Characteristics during lifting the Reactor Plant Power from MCC to 100% ASW Resonance Frequency as a Diagnostic Feature of Coolant State

Under constant operating conditions, the ASW parameters (frequencies) are the most conservative parameters. They are determined by geometry of MCC, coolant temperature, concentration of impurities and volume concentration of steam (gas) phase in MCC. When repeatedly heating up the reactor plant and switching to power operation, the dependencies of ASW frequencies on the temperature and power, which have been identified earlier, shall be reproduced in exact manner every time. If they are not reproduced as specified earlier, then it is necessary to check the coolant for presence of non-design concentration of impurities and/or steam and gas components.

The minimum controllable power level of the power plant is so small a value that the coolant state can still be considered isothermal.

Along with increasing power, coolant temperature in the "hot" leg — T_h — will be higher than coolant temperature in the "cold" pipeline — T_c. The boric acid concentration at MCC is much less than in sub-critical modes. While changing the coolant density, boric acid concentration changes sound velocity, and therefore the ASW frequencies also change. VVER vibration experience [381, 383] demonstrates this dependence to be rather weak. It can be detected only by spectral analysis of very high frequency resolution. There is still no knowledge about dependence of the sound velocity on boric acid concentration and you shall begin assessing it in this mode and continue in power operation mode throughout the campaign, with boric acid concentration decreasing to almost zero.

The so-called "sub-cooled" boiling is allowed by design of VVER-1000/1200 reactor plant, which means that early stages of boiling when the average coolant temperature is less than the saturation temperature — T_s — may take place in high powered areas of the reactor core (e.g. near-wall and nucleate boiling). The steam bubbles originate and die already on the FA surface during near-wall boiling or in the sub-cooled coolant flow at the bubble stage. In any case, certain, though a small volume fraction of the steam phase is present in the core from a certain power level of the reactor plant and it contributes to the ex-core and in-core neutron noise. While replacing the volume of water moderator, the stem volume worsens the process of fast neutron thermalization, thus forming a low magnitude but positive steam component of reactivity. The intermediate power level of reactor plant, at which the steam reactivity coefficient significantly differs from zero, has not yet been found experimentally.

Vibration and neutron diagnostic systems are traditionally used in two functions, separated in time. While heating up the unit, only thermal displacement of the main equipment is monitored by the permanent components of the relative displacement sensor signals. Vibration noise diagnostics starts at 100% stationary power level of the reactor plant. It is fundamentally based on mathematical tool of stationary random processes. Moreover, during power ascension from MCC to 100%, the neutron noise channels based on ionization chambers (IC) and SPND are not used, since it is considered that the useful neutron current components of the neutron sensors are hidden by the white noise gamma component.

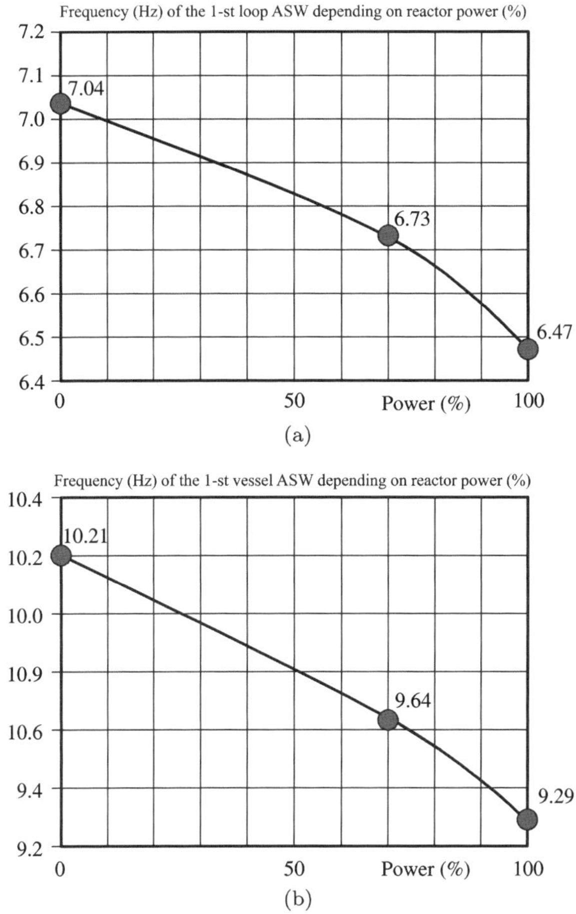

Figure 14.3: Dependence of central frequency of ASW_{1loop} (a) and $ASW_{1vessel}$ (b) on reactor power. Tianwan NPP Unit 1.

Indeed, the neutron noise of IC sensors and SPND exceeds the parasite gamma component only starting from a quite high level of reactor plant. Central frequencies ASW_{1loop} and $ASW_{1vessel}$ (Fig. 14.3) from APSD were extracted at Unit 1 of Tianwan NPP — only at 70% power level and 100% power level. If we add the corresponding ASW at Minimum Controlled Level to them, we will obtain nonlinear, convex, but close to linear dependencies of ASW from the power (Fig. 14.3).

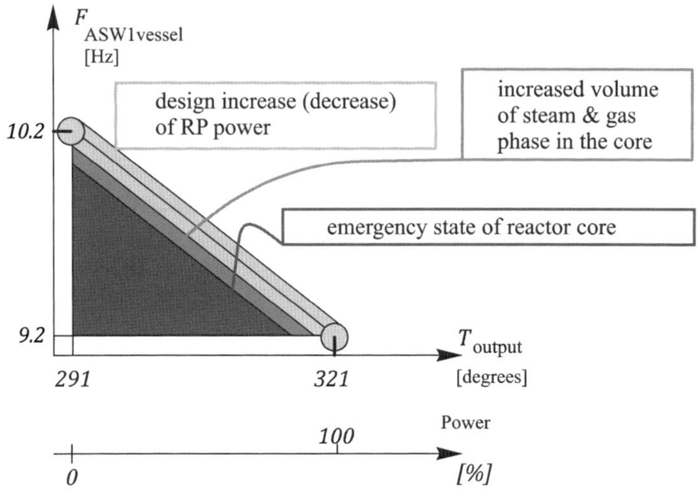

Figure 14.4: Dependence of central frequency of the first vessel ASW on coolant temperature for detecting a steam phase in VVER-1000 reactor plant core.

Based on such a small number of points and without direct witnesses of the coolant boiling, it is difficult to state that the sub-cooled boiling of the coolant causes this nonlinearity. Other reasons may be non-isothermal mode of reactor plant power operation and change of boric acid concentration.

Dependence of the resonance frequency — F_{ASW} — on temperature in Figs. 14.3 and 14.4, indicates the axis "X" can be calibrated in power units, and then we shall get another method for its estimation. It is obvious that in order to implement such an acoustic power meter for the reactor plant it is necessary to perform measurements at a large number of intermediate power levels, and, perhaps, take into account nonlinearity of ASW frequency dependence on the coolant temperature or the power. Such measurements with a small power step could be performed at the low power testing stage of Novovoronezh NPP-2 power Unit 1.

Traditional measurements of the coolant temperature using thermocouples or resistance temperature devices are fundamentally a point type of measurement due to the point type of the detector. Local properties of the temperature measurements are not always positive, since the local hydrodynamic state of the coolant is often unknown. "Hot" and "cold" jets, which are not mixed with the main layer of the coolant, or local boiling, are possible. On the contrary, acoustic measurement of the coolant temperature provides an average value of the coolant temperature for a large MCC volume.

In numerous technical applications, the frequency of a resonator filled with a certain medium (or even with several media with interfaces between them) is used to determine the medium parameters, such as density, viscosity, sound velocity, liquid level, phase composition of the medium (presence of steam and gas phase or solid inclusions in liquid) and even flow rate of the liquid, if it moves through the resonator. For this purpose, such a resonator is artificially created, or a resonator is introduced into the measurement object medium from the outside. For example, in the oil industry a special acoustic resonator is put into the oil well to measure the above-mentioned oil parameters. In case of a reactor plant, the acoustic resonator is automatically created due to the design features of MCC. At the same time, it is very important for reactor diagnostics to learn the thermal and physical states of the coolant, in particular:

- existence of coolant boiling in the core and stages of this boiling;
- change of coolant density along the height of the core, which is associated with recovery of the power generation field;
- measurement of sound velocity due to changes in the density and concentration of boric acid;
- stratification of the coolant in the connecting pipeline (between pressurizer and “hot” leg of the 4th loop);
- coolant level in pressurizer and composition of steam and gas “blanket”;
- gas voids in the so-called “stagnant” MCC areas: upper parts of SG headers, internal volume under reactor lid and volumes of Reactor Control Rod Moving System channels (Fig. 14.5);
- size of coolant leak.

The spatial extent of the studied phenomenon shall be compared with the spatial wavelength of the ASW. For example, if a gas fraction occurs in the stagnant area at the top of SG header, the first loop ASW must change its frequency. Its wavelength for single-phase coolant is very long and close to the doubled length of the loop. If there is a gas volume in the header and even if it is much less than the water volume in the loop, this, however, will significantly change the

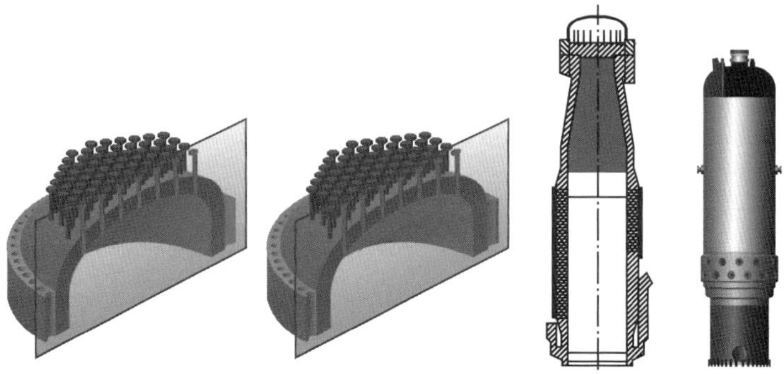

Figure 14.5: MCC stagnant areas: Reactor plant vessel lid, SG header, pressurizer.

ASW_{1loop} frequency, since the sound velocity in gas media is much less than in the liquid media.

A "hot" SG header lid is a reflective surface for the first vessel standing wave $ASW_{1vessel}$. Accumulation of gas under this lid will change its reflective properties. Pressure pulsations absorption coefficient of the gas is much higher than that of metal, so the $ASW_{1vessel}$ amplitude will decrease. The $ASW_{1vessel}$ frequency will also decrease due to decrease of sound velocity in the presence of steam. The phase relations between the incident and reflected waves will also change. When all parameters of the standing wave change, the $ASW_{1vessel}$ gets specific properties, different from that of $ASW_{1vessel}$ in the single-phase coolant medium.

The same applies to the gas volume, which can be accumulated in a stagnant area, such as the volume under the reactor vessel lid, in particular, the volume under the cover of CPS control rods. Its volume is incommensurably small compared to the volume of the reactor vessel, but, starting with a certain minimum gas volume, the vessel ASW will significantly change its frequency.

The monotonic increase of the coolant temperature along the height of the core also determines the monotonic decrease of the coolant density, which changes the sound velocity. Dimensions of the core are comparable with the dimensions of the reactor vessel. Thus, we shall expect a significant impact of the above stratification of the coolant properties along the core height on the frequency properties of the $ASW_{1vessel}$ and higher vessel ASWs.

Calibration of the ASW frequency in units of coolant temperature or reactor plant power requires high accuracy of its evaluation, which requires spectral analysis of high frequency resolution. For example, a minimum spectral resolution of 0.007 Hz shall be available to measure the coolant temperature by means of the $\text{ASW}_{1\text{vessel}}$ frequency with a minimum accuracy of 0.1°C.

The ratio of central frequencies of $\text{ASW}_{1\text{vessel}}$ and $\text{ASW}_{1\text{loop}}$ quite slightly depend on the temperature and can serve as an invariant of the single-phase coolant medium:

$$\mu(T) = \frac{F_k(T)}{F_n(T)} \approx \frac{F_{k0}(1 - \propto_k \cdot T)}{F_{n0}(1 - \propto_n \cdot T)} \approx \frac{F_{k0}}{F_{k0}}$$

$$[1 + (-\propto_k + \propto_n)T] \approx \frac{F_{k0}}{F_{n0}} \approx \text{const}\,(\%)$$

The percentage "corridor" around the constant $\mu(T)$ is set for the temperature range within (290–320°C) for VVER-1000 and (298–328°C) for VVER-1200, that means for all reactor plant modes, where it has a non-zero power level. The threshold dependence $\mu(T) \approx \text{const}$ is a diagnostic proof that there is no significant amount of steam, which prevents cases of sub-cooled coolant boiling, i.e. it is an invariant of the design coolant single-phase. This invariant of the single-phase coolant medium can not only detect, but it can also localize a steam and gas coolant phase. If steam (gas) entered the loop, the invariant denominator will decrease. If steam (gas) entered the reactor vessel, the numerator will decrease (Fig. 14.6):

$$\mu(T) = \frac{F_k(T)}{F_n(T)} = \begin{cases} > (\text{const1}) - \text{gas volume in loop,} \\ < (\text{const2}) - \text{gas volume in vessel} \end{cases}$$

The following diagnostic rules are applicable for resonance frequencies of three lower types of ASWs:

$$F_{\text{ASWvessel}} \approx F_{\text{ASWvessel0}}$$

$$- \begin{cases} \text{const} \cdot T \text{ — designed operating conditions} \\ \text{const} \cdot T + \text{const} \cdot \alpha \text{ — generation of steam} \\ \qquad \textit{gas volume in the core} \\ \text{const} \cdot T + \text{const} \cdot \alpha + \text{const} \cdot H \text{ — emergency} \\ \qquad \text{decrease of coolant level} \end{cases}$$

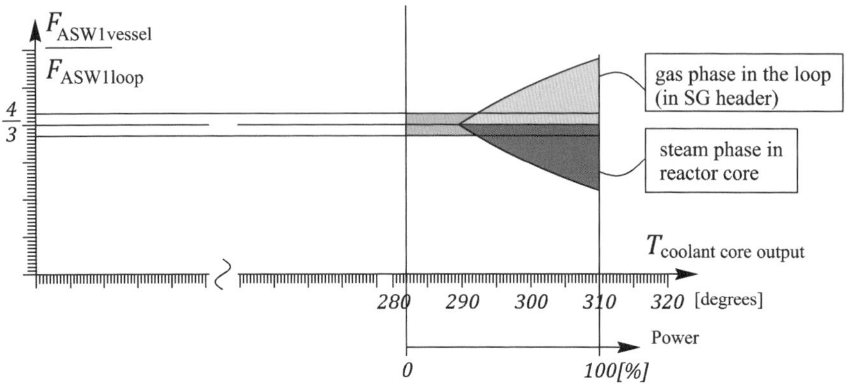

Figure 14.6: Ratios of central frequencies of $\mathrm{ASW}_{1\mathrm{vessel}}$ and $\mathrm{ASW}_{1\mathrm{loop}}$ as invariant of single-phase coolant.

$$F_{\mathrm{ASW}_{1\mathrm{loop}}} = F_{\mathrm{ASW}_{1\mathrm{loop}0}} - \begin{cases} \mathrm{const} \cdot T \quad \text{designed operating conditions} \\ \mathrm{const} \cdot T + \mathrm{const} \cdot \alpha - \text{generation of steam} \\ \qquad \textit{gas volume in the loop} \end{cases}$$

(index "0" means the starting ASW frequency).

It is essential that these expressions are valid not only for the stationary nominal operating mode of reactor plant, but also for its dynamic modes.

The coolant temperature measured in some point of the core is a parameter, which depends on numerous factors, e.g. power generation processes in fuel, heat transfer to its coolant, coolant flow rate, boiling processes in the coolant, etc. The coolant temperature can be significantly changed in emergency states of the core, such as coolant loss due to its leak from MCC, blockage of coolant flow section by foreign objects, heat exchange crisis in the core accompanied by "steaming" in a significant volume of the core, etc. It is obvious that acoustic properties of MCC volume will also undergo significant changes. The reactor plant operating procedure does not allow evolution of processes which cause non-standard temperature of the core coolant. For example, during reactor plant operation it is possible to increase the steam and gas volume in the upper "stagnant" part of reactor vessel, and the primary cause may be the coolant radiolysis process in the core or (non-emergency) evolution of the coolant boiling stage. Deviation of power generation is also possible in some core point within the specified limits. The acoustic properties of the core, expressed in terms of ASW frequencies, can be much more sensitive to the above phenomena compared to the point measurements of the coolant temperature in different locations of MCC pipelines and at the core outlet.

14.5 Vibration Measurements During Power up stages of Novovoronezh NPP-2 Unit 1-2

14.5.1 *Reactor vessel oscillations*

A new generation of VNDS, implemented at Novovoronezh NPP-2 Unit 1, is capable of automatic registration of multi-channel measurement using a preprogrammed time schedule. This new feature enabled to record multiple VNDS processes for the first time during several changes (increase and decrease) of the reactor plant power, in particular, at the following ten power levels:

Minimum Controllable Level, 9%, 17%, 32%, 41%, 57%, 77%, 94%, 99%, 100%.

The waterfalls of autospectra of accelerometer signals, installed at the upper unit of the reactor, provided a characteristic picture of monotonic decrease of the central frequencies of ASWs and the constant frequencies of vibration resonances at these power levels of reactor plant (Fig. 14.7). The vessel is susceptible to the forced oscillations at $\mathrm{ASW}_{1\mathrm{loop}}$ frequency and at frequency of its second harmonic $\mathrm{ASW}_{2\mathrm{loop}}$; the vessel $\mathrm{ASW}_{1\mathrm{vessel}}$ very slightly affects its vibration state. Oscillations at $\mathrm{ASW}_{2\mathrm{loop}}$ frequency are the dominant ones with regard to power; and its amplitude is only inferior to oscillations at the MCP rotational frequency. The next sections will show that there are also mutual oscillations of the vessel, pipelines and main MCC equipment in this frequency sub-range.

There is a coincidence of $\mathrm{ASW}_{2\mathrm{loop}}$ frequency with natural common oscillations frequency of MCC equipment, which generates an almost global common loop oscillatory process. A quite weak impact of the higher $\mathrm{ASW}_{4\mathrm{loop}}$, $\mathrm{ASW}_{6\mathrm{loop}}$ and $\mathrm{ASW}_{8\mathrm{loop}}$ can be observed

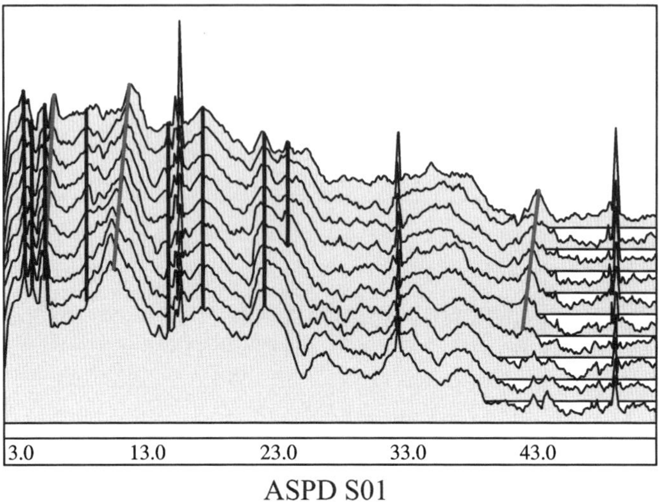

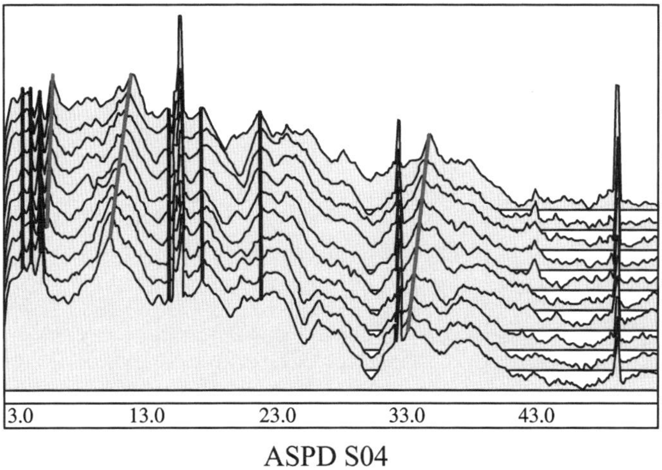

Figure 14.7: APSD families of signals of accelerometer, installed at the upper unit (S01 — upper picture, S04 — lower picture) changing power of reactor plant (top-to-down): MCPL, 9%, 17%, 32%, 41%, 57%, 77%, 94%, 99%, 100%.

here as small amplitudes of forced oscillations of the reactor vessel. Note that the ASW_{1loop} resonance almost coincides with the vibration resonance when 100% reactor power has been achieved, but no hazardous effect of the amplitude increase was observed. Thus, two dominant resonances ASW_{1loop} and ASW_{2loop} have vibration effects at natural frequencies in their vicinity.

The mutual properties of the considered accelerometer signals provide a unique picture of oscillations of VVER-1200 reactor plant vessel (Fig. 14.8). This means that there are only vertical oscillations of the vessel and their amplitude is negligible: less than 5 microns. The accumulated experience in vibration monitoring of reactor vessel oscillations at various VVER-1000 and VVER-440 units shows that there is a large number of oscillation types with combinations of characteristic phases (0°, 90°, 180°). There are pendulum-type oscillations (which is the most unfavorable in terms of exhausting the vibration resources of supporting structures), circular, vertical and so-called stochastic-types of oscillations with amplitudes exceeding 5 microns. New design solutions for anchoring of VVER-1200 reactor plant and new fixtures of VVER-1200 Reactor Internals (RI) significantly reduced vibration loading of its vessel.

The types of vessel oscillations shall be defined not only by condition (stiffness characteristics) of its own support structures, but also by stiffness characteristics of RI fixtures and in particular by three horizons of core barrel fixtures (Fig. 14.9). Therefore, the signals of vibration sensors, installed on the reactor vessel, appear to contain components of RI vibrations.

The elastic element in the upper fixture horizon is deformed in the vertical direction thereby providing also vertical clamping of core barrel. The horizontal static forces in the middle node affect the core barrel in normal vector direction, and horizontal static forces in the lower node affect the RC tangentially. The three horizontal fixtures of core barrel of VVER-1200 design have been further developed.

Thus, the tubular elastic segments of VVER-1000 in the upper fixture are replaced by elastic elements made of expanded graphite. They almost do not lose their stiffness characteristics throughout the campaign. The number of vibration dampers was increased in the lower fixture from 8 to 12 pieces. Stiffness of the middle level of core barrel fixture shall be defined by the so-called circular pressure of the core barrel at nominal temperatures of the coolant. In turn, compression is determined by the azimuthal invariance of the gap in "Core barrel–Flow separator" pairs in the "cold" state and high quality surfaces in these pairs.

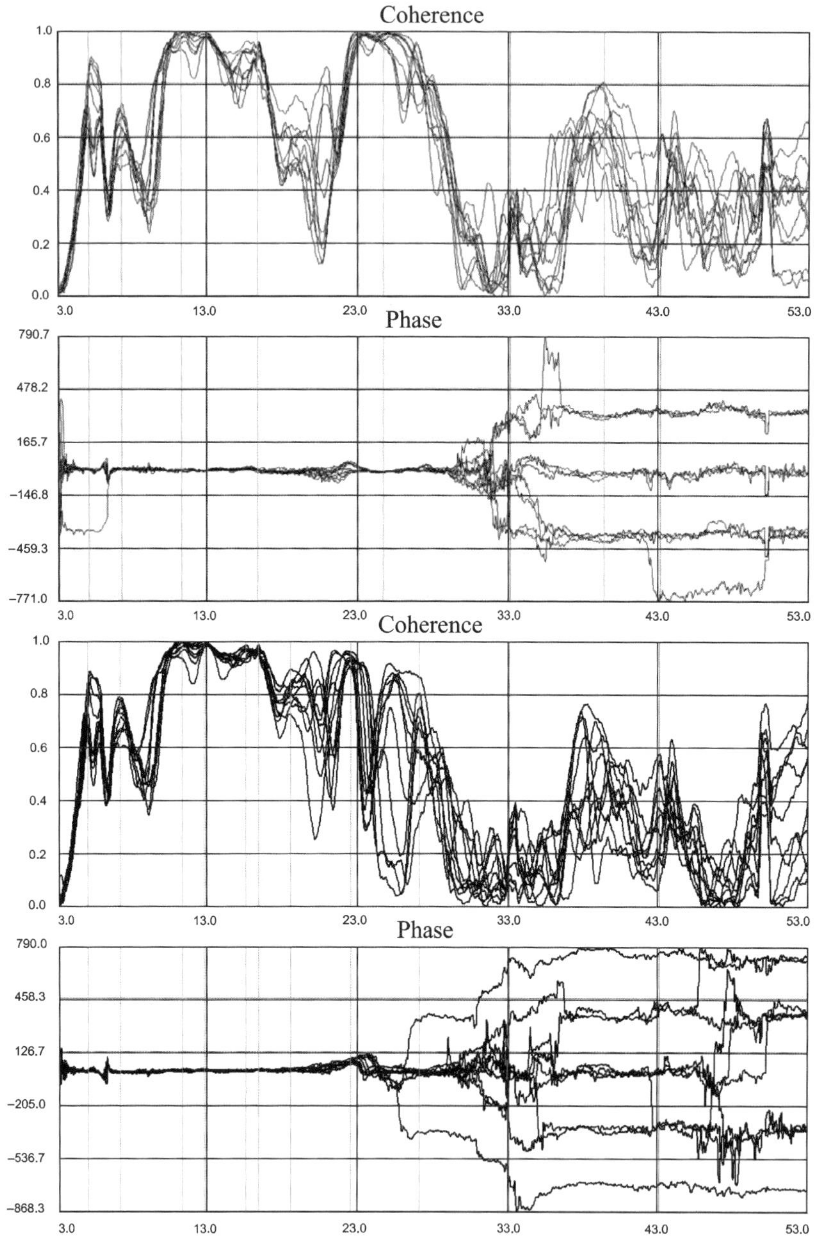

Figure 14.8: Families of coherence and phase functions between accelerometer signals, installed on the upper unit (S01–S04 — see upper two pictures and S03–S04 — see lower two pictures) during the following change of reactor power level: minimal controllable level, 9%, 17%, 32%, 41%, 57%, 77%, 94%, 99%, 100%.

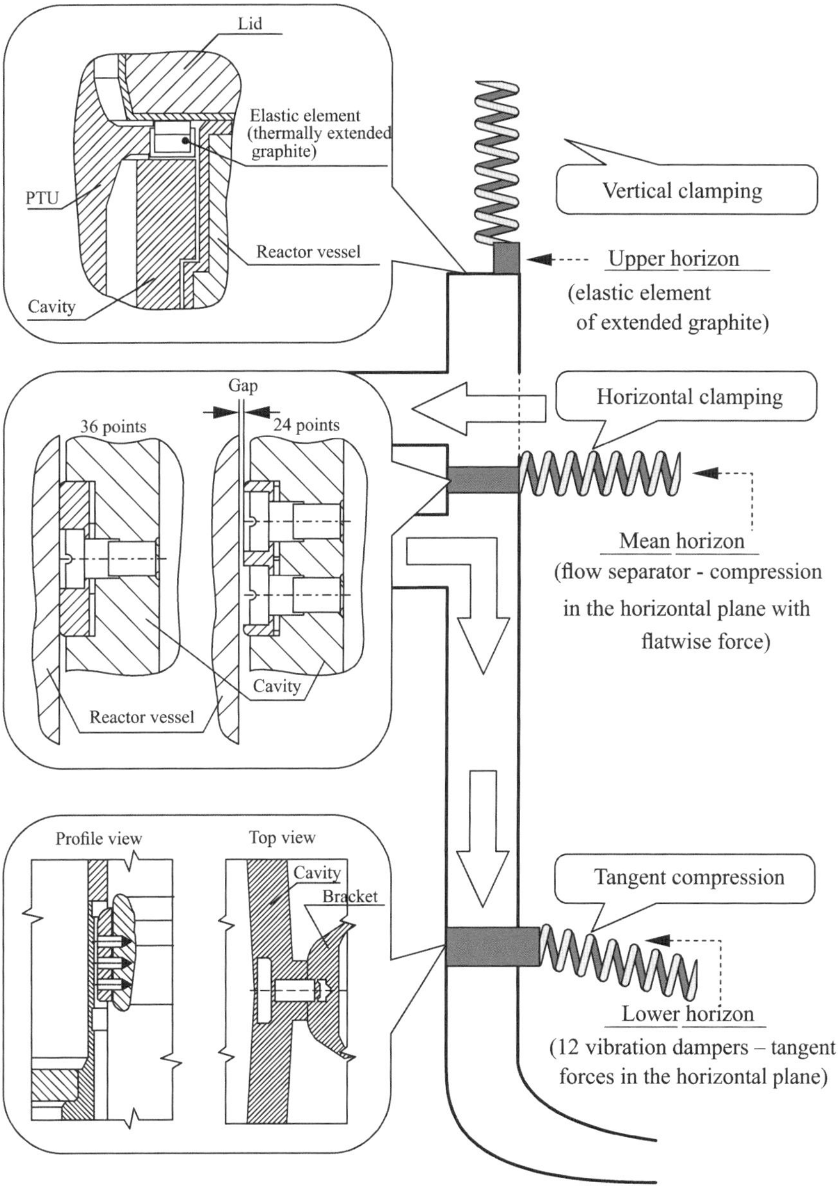

Figure 14.9: Qualitative representation of three RIC fixtures by spring stiffness elements.

A new approach to monitoring the quality of the middle level of core barrel (RIC) fixture is implemented in VVER-1200 project. Compensation plates with lining washers are installed along the perimeter of the RIC shell. The washer thickness shall take into account an actual gap between the plates and RIC wall, so that the "installation" gap between plates and separation ring corresponds to the design value during assembly. When the operating parameters of the coolant are reached, the "installation" gap is reduced to zero due to the difference in temperature expansion of reactor internal core barrel (RIC) and reactor vessel.

Insufficient stiffness of the middle fixture reduces the main lower-type of RIC oscillations, which correspond to oscillations of the beam with two fixed ends. Insufficient stiffness of the lower fixtures releases the lower pendulum-type of RIC oscillations, which is a beam oscillation-type with a single upper attachment point. Insufficient stiffness of the upper fixture can cause the so-called "floating" type of RIC oscillations, which correspond to weak attachment of the RIC top with very low natural frequency. These types of oscillations were observed during continuous development of VVER technology [381].

After a significant wear of RIC fixtures, after many years of operation, some type of natural oscillations will prevail, thereby determining the RIC attachment level, which changes by the most significant extent. In this sense, the starting spectral characteristics of vibration signals of the reactor vessel of Novovoronezh NPP-2 Unit 1, where no RIC natural oscillations were detected, appear to be very favorable.

14.6 Control of CPS Control Rods in Power Ascension Mode of Reactor Plant according to VNDS Vibration Channels

Vibration sources may have changes related to movement of CPS control rods. While moving the CPS control rod suspension changes its length and, by analogy with the pendulum, if the length of pendulum changes, the frequency of its oscillations also changes. While changing its spatial position, the CPS control rods group has an equivalent length of pendulum, which is always longer than that of the other groups of control rods which are withdrawn to the maximum upper position. By conducting a spectral high-resolution analysis, it is possible to detect the corresponding shifts in position of resonances on the frequency axis, and thereby determine a depth of CPS control rod immersion and operability of its components.

This stationary vibration effect can be detected at 100% power level of reactor plant. At low power levels, the signals of ionization chambers reflect non-stationary processes. There are upwards and downwards flashes, which significantly exceed the amplitude of the steady-state background which exists due to the parasitic gamma component (Fig. 14.10). CPS control rods movement causes the current flashes of ionization chamber. If such emission is zoomed in time, for example, in a period of 2 s, then an almost deterministic and periodically repeated waveform of the signal is detected. One can distinguish two such periods in Fig. 14.10. The time distance between the local maxima of these periodic fragments is one second. Movement of control rods shall be controlled by a deterministic sequence of switching currents of the pulling, locking and latching solenoid of a particular group of control rods. The switching sequence is always the same and it ensures safe step-by-step movement of control rods. The way that deterministic current image of ionization chamber signals is reproduced in time gives evidence of operability of control rod components. This selected time fragment of IC record is an etalon, with which the current fragment of IC record caused by movement of control rods shall be compared.

At low power levels, up to several dozens of such control rod movements are registered in form of IC signal recordings of 10 min duration (Fig. 14.10). The reactor plant is also adjusted by means of control rods at 100% power level, but these are very rare situations, which shall be firstly detected before analyzing, since not every 10-min record of IC signals includes such events.

Figure 14.11 shows APSD of IC signal for four low levels of reactor plant power. If we ignore the signal amplitude, it becomes obvious that the shape of these APSDs is the same.

Control rods were repeatedly and randomly moved during all of the four IC realizations. They also determined a periodic form of APSD: the local frequency maxima are located with 1 Hz step (meaning 1-s period), and this is a diagnostic evidence of control rod components operability. Thus, a device for spectral analysis of stationary processes is used for non-stationary processes, when non-stationary

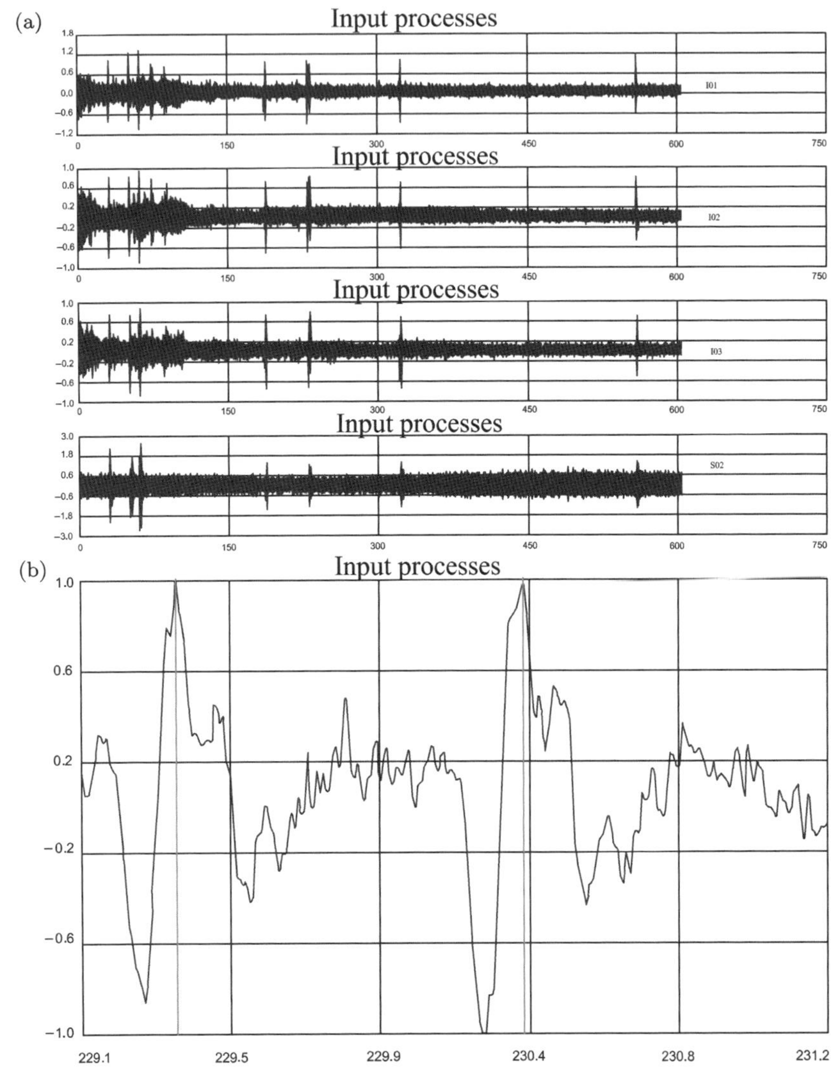

Figure 14.10: Time records of three IC signals (a) and S02 accelerometer signal during control rods movement at 9% power level. IC signal fragment consisting of Two Periods (b).

has a form of randomly repeated set with deterministic periodic structure. The mode of exiting from Minimum Controllable Level to the low power level is the first dynamic mode, in which the control rods shall be subject to diagnostics. At 100% power level, these

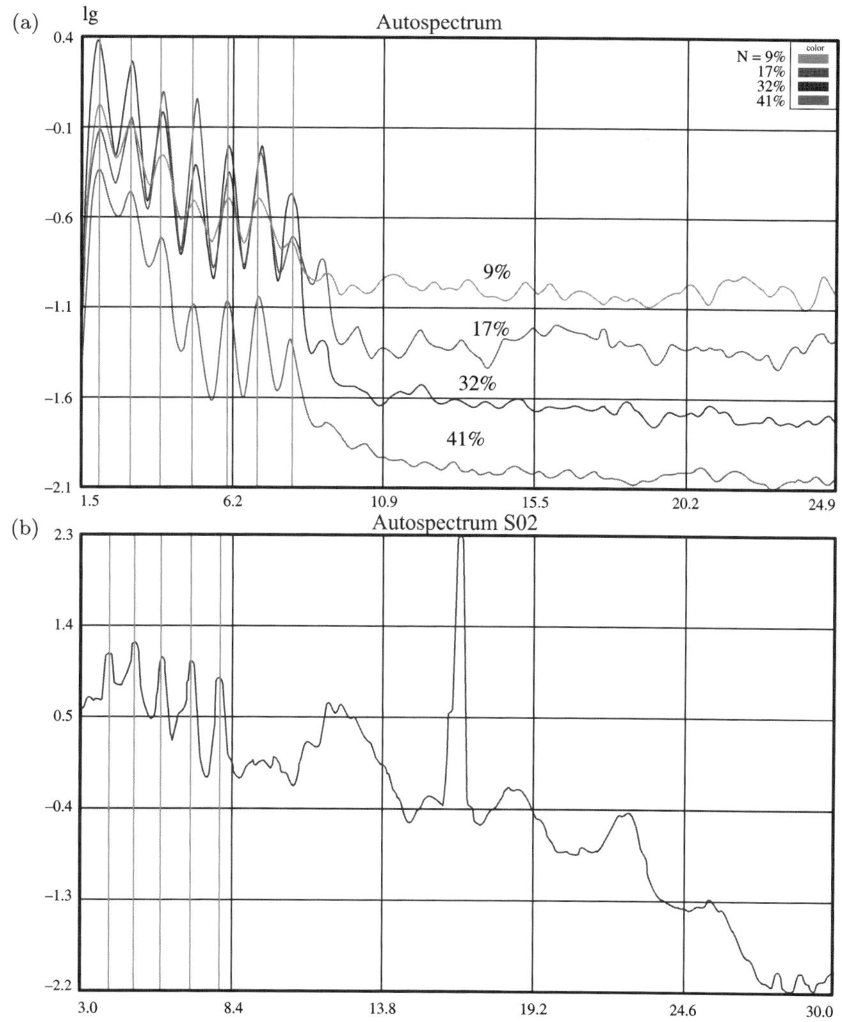

Figure 14.11: APSD of IC signal (a) at four low power levels and APSD of S02 accelerometer signal during control rods movement at 9% power level (b).

diagnostic evidences will be rare events, which shall be automatically detected and control rods shall be subject to diagnostics within the allocated time fragment in automatic mode as well.

Only one of four accelerometers, installed on the upper unit of reactor plant, namely, S02 appeared to be sensitive to control rods movement. Its signal has flashes at the same time as IC signals

(Fig. 14.10), and its APSD has the same form and frequency as APSD of IC signals (Fig. 14.11).

All three IC, installed around the reactor plant vessel at significant distance from each other, are sensitive to control rods displacement. That is, this neutron effect is global for them. Control rods displacement is a local phenomenon for vibration sensors S01, S02, S03, S04 with much less distance from each other compared to that of IC sensors. Only the accelerometer located close to the vibration source, i.e. a control rod component here, is sensitive to displacement.

In-core self-powered neutron detectors (SPND) are also sensitive to control rods displacement. Similar neutron flux flashes were registered according to these signals on Tianwan NPP Unit 1 (Fig. 14.12) at 100% power level. Strictly speaking, all SPNDs shall be sensitive to such global phenomenon, but at the lower $SPND_1$ a change of the signal in this time interval is negligible, and as the numbering of SPND increases, the amplitude of flashes grows monotonically and achieves significant high values in the signal of the upper $SPND_7$.

14.6.1 *Vibration by neutron measurements of central ASW frequencies*

To determine a power level of the reactor plant at which the neutron current component of the ionization chamber (IC) starts to exceed the gamma-ray component, VNDS implementations at 12 power levels were recorded at Novovoronezh NPP-2 Unit 1 (Fig. 14.13).

Registration of neutron noise, starting from Minimum Controllable Level and further with a small power step allowed to find that the level 40% of power, at which the neutron barometric effects of ASW start to be seen (Fig. 14.14).

No significant neutron resonance phenomena are detected at low power levels on the high amplitude white noise background. ASW_{1loop} resonance popping out from background only at the 40% power level and its frequency monotonically decreases along with monotonous increase of the power, and the amplitude monotonically grows. Let us recall that the ASW_{1loop} generates a global barometric effect on the neutron noise, as one of its half-waves covers the entire reactor core. The ASW_{1loop} is the only significant local resonance of APSD of all IC signals. Other local resonances can be detected by means of cross-spectral functions of various pairs of neutron and

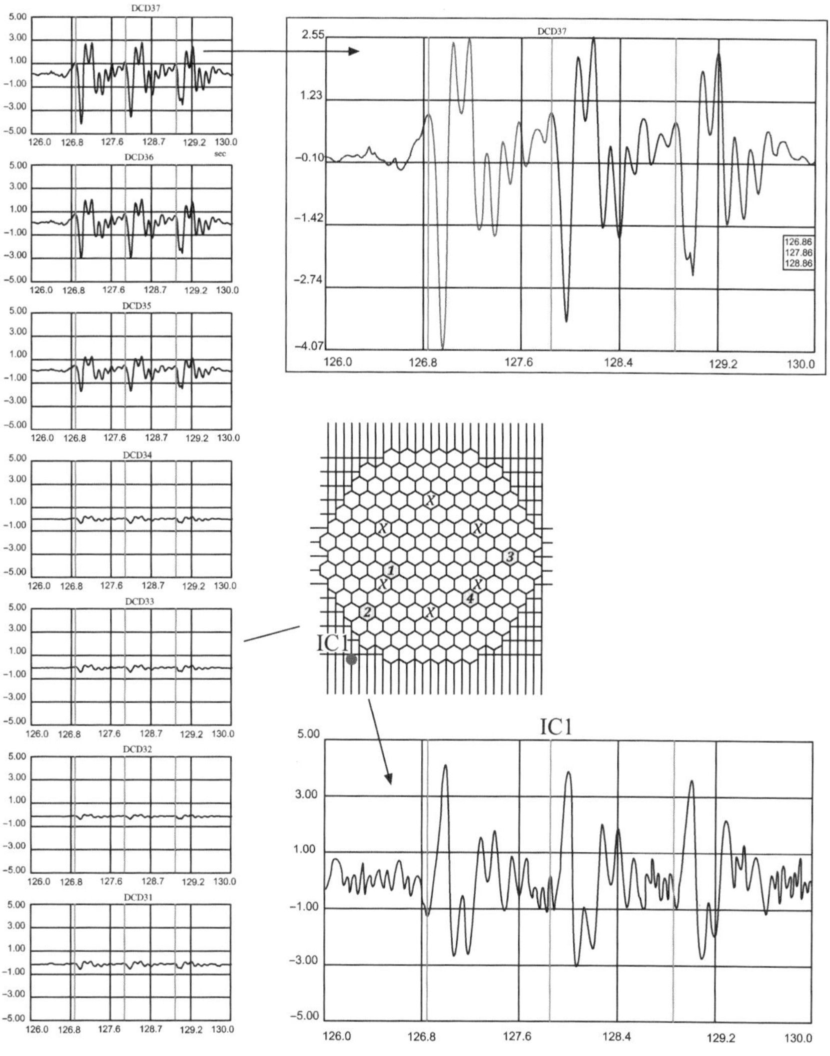

Figure 14.12: Signals of seven SPNDs of one assembly (left column of graphs), IC signal (lower graph) and three fragments of SPND7 signals of three neutron-flux measuring channel assemblies. Tianwan NPP Unit 1.

vibration signals. These resonances occur due to acoustic and vibration effects (Fig. 14.15), but in APSD of IC signal in all VVER reactor plants (VVER-440/1000/1200), the ASW_{1loop} resonance is almost the only significant local maximum.

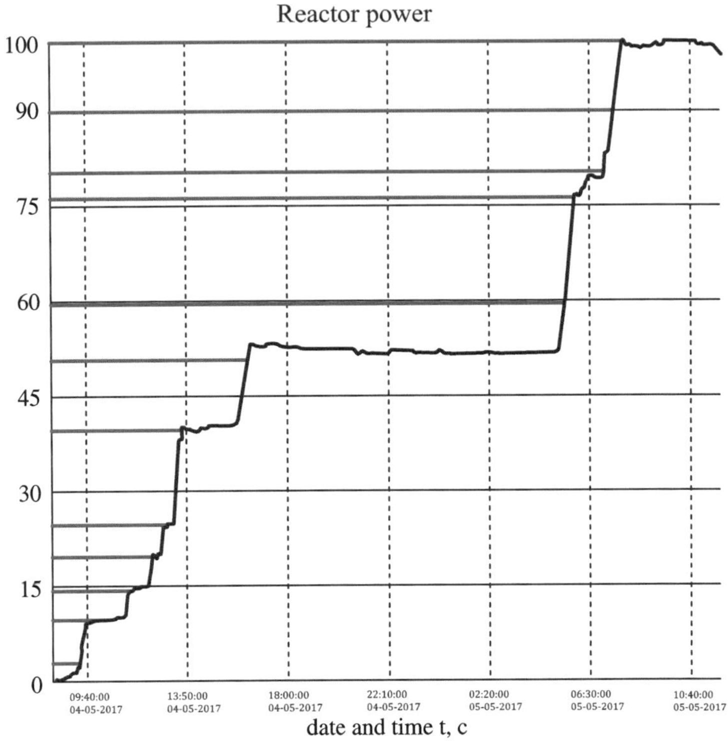

Figure 14.13: Change of Novovoronezh NPP-2 Unit 1 power from 0% to 100% during 25 h and 12 power levels, at which VNDS signals were recorded.

The lowest ASW_{1loop} having its four alternating in-sign components along the azimuth of the core is self-compensated as global barometric effect. Vibration effects of ASW_{1loop} of four loops cause its weak effect on the neutron flux. A resonance at $ASW_{1vessel}$ frequency dominates in coherence function for neutron signal pairs ("IC–IC", "SPND–SPND", "IC–SPND") as a barometric global effect. Two different physical effects correlate for the signal pairs "vibration sensor – neutron sensor": a vibration effect in accelerometer signal at $ASW_{1vessel}$ frequency and a barometric effect in a neutron sensor signal at the same frequency.

In addition to vibration and acoustic effects, spatial temperature and neutron effects [358, 360] are observed in mutual signal characteristics of remote IC. They are expressed in the so-called sink structures, which are periodic components of the spectral functions

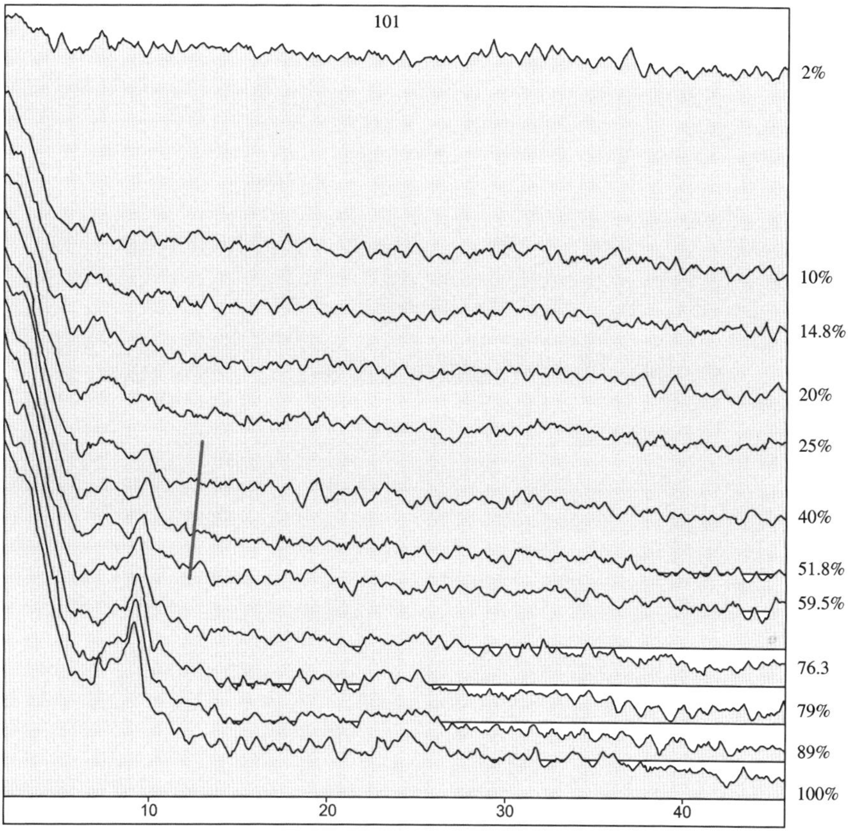

Figure 14.14: APSD of IC Signal in Novovoronezh NPP-2 Unit 1 power level change from 0% to 100% (top-down) during 25 h.

$\frac{\sin(f\tau_0)}{f\tau_0}$. The τ_0 parameter, which refers to the time of coolant passing through effective height of reactor core, can be used to assess the core coolant flow rate through the neutron and noise channels. Such a component in Fig. 14.15 has a central frequency of about 3.2 Hz and an opposite phase near this frequency in signals of the opposite side ICs.

The same effects are observed in the signals of the opposite side ICs at Novovoronezh NPP-2 Unit 1 (Fig. 14.16). As the reactor plant power increases from 40% to 100%, the $\mathrm{ASW}_{1\mathrm{vessel}}$ frequency changes from about 9.4 Hz to 8.7 Hz, while the phase evaluation is always zero (table in Fig. 14.16).

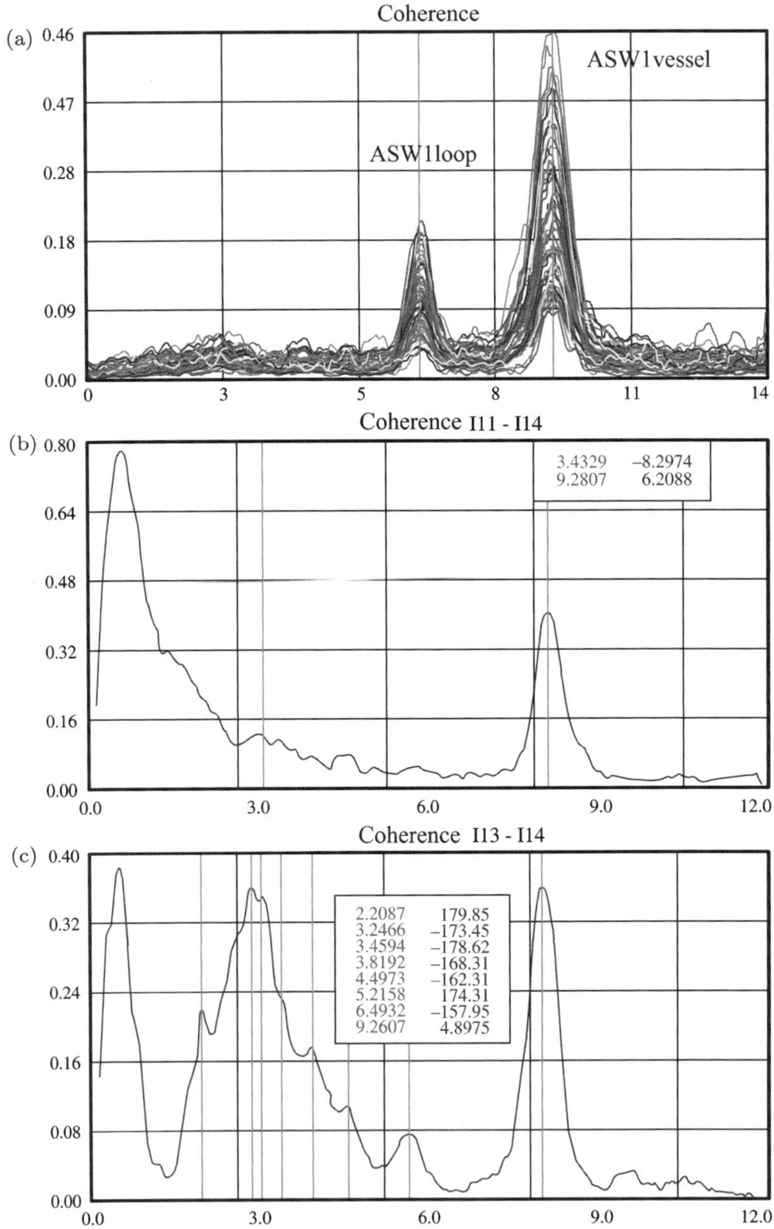

Figure 14.15: Coherence functions of signal pairs: "Accel–Accel" (a), close to each other "IC-IC" (b) and far from each other "IC-IC" (c) (Tianwan NPP Unit 1).

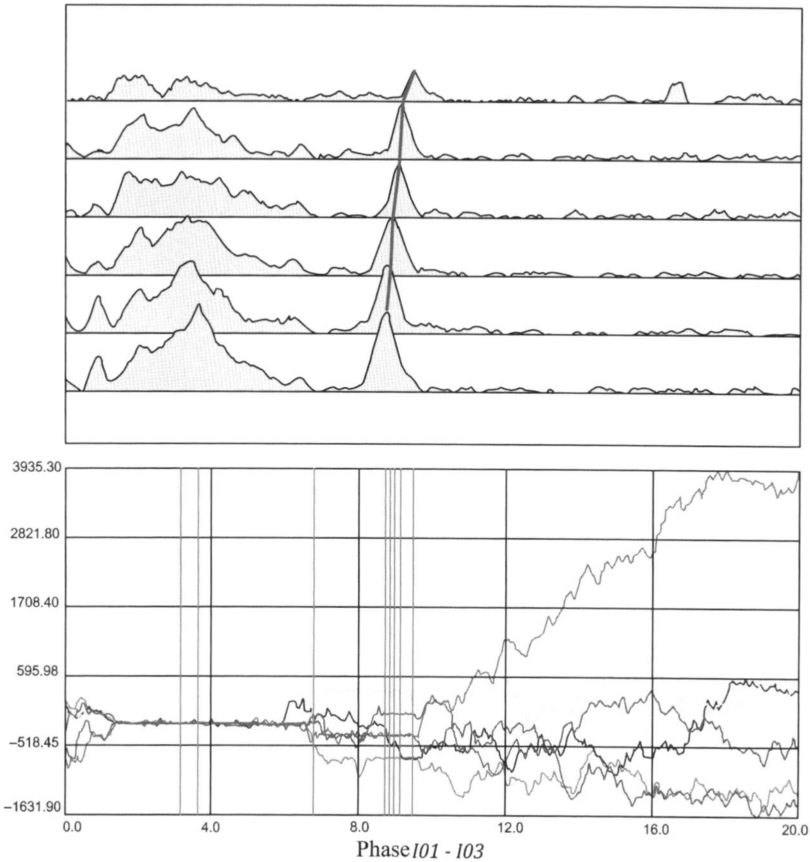

Hz	Degrees					
3.1329	−169.89	−159.7	177.55	−172.01	−173.24	−170.2
3.6191	−140.81	−179.07	−163.43	−176.47	−172.33	−164.42
6.7777	−110.27	−20.333	−161.81	−156.29	31.932	−91.909
8.7012	137.97	−30.886	14.596	4.3022	−13.52	−7.6123
8.7815	−78.819	21.104	−13.377	−0.63208	−12.265	−8.2395
8.913	−157.78	−3.175	−9.3964	−0.94934	8.8239	2.1349
9.0507	−175.81	−10.935	−2.932	−4.575	4.4542	1.2708
9.1007	158.72	−15.127	−9.012	−9.8106	−8.0816	1.8821
9.4273	3.7918	1.4879	−4.1197	23.332	112.76	19.806

Figure 14.16: Coherences and phases, obtained from signals of opposite side IC at reactor plant power increasing from 40% to 100%. Table of phases at nine characteristic frequencies. Phases at $\mathrm{ASW}_{1\mathrm{vessel}}$ frequencies are indicated from bottom upwards. Phases at frequencies of the sink-structures are indicated at the top. Novovoronezh NPP-2 Unit 1.

From the appearance of reactor power and non-zero temperature drops of the coolant in reactor core – $\Delta T = T_h - T_c$, it can be approximately seen that the coolant temperature from the core outlet to SG inlet is T_h, and from SG outlet to the core inlet — T_c. Having such long half-waves of the lower ASWs, covering both "hot" and "cold" pipelines, we obtain an anisotropic medium at one half-wave length of the lower ASW. At a quarter of the $\mathrm{ASW}_{1\mathrm{loop}}$ wavelength, the coolant temperature is T_h, and at the other quarter, it is Tc. Having one half-wave covering two "hot" pipelines and reactor volume, the $\mathrm{ASW}_{1\mathrm{vessel}}$ will be defined, on the one hand, by a medium with T_h temperature, and, on the other hand, a half-wave of the same $\mathrm{ASW}_{1\mathrm{vessel}}$, concentrated in "cold" pipelines, will be defined by a medium with Tc temperature. On the one hand, it is fair enough to assume that the same $\mathrm{ASW}_{1\mathrm{vessel}}$ has half-waves of different spatial length and different frequencies. On the other hand, the same ASW shall have a single frequency, as it is obtained as a result of averaging when being subject to "back and forth" elastic perturbation along anisotropic medium at the half-wave length.

In order to resolve this a paradox, let us again turn to the experimental data, obtained at Tianwan NPP Unit 1, where all 12 PPSs operate at 100% power level. VNDS of the previous generation is used there as intended, and vibration diagnostics is performed only at 100% power level of the reactor plant. Since three PPSs are installed in "hot" and "cold" pipelines of a single loop, we obtain statistics from four (by the number of loops) measurements of the single-type PPS. According to one hypothesis, the ASW frequency at a sufficiently high spectral resolution, obtained from APSD of $\mathrm{PPS_h}$, shall be significantly lower than the ASW frequency, obtained from $\mathrm{PPS_U}$ and $\mathrm{PPS_c}$ installed in the "cold" pipelines. According to the other hypothesis, which assumes that the standing wave has averaging properties, all ASW frequencies shall not significantly differ from each other.

Using the following formula (12.1)

$$F_{\mathrm{ASW}_{1\mathrm{loop}}} = 12.20 - 0.0175 \cdot T$$

$$F_{\mathrm{ASW}_{1\mathrm{vessel}}} = 16.60 - 0.0215 \cdot T$$

it is possible to assess the maximum offsets of $\mathrm{ASW}_{1\mathrm{oop}}$ and $\mathrm{ASW}_{1\mathrm{vessel}}$ frequencies with increasing power from Minimum Controllable Level to 100%, assuming that the coolant temperature changed from 298°C to 328°C (for VVER-1200). The ASW_{1loop} under this formula will be shifted for 0.52 Hz and the $ASW_{1vessel}$ frequency will be shifted for 0.64 Hz. It is quite possible to measure such shifts using high-resolution spectral analysis.

Figure 14.17 shows APSD of all 12 PPSs of three types in four loops. Table 14.2 lists experimental values of ASW_{1loop} frequencies.

Taking into account their standard deviation around the average (see RMS assessments in Table 14.2), it can be argued that no significant differences of ASW_{1loop} frequencies in "hot" and "cold" pipelines were observed. However, there are differences in the loops: in all three dimensions of the 4th loop the ASW_{1loop} frequency is systematically higher than in the other loops. The same effect was observed in the ASW_{1loop} at Novovoronezh NPP Unit 1. This was associated with an increase of the acoustic path due to the complex resistance introduced by connecting the pipeline nozzle. So, no frequency change is experimentally observed at the ASW_{1loop} half-wave length at 100% power level. The ASW_{1loop} with a single half-wave in anisotropic medium "averages" its properties and has a single value of the central frequency.

Now, let us consider a case of several half-waves of one ASW! Is the central frequency split in such anisotropic structure? The $ASW_{1vessel}$ is the most suitable to investigate this. One $ASW_{1vessel}$ half-wave contains only anisotropic medium with T_h temperature; the second half-wave contains anisotropic medium with T_h and T_c temperatures; the third half-wave again contains anisotropic medium, but with T_c temperature (Fig. 14.18).

As one can see from APSD-type of PPS signals in Fig. 14.17, the $ASW_{1vessel}$ resonance does not always definitely stand out from the background of other physical sources of coolant pressure pulsations. Figure 14.19 provides a detailed scaled set of APSD signals of 4 PPS_h and 4 PPS_c with $ASW_{1vessel}$ resonance frequencies. Indeed, the $ASW_{1vessel}$ resonance amplitudes are much lower than the amplitudes of two neighboring ASW_{1loop} and ASW_{2loop} resonances, but there are no significant differences in evaluations of $ASW_{1vessel}$ frequencies between the "hot" and "cold" PPSs: the average for PPS_h according to four evaluations of $ASW_{1vessel}$ frequency is 9.194 Hz and the average for PPS_c is 9.219 Hz. Their difference by 0.025 Hz is too small and lies within the error margin of evaluations. Let us also note that the $ASW_{1vessel}$ frequency in the fourth loop is again significantly

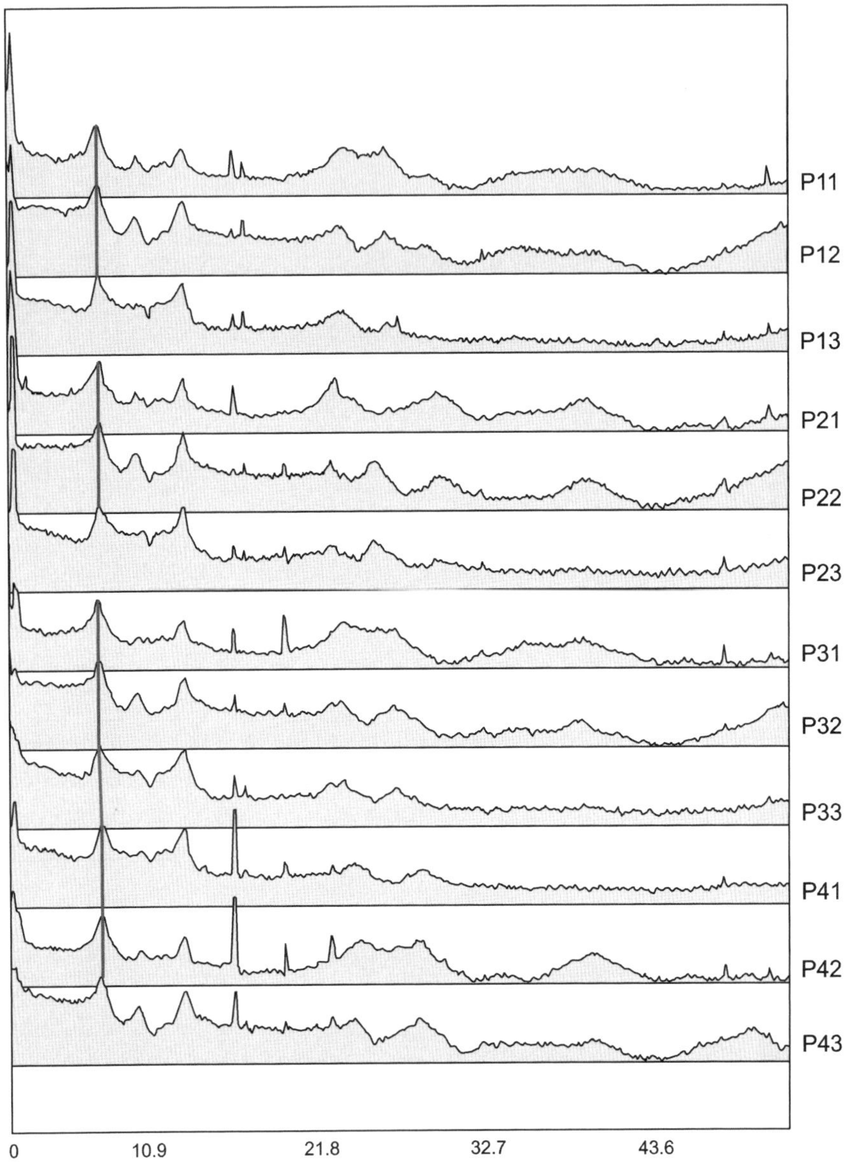

Figure 14.17: Signal APSDs of 12 PPSs of Tianwan NPP Unit 1 at 100% power level P11, P21, P31, P41$-$PPS$_h$ in four loops, P12, P22, P32, P42$-$PPS$_U$ in four loops, P13, P23, P33, P43$-$PPS$_c$ in four loops.

Table 14.2: ASW_{1loop} experimental central frequencies [Hz] of Tianwan NPP Unit 1 at 100% power level.

i-No of loop	$i = 1$	$i = 2$	$i = 3$	$i = 4$	Average of i
PPS_{hi}	6.38	6.43	6.32	6.48	6.40 ± 0.068
PPS_{Ui}	6.37	6.42	6.31	6.47	6.39 ± 0.068
PPS_{ci}	6.40	6.44	6.39	6.46	6.42 ± 0.033
Medium	6.38 ± 0.015	6.43 ± 0.010	6.34 ± 0.043	6.47 ± 0.010	6.40

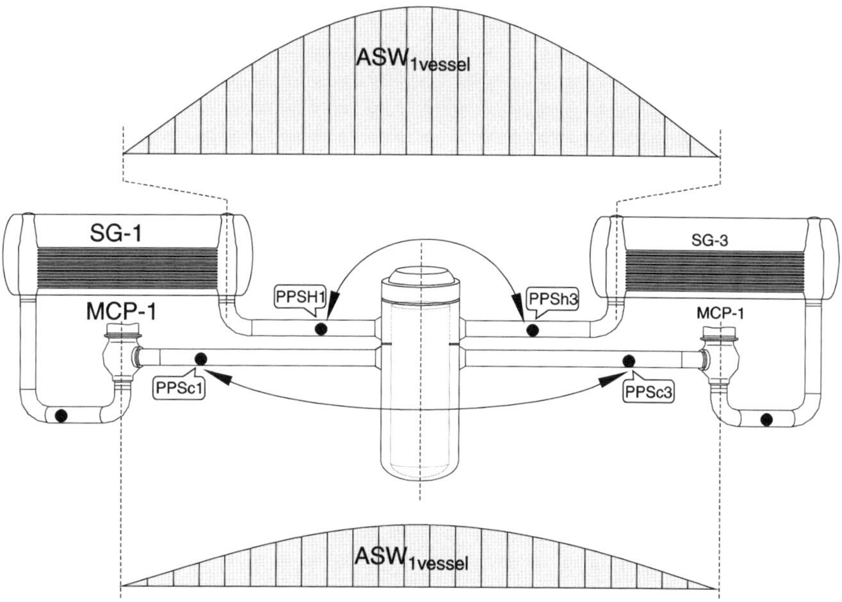

Figure 14.18: Two $ASW_{1vessel}$ half-waves in "hot" and "cold" MCC.

lower than all the others regardless of the coolant temperature, and this corresponds to the longer acoustic length of the fourth loop.

Background sources of pressure pulsations, competing with the $ASW_{1vessel}$ and existing interference during evaluation of $ASW_{1vessel}$ resonance frequencies, can be eliminated by means of cross-spectral analysis, thereby improving the signal/background ratio at the $ASW_{1vessel}$ frequency. For example, when eliminating uncorrelated background components of two signals, a coherence function will

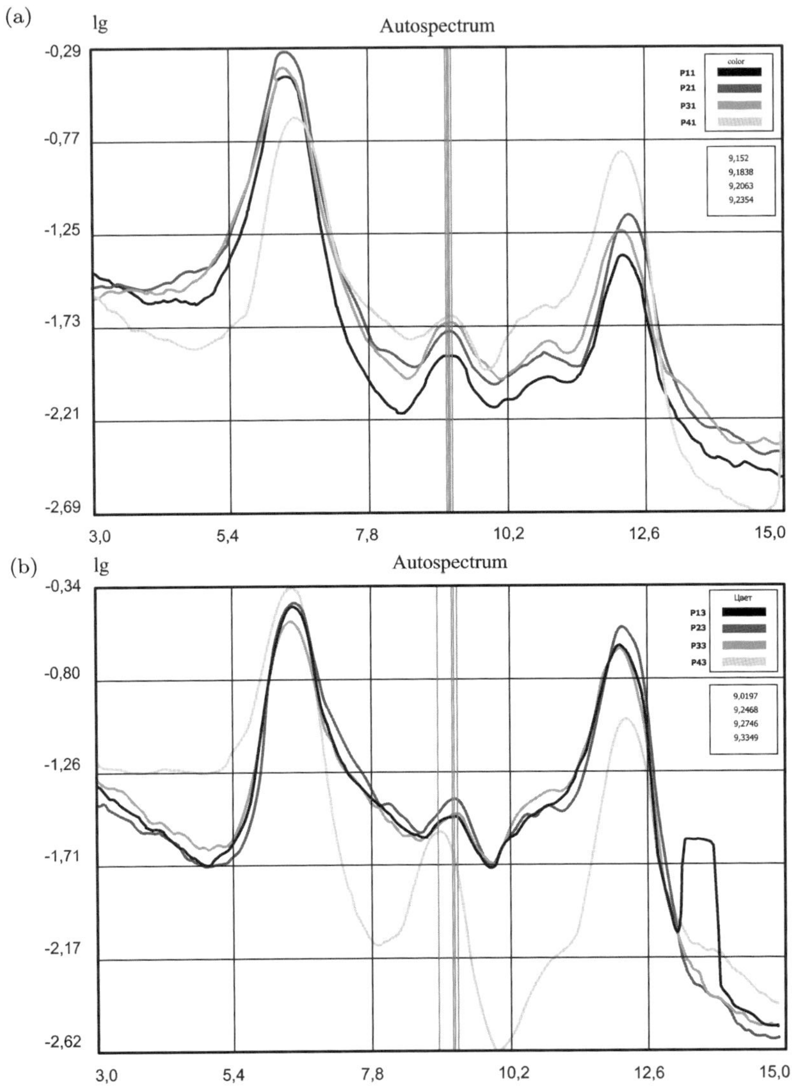

Figure 14.19: APSD of four PSSs' signals of four loops with $ASW_{1vessel}$ frequencies: PPS_h, installed in "hot" pipelines (a), PPS_c, installed in "cold" pipelines (b).

allocate the $ASW_{1vessel}$ resonance in much more contrast than two corresponding APSDs. However, a systematic shift of $ASW_{1vessel}$ resonance evaluation of coherence function is possible, as two signals with slightly differing frequencies of these resonances are involved in its generation. To avoid such systematic error we will make pairs of the same-name signals of various loops. This means we will use a globalism property of the $\mathrm{ASW}_{\mathrm{1vessel}}$ standing wave and correlate signals of the "hot" PPSs of different loops with each other. Six combinations of signal pairs are obtained from four $\mathrm{PPS}_{\mathrm{h}i}$ signals: $\mathrm{PPS}_{\mathrm{h}i} - \mathrm{PPS}_{\mathrm{h}j,i<j}, i, j = 1, 2, 3, 4$. And six combinations of signal pairs are obtained from four $\mathrm{PPS}_{\mathrm{c}i}$ signals: $\mathrm{PPS}_{\mathrm{c}i} - \mathrm{PPS}_{\mathrm{c}j,i<j}, i, j = 1, 2, 3, 4$. Figure 14.20 shows these coherence functions in various scales with marks of $\mathrm{ASW}_{\mathrm{1vessel}}$ frequencies. Indeed, the selected coherence functions show the resonance in much more contrast than APSD.

Now the $\mathrm{ASW}_{\mathrm{1vessel}}$ resonance dominates over the neighboring $\mathrm{ASW}_{\mathrm{1loop}}$ and $\mathrm{ASW}_{\mathrm{2loop}}$ resonances and its maximum value reaches 0.4, which ensures a satisfactory accuracy of central frequency evaluation. All six coherences of signal pairs of "hot" $\mathrm{PPS}_{\mathrm{h}}$ without exception are suitable for $\mathrm{ASW}_{\mathrm{1vessel}}$ allocation and provide an average evaluation of 9.344 Hz. The $\mathrm{ASW}_{\mathrm{1vessel}}$ half-waves, enclosed between the opposite MCPs (MCP_1-MCP_3 and MCP_2-MCP_4) have lower amplitude than the $\mathrm{ASW}_{\mathrm{1vessel}}$ half-waves, enclosed between the opposite SGs (Fig. 14.18). Therefore, not all six coherence functions $\mathrm{PPS}_{\mathrm{c}i}$–$\mathrm{PPS}_{\mathrm{c}j}$ were used to evaluate the $\mathrm{ASW}_{\mathrm{1vessel}}$ frequency, but only those three functions whose coherence were not lower than 0.3 (Fig. 14.20). They provided an average evaluation of 9.366 Hz. A negligible difference between 9.344 Hz and 9.366 Hz evaluations prove that there is no splitting effect of $\mathrm{ASW}_{\mathrm{1vessel}}$ frequency in the presence of temperature anisotropy. This conclusion is applicable to both simple types of ASW consisting of a single half-wave, and to ASW consisting of multiple half-waves.

Some of the higher types of local ASWs are located only in the "cold" pipeline and some of them are only in the "hot" pipeline. There is no effect of central frequency splitting in them, but the shift of this central frequency along the frequency axis will be driven by the relevant coolant temperature in the part of MCC, where this ASW is localized. Two of the higher ASWs, individually localized in the hot and cold pipelines, are affective for measuring the reactor plant power or coolant temperature drop in the core by difference of their central ASW frequencies. The higher the central frequency of the ASW, the higher the absolute temperature shift of the frequency of ASW. A 10% absolute shift of the highest simple ASW with 30 Hz

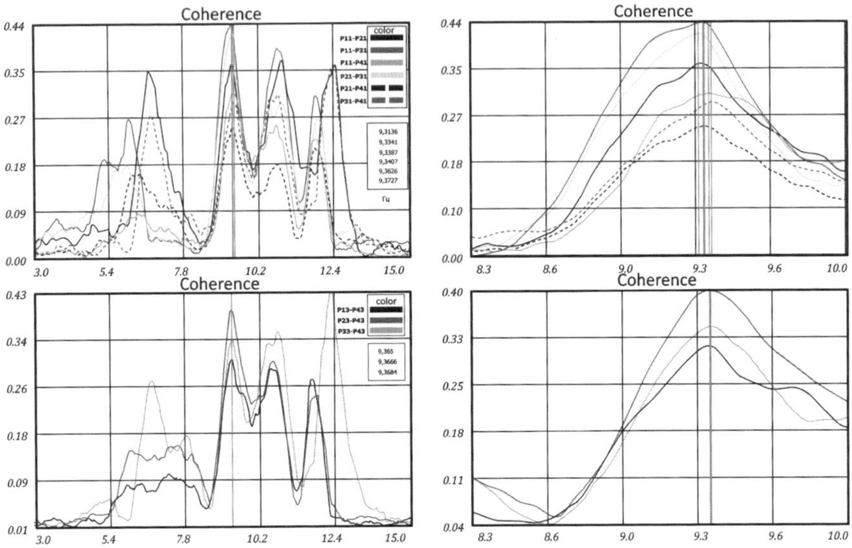

(a) (b)

Figure 14.20: Coherence functions of PPS signals of four loops with $ASW_{1vessel}$ frequencies (a) and their fragments (b): — PPS_h–PPS_c, various pairs (6 pcs.) of four PPS_h, installed in four "hot" pipelines (top line of graphs) — PPS_c–PPS_c, installed in the "cold" pipelines (3 pcs. — only significant coherences) (lower family).

central frequency, which is equal to 3 Hz, can be graduated in power units or temperature drop units of coolant in the core or only Toutput of coolant. Thus, a fairly accurate acoustic reactor plant power meter (or coolant temperature meter) can be obtained based on evaluations of frequencies of various simple (!) ASWs. The following signals can be used as the input measuring signals:

- noise components of standard IC and SPND signals;
- coolant pressure pulsations:
- signals of accelerometers, installed in MCC pipelines and in the main loop equipment;
- signals of other vibration sensors (absolute and relative displacement sensors).

It has been shown that there is a full identity of detected lower types of ASW from the PPS signals and from the signals of accelerometers, installed in MCC pipeline close to those PPSs. Now for the $ASW_{1vessel}$ reactor plant power, above 40% shall be detected also by ionization chamber signals. Let us check how the frequency evaluations of these $ASW_{1vessel}$ estimated from different signals will coincide with each other. To do this, we will use the waterfall spectra (APSD) of accelerometer signal, installed in the cold pipeline of the first loop (Fig. 14.21), and APSD of IC signals (Fig. 14.14).

The graphs of central frequencies of $ASW_{1vessel}$ in Fig. 14.22 almost coincide with each other. This experimental result is another proof of invariance of the central frequency of one ASW type at its

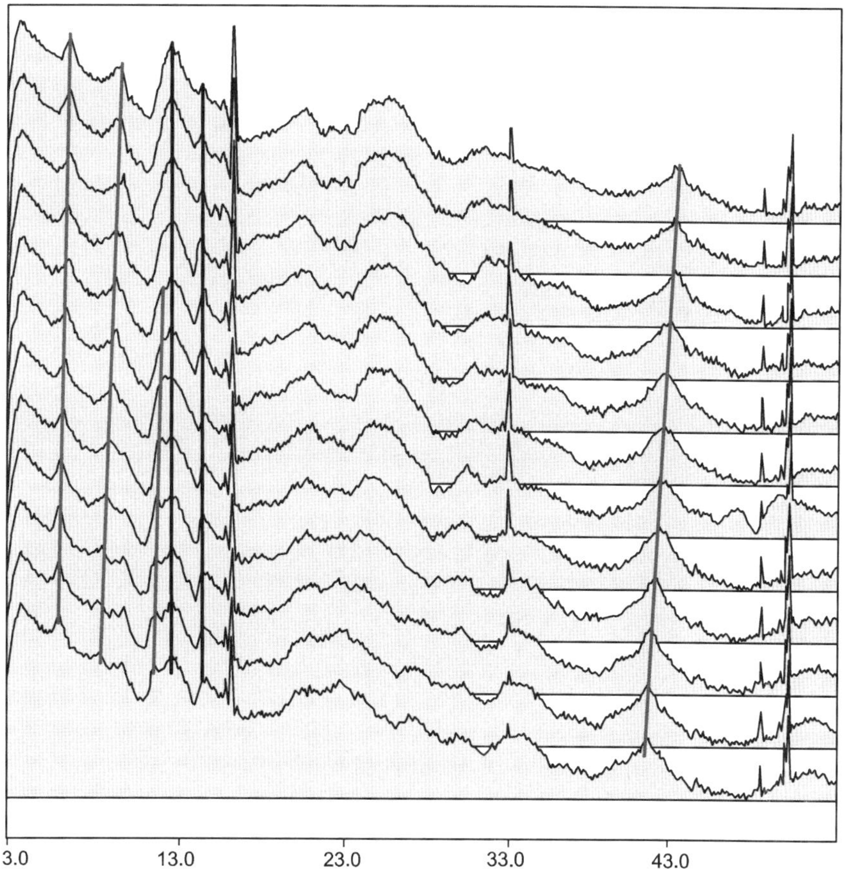

Figure 14.21: APSD of signal of Accelc1 in Novovoronezh NPP-2 Unit 1 while power level changes from 0% to 100% (12 power levels top-to-down) during 25 h.

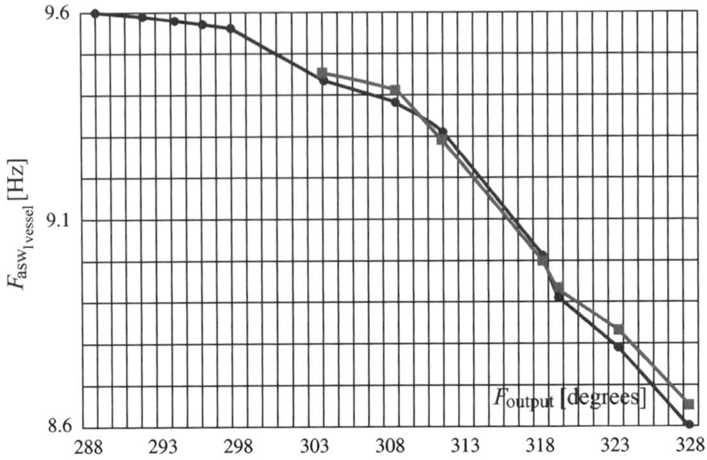

Figure 14.22: Dependencies of central $\mathrm{ASW}_{1\mathrm{vessel}}$ from coolant temperature in core outlet, obtained from IC signals (for reactor power level above 40%) and signal of accelerometer, installed in the "Cold" MCC pipeline.

different half-periods. The ionization chamber records an $\mathrm{ASW}_{1\mathrm{vessel}}$ half-wave, which covers the reactor, while the accelerometer records a half-wave, covering the "cold" pipeline.

Now, we can definitely assert that the temperature dependence of $\mathrm{ASW}_{1\mathrm{vessel}}$ frequency in the power ascension mode of reactor plant is not linear. For a result, which had been obtained in VVER-1000 reactor (Fig. 14.3), where such a curve consisted of only three points, such a conclusion would have been premature. The difference of curves for VVER-1200 and VVER-1000 reactors (Fig. 14.23) is that the $\mathrm{ASW}_{1\mathrm{vessel}}$ frequencies of VVER-1200 reactor are systematically lower than the $\mathrm{ASW}_{1\mathrm{vessel}}$ frequencies of VVER-100 reactor and the steepness of the straight lines of LSM approximations is almost the same, although it is slightly steeper for VVER-1000 reactor.

The total value of the central $\mathrm{ASW}_{1\mathrm{vessel}}$ frequency is determined mainly by two factors: the higher the average coolant temperature and the more the volume of the coolant steam phase in the core, the lower the central $\mathrm{ASW}_{1\mathrm{vessel}}$ frequency. If we compare the main operating parameters, achieved in VVER-1200 at 100% power level, with the same parameters for VVER-1000 (Table 14.3), we can conclude that an average energy density of VVER-1200 reactor core is less than the intensity of VVER-1000 reactor core.

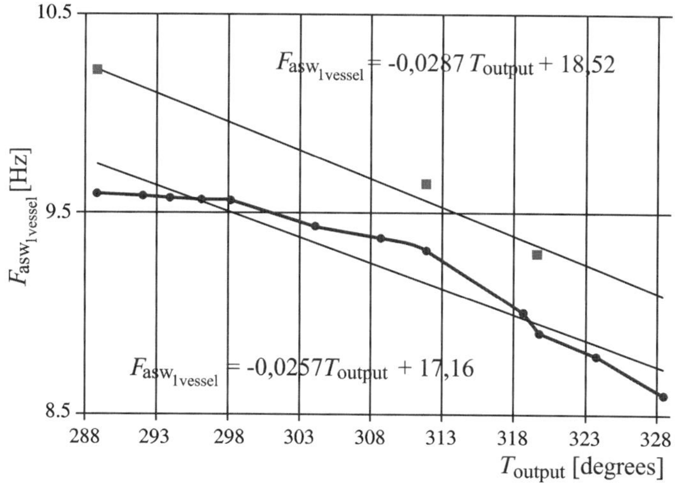

Figure 14.23: Dependencies of central frequency of $ASW_{1vessel}$ on coolant temperature in core outlet, obtained from PPS signals at Tianwan NPP Unit 1 with VVER-1000 reactor at three power levels, and signal of accelerometer, installed in the "Cold" MCC Pipeline of Novovoronezh NPP-2 Unit 1.

Table 14.3: Comparison of main mode parameters of VVER-1000 and VVER-1200 reactor plants.

Main mode parameters	VVER-1000	VVER-1200
T_{output} — Coolant temperature in the "hot" pipeline [°C]	321	328
T_{input} — Coolant temperature in the "hot" pipeline [°C]	291	298
ΔT — heating [°C]	30	30
G — coolant flow rate through reactor [m^3/h]	80000	90000
P — primary pressure [MPa]	16.0	16.2
q — linear heat flux [w/cm]	448	420

At the same heating of coolant in the core, the higher flow rate of the coolant through VVER-1200 reactor provides the lower average energy density in its core compared to that of VVER-1000 core. The volume of the steam phase in VVER-1200 core is also less than in VVER-1000 core. This is further evidenced by such parameters of Table 14.3 as pressure in the primary loop and linear heat flux.

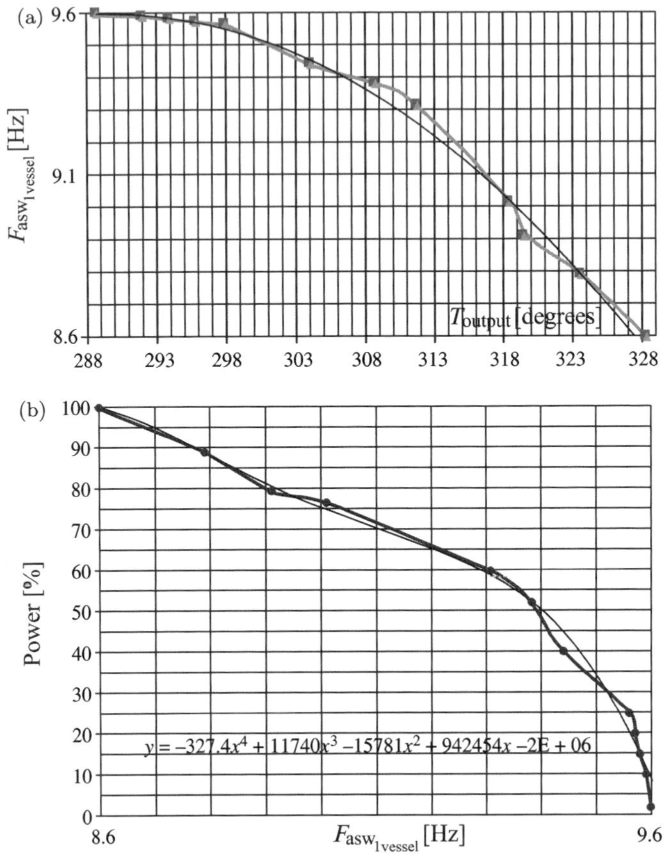

Figure 14.24: Dependence of central frequency of $ASW_{1vessel}$ on coolant temperature at reactor core outlet, obtained by the signal of accelerometer, installed in "Cold" MCC pipeline and its LSM-parabolic approximation (a). Correspondence "reactor plant power – $ASW_{1vessel}$ frequency" and its approximation by a polynomial for analytical determination of power according to $ASW_{1vessel}$ frequency measurement (b).

Dependencies of the central frequency of $ASW_{1vessel}$ from temperature at reactor outlet and from reactor plant power are shown in Fig. 14.24 together with power-law LSM approximations. These dependencies can be considered as calibration curves of acoustic measurements of coolant temperature and reactor plant power. An unexpected practical result is obtained: the main thermal and physical parameters of reactor core can be measured by measuring

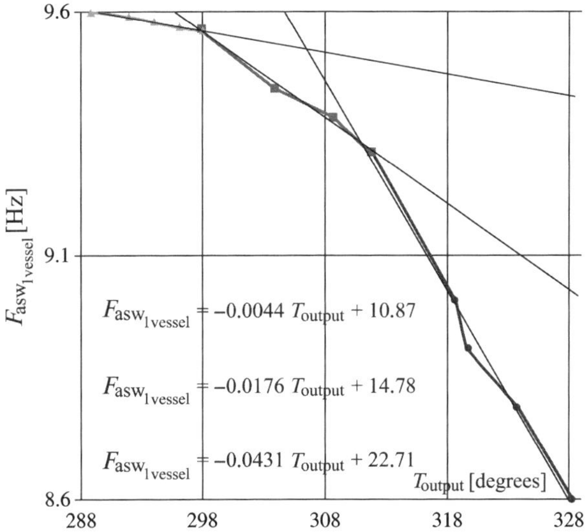

Figure 14.25: Dependence of central frequency of $\text{ASW}_{1\text{vessel}}$ on coolant temperature at reactor core outlet, obtained by the signal of accelerometer, installed in "Cold" MCC pipeline and its three LSM-piecewise linear approximations.

the forced vibrations with accelerometer, installed in the "cold" MCC pipeline.

The $\text{ASW}_{1\text{vessel}}$ frequency, considered with even a higher accuracy, is approximated by piecewise linear adjacent fragments (Fig. 14.25). They are obtained purely empirically from the shape of this curve. The first LSM straight line with the smallest slope — 0.0044 [°/Hz] — was obtained in the range of initial temperatures (288–298°C). The second one with exactly four times larger slope — 0.0176 [°/Hz] — was obtained in the next temperature range of (298–312°C). The third one with the slope of 0.0431 [°/Hz] (which is almost an order larger than that of the first straight line) was obtained in the range of 312–328°C. Most likely, the coolant steam phase had a dominant role in the formation of the last straight line. The steam phase reduces the $\text{ASW}_{1\text{vessel}}$ frequency significantly faster than a single effect of the coolant temperature increase factor starting from the $T_{\text{output}} = 312°\text{C}$ level, which corresponds to reactor plant power level of 60%.

Monotonous increase of the volume fraction of steam in the core with increasing reactor plant power results in such a dramatic reduction in the $\text{ASW}_{1\text{vessel}}$ frequency.

The first fragment of the slow linear dependence for initial coolant temperature range 288–298°C is most likely due to two multi-directional factors affecting the $ASW_{1vessel}$ frequency. The fall of the frequency of $ASW_{1vessel}$ due to the coolant temperature increase is reduced by the factor of reducing the boric acid concentration.

The $ASW_{1vessel}$ frequency at the initial section of reactor plant power increase, where decreasing of boric acid concentration dominates, and the end section with dominating effect of the steam phase volume fraction growth can be used for corresponding noise measurements. Figure 14.26 shows fragments of these dependencies, which

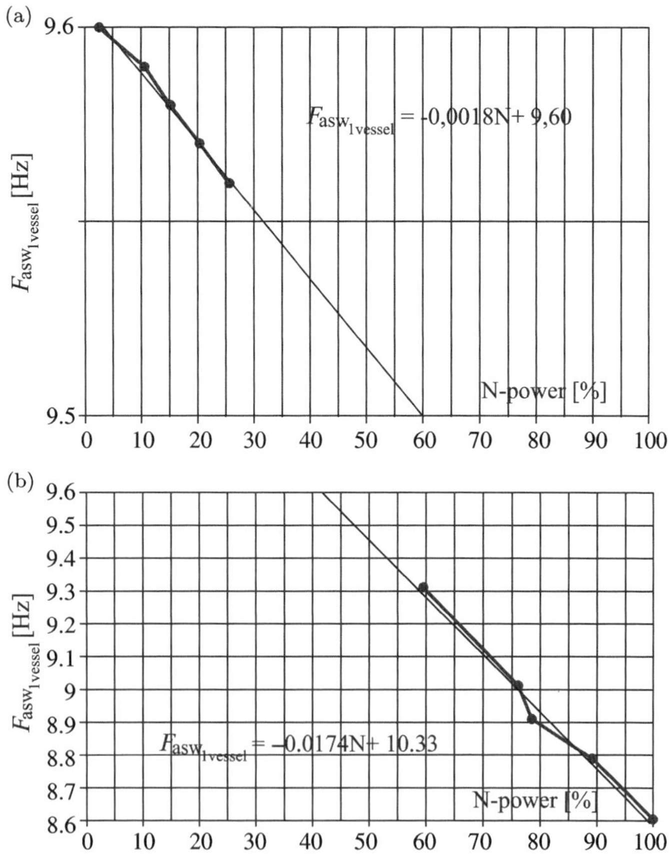

Figure 14.26: Dependence of central frequency of $ASW_{1vessel}$ on reactor plant power and its LSM-linear approximation to measure boric acid concentration (a) and volume fraction of steam phase in reactor core (b).

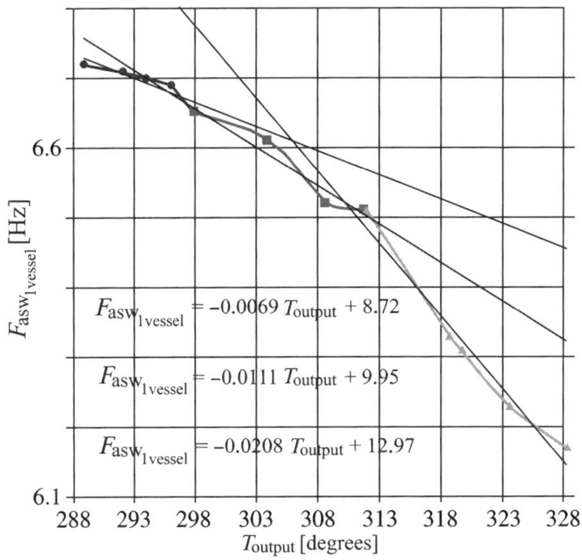

Figure 14.27: Dependence of central frequency of ASW_{1loop} on coolant temperature at reactor core outlet, obtained by the signal of accelerometer, installed in "Cold" MCC pipeline and its three LSM-piecewise linear approximation.

can be considered as calibration lineals. Now it is necessary to convert the "X" power axis into boric acid concentration, and into volume fraction of steam phase in the second graph.

The last parameter must be calculated. The boric acid concentration can be obtained either by standard I&C measurements or by more accurate chemical measurements. The same piecewise linear approximations for the same coolant temperature sub-ranges were implemented for ASW_{1loop} as well (Fig. 14.27).

Their slopes (–0.0069, –0.011, –0.0208 [°C/Hz]) slightly differ from each other compared to the respective slopes of the straight lines, built for the $ASW_{1vessel}$ (–0.0044, –0.0176, –0.0431 [°/Hz]). This is because the factor of steam phase appearance in the core has a strong influence on the $ASW_{1vessel}$, and weaker influence on the ASW_{1loop}.

Let us recall that the temperature dependencies of ASW frequencies in the isothermal heating (cooling down) modes of the reactor plant, obtained in the previous sections, were well approximated by the straight lines. Now, the ASWs are implemented at some averaged values of their frequencies in power operation, when the coolant

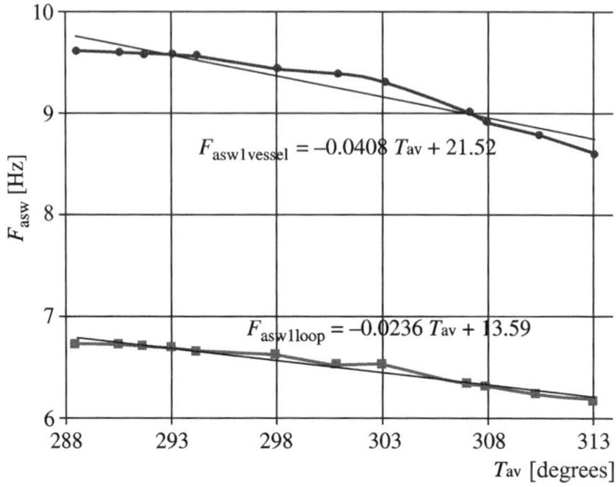

Figure 14.28: Dependencies of central frequencies of $ASW_{1vessel}$ and ASW_{1loop} on the average coolant temperature and their LSM-linear approximations

temperatures in reactor inlet and outlet differ. To obtain them, it is necessary to average the "cold" and "hot" water volumes, covered by one ASW type at the entire length of its existence. Because of rough approximation of such averaging, it is possible to consider a half-sum of the coolant input and output temperature, i.e. the average coolant temperature. Dependencies of ASW_{1loop} and $ASW_{1vessel}$ frequencies on the average temperature still significantly deviate from the LSM-straight line and have the form of convex curves, with the $ASW_{1vessel}$ curve convexity more than the ASW_{1loop} curve convexity (Fig. 14.28). The explanation for this experimental fact is the beginning of the steam phase growth at 60% power, which causes a kink of the presented dependencies.

This section provides sufficiently detailed calibrations of acoustic-neutron noise methods for measuring the coolant temperature, reactor plant power, volume steam fracture and boric acid concentration to be applied in practice.

Chapter 15

Mutual Oscillations of Reactor Vessel and Reactor Core Barrel

15.1 Structural Features of Reactor Vessel and RCB Attachment Points

Reactor vessel and reactor core barrel are a system of coupled masses which oscillate due to internal hydrodynamic forces of the coolant flow. They can be modeled easily in cylindrical coordinates by two coaxials (in technical applications they are also called shell rings), which are connected with each other by internal stiffening elements, and the reactor vessel is bound to the civil structures with external distributed ring stiffnesses in two horizons (Fig. 15.1).

Attachment points are three-dimensional elements described by the corresponding stiffness matrices. However, almost all of them have a well-defined direction, i.e. there is a coordinate vector along which stiffness reaches its extreme value. Let us also single out point (lumped) stiffness and distributed stiffness. For example, reactor vessel support ring is a stiffness element, continuously distributed in azimuth, while insert attachment points of the reactor core barrel (RCB) are point elements. In these attachment points, there may be gaps between the coupled elements, which make their modeling quite difficult.

Strictly speaking, we cannot declare that certain RCB oscillations exist independently of reactor vessel oscillations. However, it is also incorrect to say that certain RCB oscillations appear solely due to the

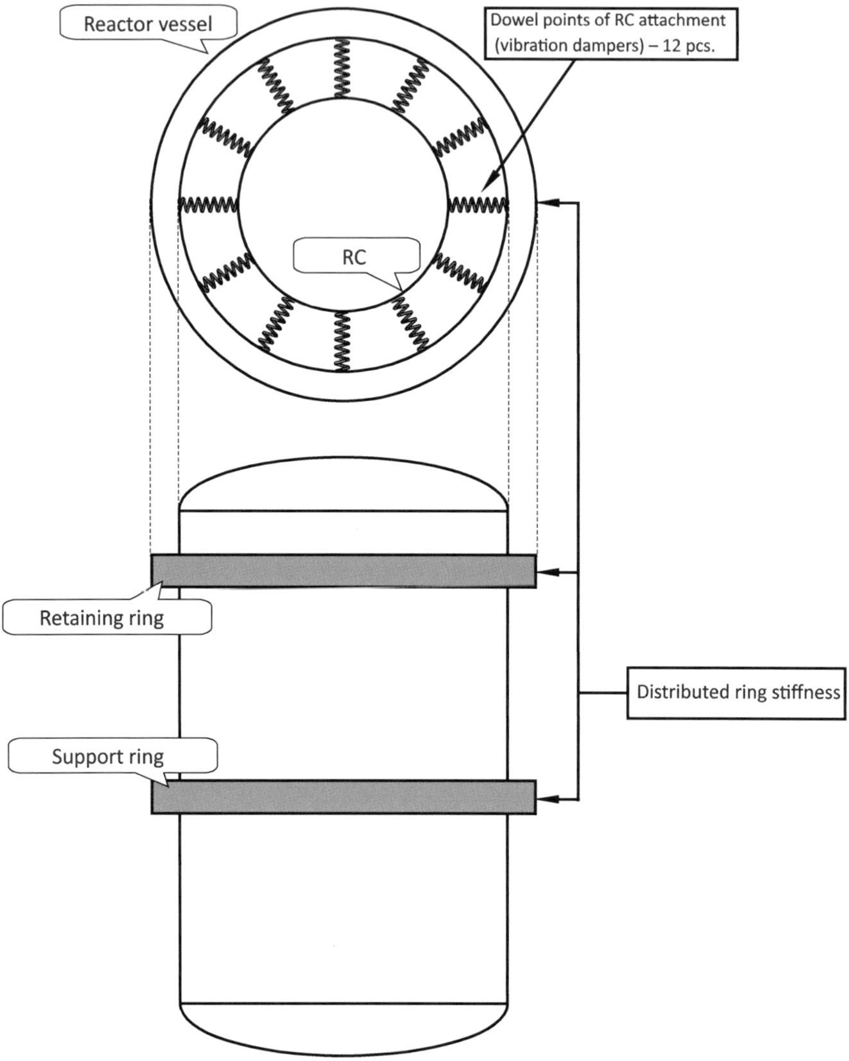

Figure 15.1: Schematic representation of reactor vessel and RC fixtures.

oscillations of the reactor vessel. In all cases, we have to speak of joint oscillations of reactor vessel and reactor core barrel. Nevertheless, in neutron-noise vibration models of the reactor core barrel, as a rule, it is possible to assume that the reactor vessel and the civil structures are rigidly fixed, and, consequently, vibration displacement of RCB in relation to the RP vessel is a kind of RCB natural oscillation. In this case, the state of external RP vessel fixtures is not taken into account *a priori*, and it is assumed that the detected neutron noise bears the information on RCB natural oscillations only — in particular, on the state of its attachment points.

Such a model, with vibration displacement of RP vessel considered to be zero, makes no sense when vibration sensors are placed directly on the RP vessel. When vibration sensors register physical vibration displacement of RP vessel, they also bear the information on vibrations of reactor internals, as their considered oscillations are mutual. The types of vessel oscillations are defined not only by the state (stiffness characteristics) of its own support structures, but also by stiffness characteristics of points of attachment of reactor internals — in particular, three horizons of RCB points of attachment (Fig. 14.9). This is why in the signals of vibration sensors installed on the RP vessel, or on the upper unit of the RP vessel, components of reactor internals vibrations can also be found.

Let us consider in detail the structure of external points of attachment of the reactor vessel and internal points of attachment of the reactor core barrel. Reactor vessel is placed in a concrete cavity and fixed in two horizons (Fig. 15.2). The upper horizon includes the flange of the RP vessel (item 1 in Fig. 15.2) and a thrust ring (item 2). It is fixed with spikes (item 3), which are inserted in the notches of the ring.

Via a support ring, loads are transferred to a thrust truss, which is a civil structure. The upper horizon of attachment enables the reactor vessel to withstand tilting forces.

The lower horizon of vessel attachment includes a thrust collar and a support ring. The thrust collar of the vessel (item 4) is a circular 80 mm rim on the vessel with 30 notches evenly distributed along the circle. In these notches, inserts (item 5) are inserted, which by means of a plate (item 6) are fixed in the support ring (item 7). The support ring, in its turn, is fixed to the thrust truss (item 8), a civil structure, with the fixture (item 9) and aligned along the horizon with support wedges (item 10). Moreover, the aggregate of elements "plate — insert — support ring — fixture — thrust truss" prevents the reactor vessels from turning along its vertical axis.

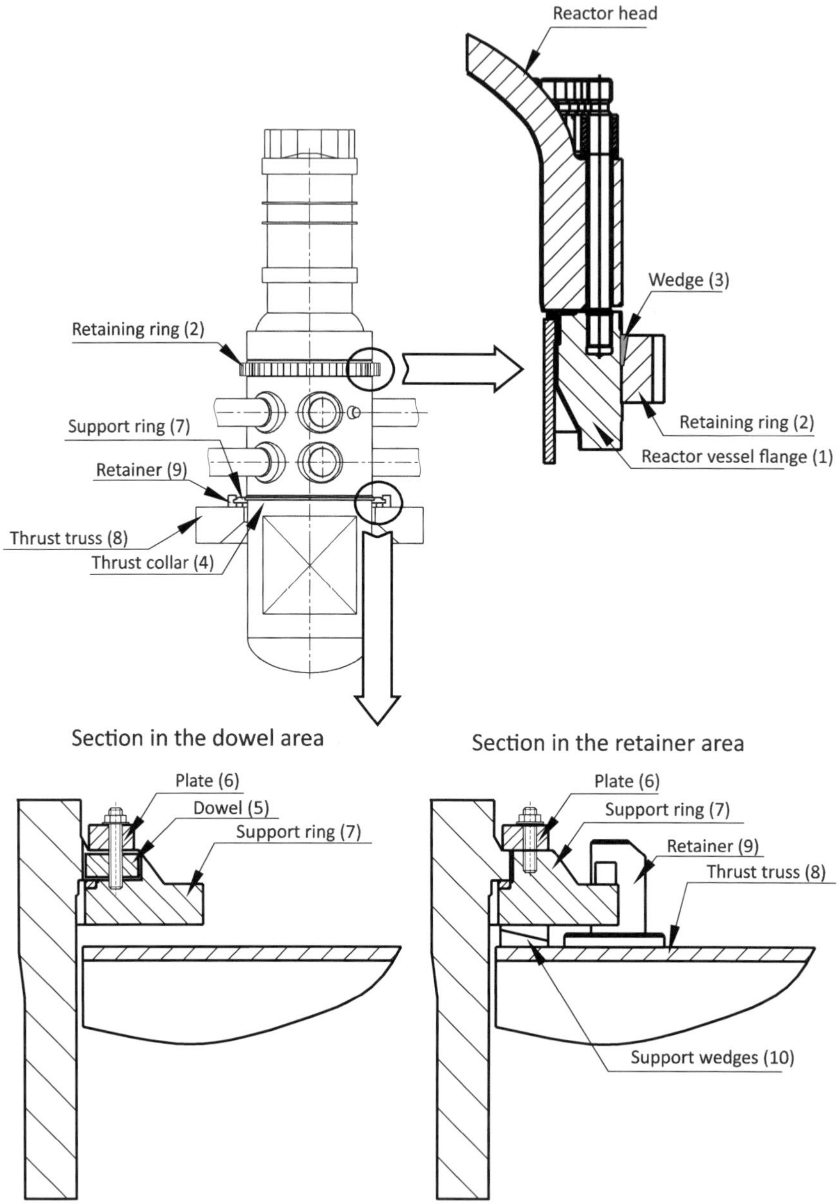

Figure 15.2: Two horizons of RP vessel attachment to civil structures.

15.2 Oscillations of Reactor Vessel and RCB at Different Numbers of MCP Sets in Operation

The first generation of vibration noise diagnostic systems monitored the vibrations of RP vessel and RCB with four absolute displacement sensors (ADS) installed on a clamp ring (VVER-440) or on the main stud-bolts (VVER-1000) in the corners of a corresponding rectangle. These vibration sensors are high by sensitive and capable of measuring vibration displacement as low as microns. Due to high operating temperatures and intensive radiation fields, they were very unreliable. Moreover, they were inconvenient in terms of maintenance, as they had to be removed every time it was necessary to remove the reactor head. That is why they were eventually replaced with four accelerometers installed on four vertical stands of the upper unit (Fig. 15.3) in the corners of the corresponding rectangle.

The operating temperature and radiation fields in these locations proved to be suitable for long-term operation of accelerometers. The equality of such replacement of vibration sensors was proven experimentally at Kalinin NPP Unit 3, and later at two power units of Tianwan NPP. It turned out that accelerometers were even more sensitive to RP vessel and RCB vibrations than ADSs, with the same number of significant resonances and all phase characteristics of signals. This means that the upper unit stands, where accelerometers are installed, are absolutely rigidly fixed to the reactor head, and they do not distort the transfer of the vibration image of oscillations from the horizon of the main head flange to the horizon of accelerometers location.

Vibration displacement of the vessel appears due to the hydrodynamic flow of the coolant and vibrations coming from MCP sets along the metal structures. At the same time, reactor internals and fuel assemblies also get involved in mutual oscillations together with the vessel. They also have an impact on the in-core neutron flux detected by self-powered neutron detectors (SPND), and on the excore neutron flux detected by ionization chambers (IC). The correlations of signals with different physical natures enable us to find the mutual oscillations of the vessel — RCB, RCB — FA, vessel — RCB — FA. Before reactor power operation begins, when only vibration sensors are involved in the measurement, it is too early to speak about definitely selected mutual oscillations.

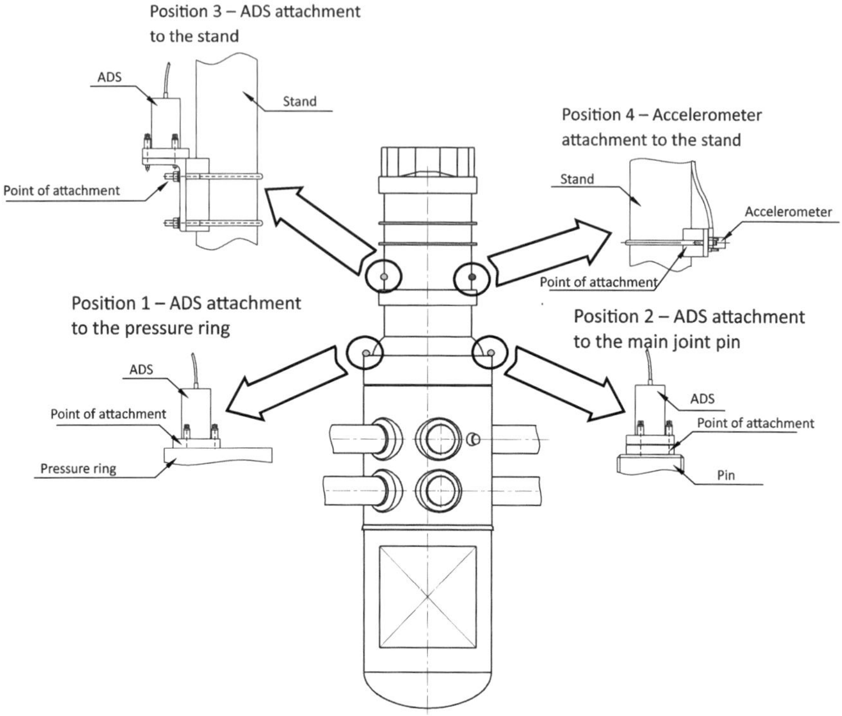

Figure 15.3: Locations of ADSs on main stud-bolts and accelerometers on the upper unit stands.

In what follows, we consider the active experiments enabling us to interpret the vibration state of the RP vessel. When not all of the MCP sets are in operation, which happens every time during the power unit heating up and cooling down, the hydrostatic impact of the coolant on RP vessel and RCB from the side of main loops may be imbalanced. Only in two cases, when two MCP sets from the opposite loops are in operation (MCP one and three, or two and four), the resulting static force, acting on the vessel and RCB from the coolant will be equal to zero. This conclusion is based on the central symmetry of the MCC and the forces from the coolant of two active and two passive loops, opposite in direction and equal by modulus, as well as on the assumed equality of four loops.

In all other cases with the incomplete set of MCPs in operation, the RP vessel and RCB will throw off the balance due to a non-zero and imbalanced static force. As a result, the core barrel is forced to come into a "rigid" contact with some of the vibration guide lugs (dampers, bumpers inserts). These are primarily the vibration dampers diametrically opposite to the active loop. This deviation of RCB from vertical is limited by another pair of diametrically opposite vibration dampers. They are perpendicular to the active loop.

Suppose that a loop starts to operate, which is opposite in direction to the first one. For example, if MCP1 was in operation at the beginning, and all three other MCPs were off, with only MCP3 set in operation, which is in the opposite loop. Now, the cavity will be tilt in the same direction, but to the opposite side. Its displacement is still limited by the same pair of perpendicularly located dowels, but by their other surfaces. In the end, summing up the modulus of RCB displacement in both sides, we will get a complete, physically possible range of RCB oscillations in one direction (here, in the direction "loop one–loop three").

Analogous switching is applied to MCP2 and MCP4 in order to estimate the range of RCB oscillations in the direction "loop 2–loop 4". This switching of four MCP sets enables us, firstly, to estimate the insert gaps in azimuthal directions and, secondly, to identify the frequencies, which may be related to RCB oscillations as a beam with two fixed ends. This type of oscillations is the most advantageous in terms of RCB fixtures durability. As far as vibration dampers get worn out and the insert gaps increase, the undesired oscillations appear — as a beam with a loose lower end. It is considered the lowest-type oscillation of the core barrel, although when the upper attachment point of the core barrel gets loose at the same time, the so-called "floating" oscillations appear. Their natural frequency is even lower than the natural frequency of oscillations of a beam with only the upper end fixed.

By such integral characteristic as root mean square (RMS) value, fibrillation of oscillation amplitudes is observed. This means that RMS values of the opposite accelerometers, "V1-V3" and "V2-V4", will reach their extremes simultaneously. For example, RMSv1 and RMSv3 will reach their maximum simultaneously, while RMSv2 and RMSv4 will reach their minimum at the same time.

The numbering of accelerometers installed on the upper unit — V1, V2, V3, V4 — is such that V1 is on the axis of loop 1, V2 is on the axis of loop 2, V3 — loop 3, V4 — loop 4. If MCP1 and

MCP3 are in operation, i.e. there is a perturbation in the direction "loop 1–loop 3", and we observe max RMS values in the pair V1 and V3 and min RMS values in the pair V2–V4, this effect is caused by forced oscillations.

Static displacement of RCB will be limited by the sizes of gaps in RCB attachment points, and the turbulent components of coolant pressure pulsations will cause RCB vibrations around the static state. If some of the gaps in attachment points are removed, i.e. become equal to zero, vibrations will be accompanied by collisions in the RP vessel and RCB attachment points.

A special role here belongs to the mode of operation of the single working MCP set. Sequential, loop-by-loop, operation of a single working MCP is considered in detail only during commissioning measurements. In standard RP heating up/cooling down modes, operation of a single MCP is not implied. The only operating MCP not only creates a coherent field of standing waves, but also unambiguously defines the direction of a static disturbing force for RCB and RP vessel. The four cases when only one MCP set is on sequentially in each loop are a sort of vibration tests, where gaps in RCB and RP vessel attachment points can be checked.

Let us consider a simplified case of vibration dampers, or the so-called insert connections of RCB and RP vessel in the lower attachment point. In VVER-1200, there are 12 such points, in VVER-1000 — 8 points. To make things simple, let us consider first the case when RCB is fixed in only four of such points (Fig. 15.4), and each point in the horizontal section is characterized by three hypothetical gaps. In particular, any of them may be equal to zero. Under an idealized assumption that all gaps are equal, we can conclude that in one direction, the extreme vibration displacements are

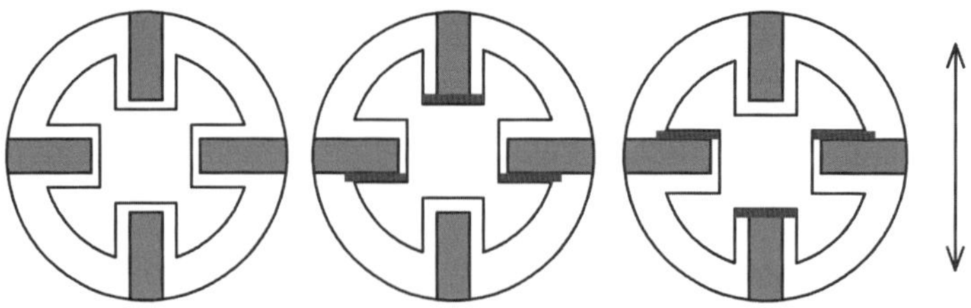

Figure 15.4: A modeling case of four-dowel attachment points of RCB and RP vessel. Equilibrium state (left picture) and two extreme states with no gaps left due to vibration. Dark gray color indicates zero gaps.

limited by three gaps only (see the dark gray color gaps in Fig. 15.4), and, in the other direction, by three other gaps.

In the process of operation, contact pairs of dowel connections get worn out, and gaps get wider. They are periodically checked during scheduled outages by direct geometrical measurements of RCB gaps and dowels on the RP vessel. Measurement of RP vessel vibrations with an incomplete number of MCP sets in operation provides a way of remote indirect monitoring of gaps in dowel attachment points, being a kind of experimental test. In the nominal mode of operation, with all four MCP sets working, the established range of vibrations of RCB cannot "decide on" the gaps, unlike in case of the forced deviation of RCB with an incomplete number of MCP sets in operation.

In the considered simplified example above, two types of oppositely directed forced oscillations of RCB along one axis have affected six dowel gaps out of 12. In the perpendicular direction of forced oscillations, other six gaps will be affected. That is, during four tests, all 12 gaps will suffer from some change. However, it does not mean that it is possible to estimate the gaps, even in relative units. The assumption that all gaps are equal is actually wrong, and it was taken solely for demonstration purposes, in order to set the task of gap monitoring. In the design-basis state, a dowel on the RP vessel comes into contact with the RCB notch with two side surfaces. Direct geometrical measurements of notch and insert give the real starting image of gaps, which may be further interpreted, via vibration measurements, from the point of view of oscillation ranges.

When there are 12 dowel connections, actually, any of the 24 gaps may limit the displacement of RCB. Let us designate the number of a certain dowel point using the analogy of a clock dial (Fig. 15.5). If the direction of forced oscillations is set along the "12–6" axis, then, assuming that all 24 side gaps are equal, vibration displacement will be limited to the points numbered 3, 6, 9, 12.

If the direction of forced oscillations is set to the first dowel point, i.e. along the axis "1–7", vibration displacement is limited by points 4, 7, 10, 1. For axis "2–8", there are points 5, 8, 11, 2. For axis "3–9", there are points 6, 9, 12, 3, etc. Now, let us imagine a case when we know *a priori* the real gaps from geometrical measurements. Then, based on a certain direction of forced oscillations, through trigonometric calculations, we can easily identify the dowel attachment points, which most probably limit the vibration displacement.

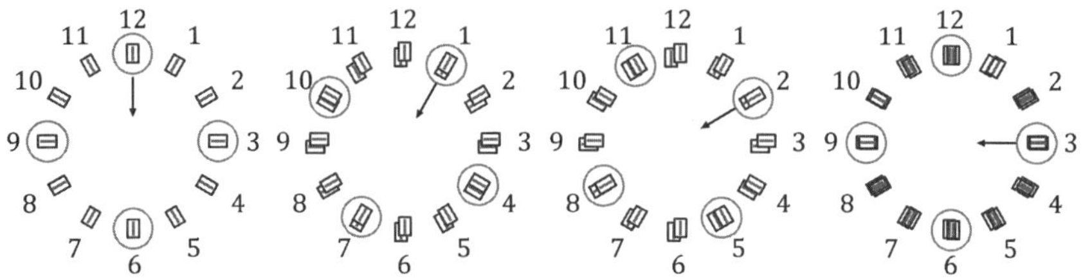

Figure 15.5: Conventional images of 12 dowel points with four different directions of forced oscillations. Direction "12–6" — left dial, direction "1–7" — the second dial on the left, direction "2–8" — the second dial on the right, direction "3–9" — the right dial.

Table 15.1: Modes with different numbers of MCP sets in operation.

Mode number	Number of MCP in operation				Tc [°C]
	1	2	3	4	
1	—	—	⊕	—	117
2	⊕	—	—	—	116
3	⊕	⊕	—		124
4	—	⊕	—	—	124
5	⊕	—	⊕	⊕	121
6	⊕	—	—	⊕	120
7	⊕	⊕	—	⊕	121
8	⊕	—	⊕	—	120
9	—	—	—	⊕	120
10	—	—	⊕	⊕	122

In commissioning measurements (Table 15.1), signals of four accelerometers installed on the RP upper unit were registered, in modes with different numbers of MCP sets in operation, with minor change of coolant temperature within the range $T_C = (116–124°)$. All in all, 10 multi-channel records were processed, which corresponded to a different number of MCP sets in operation; among them, there were four records with only one MCP on. Remember that at this temperature, there should not be more than three MCP sets in operation.

In mode 1 (Figs. 15.6 and 15.7), there was one single MCP in operation in loop No. 3. RMS values of four accelerometers formed

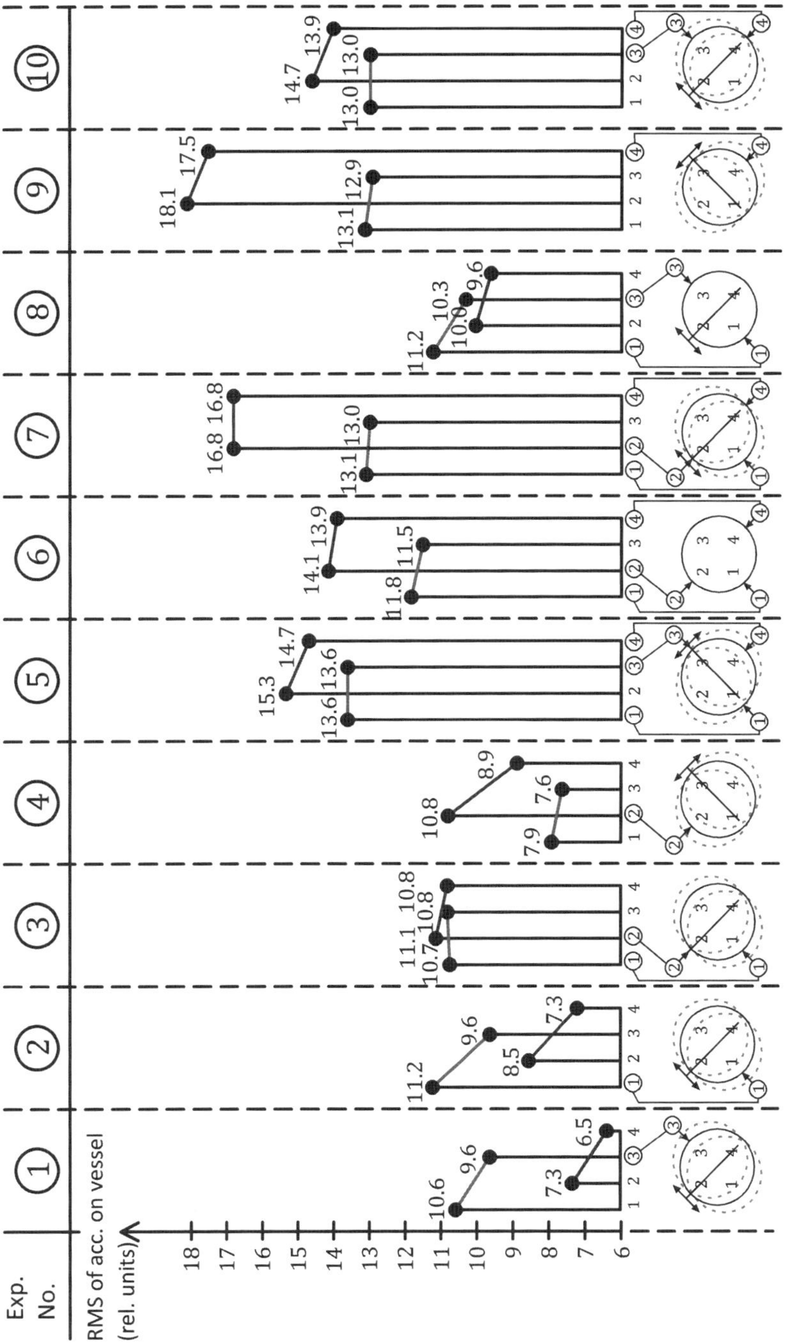

Figure 15.6: RMS values of signals of four accelerometers in ten modes with different number of MCP sets in operation.

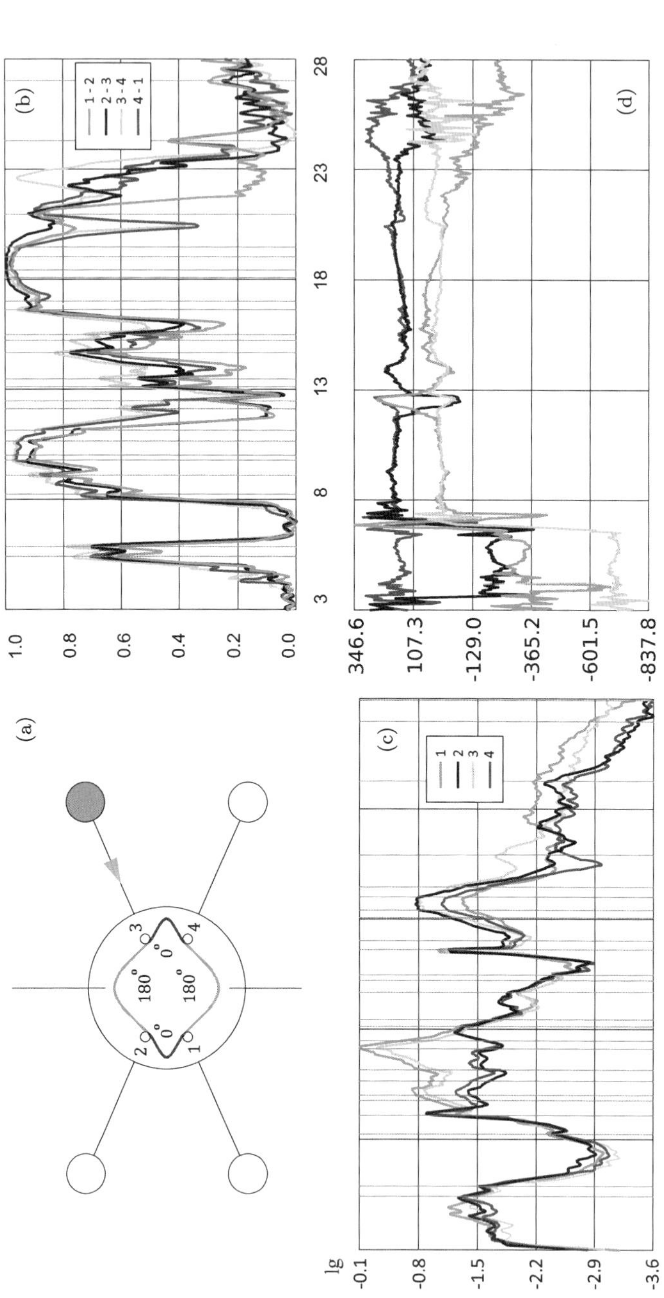

Figure 15.7: Mode 1. Conventional image of active loop and phases of signals of the adjacent sensors (top left cell of the table). APSD of signals of four accelerometers, four coherence functions and phases of signals of the adjacent sensors (1–2, 2–3, 3–4, 4–1).

two pairs: along the axis of loop No. 3, signals of accelerometers 1 and 3 had RMS values higher than the signals of accelerometers 2 and 4 located transverse to the axis of loop No. 3. Such RMS values are typical for unidirectional pendulum oscillations.

They are supposed to have typical phase images. Moving in closed sequence along the pairs of sensors — 1–2, 2–3, 3–4, 4–1 — we get the so-called phase balance, meaning that their sum is always a multiple of 360° or is equal to zero. In the given case, we have the sequence 0°, 180°, 0°, 180°, which is in reality typical of unidirectional pendulum oscillations. That is, when signals of sensors having the same phase number 1 and 2 increase, the signals of sensors numbered 3 and 4, opposite in phase, decrease, and vice versa (Figs. 15.7 and 15.8). The axis, along which pendulum oscillations occur, is defined with the accuracy of ±45°, and its average position coincides with the "II–IV" axis of the reactor.

This is the first time in vibration noise analysis when such a sequence of phases is observed in the whole range of the considered frequencies. Pendulum oscillations usually occur at some single natural frequency, covering a minor area around this frequency. However, here we have all registered frequencies with one and the same property of pendulum oscillations. Obviously, these are not natural oscillations. The practical eigenvalue problem cannot be solved at the continuum of frequencies.

Four APSDs (Fig. 15.7) have been grouped into the same pairs: 1, 3 and 2, 4. Significant difference in APSDs of the groups is observed at RP vessel oscillation frequencies from 8 to 15 Hz. Disturbing forces did not induce oscillations at new resonance frequencies, but instead redistributed their energy among themselves and imposed one and the same spatial type of pendulum oscillations.

In the second mode, the only MCP set in operation is in loop one (Figs. 15.6 and 15.9). That is, we have a pattern of disturbing forces symmetrical to mode one. Their direction along the axis "loop one–loop three" remains. RMS values are a bit higher than in case one, but the groups of sensors also remain the same. We may assume that along this axis all gaps limit the vibration displacement of RC right and left of the vertical. The phase image of forced oscillations remained the same (0°, 180°, 0°, 180°), which was expected, as the direction of forced oscillations also has not changed.

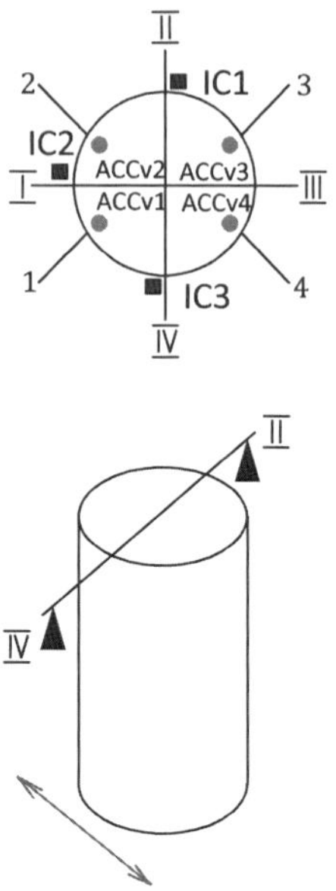

Figure 15.8: Location of ionization chambers (IC) and accelerometers on the upper unit, and direction of pendulum oscillations of RP vessel.

In mode three (Figs. 15.6 and 15.10), there were two MCP sets in operation: MCP1 and MCP2. Being in the adjacent loops, they create an imbalanced static force, which affects the RCB. However, RMS values of all 4 accelerometers became nearly equal to each other (Fig. 15.6). This experimental fact proves the existence of azimuthal asymmetry in the stiffnesses of dowel connections. The phase image remained the same again (0°, 180°, 0°, 180°). Looking ahead, we can say that it will remain unchanged in all ten modes.

In mode four, there is only one MCP set in operation, which indices RCB oscillations along the axis of pipelines of loop two. Now, max and min RMS values have swapped: the amplitudes of sensors 2 and 4 are higher than the amplitudes of sensors 1 and 3, which is an expected result.

Now, let us jump to mode nine, skipping all intermediate modes. In mode nine, there is only MCP4 in operation. Here we have the highest RMS values (Figs. 15.6 and 15.9). It is easy to notice that higher RMS values are observed in all modes with MCP4 in operation. That is, in the direction of axis from loop four, we can see the

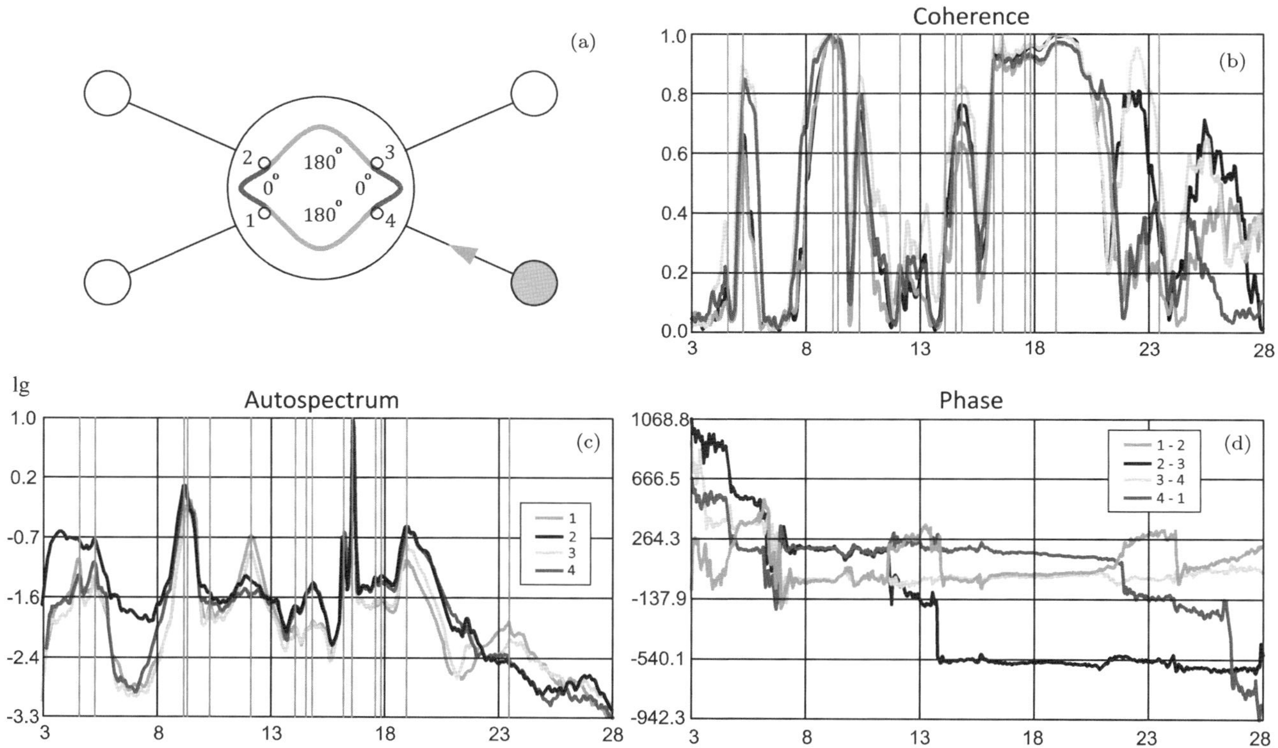

Figure 15.9: Mode 9. Conventional image of active loop and phases of signals of the adjacent sensors (b). APSD of signals of four accelerometers, four coherence functions and phases of signals of the adjacent sensors (1–2, 2–3, 3–4, 4–1).

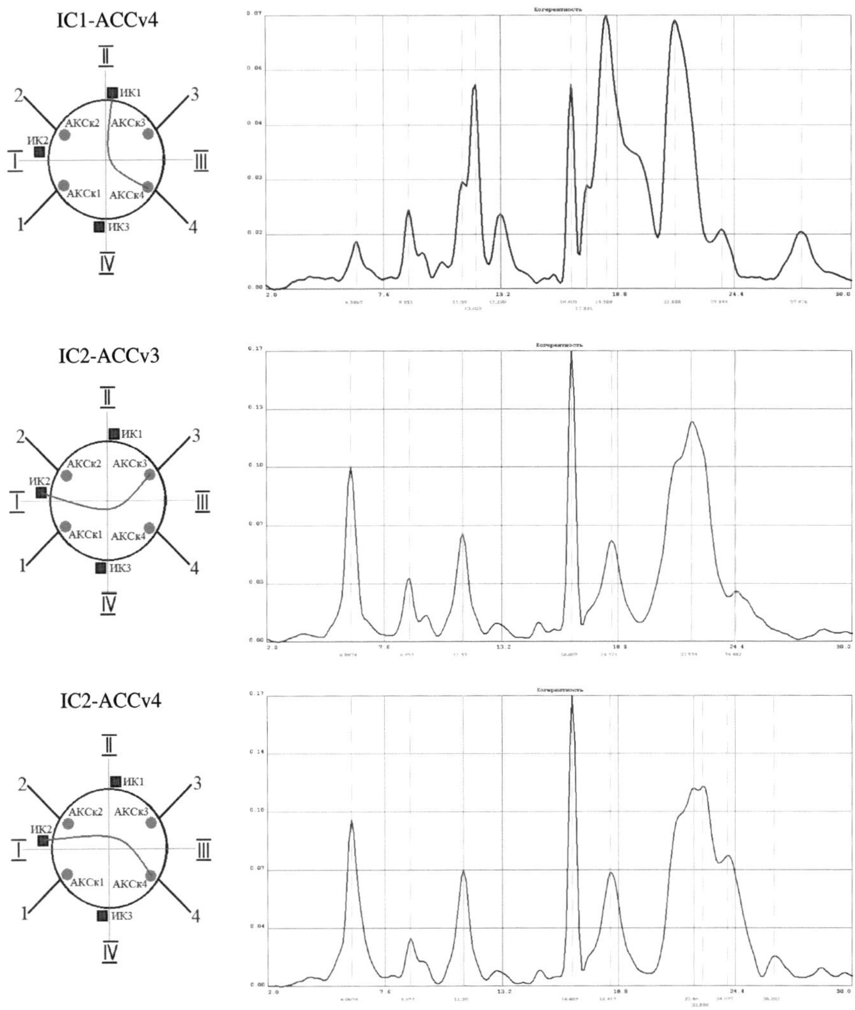

Figure 15.10: Coherence functions of "IC–ACC" signals.

worst combination of dowel gaps, while along the same line but in the opposite direction, from loop two, the gaps are nearly the same as the ones along the axis "loop one–loop three".

In the final series of spectral estimations (Fig. 15.11), similar curves are given for 100% reactor power, that is, with four MCP sets in operation. With the same resonance features and minor difference

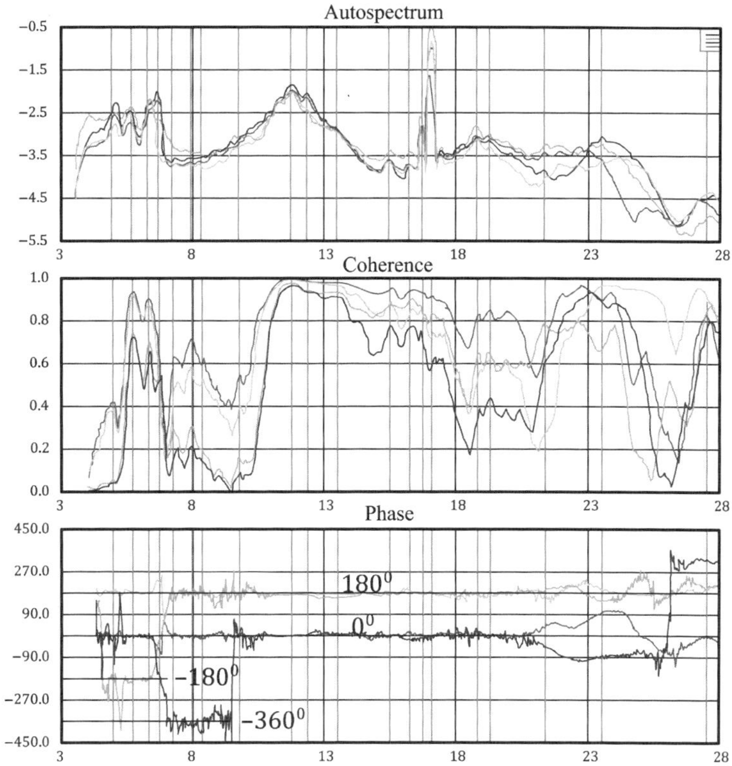

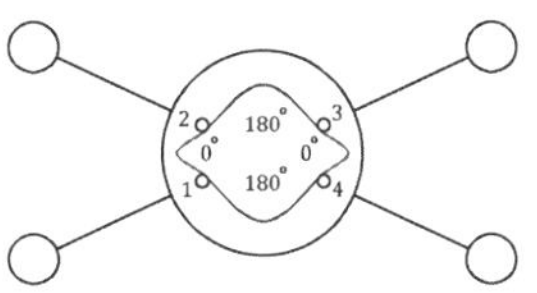

Hz	1-2	2-3	3-4	4-1
4.4303	5.589	175.22	1.3947	-178.26
5.1619	-6.3193	178.57	4.498	-176.23
5.769	-3.035	-175.18	-5.3898	-175.88
6.1582	-159.89	-21.622	-47.599	-133.03
6.7341	36.704	153.37	10.288	166.82
7.4191	5.6854	169.2	3.3817	-178.75
7.8394	-32.181	-166.17	-4.8412	-167.11
9.2474	40.52	156.12	7.3876	170.5
11.217	2.5988	173.42	2.7605	-178.75
11.84	7.4685	171.95	0.28067	-179.53
12.976	15.564	161.27	10.534	172.74
15	0.52429	175.8	3.5016	-178.98
15.763	-15.276	175.13	15.025	-175.16
16.261	-5.2358	175.61	11.028	178.6
16.619	-8.6875	176.69	10.172	-179.08
18.3	16.719	162.99	13.941	162.39
18.798	7.7703	173.73	7.861	167.23
20.884	-41.356	-163.5	42.283	158.27
23.017	-82.075	151.44	96.272	-168.59
27.034	-46.188	-156.05	0.14297	-167.81

Figure 15.11: APSDs of signals of four accelerometers installed on the upper unit (the upper family of curves), coherence functions and phases of signals of the adjacent accelerometers (1–2, 2–3, 3–4, 4–1), values of four phases at the marked frequencies. In the table, the set of phases 0°, 180°, 0°, 180° is typical for unidirectional pendulum oscillations. 100% power.

of RMS values of 4 signals from each other, we have the same phase image again. Minor systematic deviations from 0° and 180° phases at certain frequencies (black in Fig. 15.11), most often at ASW frequencies, demonstrate that forced oscillations are now accompanied by another vibration effect.

The same type of unidirectional pendular oscillations of RP vessel imposed at all frequencies, most probably, has the following explanation. The reactor vessel sort of rests upon two points of the "II–IV" symmetry axis (Fig. 15.8), and any forced oscillations from any loop swing it around this axis, not changing the type of pendular oscillations, but just changing the amplitudes of oscillations. This is a rough model, and in reality the vessel rests upon attachment points with different stiffnesses. At the same time, gaps in external attachment points of two horizons of RP vessel fixture to civil structures may be so small that it always leads to unidirectional pendular oscillations around the 2–4 symmetry axis of the reactor with very small amplitudes.

Other points of reactor internals attachment, even at most disadvantageous vertical deviation of RCB, cannot change this type of RP vessel oscillations, they only change the amplitudes of accelerometer signals. It demonstrates in practice that the type of RP vessel oscillations does not depend on the stiffness of internal points of attachment. This means that one and the same spatial type of oscillations is imposed from the outside by stiffnesses of external points of RP vessel attachment to civil structures. It masks all mutual oscillations of RP vessel and RCB at natural and forced frequencies in such a way that the corresponding phases have insignificant increments. This is the major difference of the vibration state of VVER-1200 reactor vessel of Novovoronezh NPP-2 Unit 1 from the known vibration states of VVER-1000 and VVER-440 reactor vessels.

15.3 Neutron-Vibration Measurements at 100% Power, NovoVoronezh NPP Unit 6

VMS, apart from direct vibration measurement channels, includes instrument channels, which are indirectly sensitive to vibration displacement of the core and, consequently, to vibration displacement of the reactor core barrel. These are the ex-core and in-core neutron channels based on ionization chambers and self-powered neutron detectors. Autospectra of IC signals at 100% power of VVER-1200 reactor plant (Fig. 15.12) have a single significant local maximum at $ASW_{1vessel}$ frequency (8.78 Hz). However, while it is possible to conduct multiple measurements, almost unlimited in time, at the constant stationary state of the reactor, other resonances may be singled out, insignificant in amplitude (see the lower curve in Fig. 15.12). The functions of coherence of "IC–IC" signals show them as minor (Fig. 15.13), since vibration effects are so insignificant that they are masked by global neutron and temperature effects at low frequencies, while at high frequencies they are masked by the "white" noise of ICs.

Searching further for RCB vibration resonances, let us consider the cross-characteristics of "IC–accelerometer" signals. The coherence between the signal of an accelerometer at the reactor upper unit and the signal of an ionization chamber shows the mutual oscillations of the RP vessel and RC (Fig. 15.14). Here, we face the same problem of extremely low coherence. However, upon averaging of a large number of estimations, extremely small values of the coherence function become significant.

Having three ionization chambers (IC1, IC2, IC3) and four accelerometers at the RP vessel (ACCv1, ACCv2, ACCv3, ACCv4), we get 12 coherences of pairs of signals with different physical natures, "IC–ACCv", which have the same frequencies of ASW resonances. Figure 15.10 shows not all 12 coherence functions, but only three coherence functions averaged in time. They enable us to single out from the background only resonances at ASW frequencies. These are the mutual oscillations of RP vessel and RCB induced by standing waves of different types.

RCB is also involved in mutual oscillations with the equipment and pipelines of main cooling loops. Possibly, vibration resonances at natural frequencies will be found there. Figure 15.15 shows the coherence function of such a pair of signals, and again we can see only oscillations at ASW frequencies. Thus, natural oscillations of RCB were not found for any of the signal pairs: "IC–IC", "IC–ACCv", "ACCv–ACCv", which proves that RCB vibration displacement is really small, and, therefore, the vibration state of RCB is very good.

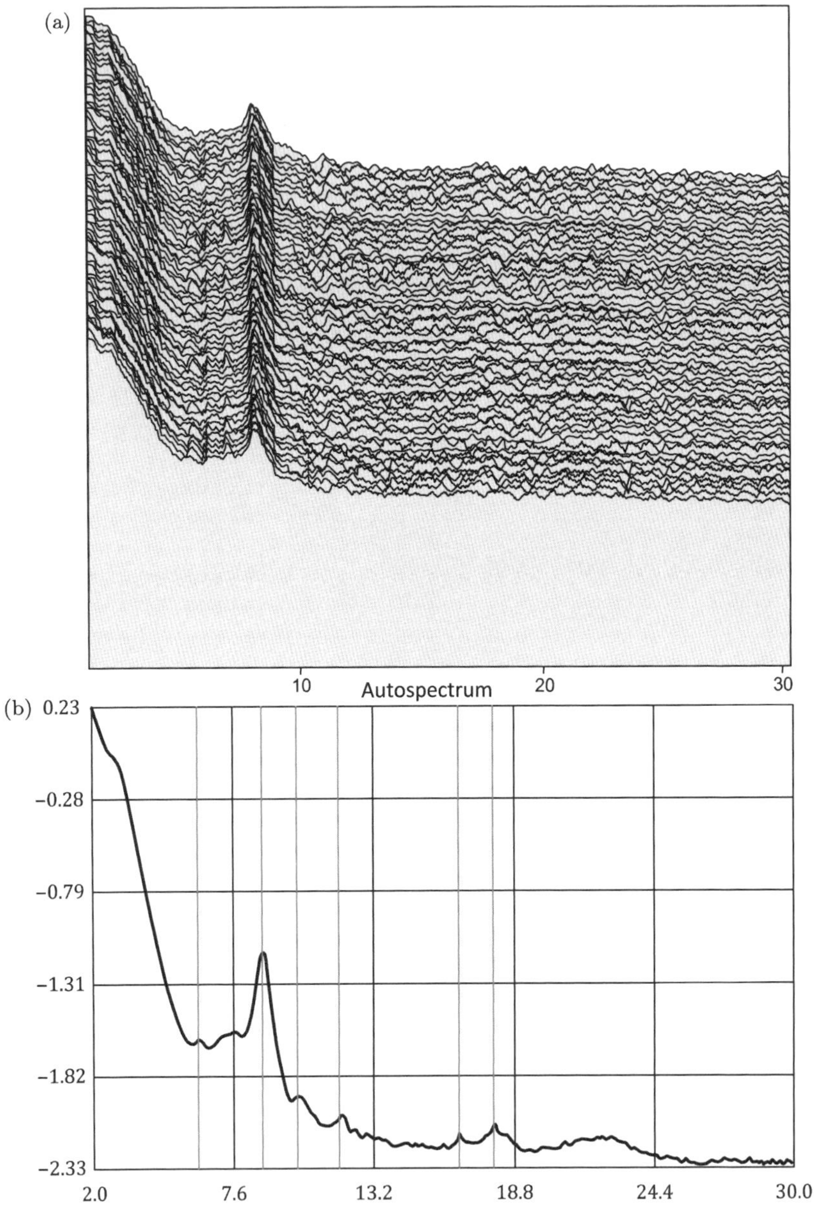

Figure 15.12: APSD family of IC1 ionization chamber signal from multiple measurements at 100% reactor power (a), and their average APSD (b).

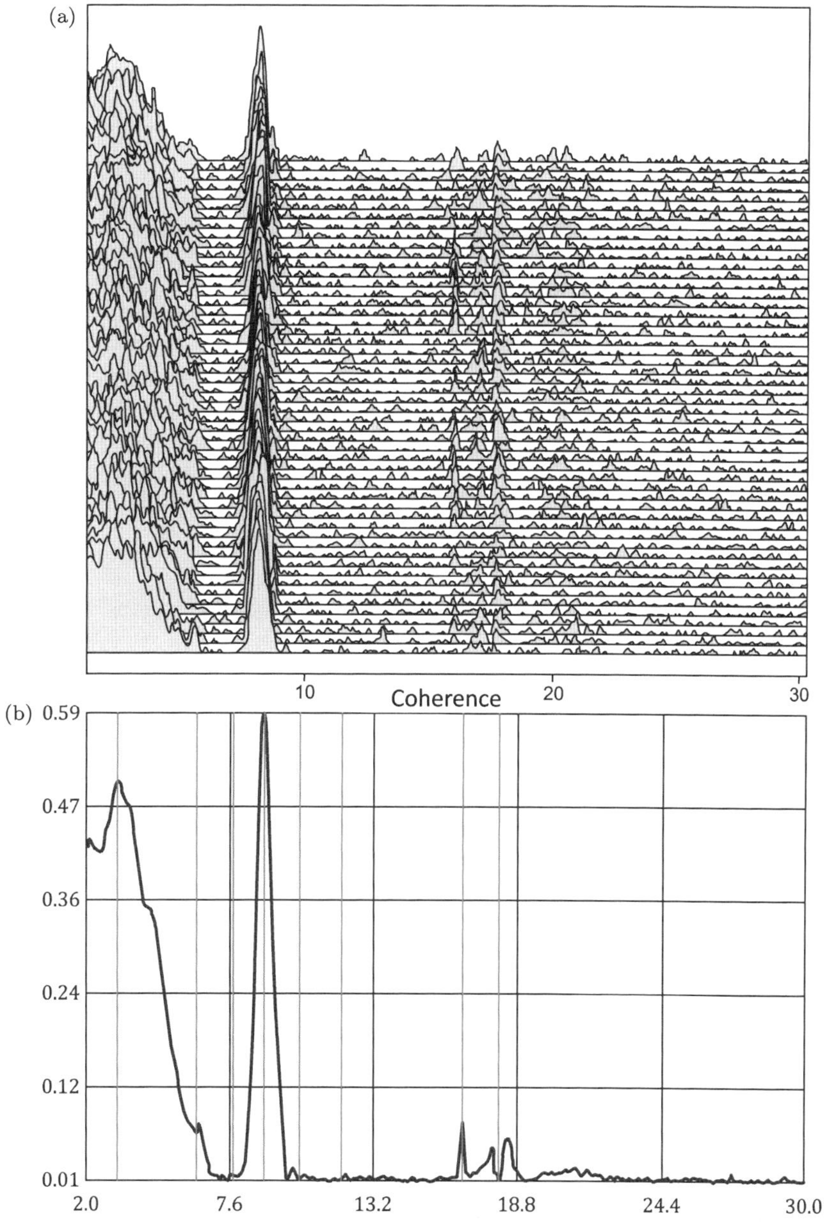

Figure 15.13: Family of coherence functions of opposite ICs (IC1-IC3) from multiple measurements at 100% reactor power (a), and their average coherence (b).

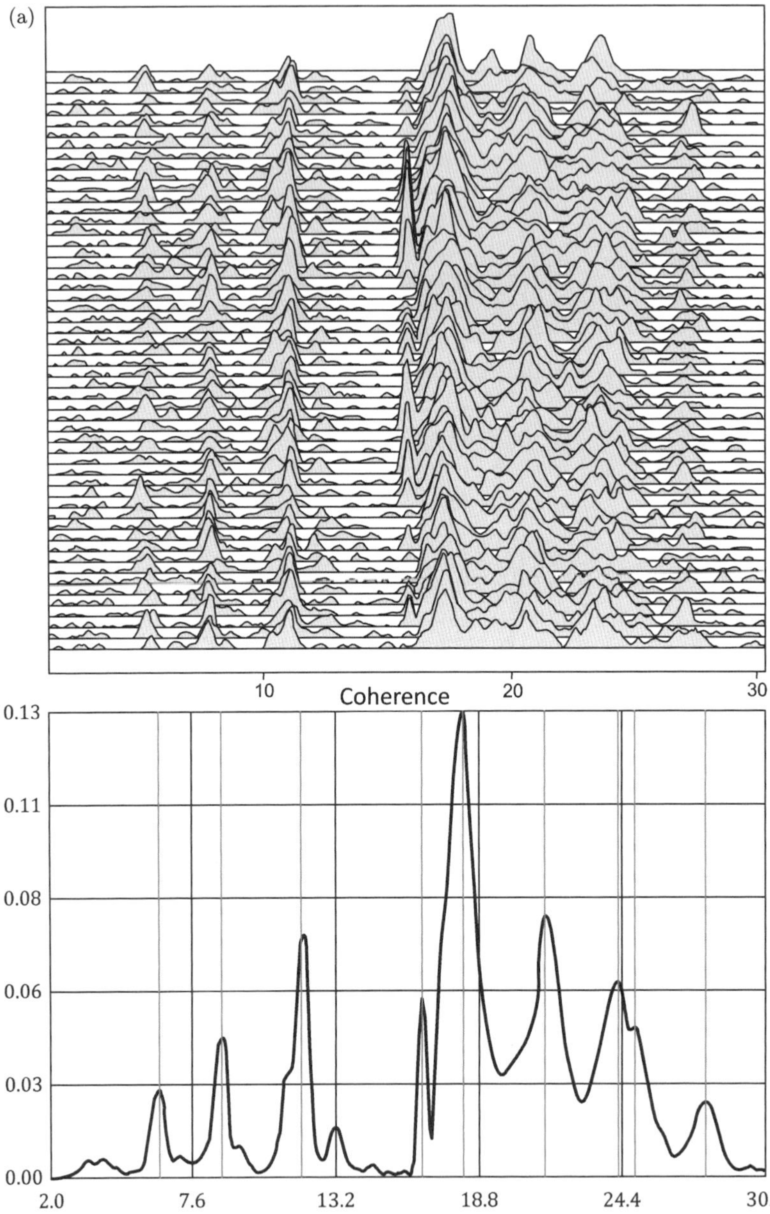

Figure 15.14: Family of coherence functions of IC signal (IC1) and accelerometer signal (ACCv3) from multiple measurements at 100% reactor power (a), and their average coherence (b).

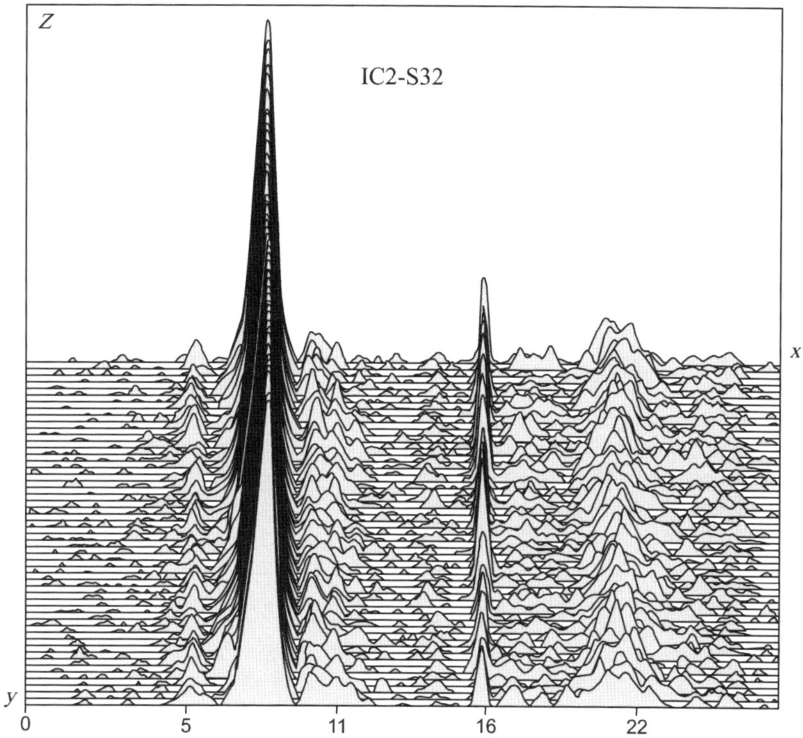

Figure 15.15: Family of coherence functions of the signal of the accelerometer installed on the "cold" leg of loop three and IC2 signal from multiple measurements at 100% reactor power.

RCB vibration monitoring is also possible at the frequencies of forced oscillations. In this case, there is no need to wait for the change of frequencies of forced oscillations (here, ASW frequencies), and it is enough to monitor only amplitudes and Q-factors of ASW resonances.

References

1. Fizestone, F.A. New analogy between mechanical and electrical systems. *JASA*. 1933, Vol. 3, No. 4, pp. 249–267.
2. Gamburtsev, G.A. On building up analogies between mechanical and electrical systems. *DAN*. 1935, Vol. 4, Nos. 8–9, pp. 303–306.
3. Belov, A.I. Drawing electrical circuits equivalent to mechanical oscillatory systems. *Technical Physics. The Russian Journal of Applied Physics*. 1935, Vol. 5, No. 9, pp. 1545–1551.
4. Koshlyakov, N.S., Svetlov, A.V., Stroganov, V.G. *Introduction to the Theory of Small Oscillations Applied to Acoustics*. Moscow, Leningrad: GONTI, 1937, 121 p.
5. Theodorchik, K.F. Two systems of electrical and mechanical analogies in terms of lagrange motion equations. *Technical Physics. The Russian Journal of Applied Physics*. 1938, Vol. 8. No. 18, pp. 1652–1658.
6. Strutt, J.W. (Baron Rayleigh). *The Theory of Sound*. Vol. 1 (translation of the 3rd English edition by P.N. Uspensky and S.A. Kamenetsky, endorsed by S.M. Rytov and K.F. Theodorchik). Moscow, Leningrad: Gostechizdat, 1940, 503 p.
7. Strutt, J.W. (Baron Rayleigh). *The Theory of Sound*, Vol. 2 (translation of the 3rd English edition by P.N. Uspensky and S.A. Kamenetsky) Moscow, Leningrad: Gostechizdat, 1944, 476 p.
8. Blokhintsev, D.I. *Acoustics of Inhomogeneous Moving Medium*. Moscow, Leningrad: Gostechizdat, 1946, 220 p.
9. Olson, G.F. *Dynamical Analogies* (endorsed by M.A. Aizerman). Moscow, Foreign Literature Publishing House, 1947, 224 p.
10. Lamb, G. *Hydrodynamics* (translation of the 6th English edition by A.V. Germogenov and V.A. Kudryavtsev) edited by N.A. Slyozkin). Moscow, Leningrad: Gostechizdat, 1947, 928 p.

11. Furduev, V.V. *Reciprocity Theorems in Mechanical, Acoustical and Electromechanical Quadrupoles.* Moscow, Leningrad: Gostechizdat, 1948, 92 p.
12. Lebedev, I.V. *SHF Equipment and Devices*, Vol. 1. Moscow. Vysshaya shkola Publ., 1970, 440 p.
13. Lebedev, I.V. *SHF Equipment and Devices*, Vol. 2. Moscow. Vysshaya shkola Publ., 1972, 376 p.
14. Atabekov, G.I. *Theoretical Foundations of Electrical Engineering.* Moscow, Energiya, 1978, 592 p.
15. Morz, F. *Vibrations and Sound.* (translation of the 2nd English edition, endorsed by S.N. Rzhevkin). Moscow, Leningrad: Gostechizdat, 1949, 496 p.
16. Zhukovsky, N.E. *On Hydraulic Shock in Water Pipes.* Moscow, Leningrad: Gostechizdat, 1949, 103 p.
17. Bulgakov, B.V. *Vibrations.* (Parts 1–3). Gostechizdat, 1954, 892 p.
18. Kudryavtsev, B.B. The speed of sound in liquids and chemical&physical properties of liquids. *The Journal of Physical Chemistry.* 1954, Vol. 28, No. 5, pp. 930–935.
19. Shalnev, K.K. Cavitation in hydrodynamics. *Izvestiya Akad. Nauk. SSSR.* 1956, No. 8, pp. 12–16.
20. Gorelik, G.S. Oscillations and Waves. An Introduction in Acoustics, Radiophysics, and Optics. In A Textbook for High Schools, edited by S.M. Rytov, 2nd edition. Moscow, Fizmatgiz, 1959, 572 p.
21. Timoshenko, S.P. *Vibration Problems in Engineering.* Moscow, Fizmatgiz, 1959, 439 p.
22. Den Hartog, J.P. *Mechanical Vibrations* (translation of the 4th American edition by A.N. Obmorshev). Moscow, Fizmatgiz, 1960, 580 p.
23. Rzhevkin, S.N. *A Course of Lectures on the Theory of Sound.* Moscow, MSU Publications, 1960, 336 p.
24. Timoshenko, S.P., Woinowsky-Krieger S. *Theory of Plates and Shells.* Moscow, Fizmatgiz, 1963, 635 p.
25. Strelkov, S.P. *Introduction to the Theory of Vibrations.* Moscow: Nauka, 1964, 437 p.
26. Babakov, I.M. *Theory of Vibrations.* Moscow: Nauka, 1965, 559 p.
27. Panovko, Y.G. *Fundamentals of Applied Theory of Elastic Oscillations.* Moscow. Mashinostroenie Publ., 1967, 316 p.
28. Strength, *Stability and Vibrations* (reference book in three volumes), endorsed by I.A. Birger and Y.G. Panovko. Moscow: Mechanical engineering, 1968. Vol. 1, pp. 831, Vol. 2, pp. 464, Vol. 3, pp. 567.
29. Skuchik, E. *Simple and Complex Oscillatory Systems* (translated from English edited by L.M. Lyamshev). Moscow: Mir, 1971, 557 p.
30. Isakovich, M.A. *General Acoustics.* Moscow: Nauka, 1973, 495 p.

31. Timoshenko, S.P. *Strength and Vibrations of Structural Elements* (selected works), E.I. Grigolyuk. Moscow: Nauka, 1975, 704 p.
32. Skuchik, E. *Fundamentals of Acoustics*, Vol. 1, (translated from English) edited by L.M. Lyamshev. Moscow: Mir, 1976, 520 p.
33. Skuchik, E. *Fundamentals of Acoustics*, Vol. 2, (translated from English) edited by L.M. Lyamshev. Moscow: Mir, 1976, 542 p.
34. Lependin, L.F. *Acoustics*. Moscow: Vysshaya shkola Publ., 1978. 448 p.
35. Migulin, V.V., Medvedev, V.I., Mustel, E.R., Parygin, V.N. *Fundamentals of Vibration Theory*. Moscow: Nauka, 1978, 392 p.
36. Andronov, A.A., Vitt, A.A., Khaikin, S.E. *Theory of Vibrations*. Moscow: Nauka, 1981, 586 p.
37. Prokhorov, A.M. *Encyclopaedic Dictionary of Physics*. Moscow: Soviet Encyclopedia, 1983, 928 p.
38. Krasilnikov, V.A. and Krylov, V.V. *Introduction to Physical Acoustics*. Moscow: Nauka, 1984, 400 p.
39. Rabinovich, M.I. and Trubetskov, D.I. *Introduction to the Theory of Vibrations and Waves*. Moscow: Nauka, 1984, 432 p.
40. Panovko, Y.G. and Gubanova, I.I. *Stability and Oscillations of Elastic Systems*. Moscow: Nauka, 1987, 352 p.
41. Butenin, N.V., Neimark, Y.I., Fufaev, N.A. *Introduction to the Theory of Non-linear Oscillations*. Moscow: Nauka, 1987, 384 p.
42. Pavlenko, V.G. *Fundamentals of Fluid Mechanics*. Leningrad: Sudostroenie, 1988, 240 p.
43. Levitsky, N.I. *Oscillations in Mechanisms*. Moscow: Nauka, 1988, 336 p.
44. Ryskin, N.M. and Trubetskov, D.I. *Nonlinear Waves*. Moscow: Nauka, Fizmatlit, 2000, 272 p.
45. Trubetskov, D.I. and Rozhnev, A.G. *Linear Oscillations and Waves*. Moscow: Fizmatlit, 2001, 415 p.
46. Kuznetsov, A.P., Rozhnev A.G., Trubetskov D.I. *Linear Oscillations and Waves. A Collection of Problems*. Moscow: Fizmatlit, 2001, 128 p.
47. Landau, L.D. and Lifschitz E.M. *Fluid Mechanics* (2nd edition, revised and enlarged). Moscow: Gostechizdat, 1953, 788 p.
48. Mandelshtam, L.I. *Lectures on Oscillations* (collected works), Vol. 4. Moscow: Akad. Nauk. SSSR Publication, 1957, 511 p.
49. Milne Thomson, L.N. *Theoretical Hydrodynamics*. Moscow: Mir, 1964, 655 p.
50. Fabricant, N.Y. *Aerodynamics*. Moscow: Nauka, 1964, 814 p.
51. Daily, J.W. and Harleman, D. *Fluid Dynamics*. Moscow: Energiya, 1971, 480 p.

52. Crawford, F. *Waves* (Berkeley Physics Course. T.3). Moscow: Nauka, 1974, 527 p.
53. Bogoliubov, N.N. and Mitropolsky, Y.A. *Asymptotic Methods in the Theory of Nonlinear Oscillations.* Moscow: Nauka, 1974, 504 p.
54. Lavrentiev, M.A. and Shabat, B.V. *Problems of Hydrodynamics and their Mathematical Models.* Moscow: Nauka, 1977, 407 p.
55. Magnus, K. *Vibrations.* Moscow: Mir, 1982, 303 p.
56. Crandall, I.B. *Acoustics.* Moscow: Comkniga, 2007, 168 p.
57. Krylov, A.N. *On some Differential Equations of Mathematical Physics with Applications to Technical Questions.* Moscow: Academkniga, 1949, 268 p.
58. Petrovsky, I.G. *Lectures on Partial Differential Equations.* Moscow: Gostechizdat, 1961, 402 p.
59. Kollats, L. *Eigenvalue Problems.* Moscow: Nauka, 1968, 503 p.
60. Cole, J. *Perturbation Methods in Applied Mathematics.* Moscow: Mir, 1972, 274 p.
61. Tikhonov, A.N. and Samarskii A.A. *Equations of Mathematical Physics.* Moscow: Nauka, 1977, 735 p.
62. Vladimirov, V.S. *Equations of Mathematical Physics.* Moscow: Nauka, 1981, 512 p.
63. Lau, P.C.M. Finite Difference Approximation for Ordinary Derivatives. *International Journal for Numerical Methods in Engineering.* 1981, Vol. 17, pp. 663–678.
64. Aly, R. Finite Difference methods in vibration analysis. *Shock and Vibration.* 1983, Vol. 15, pp. 3–7.
65. Arnold, V.I. *Ordinary Differential Equations.* Moscow: Nauka, 1984, 271 p.
66. Lavrentiev, M.A. and Shabat B.V. *Methods of the Theory of Functions of Complex Variable.* Moscow: Nauka, 1987, 688 p.
67. Arnold, V.I. *Mathematical Methods of Classical Mechanics.* Moscow: Nauka, 1989, 472 p.
68. Zavaljevski, N., Gross, K.C., Wegerich, S.W. Regularization Methods for the Multivariate State Estimation Technique (MSET). In *Proc. of the Int. Conf. on Mathematics and Computations, Reactor Physics and Environmental Analysis in Nuclear Applications.* Madrid Spain, 1999, pp. 720–729.
69. Whittle, P. On the fitting of multivariate autoregressions and the approximate canonical factorization of a spectral density matrix. *Biometrika.* 1963, Vol. 50, pp. 129–134.
70. Bendat, J. and Piersol A. *Random Data: Analysis and Measurement Procedures.* Moscow: Mir, 1974, 464 p.

71. Max. J., *Méthodes es techniques de traitement du signal et applications aux mesures physiques* (in 2 volumes). Moscow: Mir, 1983, Vol. 1, p. 311; Vol. 2, p. 256.
72. Goldenberg, L.M., Matyushkin, B. D., Polyak, M. N. *Digital Signal Processing: Reference Book.* Moscow: Radio and Svyaz Publ., 1985, 312 p.
73. Bendat, J. and Piersol A. *Engineering Applications of Correlation and Spectral Analysis* (translated from English) edited by I.N. Kovalenko. Moscow: Mir, 1987, 312 p.
74. Bendat, J. and Piersol A. *Random Data: Analysis and Measurement Procedures.* Moscow: Mir, 1989, 540 p.
75. George, R.C. and Clare McGillem, D. *Probabilistic Methods of Signal and System Analysis.* Moscow: Mir, 1989, 376 p.
76. Randall, R.B. *Frequency Analysis.* Denmark, K. Larsen & Son, 1989, 379 p.
77. Ole Døssing. Mechanical Mobility Measurements. *Structural Testing. Part 1.* Denmark, Nerum: Brüel & Kjær, 1988, 47 p.
78. Ole Døssing. Modal Analysis and Simulation. *Structural Testing. Part 2.* Denmark, Nerum: Brüel & Kjær, 1989, 72 p.
79. Marple, S.L. Jr. *Digital Spectral Analysis with Applications.* Moscow: Mir, 1990, 584 p.
80. *Equipment for Acoustics, Electro-acoustics, Vibration Measurements, Signal Analysis and Medicinal Diagnosis.* Brüel & Kjær Catalogue, 1990, 461 p.
81. Dremin, I.L., Ivanov, O.V., and Nechitailo, V.A. Wavelets and their application. *Advances in Physical Sciences.* 2001, Vol. 171, No. 5, pp. 465–501.
82. Kaplun, A.B., Morozov, E.M., and Olferyeva, M.A. *ANSYS in the Hands of an Engineer: A Practical Guide.* Moscow: Editorial URSS, 2003, 272 p.
83. Sergienko, A.B. *Digital Signal Processing: A Textbook for High Schools*, 3rd edition. Saint Petersburg: BHV–Petersburg, 2011, 768 p.
84. Birger, I.A. Determining Diagnostic Value of Indicators. *Cybernetics.* 1968, No. 3, pp. 80–85.
85. Parkhomenko, P.P. *On Technical Diagnostics.* Moscow: Znanie, 1969, 64 p.
86. Lints, V.P. *Technical Diagnostics of Machines.* Moscow: Znanie, 1971, 46 p.
87. Korenev, B.G. *Reference Book on the Dynamics of Structures.* Rabinovich I.M. (eds.). Moscow: Strojizdat, 1972, 511 p.

88. Mozgalevsky, A.V. and Gaskarov D.V. *Technical Diagnostics: Continuous Objects.* Moscow: Vysshaya shkola Publ., 1975, 207 p.
89. Karibsky, V.V., Parkhomenko P.P., Sogomonian E.S. *et al. Technical Diagnostics Fundamentals.* Moscow: Energiya, 1976, 464 p.
90. Deich, A.M. *Methods for Identification of Dynamic Objects.* Moscow: Energiya, 1979, 240 p.
91. Mironovsky, L.A. Functional diagnosis of dynamic systems (A survey). *Automation and Remote Control.* 1980, No. 8, pp. 96–121.
92. Birger, I.A. *Technical Diagnostics.* Moscow: Mashinostroenie Publ., 1978, 240 p.
93. Klyuev. V.V. (ed.). *Equipments for Non-destructive Testing of Materials and Products. Hand-book.* Moscow: Mashinostroenie Publ., 1986, Vol. 1, p. 487, Vol. 2, 351 p.
94. Mozgalevsky, A.V., Parkhomenko, P.P., Sogomonian, E.S. *et al.* Technical Diagnostics. *Reliability and Efficiency in Technique.* Reference Book, Vol. 9. Moscow: Mashinostroenie Publ., 1987, 352 p.
95. *Engineering Tools of Diagnosis: Reference Book,* endorsed by V.V. Klyuev. Moscow: Mashinostroenie Publ., 1989, 672 p.
96. Frank, P.M. Fault diagnosis in dynamic systems using analytical and knowledge-based redundancy — A survey and some new results. *Automatica.* 1990, Vol. 26, No. 3, pp. 459–474.
97. Yorish, Y.I. *Vibrometry.* Moscow: Mashinostroenie Publ., 1965, 773 p.
98. *Vibration of Power Mechanics. Handbook,* (endorsed by Grigoriev N.V). Leningrad: Mashinostroenie Publ., 1974, 463 p.
99. Vibrations in Linear Systems *Vibrations in Engineering.* Reference book in six volumes, Vol. 1. Bolotin V.V. (ed.). Moscow: Mashinostroenie Publ., 1978, 352 p.
100. Balitskii, F.Y., Ivanova M.A., Sokolova A.G., Khomyakov E.I. *Vibroacoustic Diagnostics of Nascent Defects.* Moscow: Nauka, 1984, 120 p.
101. Ahid, D. Nashif, David, I.G. Jones, John, and P. Henderson. *Vibration Damping.* Moscow: Mir, 1988, 448 p.
102. Aleksandrov, A.A., Barkov, A.V., Barkova, N.A., Shafranskij, V.A. *Vibration and Vibration-based Diagnostics of the Marine Electric Equipment.* Leningrad: Sudostroenie, 1986, 276 p.
103. Genkin, M.D. and Sokolova, A.G. *Vibroacoustic Diagnostic of Machines and Mechanisms.* Moscow: Mashinostroenie Publ., 1987, 228 p.
104. McNulty, P.J. Acoustics in Diagnosing and Vibration in Hydraulic Turbomachinery. *Proc. Spring Conf. Acoustics-80.* London, 1980, pp. 209–212.
105. KIiman, G.B. and Stein, J. Methods of motor current signature analysis. *Electric Machines and Power Systems.* 1992, Vol. 20, pp. 463–474.

106. Heylen, W., Lammens, S., Sas, P. *Modal Analysis: Theory and Testing.* Moscow: Novatest, 2010, 319 p.
107. Alster, M. Improved calculation of resonant frequencies of helmholtz resonator. *Journal of Sound Vibration.* 1972, Vol. 24, No. 1, pp. 63–85.
108. Lebedev, N.N. *et al.* Synthesis of optimal in response time algorithms of defect diagnosis in nuclear facilities. *Problems of Atomic Science and Technology. Dynamics of Nuclear Facility.* 1975, Vol. 2, No. 8, pp. 74–78.
109. Mogilner, A.I. and Skomorokhov, A.O. Computer-based studies of diagnostic method for thermal state of the nuclear reactor core. *Problems of Atomic Science and Technology.* 1979, Vol. 4, No. 8, pp. 5–14.
110. Mogilner, A.I., Skomorokhov, A.O., Shvetsov, D.M. On the problem of noise spectra classification in nuclear power plant operation diagnostics. *Nuclear Technology,* 1981, Vol. 53, No. 1, pp. 8–18.
111. Mitelman, M.G. and Rosenblum, N.D. *Charge-Coupled Detectors of Ionizing Radiation.* Moscow: Energoizdat, 1982, 77 p.
112. Abagyan, A.A., Dmitriev, V.M., Klebanov, L.A., *et al.* Monitoring and Diagnostic Systems for NPP Operating Regimes. *Atomnaya Energiya.* 1987, Vol. 63, No. 5, pp. 311–315.
113. Rakitin, I.D. and Son, V.V. Development of diagnostic systems for NPP. *Atomnaya Technika Za Rubezhom.* 1987, No. 5, pp. 3–10.
114. Ilkov, V.D., Molchanov, S.A., Mitin, V.I., *et al.* Current State and Prospects of NPP Diagnostic Issues. A Survey. *Atomnye Electrostancii Publ.*, 1988, Edition 8. Moscow: Informenergo, p. 52.
115. Komarevsky, I.V. Expert systems of operator support in controlling nuclear facilities. *Atomnaya Technika Za Rubezhom.* 1988, No. 7, pp. 3–9.
116. Fomichev, M.S. *Experimental Hydrodynamics of Nuclear Power.* Moscow: Energia Atomizdat, 1989, 246 p.
117. Fedorovich, E.D., Fokin, B.S., Akselrod, A.F., Goldberg, E.N. *Vibrations of NPP Equipment Components.* Moscow: Energia Atomizdat, 1989, 168 p.
118. Babkin, N.A., Gorlin, A.I., Kolgan, S.V., Kroshilin, A.E., Mironova, G.V. Application of Anomalous Dynamic Characteristics for Online Diagnosis of NPP Emergency Operating Modes. *Problems of Atomic Science and Technology.* Series: Physics of Nuclear Reactors, 1989, Edition 3, pp. 47–52.
119. Kalinushkin, A.E., Mitin, V.I., Semchenkov Y.M. Development of expert systems for nuclear power plants. *Atomnaya Technika Za Rubezhom,* 1990, No. 7, pp. 3–8.

120. Accidents and Incidents on Nuclear Power Plants. *A Textbook for High Schools.* Edited by Soloviev S.P. Obninsk: IATE, 1992, 300 p.
121. Kazakov, V.A. and Levadny V.A. Noise measurement method for computing reactivity coefficients and monitoring gas-steam volume in VVER-1000 reactor. *Atomic Energy.* 1993, Vol. 75, No. 5, pp. 341–345.
122. Antonov, B.V. Results of Russian NPP operation in 1995. *Izvestiya Vuzov. Nuclear Industry.* 1996, No. 3, pp. 4–8.
123. Gerasimov, V.V., Gorbatykh, V.P., Morozov, A.V. Several aspects of thermal power equipment metal lifetime management at nuclear power stations. *Thermal Engineering.* 2000, No. 8, pp. 47–51.
124. Gorbatykh, V.P., Morozov, A.V., Dubar, A., Ivanov, S.O., Parfenov, V.V. Analysis and Diagnosis of Metal State and Techniques of Managing Equipment Residual Life of NPP with VEER. In *Safety Assurance of NPP with VVER: The III-rd International Scientific and Technical Conference.* Podolsk, OKB Gidropress, Vol. 4, May 26–30, 2003, pp. 88–96.
125. Solonin, V.I., Perevezentsev, V.V., Rekshnya, N.F. Computational and experimental studies of vibration behavior of fuel assemblies of water-cooled reactors under combined hydro-mechanical loading. In *The V-th International Conference on Vibration. ICOVP-2001.* Moscow: Mechanical Engineering Research Institute of Russian Academy of Sciences, 2001, pp. 433–437.
126. Boltenkov, V.A. and Molina, T.O. Smart technologies in NPP diagnostic systems. *Artificial Intellect.* 2004, No. 3, pp. 274–278.
127. Kalyavin, V.P. and Pankin, A.M. *Fundamentals of Reliability and Technical Diagnostics of NPP Components and Systems.* Saint Petersburg: Polytechnic University Publ., 2007, 213 p.
128. Ryasny, S.I. Natural Experimental Substantiation and Maintenance of Operation of the Nuclear Power Plant. In *Safety Assurance of NPP with WWER: The VII-th International Scientific and Technical Conference.* Podolsk, OKB Gidropress, May 17–20, 2011, pp. 123–126.
129. Ananyev, A.N., Kaznovsky P.S., Kaznovsky S.P. *et al.* Seismic Safety of Nuclear Power Plants. Moscow: Bauman Press, 2011, 229 p.
130. Kaznovsky, P.S. *et al.* Analyses of Modern Russian and Foreign Approaches to Estimation of NPP Equipment Seismic Resistance under Operation Conditions. In *Safety Assurance of NPP with VVER: The VIII-th International Scientific and Technical Conference.* Podolsk, OKB Gidropress, May 28–31, 2013, pp. 173–176.
131. Sargsyan, A.E. *Dynamics and Seismic Stability of Nuclear Power Plants.* Sarov: Izd-vo RFYaTs–VNIIEF, 2013, p. 550.
132. Asmolov, V.G. Operation Experience of NPPs of JSC "Concern Rosenergoatom". Ensuring Safety and Enhancement of Efficiency

of Russian Nuclear Power. In *Eighth International Scientific and Technical Conference "Safety, Efficiency and Economy of Nuclear Power Industry"*. Moscow, Russia, REA Concern, May 23–25, 2012, pp. 31–37.

133. Limarenko, V.I. Life Cycle Management at the Stage of Construction. Construction Control Information System. In *Proceedings of the 2nd International Research and Practice Forum "Life Cycle Management of Complex Engineering Facilities.* Nizhniy Novgorod, Russia, Afrikantov OKB Mechanical Engineering, June 14–15, 2012, pp. 12–15.
134. Samoilov, S.N., Golubev, L.I., Samarin, A.A. *et al.* Vibration acoustic monitoring of primary circuit equipment at novovoronezh nuclear power plant. *Teploenergetika*, 1973, No. 2, pp. 21–22.
135. Samarin, A.A. and Nozdrin, G.N. Analysis of the Causes of Vibration of Equipment in the First Loop of Nuclear Power Stations with Water-Modulated Water-Cooled Reactors. *Teploenergetika.* 1973, No. 6, pp. 40–42.
136. Ten-year Experience of Operation of Novovoronezh NPP. In *Proceedings of Scientific and Technical Conference.* Novovoronezh, September, 1974, p. 178.
137. Smirnov L.V. and Ovchinnikov V.F. Flow-induced Fluctuations in the Structural Components of the Nuclear Power Plant (Part 1). In *Problems of Atomic Science and Technology.* Series Dynamics of Nuclear Facility, 1975. Edition 2(8), pp. 3–22; Part 2. — *Ibid.*, 1976, Edition 1(9), pp. 3–24.
138. Verezemsky V.G., Grudev I.D., Korneeva S.N. Free Oscillations of Heat Exchange Loop of VVER-1000 Primary Circuit. *Dynamic Deformations in Power Equipment.* Moscow: IMASH Akad. Nauk. SSSR Publication, 1978, pp. 28–36.
139. Samarin A.A. *Vibrations of the Pipes of Power Installations and Methods of Controlling them.* Moscow: Energiya, 1979, p. 288.
140. Smirnov L.V. Studies of Dynamic Properties of a Straight Pipeline with a Fluid Flow. *Dynamic Characteristics and Oscillations of Power Equipment.* Moscow: Nauka, 1980, pp. 58–66.
141. Fyodorov V.G. and Sinyavsky V.F. Main Problems of Substantiation of Vibration Strength of Nuclear Reactor and Steam Generator Internals. *Dynamic Characteristics and Oscillations of Power Equipment.* Moscow: Nauka, 1980, C.53–71.
142. Verezemsky V.G. and Kaplounov S.M. Dynamic Experimental Studies on a Large-scale Model of the NPP Primary Circuit. *Dynamic Characteristics and Oscillations of Power Equipment.* Moscow: Nauka, 1980, pp. 103–115.

143. Ovchinnikov, V.F. and Smirnov, L.V. Fluctuations of Pipelines with Non-stationary Fluid Flow. *Problems of Atomic Science and Technology.* Series: Physics of Nuclear Reactors. 1985, Edition 2, pp. 3–11.
144. Frolov, K.V. *et al.* Vibration stability of main circulation pipelines in nuclear power plants. *Strength of Materials*, 1985, No. 10, pp. 3–12.
145. Melnikov, V.I. and Usynin, G.B. *Acoustic Diagnostic Methods for Two-phase Coolants of a Nuclear Energy Station.* Moscow: Energia Atomizdat, 1987, 160 p.
146. Smirnov, L.V., Kaplounov, S.M., Ovchinnikov, V.F. *et al.* Identification of NPP Steam Piping Vibration. *Problems of Atomic Science and Technology.* Series: Physics of Nuclear Reactors, 1989, Edition 3, pp. 36–40.
147. Rudenko, O.V. and Khirnykh, K.L. Model of Helmholtz resonator for absorption of intense sound. *Acoustic Journal*, 1990, Vol. 36, No. 3, pp. 527–534.
148. Smirnov, L.V., Ovchinnikov, V.F., Usanov, A.I. Dynamic Properties of NPP Primary Circuit Coolant Circulation. *Problems of Atomic Sciencc and Technology.* Series: Physics of Nuclear Reactors, 1991, Edition 3, pp. 25–31.
149. Verezemsky, V.G. Vibration as a factor contributing to cracking of PGV-100 cold headers. *Atomnaya Energiya.* 1997, Vol. 83, Edition 1, pp. 66–69.
150. Perevezentsev, V.V. *et al.* Measurement of Pressure Pulsations and Vibrations in Fuel Rod Bundles. *Research in the Nuclear Power Field in Technical Universities of Russia.* Moscow: MPEI Publ., 1999, pp. 61–62.
151. Abramov, V.V., Valshonok, L.S., Dodonov, V.A., *et al.* Dynamic Stresses in Structural Components in the Fluid Flow. *Experimental Studies and Calculation of Stresses in Structures.* Moscow: Nauka, 1975, pp. 149–160.
152. Abramov, V.V., Gusarov, A.A., Kaplounov, S.M. Current State and Area of Research of Dynamic Stresses and Deformations in Nuclear Reactors. *Dynamic Stresses and Deformations in Structural Elements of Power Equipment.* Moscow: Nauka, 1977, C.18–22.
153. Daichik, M.L., Makhutov, N.A., Mikhalev, Y.K., *et al.* Study of Vibration Stresses in Reactor Internals and Pressure Pulsations in VVER-440 Reactor Vessel. *Study of Structural Stresses.* Moscow: Nauka, 1980, pp. 35–49.
154. Frolov, K.V. *et al.* Vibration stability of main circulation pipelines in nuclear power plants. *Strength of Materials.* 1985, No. 10, pp. 3–12.

155. Stekolnikov, V.V., Fyodorov, V.A., Lyashenko, V.N. Studies of Control Rod Oscillations. *Dynamic Deformations in Power Equipment.* Moscow: Nauka, 1987, pp. 85–89.
156. Makhutov, N.A., Gusarov, A.A., Kaplounov, S.M., Valles, N.G., Usanov, A.I. Vibrational strength of NPP equipment. *Engineering and Automation Problems.* 1988, Vol. 22, pp. 68–80.
157. Birbraer, A.N. and Shulman, S.G. *Strength and Reliability of NPP Structures in Special Dynamic Forces.* Moscow: Energoatomizdat, 1989, 304 p.
158. Makhutov, N.A., Frolov, K.V., Stekolnikov, V.V., *et al.* Experimental Investigations of Strains and Stresses in Water-moderated Power Reactors. Moscow: Nauka, 1990, 296 p.
159. Usanov, A.I., Afrov, A.M., Dranchenko, B.N. Issues of Vibration Diagnostics of VVER Reactors. *Applied Problems of Theory of Oscillations: Interuniversity Collection of Scientific Works.* Nizhny Novgorod: NNGU Publ., 1993, pp. 90–107.
160. Birbraer, A.N. *Estimate of Structure Seismic Resistance.* Saint Petersburg: Nauka, 1998, 255 p.
161. Repin, A.I., Ryzhov, S.B., Bocharov, Yu. L., *et al.* Basic Technical Solutions to Increase Safety and Efficiency of Reactor Plant Equipment for Tianwan NPP, Bushehr NPP and NVNPP-2. *Safety Assurance of NPP with VVER. Proceedings of the II All-Russian Scientific and Technical Conference.* November 19–23, 2001, Podolsk, JSC OKB Gidropress, pp. 42–46.
162. Denisov, V.P. and Dragunov, E.G. *VVER Reactor Facilities for Nuclear Power Plants.* Moscow: IzdAT, 2002, 480 p.
163. Evdokimenko, V.V., Shariy, N.V., Yuremenko, V.P. Development of Methodology for Dynamic Characteristics of VVER Internals. *Safety Assurance of NPP with VVER: The III-rd International Scientific and Technical Conference.* Podolsk, OKB Gidropress, May 26–30, 2003, pp. 17–24.
164. Rezepov, V.K., Denisov, V.P., Kirilyuk, N.A., Dragunov, Yu.G., Ryzhov, S.B. *VVER-1000 Reactors for Nuclear Power Plants.* Moscow: Academkniga, 2004, 333 p.
165. Sharyi, N.V., Semishkin, V.P., Pimenov, V.A., Dragunov, Yu.G. *Strength of Main Equipment and Pipelines in VVER Reactor Facilities.* Moscow: IzdAT, 2004, 496 p.
166. Dranchenko, B.N., Dragunov, Yu.G., Portnov, B.B., Seleznev, A.V. *Experimental Studies of Stress and Strength of VVER Equipment.* Moscow: Academkniga, 2004, 640 p.
167. Berkovich, V.M., Kopytov I.I., Taranov, G.S. *et al.* Salient features of the design for a new generation nuclear power station equipped with

a VVER-1000 reactor having enhanced safety. *Teploenergetika.* 2005, No. 1, pp. 9–15.

168. Abramov, V.V. and Yuremenko, V.P. Development of 3D Dynamic Models of VVER Internals. *Problems of Atomic Science and Technology.* Series The NPP Safety Assurance. 2006, No. 13, pp. 83–94.
169. Ryzhov, S.B., Mokhov, V.A., Podshibyakin, A.K., *et al.* New Designs of VVER Reactor Plants at Modern Stage of Nuclear Engineering Development. In *Safety Assurance of NPP with VVER: Proceedings of the VI International Scientific and Technical Conference.* Podolsk, OKB Gidropress, May 26–29, 2009, 96–101.
170. Bezrukov, Yu.A., Lysenkov, E.A., Seleznev, A.V. Analysis of Possibility of Water Hammer in Primary Circuit of VVER Reactors. In *Safety Assurance of NPP with VVER: Proceedings of the VI International Scientific and Technical Conference.* Podolsk, OKB Gidropress, May 26–29, 2009, pp. 133–136.
171. Shariy, N.V. Current Issues of Computational Studies of VVER Reactor Plant Strength and Ways of their Solution. *Problems of Atomic Science and Technology.* Series The NPP Safety Assurance, 2010, No. 27, pp. 5–21.
172. Makhutov, N.A., Frolov, K.V., Stekolnikov, V.V. *et al. Strength and Life of Water Moderated Power Reactors.* Moscow: Nauka, 1988, 312 p.
173. Makhutov, N.A., Frolov, K.V., Dragunov, Y.G. *et al. Stress and Strength of Nuclear Reactors: Simulation and Tensometry for Power Reactors.* Moscow: Nauka, 2001, 293 p.
174. Frolov K.V., Makhutov N.A., Kaplounov S.M. *et al. Dynamics of Hydroaeroelastic System Structures.* Moscow: Nauka, 2002, 399 p.
175. Makhutov N.A., Dragunov Yu.G., Frolov K.V. *et al. Dynamics and Strength of Water Moderated Power Reactors.* Moscow: Nauka, 2004, 440 p.
176. Makhutov N.A., Frolov K.V., Gadenin M.M. *et al. Scientific Principles of Increasing Low-Cycle Strength.* Moscow: Nauka, 2006, 584 p.
177. Makhutov, N.A., Frolov, K.V., Dragunov, Yu.G., *et al. Problems of Strength and Safety of Water Moderated and Cooled Power Reactors*, edited by Makhutov N.A., Gadenin M.M. Moscow: Nauka, 2008, 464 p.
178. Barkov, A.V. *Vibration and Vibration-based Diagnostics of the Marine Electric Equipment.* Leningrad: Sudostroenie, 1986, 276 p.
179. Barkova, N.A. *Vibroacoustic Methods for Diagnostics of Marine Power Equipment: A Handbook.* Leningrad: Saint Petersburg State Marine Technical University Publ., 1986, 203 p.
180. Barkov, A.V., Barkova, N.A., Rogov, S.N. Vibroacoustic Diagnostics of Bearings. *Technical Diagnostics and Operational Control at NPP:*

State and Prospects of Development. Collection of Abstracts. 1990, pp. 31–32.

181. Barkov, A.V., Barkova, N.A., Rogov, S.N. Vibroacoustic Diagnostics of Pumps. *Technical Diagnostics and Operational Control at NPP: State and Prospects of Development.* Collection of Abstracts. 1990, pp. 32–33.
182. Barkov, A.V. and Barkova, N.A. Diagnostics of Gearings and Geared Couplings Using Envelope Spectrum Methods. *Technical Diagnostics and Operational Control at Nuclear Power Plants: State and Prospects of Development.* Collection of Abstracts. 1990, pp. 29–31.
183. Mitchel, J.S. *An Introduction to Machinery Analysis and Monitoring.* Tulsa: Penn Well Books, 1993, 322 p.
184. Barkov, A.V., Barkova, N.A., Mitchel, J.S. Condition assessment and life prediction of rolling element bearings, Part 1. *Sound and Vibration.* 1995, June, No. 6, pp. 10–17.
185. Barkov, A.V., Barkova, N.A., Mitchel, J.S. Condition assessment and life prediction of rolling element bearings, Part 1. *Sound and Vibration.* September, 1995, No. 9, pp. 27–31.
186. Barkov, A.V., Barkova, N.A., Azovtsev, A.Y. *Monitoring and Diagnostics of Rotor Machines by Vibration Level: A Textbook for High Schools.* Saint Petersburg: Sevzapuchcenter, 2013, 158 p.
187. Goldin, A.S. *Vibration of Rotary Machines.* Moscow: Mashinostroenie Publ., 2000, 344 p.
188. Prigorovsky, N.I., Khurshudov, G.K., Daichik, M.L., *et al.* Tensometry of Reactor Internals in Trial Run of Primary Circuit. *Methods for Structural Stress Studies.* Moscow: Nauka, 1976, pp. 113–116.
189. Dranchenko, B.N., Usanov, A.I., Khairetdinov, V.U. Studies of VVER-1000 operational vibration reliability based on the results of commissioning measurements. *Energetika.* 1991, Vol. 2, No. 6, pp. 177–179.
190. Dragunov, Yu. G., Seleznev, A.V., Khairetdinov, V.U. Methods and Means Ensuring Operational Reliability of VVER-1000 In-core Equipment based on the Results of Commissioning Tests and Measurements. In *Proceedings of XII NTK "Sensor-2000", Sensors and Information Converters of Measurement, Monitoring and Control Systems*". Sudak, OKB Gidropress, 2000, pp. 221–222.
191. Dragunov, Yu. G., Seleznev, A.V., Khairetdinov, V.U. *et al.* Commissioning Measurement System for Unit 1 Rostov NPP. In *Safety Assurance of NPP with VVER. Proceedings of II All-Russian Scientific and Technical Conference.* November 19–23, 2001, Podolsk, OKB Gidropress, pp. 33–36.
192. Khairetdinov, V.U. and Andropova, L.S. Acceptance Criteria of the Results of Commissioning Tests and Measurements at VVER-1000

Reactor Plant. In *Problems of Atomic Science and Technology.* Series: The NPP Safety Assurance, 2000, Edition 1, pp. 53–61.

193. Seleznev, A.V., Khairetdinov, V.U., Vasilevsky, G.M., Rodionov, A.B., Ryzhov, A.S. Methods and Means of Experimental Validation of Seismic Resistance of Control Channel Items for VVER-1000 Reactivity. In *Problems of Atomic Science and Technology.* Series: The NPP Safety Assurance, 2004, Edition 5, pp. 61–64.
194. Khairetdinov, V.U., Paderov, M.M., Kolesnikova, L.A., Abramov, V.V., Yuremenko, V.P. Methods and Means of Substantiation of Vibration Strength of Nuclear Reactor Internals. In *Sensors and Systems - 2006: Proceedings of Scientific and Technical Conference.* Moscow, 2006, pp. 253–262.
195. Khairetdinov, V.U., Tarhanov V.V., Abramov V.V. Studies of vibrational load carrying structures of reactor core at dukovany NPP during VVER-440 operation at increased power. *Problems of Atomic Science and Technology.* 2007, No. 9, pp. 93–108.
196. Saakov, E.S., Ryasnyi, S.I., Khairetdinov, V.U. Improving the efficiency of in-situ tests in commissioning NPP power units. *Electric Stations.* 2007, No. 9, pp. 10–14.
197. Saakov, E.S. Controlling NPP unit commissioning processes. *Electric Stations.* 2007, No. 10, pp. 2–6.
198. Dragunov, Yu.G., Dranchenko, B.N., Abramov, V.V., Khairetdinov, V.U. Vibrodynamic Studies to Validate VVER Designs. In *Safety Assurance of NPP with VVER. Proceedings of the V International Scientific and Technical Conference* May 29–June 1, 2007; Podolsk, OKB Gidropress, 2007, Vol. 2, pp. 356–375.
199. Khairetdinov, V.U., Tarkhanov, V.V., Abramov, V.V. Experimental and Calculational Justification of Reactor Equipment Vibration Strength at Dukovany NPP during VVER-440 Operation at Increased Power. In *Safety Assurance of NPP with VVER. Proceedings of the V International Scientific and Technical Conference*, May 29–June 1, 2007, Podolsk, OKB Gidropress, 2007, Vol. 2, pp. 376–379.
200. Khairetdinov, V.U. and Ryasny, S.I. Full-scale justification of operating conditions of reactor equipment using commissioning measurement systems at Tianwan NPP units. *Tyazheloe Mashinostroenie Publ.* 2008, No. 11, pp. 27–32.
201. Altstadt, E., Grunwald, G., Liewers, P., Schumann, P., Weiss, F.-P. Analytical Model for the Description of Control Element Vibration Occurring During Abnormal Core Barrel Motion at WWER-440 Type Reactors and for the Interpretation of phase Relations between Incore and Ex-core Neutron Noise. In *Proceedings of Third International*

congress on condition monitoring and diagnostic engineering management held at Southampton Institute, July 2–4, 1991, pp. 57–62.

202. Kinelev, V., Perov, S., Sulimov, V. The Mathematical Modeling of Free Vibrations of VVER-1000 Plant Primary Circuit for the Equipment Condition Diagnostics. In *A Symposium on Nuclear Reactor Surveillance and Diagnostics. SMORN-7.* June 19–23, 1995, Avignon, France. Session 9 — Vibrations Structures: Modeling and Monitoring, Vol. 2, pp. 108–116.
203. Altstadt, E., Grunwald, G., Weiss, F.-P. Theoretical vibration model of VVER reactors considering fluid-structure interaction. *A symposium on nuclear reactor surveillance and diagnostics.* SMORN 7, June 19–23, 1995, Avignon, France. Session 9 — Vibrations structures: modeling and monitoring. Vol. 2, pp. 111–114.
204. Kinelev, V., Perov, S., Sulimov, V. Theoretical modeling of fuel assembly vibrations for VVER-type reactors. *Specialists' meeting on in-core instrumentation and reactor core assessment.* INCORE-96. October 14–23, 1996, Mito, Japan. pp. 6.4.1–6.4.9.
205. Perov, S.L. Finite-element simulation of fluid dynamics. *Natural Sciences*, 1999, No. 2, pp. 48–58.
206. Perov, S.L. Three-dimensional Model of VVER-1000 Reactor Coolant Acoustic Oscillations. *Non-Destructive Testing and Diagnostics. Proceedings of 15th International Conference.* Moscow, 1999, pp. 21–26.
207. Hazi, G. and Por, G. Numerical simulator for noise diagnostic investigations in NPPs. *Annals of Nuclear Energy*, 1999, Vol. 26, pp. 1113–1130.
208. Fedotovsky, V.S. *Vibroacoustic Dynamics of Heterogeneous Media and NPP Structures. Thermophysical Studies.* Collection of Works dedicated to the 80th anniversary of Academician Subbotin V.I. Obninsk: SCC RF IPPE Publ., 1999, pp. 66–79.
209. Lapin, A.D. Scattering, absorption cross-sections of the helmholtz resonator in a multimode waveguide. *Acoustic Journal* 1999, Vol. 45, No. 3, pp. 376–379.
210. Perov, S., Altstadt, E., Weiss, F.P. Analysis of fluid-structure interaction at WWER-1000 type reactors. In *Proceedings of ICONE 8*, April 2–6, 2000, Baltimore, USA, ICONE 8446, pp. 87–92.
211. Perov, S., Altsdat, E., Werner, M. Vibration analysis of the pressure vessel internals of WWER-1000 type reactors with consideration of fluid-structure interaction. *Annals of Nuclear Energy.* 2000, Vol. 27, No. 16, pp. 1441–1457.
212. Fedotovsky, V.S. and Vereshchagina, T.N. On Natural Frequencies and Forms of Hydrodynamically Coupled Vibrations of Fuel Rod Bundles

of the Fuel Assembly of VVER Reactors. In *Safety Assurance of NPP with VVER. Proceedings of the II All-Russian Scientific and Technical Conference*, November 19–23, 2001, Podolsk, JSC OKB Gidropress, pp. 97–99.

213. Fedotovsky, V.S. and Vereshchagina, T.N. On Natural Vibrations of Cylindrical Shells Containing Fluid and Rod Bundles. In *Safety Assurance of NPP with VVER: Proceedings of the III All-Russian Scientific and Technical Conference*, May 26–30, 2003, Podolsk, JSC OKB Gidropress, pp. 55–59.
214. Pecinka, L., Dupal J., Zeman V. Lesson learnt from mathematical modeling of primary circuit of VVER-440\213 (SMORN-VIII). *Progress in Nuclear Energy*. 2003, Vol. 43, No. 1–4, pp. 167–175.
215. Abramov, V.V., Shariy N.V., Yuremenko V.P. Development of Dynamic Models of VVER-1000 Internals Using ANSYS. In *Safety Assurance of NPP with VVER: Proceedings of IV International Scientific and Technical Conference in 7 Volumes*, Vol. 1, Podolsk, OKB Gidropress, May 23–26, 2005, pp. 103–116.
216. Komkin, A.I. On Selection of Helmholtz Resonator Parameters in a Narrow Channel. In *Acoustics of Inhomogeneous Media. Annual Journal of Russian Acoustical Society. Proceedings of Prof. S.A. Rybak's Seminar*. Moscow: GEOS, 2012, Edition 12, pp. 77–83.
217. Komkin, A.I., Mironov, M.A., Yudin, S.I. On the attached length of orifices. *Acoustic Journal* 2012, Vol. 58, No. 6, pp. 677–682.
218. Komkin, A.I., Mironov, M.A., Yudin, S.I. Eigenfrequency of a Helmholtz Resonator at the Wall of a Rectangular Duct. *Acoustical Physics*. 2014, Vol. 60, No. 2, pp. 145–151.
219. SMORN-II. *Progress in Nuclear Energy*. 1977, Vol. 1, 374.
220. SMORN-II. *Progress in Nuclear Energy*. 1982, Vol. 9, 714.
221. SMORN-IV. *Progress in Nuclear Energy*. 1985, Vol. 15, 1004.
222. SMORN-V. *Progress in Nuclear Energy*. 1988, Vol. 21, 846.
223. SMORN-VI. *A Symposium on Nuclear Reactor Surveillance and Diagnostics*, May 19–24, 1991, Gatlinburg, Tennessee, USA, 754.
224. SMORN-VII. *Symposium on Nuclear Reactor Surveillance and Diagnostics*. June 19–23, 1995, Avignon, France. p. 599.
225. SMORN-VIII. *Progress in Nuclear Energy*. 2003, Vol. 43, No. 1–4, p. 682.
226. Bauernfeind, V. Vibration and pressure signals as sources of an on-line vibration monitoring system in PWR power plant. *Nuclear Engineering and Design*. 1977, Vol. 40, pp. 403–407.
227. Bernard, P., Brillon, A., Carre, J.C. Neutron noise measurements of PWR's. *Progress in Nuclear Energy*. 1977, Vol. 1, pp. 333–346.

228. Dragt, J.B. and Turkan, E. Borssele PWR noise measurements, analysis and interpretation. *Progress in Nuclear Energy*. 1977, Vol. 1, pp. 293–296.
229. Fuge R., Valko, J., Gribon, T. Measurement of the local and global effect of bubbles in a water moderated reactor. *Annals of Nuclear Energy*. 1977, Vol. 4, pp. 161–165.
230. Grabner, A., Liewers, P., Schuman, P. Decompositions of noise signals composed of many similar components. *Progress in Nuclear Energy*. 1977, Vol. 1, pp. 615–628.
231. Matis, M.W., Smith, C.M., Fry, D.N. Characterization studies of BWR-4 neutron noise analysis spectral. *Progress in Nuclear Energy*. 1977, Vol. 1, pp. 175–179.
232. Mayo, C.W. and Currie, R.L. Neutron noise monitoring of PWR internal vibrations. *Progress in Nuclear Energy*. 1977, Vol. 1, pp. 363–368.
233. Wach, D. and Sunder, R. Improved PWR neutron noise interpretation on detailed vibration analysis. *Progress in Nuclear Energy*. 1977, Vol. 1, pp. 309–315.
234. Akerhielm, F., Espefaelt, R., Lorenren, J. Surveillance of vibrations in PWR. *Progress in Nuclear Energy*. 1982, Vol. 9, pp. 453–457.
235. Albrecht, R.W., Crowe, R.D., Dailey, D.J., Damborg, M.J., Koslay, G. Measurement of two-phase flow properties using the nuclear reactor instrument. *Progress in Nuclear Energy*. 1982, Vol. 9, pp. 37–39.
236. Bernard, P., Messainguiral, C., Carre, J.C. *et al.* Quantitative monitoring and diagnosis of French PWR's – internal structures vibrations by ex-core neutron noise and accelerometers analysis. *Progress in Nuclear Energy*. 1982, Vol. 9, pp. 465–468.
237. Bernard, P., Cloue, J., Messainguiral, C. PWR core monitoring by in-core noise analysis. *Progress in Nuclear Energy*. 1982, Vol. 9, pp. 541–546.
238. Federico, A., Galli, C., Parmeggiani, C., Ragona, R., Tosi, V. Second transit time in a BWR power plant: experimental evidence. *Progress in Nuclear Energy*. 1982, Vol. 9, pp. 631–633.
239. Fujita, Y. and Oraki, H. Neutron noise monitoring of reactor core PWR internal vibrations at PWR's in Japan. *Progress in Nuclear Energy*. 1982, Vol. 9, pp. 531–533.
240. Grunwald, G., Hessel, G., Liewers, P., Schmitt, W. Low Frequency Pressure Oscillations in a PWR. *Progress in Nuclear Energy*. 1982, Vol. 9, pp. 569–579.
241. Katona, T., Mesko, L., Por, G., Valko, J. Same aspects of the theory of neutron noise due to propagating disturbances. *Progress in Nuclear Energy*. 1982, Vol. 9, pp. 209–222.

242. Miteff, L. and Behringer, K. Investigation of the relation between the Rms-neutron-noise and local void fraction in the bubbly flow region of a BWR. *Progress in Nuclear Energy*. 1982, Vol. 9, pp. 37–39.
243. Oguma, R. Coherence analysis of systems with feedback and it's application to a BWR noise investigation. *Progress in Nuclear Energy*. 1982, Vol. 9, pp. 137–148.
244. Por, G. and Katona, T. Some aspects of the theory of neutron noise due to propagating disturbances. *Progress in Nuclear Energy*. 1982, Vol. 9, pp. 209–216.
245. Turkcan, E. Review of Borssele PWR noise experiments analysis and instrumentation. *Progress in Nuclear Energy*. 1982, Vol. 9, pp. 437–441.
246. Bastle, W., Sunder, R., Wach, D. Experiences and results with COMOC — an on-line vibration analysis and monitoring system. *Progress in Nuclear Energy*. 1985, Vol. 15, pp. 88–92.
247. Bauernfeind, V. Vibration monitoring of four-loop PWR: model-investigations of the sensitivity of the monitored signals on mechanical failures. *Proceedings of SMORN IV*. Dijon, France, 1985, pp. 285–288.
248. Dach, K., Krett, V., Vavrin, J. *et al.* The experiences of using neutron noise analysis for vibration behaviour of internal structures. *Progress in Nuclear Energy*. 1985, Vol. 15, pp. 243–250.
249. Fry, D.N., Horne, G.P., Mayo, C.W. Report of the first United States conference on utility experience with neutron noise analysis. *Progress in Nuclear Energy*. 1985, Vol. 15, pp. 503–511.
250. Grunwald, G., Junghanss, K., Liewers, P. Investigation of pressure oscillation in PWR primary circuit. *Progress in Nuclear Energy*. 1985, Vol. 15, pp. 651–659.
251. Kostic, L.J., Runkel, J., Stegemann, D. Thermohydraulics surveillance of pressurized water reactors by experimental and theoretical investigations of the low frequency noise field. *Progress in Nuclear Energy*. 1985, Vol. 15, pp. 879–887.
252. Katona T. and Mesko L. Investigation of the noise source characteristics. *Progress in Nuclear Energy*. 1985, Vol. 15, pp. 612.
253. Katona, T. and Kozma R. Problems of estimation of the thermohydraulic parameters using neutron and temperature noise signals in PWR's. *Progress in Nuclear Energy*. 1985, Vol. 15, pp. 359–364.
254. Mullers, J.A. and Thie, J.A. Modeling and diagnostic techniques applicable to the analysis of pressure noise in pressurized water reactor and pressure-using system. *Progress in Nuclear Energy*. 1985, Vol. 15, pp. 217–224.
255. Nagy, I. and Katona, T.J. Theoretical investigation of the low-frequency pressure fluctuation in PWRs. *Progress in Nuclear Energy*. 1985, Vol. 15, No. 1–3, pp. 671–683.

256. Por, G., Izsak, E.T., Valka, S. Some results of noise measurements in PWR NPP. *Progress in Nuclear Energy.* 1985, Vol. 15, pp. 387.
257. Smith, C.M. and Gonzalez, R.C. Long-term automated surveillance of a commercial nuclear power plant. *Progress in Nuclear Energy.* 1985, Vol. 15, pp. 17–26.
258. Sweeney, F.J., Upadhyaya, B.R., Shieh, D.J. In-core coolant flow monitoring of PWR using temperature and neutron noise. *Progress in Nuclear Energy.* 1985, Vol. 15, pp. 201–208.
259. Türkcan E. On-line monitoring of a PWR for plant surveillance by noise analysis. *Progress in Nuclear Energy.* 1985, Vol. 15, pp. 365.
260. Schumann, P., Liewers, P., Schmitt, W., Weiss F.-P. Detection of core barrel motion at WWER-440 type reactors. *Progress in Nuclear Energy.* 1988, Vol. 21, pp. 89–96.
261. Por, G., Glogter, O., Rindelhardt, U. Boiling detection in PWRs by noise measurement. *Progress in Nuclear Energy.* 1988, Vol. 21, pp. 555–563.
262. Wehling, H.-J., Kingler, K., Stolben, H. Vibration Monitoring of KWU Pressurized Water Reactor — Review, Present Status and Further Development. *Nuclear Technology* 1988, Vol. 80, pp. 367.
263. Kostic, L., Husemann, A., Runkel, J., Stegemann, D., Kahlastatt, H. Estimation of MTC in a PWR as Function of Neutron Noise Amplitude. *Proc. SMORN-VI*, Gatlinburg, Tennessee, USA, May 19–24, 1991.
264. Sunder, R., Baleanu, M., Kieninger, K. e.a. Experiences and Results with COMOS — an On-line Vibration Analysis and Monitoring System. *Proc. SMORN-VI*, 1991, Vol. 1, pp. 5.01–5.12.
265. Bogard, T., Kozlosky, T., Lowenfeld, S., Singleton, N. Reactor Internals and Core Support Monitoring System. *SMORN-VI*, 1991, Vol. 1, Gatlinburg, Tennessee, USA, May 19–24, 1991, pp. 19.01.
266. Trenty, A., Puyal C., Klajnmic, H. SINBAD, A Data Base for PWR Internals Vibratory Monitoring. *Proc. SMORN-VI*, 1991, Vol. 1, pp. 253–264.
267. Vavrin, J. Noise Analysis of Reactor Core Barrel Impact Oscillations. *Proc. SMORN-VI*, 1991, Vol. 1, pp. 16.01.
268. Upadhyaya, B.R. and Glockler, O. Estimation of Feedback Parameters in Pressurized Water Reactors Using Neural Networks. *Proc. SMORN-VI*, 1991, Vol. 2, pp. 175–189.
269. Uchiyama, J. and Sakai, K. A Reactor Equipment Monitoring System for Japanese PWRs. *Proc. SMORN VII*, 1995.
270. Glockler, O. and Tulett, M.V. Application of reactor noise analysis in the CANDU reactors of Ontario Hydro. *Progress in Nuclear Energy.* 1995, Vol. 29, pp. 171–191.

271. Runkel, J., Laggiard, E., Stegemann, D., *et al.* Application of Noise Analysis in Two BWR Units of Nuclear Power Plants Gundremmingen. *Proc. SMORN VII*, 1995. Avignon, France, June 19–23, 1995.
272. Jung, S.H., Kim, T.R., Park, J.N., Park, J.H. Development of Fault Diagnostic PC-based Software for Reactor Internals. *Proc. SMORN VII*, 1995. Avignon, France, June 19–23, 1995. Vol. 2, pp. 9.8.
273. Laggiard, E., Runkel, J., Stegemann, D., Fiedler, J. Determination of Vibration Amplitudes and Neutron-Mechanical Scale Factors Using In-core Accelerometers in NPP Obrigheim. *Proc. SMORN VII*, 1995. Avignon, France, June 19–23, 1995.
274. Laggiard, E., Fiedler, J., Runkel, J., *et al.* Vibration measurements in PWR obrigheim by use of in-core accelerometers. *Progress in Nuclear Energy.* 1995, Vol. 29, No. 3–4, pp. 229–238.
275. Gerardin, J.P. and Donnette, J.E. Method for Evaluating the System Instrumentation for Loose Part Detection in the Primary Cooling Circuit of French PWRs. *Proc. SMORN VII*, 1995, Avignon, France, June 19–23, 1995.
276. Hrosso, P., Maudry, J., Korec, J. Technical diagnostics of V1 and V2 NPP at Bohunice. *Proc. SMORN VII*, 1995. Avignon, France, June 19–23, 1995, pp. 561–571.
277. Laggiard, E., Rumkel, J., Stegemann, D., *et al.* Determination of Vibration Amplitudes and Neutron-Mechanical Scale Factors Using In-core Accelerometers in NPP Obrigheim. *Proc. SMORN VII*, 1995. Avignon, France, June 19–23, 1995, Vol. 2, No. 9.1, pp. 547–552.
278. Trenty, A. Operational feedback on internal structure vibration in 54 French PWRs during 300 fuel cycles. *Progress in Nuclear Energy.* 1995, Vol. 29, No. 3/4, pp. 347–356.
279. Kunze, U. and Bechtold, B. New generation of monitoring systems with on-line diagnostics. *Progress in Nuclear Energy.* 1995, Vol. 29, 215 p.
280. Uchiyama, J. and Skai, K. A reactor equipment monitoring system for Japanese PWRs. Systeme de surveillance pour les REP Japonais. *Proc. SMORN VII*, 1995. Avignon, France, 19–23 June, 1995, Vol. 2, p. 7.6.
281. Laggiard, E., Fiedler, J., Runkel, J., *et al.* Vibration measurements in PWR Obrigheim by use of in-core accelerometers. *Progress in Nuclear Energy.* 1995, Vol. 29, pp. 229–237.
282. Glockler, O. Reactor noise measurements in the safety and regulating systems of CANDU stations. *Progress in Nuclear Energy.* 2003, Vol. 43, pp. 75.
283. Andersson, T., Demaziere, C., Nagy, A. *et al.* Development and application of core diagnostics and monitoring for the ringhals PWRs. *Progress in Nuclear Energy.* 2003, Vol. 43, No. 1, pp. 35–41.

284. Czibok, T., Kiss, G., Kiss, S., Krinizs, K., Vegh, J. Regular neutron noise diagnostics measurements at the Hungarian Paks NPP. *Progress in Nuclear Energy*. 2003, Vol. 43, pp. 67.
285. Thie, J.A. *Reactor noise*. New York: Rowman and Littlefield, 1963.
286. Thie, J.A. Noise Sources in Power Reactor. *Symp. on noise analysis in nuclear systems*. June, 1964, pp. 357–365.
287. Thie, J.A. *Stochastic Processes in Nuclear Reactors and Measurements of Dynamic Characteristics*. Moscow: Atomizdat, 1966, p. 242.
288. Wambsganss, M.W. Vibrations of reactor core components. *Reactor-Fuel Process Technology*. 1967, Vol. 10, No. 3, pp. 208–219.
289. Albrecht, R. and Seifritz, W. The information in neutron fluctuations. *Nuclear Science and Engineering*. 1970, Vol. 41, pp. 417–420.
290. Grolmes, M.A. and Fauske, H.K. Propagation characteristics of compression and rarefaction pressure pulses in one-component vapor-liquid mixtures. *Nuclear Engineering and Design*. 1970, Vol. 11, No. 1, pp. 137–142.
291. Thie, J.A. Reactor noise monitoring for malfunctions. *Reactor Technology*. 1971, Vol. 14, No. 4, pp. 354–365.
292. Seifritz, W. and Stegemann, D. Reactor noise analysis. *Atomic Energy Rev*. 1971, Vol. 9, No. 1, pp. 129–135.
293. Bohm, G.J. and Nahavandi, A.N. Dynamic analysis of reactor internal structures with impact between component. *Nuclear Science and Engineering*. 1972, No. 47, pp. 391–408.
294. Kosaly, G. and Mesko, L. To question of a transfer function which connects the entrance coolant temperature fluctuations with the temperature noise. *Atomkernenergie*. 1972, Vol. 20, No. 1, pp. 12–17.
295. Uhric, R.E. Noise analysis in power reactor. *Electrical World*. 1973, Vol. 180, No. 11, pp. 44–56.
296. Williams, M.M.R. *Random Processes in Nuclear Reactors*. Oxford: Pergamon Press, Ltd., 1974, 258 p.
297. Saito, K. On the theory of power reactor noise (I, II, III). *Ann. of Nuclear Science and Engineering*. 1974, Vol. 1, pp. 31, 107, 209.
298. Robert, E. *Uhrig Statistical Methods in Reactor Physics*: Translated from English/Edited by Mogilner A.I. Moscow: Atomizdat, 1974. 400 p.
299. Fry, N.D., Kryter, R.C., Robinson, J.C. *Analysis of Neutron density Oscillations resulting from core barrel Motion of the Palizades Nuclear Power Plant Report ORNL-TM-4570*. Oak Ridge, Tennessee, 1974. 26 p.
300. Wach, D. The analysis at-power neutron flux noise in the frequency range of vibrating reactor structures. *Annals of Nuclear Energy*. 1975, Vol. 2, pp. 353–359.

301. Kosaly, G. and Wach, D. Investigation in the joint effect of local and global driving sources in in-core neutron noise measurement. *Atomkernenergie.* 1975, Bd. 23, pp. 244–248.
302. Kosaly, G., Maroti, L., Mesko, L. A simple space dependent theory of the neutron noise in a BWR. *Annals in Nuclear Energy.* 1975, Vol. 2, pp. 315–317.
303. Thie, J.A. Theoretical considerations and their application to experimental data in the determination of reactor internals' motion from stochastic signals. *Annals of Nuclear Energy.* 1975, Vol. 2, pp. 253–259.
304. Kosaly, G. and Mesko, L. Theory of the auto-spectrum of the local components of PWR. *Annals of Nuclear Energy.* 1976, Vol. 3, pp. 233–238.
305. Van Dam H. Neutron noise in BWR. *Atomkernenergie.* 1976, Bd. 27, Lfg. 1, pp. 8–27.
306. Willsky, A.S. A survey of design methods for failure detection in dynamic systems. *Automatica.* 1976, Vol. 12, pp. 601–611.
307. Mayo, C.W. Detailed neutron noise analysis of PWR internal vibrations. *Atomkernenergie.* 1977, Bd. 29, H. 9, pp. 177–181.
308. Atta, M.A., Fry, D.N., Mott, J.E., King, W.T. Determination of void fraction profile in a BWR channel using neutron noise analysis. *Nuclear Science and Engineering.* 1978, Vol. 66, pp. 264–268.
309. Robinson, J.C., Shahrokhi, F., Kryter, R.C. Calculation of the scale factor for inference of PWR core barrel motion from neutron noise spectral density. *Nuclear Technology.* 1978, Vol. 40, pp. 35–46.
310. Robinson, J.C., Shahrokhi, F., Kryter, R.C. Quantification of core barrel motion using an analytically derived scale factor and statistical reactor noise descriptors. *Nuclear Technology.* 1978, Vol. 40, pp. 47–51.
311. Grunwald, G. *et al.* Experimental investigation of flow-induced control-element movements by noise analysis. *Proc. of symposium for nuclear power plant control and instrumentation.* 1978, Vol. 1, pp. 291–292.
312. Van Blarcom, P.P. and Smitt, R.D. Flashing Fluids at Low Pressures. *Proc. ISA Conf. and Exhib.* Chicago, 1979, pp. 391–440.
313. Saito, K. Source papers in reactor noise. *Progress in Nuclear Energy.* 1979, Vol. 3, pp. 157–168.
314. Naito, N., Ando, Y., Yamamoto, F., Takeuchi, E. Estimation of fuel inlet flow rate by noise analysis. *Nuclear Science and Engineering.* 1980, Vol. 17, No. 5, pp. 351–358.
315. Kosaly, G. Noise investigation in BWR and PWR. *Progress in Nuclear Energy.* 1980, Vol. 5, pp. 145–199.

316. Thompson, J.P., McCoy, G.R., Lubin, B.T. Experimental value of percent variation in root-mean-square ex-core detector signal to the core barrel amplitude scale factor. *Nuclear Technology.* 1980, Vol. 48, pp. 122–127.
317. Nguyen, D.L., Winter, E.R.F., Greiner, M. Sonic velocity in two-phase systems. *International Journal of Multiphase Flow.* 1981, Vol. 7, No. 3, pp. 311–320.
318. Thie, J.A. and Mullens, J.A. Utilization of Pressure Noise for Reactor Anomaly Monitoring. *Proceeding of Specialist's Meeting on Early Diagnosis of Failures in Primary System Components of Nuclear Power Plants.* IAEA Report. Vienna, 1982, pp. 16–19.
319. Mullers, J.A. and Thie, J.A. Understanding Pressure Dynamic Phenomena in PWRs for Surveillance and Diagnostic Applications. *Proceeding of 5th Power Plant Dynamics, Controls and Testing Symposium.* University of Tennesse, Knoxville, March 1983, pp. 47–49.
320. Herbert, M.R. A review of on-line diagnostic aids for nuclear power plant operators. *Nuclear Energy.* 1984, Vol. 23, No. 4, pp. 259–264.
321. Grunwald, G., Junghans, K., Liewers, P. Investigation of pressure oscillation in PWR primary circuit. *Progress in Nuclear Energy.* 1985, Vol. 15, pp. 651–659.
322. Thie, J.A. A review of the OECD specialist meeting on continuous monitoring techniques for assuring coolant circuit integrity. *Nuclear Safety.* 1986, Vol. 27, No. 2, pp. 193–198.
323. Vibration Monitoring System. SUS. Siemens AG KWU Group, 1989, 362 p.
324. Patton, R., Frank, P., Clark, R. (eds.). *Fault Diagnosis in Dynamic Systems. Theory and applications* N.J., Englewood Cliffs: Prentice Hall Inc., 1989, 624 p.
325. Runkel, J. and Stegemann, D. Vibration and noise analysis at the primary loop of nuclear power plants. *VGB Kraftwerkstechnik.* 1989, Vol. 8, pp. 655.
326. Kozma, R., Dam, H., Hoogenboom, J.E. Identification of flow patterns by neutron noise analysis during actual coolant boiling in thin rectangular channels. *Nuclear Technology.* 1992, Vol. 100, pp. 97–110.
327. Bond, C.B., Bogard, W.T., Palusamy, S.S., Kozlovsky, T.A. Advanced Monitoring and Diagnostics for Nuclear Systems. *Modern Condition Monitoring in Power Systems Conference*, May 1993, London, UK, pp. 137–139.
328. Laggiard, E., Runkel, J., Starke, D., Stegemann, D., Heidemann, P. In-core Measurement of Reactors Internals Vibrations by use of

accelerometers and neutron detectors. *Specialists' meeting on in-core instrumentation and reactor core assessment. INCORE-96,* Mito, Japan. October 14–23, 1996, pp. 7–11.

329. Xu J.L. and Chen T.H. Acoustic wave prediction in flowing steam-water two-phase mixture. *International Journal of Heat and Mass Transfer.* 2000, Vol. 43, No. 7, pp. 1079–1088.
330. Wach, D. Condition Monitoring for Improved Performance in German NPPs. In *4th International Topical Meeting on Nuclear Plant Instrumentation, Control and Human Machine Interface Technology (NPIC&HMIT'04)*, Columbus Ohio, 2004, pp. 47–49.
331. Pecinka, L., Stulik, P., Zeman, V. Influence of the Operational Vibrations on the WWER 1000/320 Reactor Internals on Temelin NPP Core Barrel Stability. *Safety Assurance of NPP with WWER: Proceedings of 5th International Conference*, Podolsk, OKB Gidropress, 2007, May 29–June 1, 2007, pp. 36–48.
332. Katona, T.J. Possibility of use of noise analysis for identification of reactor conditions during accidents. *World Journal of Nuclear Science and Technology.* 2013, Vol. 3, No. 3, pp. 96–105.
333. Leskin, S.T. Classification algorithms for analysis of the core states according to ICIS data. *Izvestiya Vuzov. Nuclear Industry.* 1996, No. 4, pp. 20–26.
334. Leskin, S.T. Statistical model for VVER core diagnostics. *Izvestiya Vuzov. Nuclear Industry.* 1996, No. 6, pp. 33–39.
335. Leskin, S.T. Development of algorithms for identification of NPP equipment malfunction based on analysis of operational process control data. *Izvestiya Vuzov. Nuclear Industry.* 1997, No. 4, pp. 4–11.
336. Leskin, S.T. and Zhidkov, S.V. Representation of state of VVER-1000 core by a minimal connected graph. *Izvestiya Vuzov. Nuclear Industry.* 1997, No. 4, pp. 9–14.
337. Leskin, S.T. and Zhidkov, S.V. System for assessing VVER-1000 core status based on data of in-core monitoring in operation of the power unit. *Izvestiya Vuzov. Nuclear Industry.* 1998, No. 3, pp. 4–10.
338. Leskin, S.T. and Zarjugin, D.G. Program complex DINA I for VVER main circulation pumps diagnostics based on technological monitoring data analysis. *Izvestiya Vuzov. Nuclear Industry.* 2001, No. 1, pp. 3–12.
339. Leskin, S.T. and Valuy, V.V. Application of principle component analysis for the diagnostics of high and low pressure pumps based on test results in VVER NPP operation. *Izvestiya Vuzov. Nuclear Industry.* 2002, No. 4, pp. 38–45.
340. Leskin, S.T., Slobodchuk, V.I., Shelegov, A.S. Analysis of VVER-1000 main circulation pump state in operation. *Izvestiya Vuzov. Nuclear Industry.* 2016, No. 4, pp. 12–22.

341. Proskuryakov, K.N. Self-oscillations in the individual steam-generating channel. *Teploenergetika.* 1965, No. 12, pp. 75–77.
342. Proskuryakov, K.N. Electric model of steam generating channel. *MEI.* 1972, No. 126, pp. 64–67.
343. Proskuryakov K.N. *et al.* Theoretical determination of frequencies of coolant natural oscillations in the primary circuit of NPP. *MEI.* 1979, No. 407, pp. 87–92.
344. Proskuryakov K.N. Calculating method of coolant resonant frequencies oscillation in normal and emergency operating of NPP with WWER. *Kernenergie* 1983, Vol. 3, pp. 102–104.
345. Proskuryakov, K.N. *Heat Hydraulic Excitation of Coolant Oscillations in Internals of Nuclear Power Stations.* Moscow: MEI, 1984, 68 p.
346. Proskuryakov, K.N. Early Boiling Detection Method OF Pre- or Post-Accident Situation on WWER and RBMK. *Proc. SMORN VII. Avignon (France)*, 1995, Vol. 1, pp. 426–424.
347. Proskuryakov, K.N. *Application of Vibroacoustic Noise for Diagnostics of Technological Process in Nuclear Power Plants.* Moscow: *MEI*, 1999, 68 p.
348. Proskuryakov, K.N. Modeling Parametric Oscillations in Cooling Systems of Nuclear Reactors. *Izvestiya Vuzov. Nuclear Industry.* 2000, No. 3, pp. 98–102.
349. Proskuryakov, K.N. Vibroacoustic Certification of Nuclear Power Plants is a Way to Increase their Reliability and Safety. *Teploenergetika*, 2005, No. 12, pp. 30–34.
350. Proskuryakov, K.N., Shakirzyanov F.N., Karataev V.V. *et al.* Procedure for prevention of resonant vibrations of VVER-1000 fuel rods and assemblies. *Problems of Atomic Science and Technology.* Series: The NPP Safety Assurance. 2008, No. 23, pp. 18–25.
351. Proskuryakov, K.N. and Novikov, K.S. The Causes of High Cyclic Dynamical Loadings in VVER-1000 Reactors. In *Proceedings of 6th International Scientific and Technical Conference "Safety, Efficiency and Economy of Nuclear Power Industry". MNTK-2008.* Moscow, Concern Rosenergoatom JSC, May 21–23, 2008, pp. 77–79.
352. Proskuryakov, K.N. and Novikov, K.S. Area of resonance interaction of vibrations of VVER-1000 internals and pressure pulsations at supercritical parameters. In *Safety Assurance of NPP with VVER. Proceedings of 6th International Scientific and Technical Conference Podolsk*, OKB Gidropress, May 26–29, 2009, pp. 42–46.
353. Proskuryakov, K.N. and Novikov, K.S. Determining Area of Coolant and Fuel Assembly Vibroacoustic Resonances in Advanced Reactors with Increased Capacity. *Atomnaya Energiya.* 2010, Edition 3, pp. 151–155.

354. Pavelko, V.I. New Spectral Methods of Estimating Delay Time in Reactor Noise Studies. *Atomnaya Energiya.* 1987, Vol. 63, Edition 4, pp. 266–288.
355. Pavelko, V.I. Spectral Methods of Estimating Delay Time in Reactor Noise Studies. *Problems of Atomic Science and Technology,* Series: Physics of Nuclear Reactors. 1989, Edition 2, pp. 58–65.
356. Pavelko, V.I. A Review of Application of Expert System Methodology in Nuclear Power Engineering. *Atomnaya Energiya.* 1990, Edition 11, pp. 1–8.
357. Pavelko, V.I. Statistical methods at the preliminary stage of operational diagnostics of nuclear power plants. *Problems of Atomic Science and Technology.* Series: Physics of Nuclear Reactors, 1990, Edition 2, pp. 23–30.
358. Pavelko, V.I. Neutron-temperature noise models of VVER core. *Atomnaya Energiya.* 1992, Vol. 72, No. 6, pp. 66–81.
359. Pavelko, V.I. Methods of signature diagnostics by cross-spectral functions. *Progress in Nuclear Energy.* 1995, Vol. 29, No. 3/4, pp. 239–246.
360. Bulavin, V.V., Gutsev, D.F., Pavelko, V.I. Investigation of characteristics of vibrational diagnostics of VVER-1000 reactors under operating conditions. *Atomnaya Energiya.* 1995, Vol. 79, No. 5, pp. 343–349.
361. Bulavin, V.V., Gutsev, D.F., Pavelko, V.I. The experimental definition of the acoustic standing waves series shapes, formed in the coolant of the primary circuit of VVER-440 type reactor. *Progress in Nuclear Energy.* 1995, Vol. 29, No. 3/4, pp. 153–170.
362. Bulavin, V.V., Gutsev, D.F., Pavelko, V.I. Some Results of the Vibrations Analysis on the VVER-440 Type Reactor Vessel Top Head and on the Facilities of its Primary Circuit Six Loops. *A symposium on Nuclear Reactor Surveillance and Diagnostics. SMORN VII,* June 19–23, 1995, Avignon, France. Session 9 — Vibrations Structures: Modeling and Monitoring. Vol. 2, No. 9.5, pp. 168–172.
363. Bulavin, V.V., Gutsev, D.F., Pavelko, V.I. Neutron-temperature noise methods and their experimental check on the reactor VVER-1000. *Specialists' meeting on in-core instrumentation and reactor core assessment. INCORE-96.* October 14–23, 1996, Settlement Account 26–29.
364. Gutsev, D.F. and Pavelko, V.I. VVER noise diagnostic methods. *Atomic Energy.* 1997, Vol. 82, No. 4, pp. 264–271.
365. Anikin, G.G., Gutsev, D.F., Pavelko, V.I. Physical interpretation of neutron-noise specimens of VVER-1000 reactor. *Atomnaya Energiya.* 1997, Vol. 82, No. 4, pp. 271–277.

366. Anikin, G.G. and Pavelko, V.I. Experience of implementing vibration monitoring systems at VVER-440 reactor plants. *Teplotechnika.* 1999, No. 6, pp. 12–17.
367. Skomorokhov, A.O. and Slepov, M.T. Pattern recognition in APL with application to reactor diagnostics. *APL Quote Quad.* 1999, Vol. 29, No. 3, pp. 164–172.
368. Skomorokhov, A.O. and Slepov, M.T. Data verification in the system of vibration diagnostics at the novovoronezhskaya nuclear power plant. *Izvestiya Vuzov. Nuclear Industry.* 1999, No. 1, pp. 56–65.
369. Arkadov, G.V., Perov, S.L., Usanov, A.I. Computational Modeling of Vibration Characteristics of Equipment and Fuel Assemblies of VVER Reactors. In *Safety Assurance of NPP with VVER. Proceedings of the II All-Russian Scientific and Technical Conference*, November 19–23, 2001, Podolsk, JSC OKB Gidropress, pp. 29–32.
370. Arkadov, G.V., Ovcharov, O.V., Pavelko, V.I. *et al.* Measurement of Coolant Flow through the VVER-440 Fuel Channel by Fluctuations of Direct Charge Detector Signals. *Atomnaya Energiya.* 2001, Vol. 91, No. 3, pp. 167–174.
371. Arkadov, G.V., Pavelko, V.I., Titov, S.I. Neutronic and noise vibration diagnostics of the VVER-440 reactor cavity. *Atomnaya Energiya.* 2001, Vol. 91, Edition 4, pp. 247–251.
372. Arkadov, G.V., Dembovsky, A.V., Usanov, A.I. *et al.* Computational and Experimental Study of Sources of Thermal and Mechanical Loading of the Welding Unit of the PGV-1000 Hot Header. In *Safety Assurance of NPP with VVER. Proceedings of the II All-Russian Scientific and Technical Conference.* November 19–23, 2001, Podolsk. JSC OKB Gidropress, pp. 87–90.
373. Arkadov, G.V., Anikeev, Yu.A., Pavelko, V.I. *et al.* Evaluation of the Vibration States of Reactor Internals and Fuel Assemblies of VVER-440 in Operating Conditions. Safety Assurance of NPP with VVER. In *Proceedings of the II All-Russian Scientific and Technical Conference.* November 19–23, 2001, Podolsk. JSC OKB Gidropress, pp. 7–11.
374. Arkadov, G.V., Pavelko, V.I., Finkel, B.M. *et al.* Development of Software and Hardware Systems to Diagnose Equipment of NPP Units under Construction. In *Safety Assurance of NPP with VVER. Proceedings of the II All-Russian Scientific and Technical Conference*, November 19–23, 2001, Podolsk, JSC OKB Gidropress, pp. 72–75.
375. Arkadov, G.V., Pavelko, V.I., Finkel, B.M. *et al.* Status and Prospects of Application of Operational Diagnostics Systems for Maintenance of Safety of Power Units with VVER. In *Safety Assurance of NPP with VVER. Proceedings of the II All-Russian Scientific and Technical*

Conference, November 19–23, 2001, Podolsk, JSC OKB Gidropress, pp. 29–31.
376. Arkadov, G.V., Matveev, V.P., Pavelko, V.I., Finkel, B.M. Software and Hardware Complex of Vibration Diagnostics System of VVER. *Problems of Atomic Science and Technology.* Series: Physics of Nuclear Reactors, 2002, Edition 3, pp. 37–45.
377. Arkadov, G.V., Matveev, V.P., Ovcharov, O.V. *et al.* The Main Spectral Characteristics of the Vibration State of VVER-1000. *Problems of Atomic Science and Technology.* Series: Physics of Nuclear Reactors, 2002, Edition 3, pp. 46–52.
378. Solonin, V.I., Pavelko, V.I., Perevezentsev V.V. Influence of Hydrodynamic Loads on Vibration of Fuel Rod Bundles Used in Fuel Assemblies of VVER-440 Reactors. In *Safety Assurance of NPP with VVER: The III-rd International Scientific and Technical Conference*, Podolsk. OKB Gidropress, May 26–30, 2003, p. 54.
379. Arkadov, G.V., Pavelko, V.I., Usanov, A.I. *Vibration and Noise Diagnostics of VVER.* Moscow: Energoatomizdat, 2004, p. 344.
380. Arkadov, G.V., Pavelko, V.I., Usanov, A.I. Development of Vibration Regulations of VVER-440 Internals. In *Safety Assurance of NPP with VVER: Proceedings of IV International Scientific and Technical Conference*, Podolsk, OKB Gidropress, 2005, pp. 62–65.
381. Arkadov, G.V., Pavelko, V.I., Finkel, B.M. *Diagnostic Systems for VVER.* Moscow: Energoatomizdat, 2010, 391 p.
382. Baranova, Yu.A. and Slepov, M.T. AES-2006 with VVER-1200 — new approach to information display from technical diagnostics systems. *Izvestiya Vuzov. Nuclear Industry.* 2014, No. 4, pp. 11–20.
383. Slepov, M.T. and Sysoev, N.P. Diagnostics of motor-operated valves — operating experience of novovoronezh NPP. *Global Nuclear Safety.* 2014, No. 2, pp. 79–85.
384. Pavelko, V.I., Slepov, M.T., Khairetdinov, V.U. Experience of complex measurements with various systems at different stages of commissioning of VVER-1200 power unit. *Izvestiya Vuzov. Nuclear Industry.* 2016, No. 4, pp. 44–54.
385. Fedorov, A. I. and Slepov, M.T. Complex measurements of diagnostic parameters of the equipment of unit 1, Novovoronezh NPP-2 in trial operation. *Izvestiya Vuzov. Nuclear Industry.* 2017, No. 3, pp. 77–87.
386. Nuclear safety rules for reactor installations of nuclear power plants. NP-082–07. *Nuclear and Radiation Safety.* 2008, No. 1(47), pp. 52–77.
387. PNAE G-7-002-86. Strength Calculation Rate for Equipment and Pipelines of Nuclear Facilities PNAE G-7-002-86. Moscow: Energoizdat, 1989, 525 p.
388. *GOST 16819-71. Vibrometers. Terms and Definitions.* Moscow: Standardinform, 2010, 8 p.

389. *GOST 18955-73. Linear Low Frequency Accelerometers.* Moscow: Publishing and Printing Complex IPK Standards Publishing 1973, 14 p.
390. *GOST 20417-75. Technical Diagnostics. General Requirements for Elaboration of Diagnostic Systems.* Moscow: Standards Publishing, 1975, 4 p.
391. *GOST 12.1.012–78. Occupational Safety Standards System. Vibration. General Safety Requirements.* Moscow: Standards Publishing, 1989, 32 p.
392. *GOST 24346-80. Vibration. Terms and Definitions.* Moscow: Standardinform, 2010, 26 p.
393. *GOST 24347-80. Vibration. Designations and Units of Quantities.* Moscow: Standards Publishing, 1986, 6 p.
394. *GOST 25275-82. Standards System on Vibration. Devices for Measuring the Vibration of Rotary Machines. General Technical Requirements.* Moscow: Standards Publishing, 1993, 11 p.
395. *GOST 26043-83. Vibration. Dynamic Characteristics of Stationary Machines. General Provisions.* Moscow: Publishing and Printing Complex IPK Standards Publishing 1984, 16 p.
396. *GOST 26656–85. Technical Diagnostics. Testability. General Requirements.* Moscow: Standardinform, 2009, 10 p.
397. *GOST 4.304-85. Production Quality Index System. Vibration Measurement Equipment and Apparatus. Index Nomenclature.* Moscow: Standards Publishing, 1994, 16 p.
398. *GOST 26875-86. Vibration. Portable Balancing Equipment. Technical Requirements.* Moscow: Standards Publishing, 1986, 10 p.
399. *GOST 27518-87. Diagnostics of Items. General Requirements.* Moscow: Standardinform, 2009, 5 p.
400. *GOST 20911-89. Technical Diagnostics. Terms and Definitions.* Moscow: Standardinform, 2009, 12 p.
401. *GOST 30296-95. General-purpose Measuring Instruments for Determination of Main Vibration Parameters. General Technical Requirements.* Moscow: Publishing and Printing Complex IPK Standards Publishing, 1998, 20 p.
402. *GOST R ISO 10816-1-97. Mechanical Vibration. Evaluation of Machine Vibration by Measurements on Non-rotating Parts. Part 1. General Requirements.* Moscow: Publishing and Printing Complex IPK Standards Publishing, 1998, 18 p.
403. *GOST R 51498-99. (ISO 10814-69). Mechanical Vibration. Susceptibility and Sensitivity of Machines to Unbalance.* Moscow: Publishing and Printing Complex IPK Standards Publishing, 2000, 16 p.

404. *GOST R ISO 7919-3-99. Mechanical Vibration. Measurements on Rotating Shafts and Evaluation Criteria. Coupled Industrial Machines.* Moscow: Publishing and Printing Complex IPK Standards Publishing, 2000, 16 p.
405. *GOST R ISO 10817-1-99. Vibration. Rotating shaft vibration measuring systems. Part 1. Relative and absolute sensing of radial vibration.* Moscow: Publishing and Printing Complex IPK Standards Publishing, 2000, 19 p.
406. *GOST ISO 5348-2002. Vibration and Shock. Mechanical mounting of accelerometers.* Moscow: Standardinform, 2007, 16 p.
407. *GOST R 53565–2009. Condition monitoring and diagnostics of machines. Hazardous equipment monitoring. Vibration generated by rotodynamic pump and compressor units.* M.; Standardinform, 2010.
408. *IEC 61502:1999. Nuclear Plants. Pressurized Water Reactors. Vibration Monitoring of Internal Structures*, 1999.
409. *Instrumentation and Control Systems Important to Safety in Nuclear Power Plants. Safety Guide. IAEA Safety Standards Series No. NS—G—1.3.* Vienna, Austria, 2002, 99 p.
410. *IEC 60988:1990. Acoustic Monitoring Systems for Loose Parts Detection — Characteristics, Design Criteria and Operational Procedures.* 1990, 46 p.

Index

G

H

I

K

L

M

W

Z